Lecture Notes on Data Engineering and Communications Technologies

Volume 110

Series Editor

Fatos Xhafa, Technical University of Catalonia, Barcelona, Spain

The aim of the book series is to present cutting edge engineering approaches to data technologies and communications. It will publish latest advances on the engineering task of building and deploying distributed, scalable and reliable data infrastructures and communication systems.

The series will have a prominent applied focus on data technologies and communications with aim to promote the bridging from fundamental research on data science and networking to data engineering and communications that lead to industry products, business knowledge and standardisation.

Indexed by SCOPUS, INSPEC, EI Compendex.

All books published in the series are submitted for consideration in Web of Science.

More information about this series at https://link.springer.com/bookseries/15362

Rajaa Saidi · Brahim El Bhiri ·
Yassine Maleh · Ayman Mosallam ·
Mohammed Essaaidi
Editors

Advanced Technologies for Humanity

Proceedings of International Conference on Advanced Technologies for Humanity (ICATH'2021)

Springer

Editors
Rajaa Saidi
INSEA
Rabat, Morocco

Brahim El Bhiri
EMSI
Rabat, Morocco

Yassine Maleh
USMS-ENSAK
Khouribga, Morocco

Ayman Mosallam
University of California
California, USA

Mohammed Essaaidi
ENSIAS
Rabat, Morocco

ISSN 2367-4512 ISSN 2367-4520 (electronic)
Lecture Notes on Data Engineering and Communications Technologies
ISBN 978-3-030-94187-1 ISBN 978-3-030-94188-8 (eBook)
https://doi.org/10.1007/978-3-030-94188-8

This Springer imprint is published by the registered company Springer Nature Switzerland AG
The registered company address is: Gewerbestrasse 11, 6330 Cham, Switzerland

Preface

Advances in science and engineering have a lower impact when the human side is ignored. For this reason, the International Conference on Advanced Technologies for Humanity (ICATH'2021) aims at establishing this vital link to magnify such benefits. After all, such advances serve humanity, and both scientists and engineers must assess the impact of their research on society and humanity. In designing the scope and theme of this conference, a broad spectrum of topics and disciplines, including different engineering and social science areas, was included.

ICATH'2021 is thus a cross-disciplinary conference that provides the critical issues for the benefit of the resource-constrained and vulnerable populations in the world. In addition, it provides an opportunity to identify the most pressing needs and discover the highlight humanitarian technologies that promote the successful practice and attract humanitarian and emergency management practitioners in order to learn from their successes and better guide future research.

This book of Lecture Notes on Data Engineering and Communications Technologies contains the papers presented in the main tracks of ICATH'2021, held in INSEA Rabat, Morocco, on November 26–27, 2021. ICATH'2021 was organized by the National Institute of Statistics and Applied Economics (INSEA) in collaboration with the Moroccan School of Engineering Sciences (EMSI), the Hassan II Institute of Agronomy and Veterinary Medicine (IAV-Hassan II), the National Institute of Posts and Telecommunications (INPT), the National School of Mineral Industry (ENSMR), the Faculty of Sciences of Rabat (UM5-FSR), the National School of Applied Sciences of Kenitra (ENSAK) and the Future University in Egypt (FUE).

ICATH'2021 was devoted to practical models and industrial applications related to advanced technologies for humanity. It was considered as a meeting point for researchers and practitioners to enable the implementation of advanced information technologies into various industries.

There were 105 submissions from 12 countries. Each submission was reviewed by at least three chairs or PC members. We accepted 48 regular papers (45%). Unfortunately, due to limitations of conference topics and edited volumes, the program committee was forced to reject some interesting papers, which did not satisfy these topics or publisher requirements. We would like to thank all authors and reviewers for their work and valuable contributions. The friendly and welcoming attitude of conference supporters and contributors made this event a success.

The papers presented in the volume are organized in topical sections on (i) Smart and Sustainable Cities, (ii) Communication Systems, Signal and Image Processing for Humanity, (iii) Cyber Security, Database and Language Processing for Human Applications, (iV) Renewable and Sustainable Energies, (V) Civil Engineering and Structures for Sustainable Constructions, (Vi) Materials and Smart Buildings and (Vii) Industry 4.0 for Smart Factories. The review process was highly competitive. We had to review 105 submissions. A team of over 100 program committee members and reviewers did this terrific job. Our special thanks go to all of them.

ICATH'2021 offered again an exciting technical program as well as networking opportunities. Distinguished scientists accepted the invitation for keynote speeches:

- Saifur RAHMAN
  Virginia Tech Advanced Research Institute, USA
- Naima KAABOUCH
  University of North Dakota, USA
- Ebada SARHAN
  Future University in Egypt, Egypt
- Mounir BOUAZAR
  UNICEF, Copenhagen Area, Denmark

We want to take this opportunity and express our thanks to the contributing authors, the members of the program committee, the reviewers, the sponsors and the organizing committee for their hard and precious work. Thanks for your help—ICATH'2021 would not exist without your contribution.

Rajaa Saidi
Brahim El Bhiri
Yassine Maleh
Ayman Mosallam
Mohammed Essaaidi

Organization

General Chairs

Ayman Mosallam	UCI, USA
Brahim El Bhiri	EMSI, Morocco
Rajaa Saidi	INSEA, Morocco

Steering Committee

Aniss Moumen	ENSAK, Morocco
Ashraf Aboshosha Eic	ICGST, Egypt
Ayman Mosallam	UCI, USA
Brahim El Bhiri	EMSI, Morocco
Moha Cherkaoui	ENSMR, Morocco
Mohammed Raiss El Fenni	INPT, Morocco
Rafika Hajji	IAV, Morocco
Rajaa Saidi	INSEA, Morocco
Yann Ben Maissa	INPT, Morocco
Younes Karfa Bekali	FSR-UM5, Morocco

Technical Program Chairs

Mohammed Essaaidi	ENSIAS, Morocco
Pierre-Martin Tardif	UdeS, Canada

Publication Chairs

Imade Benelallam	INSEA, Morocco
Yassine Maleh	USMS, Morocco

Sponsoring Chairs

Kaoutar Elhari	INSEA, Morocco
Rafika Hajji	IAV, Morocco
Younes Karfa Bekali	FSR-UM5, Morocco

Publicity Chairs

Idris Oumoussa	INSEA, Morocco
Yves Frédéric Ebobissé Djéné	SMARTiLAB, EMSI, Morocco

Organizing Committee

Abdeslam Kadrani	INSEA, Morocco
Adil Kabbaj	INSEA, Morocco
Ahmed Doghmi	INSEA, Morocco
Amina Radgui	INPT, Morocco
Aniss Moumen	ENSAK, Morocco
Aouatif Amine	ENSAK, Morocco
Fadoua Badaoui	INSEA, Morocco
Hanaa Hachimi	USMS, Morocco
Maryam Radgui	INSEA, Morocco
Mohamed Nabil Saidi	INSEA, Morocco
Mohamed Ouzineb	INSEA, Morocco
Mohammed Raiss El Fenni	INPT, Morocco
Rachid Benmansour	INSEA, Morocco
Yaacoub Hannad	INSEA, Morocco
Yann Ben Maissa	INPT, Morocco

Scientific Committee

Abdelkaher Ait Abdelouahad	FSJ, Morocco
Abdeslam Kadrani	INSEA, Morocco
Abderrahim Elqadi	EST, Morocco
Adil Kabbaj	INSEA, Morocco
Adil Salbi	EMSI, Morocco
Ahmed Essadki	ENSET, Morocco
Alaa Sheta	SCSU, USA
Ali Jorio	EMSI, Morocco
Amina Radgui	INPT, Morocco
Amine Amar	MASEN, Morocco
Anis Moumen	ENSAK, Morocco
Aouatif Amine	ENSAK, Morocco
Ashraf Aboshosha	UT, Germany
Ashraf K. Helmy	NARSS, Egypt

Atman Jbari	ENSET, Morocco
Awatif Rouijel	ISMAC, Morocco
Ayman Haggag	HU, Egypt
Ayman Mosallam	UCI, USA
Ayoub Karine	IYO, France
Azhar Hadmi	ISMAC, Morocco
Azzeddine Mazroui	UMF, Morocco
Bouchaib Cherradi	CRMEF, Morocco
Bouchra Nassih	ENSAK, Morocco
Boutaina Bouzouf	FM Global, Canada
Brahim El Bhiri	EMSI, Morocco
Btihal El Ghali	ESI, Morocco
Chaimaa Essayeh	INPT, Morocco
Chaimae Saadi	EMSI, Morocco
Chakib Elmokhi	USMS, Morocco
Fadoua Ataa Allah	IRCAM, Morocco
Fadoua Badaoui	INSEA, Morocco
Fahd Idrissi Khamlichi	DEVOTEAM, Morocco
Fayçal Mimoun	IUIT, Kenitra
Hafida Bouloiz	ENSA-Agadir, Morocco
Hafsa Abouadane	FSTM, Morocco
Hanaa Hachimi	USMS, Morocco
Hassan Silkan	MFJ, Morocco
Hind Lamharhar	INSEA, Morocco
Houssaine Ziyati	ESTC, Morocco
Houssam Eddine Chakir	UH2, Morocco
Ibtissam Medarhri	ENSMR, Morocco
Imad Benelallam	INSEA, Morocco
Ismail El-Kafazi	EMSI, Morocco
Jihad Zahir	CAU, Morocco
Kamal Anoune	AEU, Morocco
Kaoutar Elhari	INSEA, Morocco
Khalid Elatife	EMSI, Morocco
Laila El Abbadi	ENSAK, Morocco
Luciano Feo	US, Italy
Manal El Bajta	INSEA, Morocco
Maryam Gallab	EMI, Morocco
Mohamed Bouhaddioui	ENSMR, Morocco
Mohamed Et-Tolba	INPT, Morocco
Mohamed Nabil Saidi	INSEA, Morocco
Mohamed Ouzineb	INSEA, Morocco
Mohammed Raiss El Fenni	INPT, Morocco
Mostafa Khabouze	ENSMR, Morocco
Mourad Ouadou	UMV, Morocco
Naouel Moha	ÉTS, Canada

Nezha Mejjad	FSB, Morocco
Nisrine Ibadah	LRIT, Morocco
Pierre-Martin Tardif	UdeS, Canada
Rachid Benmansour	INSEA, Morocco
Rachid Saadane	EHTP, Morocco
Rafika Hajji	IAV, Morocco
Rajaa Saidi	INSEA, Morocco
Rami Hawileh	AUS, UAE
Rosa Penna	US, Italy
Safae Merzouk	EMSI, Morocco
Said Jabbour	CRIL, France
Shaohua He	GUT, Morocco
Siham Boulknadel	IRCAM, Morocco
Slimane Bah	EMI, Morocco
Soumaya El Mamoune	FSTT
Volkan Kahya	KTU, Turkey
Walid Cherif	ESI, Morocco
Willy Hermann Juimo Tchamdjou	UM, Cameroun
Yaacoub Hannad	INSEA, Morocco
Yann Ben Maissa	INPT, Morocco
Yassine Maleh	USMS, Morocco
Youness Chaabi	IRCAM, Morocco
Youness Chater	ENSAT, Morocco
Younes Karfa Bekali	UMV, Morocco
Younès Raoui	UMV, Morocco
Zakarya Erraji	INSEA, Morocco
Zaynab El Khattabi	FST, Morocco
Zineb Aarab	EMSI, Morocco

Contents

Smart and Sustainable Cities

Towards for an Agent-Based Model to Simulate Daily Mobility in Rabat Region 3
Khalid Qbouche and Khadija Rhoulami

Scenes Segmentation in Self-driving Car Perception System Based U-Net and FCN Models 10
Aimad Lahbas, Azhar Hadmi, and Amina Radgui

Autonomous Parallel Parking Using Raspberry-PI 19
Mahmoud Wagdy, Ahmed Roushdy, Ahmed Sabry, Abdallah Mostafa, Mahmoud Magdy, and Amal S. Mehanna

A Survey on the Application of the Internet of Things in the Diagnosis of Autism Spectrum Disorder 29
Fatima Ez Zahra El Arbaoui, Kaoutar El Hari, and Rajaa Saidi

Cost Reduction in Smart Grid Considering Greenhouse Gas Emissions Using Genetic Algorithm 42
F. Z. Zahraoui, H. E. Chakir, and H. Ouadi

Modelling of Cavitation in Transient Flow in Pipe-Water Hammer 54
Ouafae Rkibi, Nawal Achak, Bennasser Bahrar, and Kamal Gueraoui

The Contribution of GIS for the Modeling of Water Erosion, Using Two Spatial Approaches: The Wischmeier Model and SWAT in the High Oum Er-Rbia Watershed (Middle Atlas Region) 61
Younes Oularbi, Jamila Dahmani, and Fouad Mounir

Communication Systems, Signal and Image Processing for Humanity

A Compact Frequency Re-configurable Antenna for Many Wireless Applications 75
Younes Karfa Bekali, Asmaa Zugari, Brahim El Bhiri, Mohammed Ali Ennasar, and Mohsine Khalladi

A Review on Visible Light Communication System for 5G 84
Mohamed El Jbari, Mohamed Moussaoui, and Noha Chahboun

Sleep Stages Detection Based BCI: A Novel Single-Channel EEG Classification Based on Optimized Bandpass Filter 96
Said Abenna, Mohammed Nahid, and Hamid Bouyghf

3D Numerical Study of Sound Waves Behavior in the Presence of Obstacles Using the D3Q15-Lattice Boltzmann Model 106
Jaouad Benhamou, Salaheddine Channouf, and Mohammed Jami

Deep Learning for Building Extraction from High-Resolution Remote Sensing Images 116
Abderrahim Norelyaqine and Abderrahim Saadane

Automatic Searching of Deep Neural Networks for Medical Imaging Diagnostic 129
Zakaria Rguibi, Abdelmajid Hajami, and Zitouni Dya

PCA SVM and Xgboost Algorithms for Covid-19 Recognition in Chest X-Ray Images 141
R. Assawab, Abdellah Elzaar, Abderrahim El Allati, Nabil Benaya, and B. Benyacoub

Machine Learning Algorithms for Forest Stand Delineation Using Yearly Sentinel 2MSI Time Series 149
Anass Legdou, Aouatif Amine, Said Lahssini, Hassan Chafik, and Mohamed Berada

Cyber Security, Database and Language Processing for Human Applications

When Microservices Architecture and Blockchain Technology Meet: Challenges and Design Concepts 161
Idris Oumoussa, Soufiane Faieq, and Rajaa Saidi

Ensuring the Integrity of Cloud Computing Against Account Hijacking Using Blockchain Technology 173
Assia Akamri and Chaimae Saadi

Lightweight-Blockchain for Secured Wireless Sensor Networks: Energy Consumption of MAC Address-Based Proof-of-Authentication 182
Yves Frédéric Ebobissé Djéné, Mohammed Sbai EL Idrissi, Pierre-Martin Tardif, Brahim El Bhiri, Youssef Fakhri, and Younes Karfa Bekali

Virtual OBDA Mechanism Ontop for Answering SPARQL Queries Over Couchbase 193
Hakim El Massari, Sajida Mhammedi, Noreddine Gherabi, and Mohammed Nasri

Toward a Smart Approach of Migration from Relational Database System to NoSQL System: Analyzing and Modeling 206
Abdelhak Erraji, Abderrahim Maizate, and Mohamed Ozzif

Sentence Generation from Conceptual Graph Using Deep Learning . . . 218
Mohammed Bennani and Adil Kabbaj

LSTM-CNN Deep Learning Model for French Online Product Reviews Classification 228
Nassera Habbat, Houda Anoun, and Larbi Hassouni

Amazigh Handwriting Recognition System—Multiple DCNN Strategies Applied to New Wide and Challenging Database 241
Abdellah Elzaar, Rachida Assawab, Ayoub Aoulalay, Lahcen Oukhoya Ali, Nabil Benaya, Abderrahim El Mhouti, Mohammed Massar, and Abderrahim El Allati

Toward an End-to-End Voice to Sign Recognition for Dialect Moroccan Language 253
Anass Allak, Imade Benelallam, Hamdi Habbouza, and Mohamed Amallah

Renewable and Sustainable Energies

A Comparative Study of LSTM and RNN for Photovoltaic Power Forecasting 265
Mohammed Sabri and Mohammed El Hassouni

Solar Energy Resource Assessment Using GHI and DNI Satellite Data for Moroccan Climate 275
Omaima El Alani, Hicham Ghennioui, Mounir Abraim, Abdellatif Ghennioui, Philippe Blanc, Yves-Marie Saint-Drenan, and Zakaria Naimi

Experimental Validation of Different PV Power Prediction Models Under Beni Mellal Climate 286
Mustapha Adar, Mohamed-Amin Babay, Souad Taouiri, Abdelmounaim Alioui, Yousef Najih, Zakaria Khaouch, and Mustapha Mabrouki

Comparative Study of MPPT Controllers for a Wind Energy Conversion System ... 300
Hamid Chojaa, Aziz Derouich, Yassine Bourkhime, Elmostafa Chetouani, Billel Meghni, Seif Eddine Chehaidia, and Mourad Yessef

Optimization of Energy Consumption of a Thermal Installation Based on the Energy Management System EnMS ... 311
Ali Elkihel, Amar Bakdid, Yousra Elkihel, and Hassan Gziri

Legendre Polynomial Modeling of a Piezoelectric Transformer ... 320
D. J. Falimiaramanana, H. Khalfi, J. Randrianarivelo, F. E. Ratolojanahary, L. Elmaimouni, I. Naciri, and M. Rguiti

Civil Engineering and Structures for Sustainable Constructions

Numerical Simulation of Fatigue Crack Propagation of S355 Steel Under Mode I Loading ... 333
Jielin Liu, Haohui Xin, and Ayman Mosallam

The Screening Efficiency of Open and Infill Trenches: A Review ... 346
Hinde Laghfiri and Nouzha Lamdouar

Relationship Between Eight-Fold Star and Other Tiles in Traditional Method 'Tastir' ... 352
Zouhair Ouazene, Aziz Khamjane, and Rachid Benslimane

Periodic Structures as a Countermeasure of Traffic Vibration and Earthquake: A Review ... 359
Hinde Laghfiri and Nouzha Lamdouar

Numerical Evaluation of Channel-Beam Railway Bridge with Hollow Section Concrete Deck ... 374
Shaohua He, Luoquan Zou, and Ayman Mosallam

Seismic and Energy Upgrading of Existing RC Buildings: Methodological Aspects and Application to a Case-Study on the Italian Experience ... 387
Luciano Feo, Enzo Martinelli, Rosa Penna, and Marco Pepe

Impact of Coronavirus Pandemic Crisis on Construction Control Processes in Egypt ... 404
Nora Magdy Essa, Hassan Mohamed Ibrahim, and Ibrahim Mahmoud Mahdi

Structural Evaluation of FRP Composite Systems for Repair Upgrade of Reinforced Concrete Beams ... 428
Ashraf A-K Mostafa Agwa

Materials and Smart Buildings

The Macroscopic Effect of COVID 19 on Flexible Pavement Condition Indicators Based on Analysis of Road Inspection Results 441
Mohammed Amine Mehdi, Toufik Cherradi, Said Elkarkouri, and Ahmed Qachar

Effect of Control Partitions on Drag Reduction and Suppression of Vortex Shedding Around a Bluff Body Cylinder 453
Youssef Admi, Mohammed Amine Moussaoui, and Ahmed Mezrhab

Probabilistic Fatigue Life Analysis of Fiber Reinforced Polymer (FRP) Composites Made of ±45° Layers 464
Qinglin Gao, Haohui Xin, and Ayman Mosallam

Evaluation for the Punching Shear Design of Concrete Slabs Subjected to Compressive Forces 474
A. Deifalla

Random Lattice Modeling of Fracture in Structural Glass-Fiber Reinforced Polymers 482
Alessandro Fascetti, Luciano Feo, Rosa Penna, and Yingbo Zhu

Industry 4.0 for Smart Factories

Qualitative Functional and Dysfunctional Analysis and Physical Modeling of an Eco-Designed Mechatronics System Using Coloured Petri-nets: Application on a Regenerative Braking System........... 495
Imane Mehdi and El Mostapha Boudi

On Permutation Flow Shop Scheduling Problem with Sequence-Independent Setup Time and Total Flow Time 507
Hajar Sadki, Jabrane Belabid, Said Aqil, and Karam Allali

A Novel Hybrid Heuristic Based on Ant Colony Algorithm for Solving Multi-product Inventory Routing Problem.............. 519
Fadoua Oudouar and El Miloud Zaoui

Overview of Lean Management Within PLM 530
Nada El Faydy and Laila El Abbadi

Comparison Study Between CB-SEM and PLS-SEM for Sustainable Supply Chain Innovation Model.......................... 537
Ahmed El Maalmi, Kaoutar Jenoui, and Laila El Abbadi

Correction to: Modelling of Cavitation in Transient Flow in Pipe-Water Hammer C1
Ouafae Rkibi, Nawal Achak, Bennasser Bahrar, and Kamal Gueraoui

Author Index.. 553

Smart and Sustainable Cities

Towards for an Agent-Based Model to Simulate Daily Mobility in Rabat Region

Khalid Qbouche[1(✉)] and Khadija Rhoulami[2]

[1] LRIT Associated Unit to CNRST (URAC°29), Faculty of Science, Mohammed V University in Rabat, 4 Av.Ibn Battouta, B.P. 1014 RP, 10006 Rabat, Morocco
qbouchekhalid@gmail.com
[2] DESTEC, FLSHR Mohammed V University in Rabat, Rabat, Morocco
rhoualmikhadija@yahoo.fr

Abstract. The issue of daily movements has become very important as a result of the world's urbanization, as it is concerned with studying the daily movements of the population and analyzing the behavior of individuals between the starting point, which is a person's place of residence, and the target, which is a person's place of work. This system is directly related to the urban field, especially traffic. In this work, we'll provide a mixed model of daily mobility and a person's shifting condition. The model is built on bottom-up techniques such as Multi-Agent Systems and Chain Markov, which allow for the creation of individual and individual group displacements. The model combines Bayesian Belief Networks and Markov Chains, allowing for the design of individual behavior displacement and management of the person's forecasting position, respectively.

Keywords: Daily mobility · Markov Chain (MC) · Multi-Agent System (MAS) · Bayesian Belief Network (BBN)

1 Introduction

Daily mobility is described as a collection of practices of population displacements in their normal context, where these displacements are carried out by individuals over some time and are also characterized by (motives, a moment of the day, origin, destination, duration, speed, means of transport used, etc.). As a result, these movements pertain to a variety of issues and analyses: geography, sociology, economics, urbanism, and transportation. The significance of the topic stems mostly from the difficulty of comprehending human behavior in cities, modeling and anticipating the consequences of mobility, and meeting societal transportation and equipment infrastructure demands. The expansion of the population can be viewed as a driving force behind daily mobility. As the population grows, so does the number of people who are displaced, which means that people are leaving their homes regularly for various reasons (work, school, etc.). Furthermore,

R. Saidi et al. (Eds.): ICATH 2021, LNDECT 110, pp. 3–9, 2022.
https://doi.org/10.1007/978-3-030-94188-8_1

it generates congestion, which has long been seen as a significant issue in daily mobility. Such everyday mobility phenomena are gaining in popularity, and they appear to be posing a significant problem for sustainable cities due to both their unforeseen negative consequences and the new societal demands they generate. Computer models are an effective tool for dealing with such issues. These models, with their prediction and simulation capabilities, might foresee the consequences of these phenomena and assist planners and decision-makers in evaluating the success of urban policies based on scenarios previously anticipated by models. For this paper, we used this models mathematics represented by: Markov Chain, Multi-agent System, and Bayesian Belief Network to study displacement of persons in Rabat region. The remainder of this paper is structured as follows. In Sect. 2, we will have the works in the context of daily mobility and the methods of our proposed model. In Sect. 3, we will detail the methods that we have applied and are: Chain Markov (CM), Multi-agent System (MAS) and Bayesian Belief Network (BBN). In Sect. 4, we describe our experiments before to conclude.

2 Related Works

Daily mobility is considered a social process that allows making changes in the urban area and the lifestyles of populations of different categories. To model this phenomenon and to predict future scenarios, researchers have used multi-agent systems and mathematical models to predict. Among these models, we could find Cellular Automata, Markov chain, and Bayesian belief network. For multi-agent systems, we could find the Gama platform to simulate the movements of the person [5]. Used system multi-agent Gama platform for simulating daily mobility in Rabat region using census data [2]. BBN was used to create a spatial model that simulates land-use change as a result of human land-use decision-making. [3] they used CA-Markov to predict urban growth in 2059 and examined the urban area at the base of the 1984–2020 dates, they found that urban growth will continue in the future in Eskişehir, which is experiencing a rapid urbanization process. [1] they extended a mobility model called Mobility Markov Chain (MMC) that allows incorporating the n previously visited locations and also they developed a new algorithm for the prediction of the next location based on this mobility model that we called n-MMC. In addition, we found using mobile data traces to study mobility prediction technique for movement in wireless network, such as [7]. [11] worked on urban transportation system simulations. And, utilizing a layer of roads, lighting, parking, and modes of transportation, create a multi-agent system paradigm. This paradigm is also being applied in the city of La Rochelle, based on INSEE statistics data (the French National Institute of Statistics and Economic Studies).

3 Methods

The total system consists of a collection of houses, urban space, roadways, green areas, and people's activities depicted between their places of employment and dwelling. We'll look at how the MC, MAS, gama-platform, and BNN methods were utilized to model the complete system based on the system's components.

3.1 Chain Markov

The Markov Chain (MC) is a mathematical technique for forecasting future land use dynamics. Informally, a Markov Chain is a set of random variables (X_n; n = 0; 1; ...) with values in a finite set known as the MC state space [4].
(X_n; n$\succ$ 0) is referred to be a Markov Chain if:

$$\forall i_0, i_1, .., i_{n-1}, i, j \in S, \forall n \geq 0,$$
$$P(X_{n+1} = j/X_0 = i_0, X_1 = i_1, .., X_n = i_n) = P(X_{n+1} = i_{n+1}/X_n = i_n)$$

At random, variable The state of the system at step n is represented by the symbol X_n. The MC process starts in one of the states and moves through the others. At the next phase, the process moves from i to j with a probability represented by P_{ij}. State X_{n+1} at stage n + 1 in the system might be calculated using syntaxes from state X_n at stage n.

$$X_{n+1} = \mathbf{P}X_n$$

P_{ij} the transition probability matrix, which is calculated using the formula, is made up of transition probabilities. The state vector at step n is X_n.

$$P = \begin{bmatrix} P_{11} & \dots & P_{1m} \\ . & \dots & . \\ . & \dots & . \\ P_{m1} & \dots & P_{mm} \end{bmatrix}$$

where P_ij is the chance of converting from type i to type j, and m is the number of various types of soil occupation. Using the Chapman-Kolmogorov Eq., which is based on the Markov stochastic process theory, we can determine the transition probability from state i to state j in precisely n steps.

$$P_{ij}^n = \sum\nolimits_{k \in S} P_{ik} P_{kj}^{(n-1)}$$

The initial transition probability matrix might be used to create the n transition probability matrix, where S is the MC state space.

3.2 Multi-agent System

A multi-agent system consists of a collection of agents that act and interact in a shared environment. A self-contained agent is sociable, reactive, and proactive. The goal of the MAS method is to design the city's population dynamics at the size of each individual actor [8]. Each person in our paradigm is represented by a cognitive agent. Agents are supplied information based on attributes (such as age, educational level, professional position, and kind of activity section). We used information from HCP 2004 [6] to determine each agent's housing and employment location. In this paper, we used the Gama-platform to simulate the movement of people in the Rabat region, it is a simulation platform with a complete integrated modeling and simulation development environment to create spatially explicit agent-based models [10]. In addition, it specializes in simulating individuals in metropolitan networks, such as [5].

3.3 Bayesian Belief Network

Bayesian Belief Network (BBN) is a probabilistic graphical model that uses a directed acyclic graph to describe a set of random variables and their conditional relationships [9]. We utilized BBN to design the displacement process in our model, which is based on explicit variables that allow individuals to choose whether or not to go to work (Fig. 1).

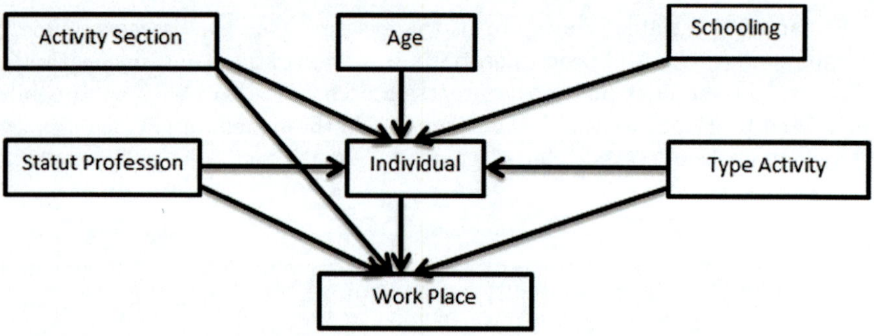

Fig. 1. The Bayesian belief network's representation of explanatory variables.

The displacement propensity of a person is calculated using Eq.

$$P(X_1..X_n) = \prod_{i=1}^{n} P(X_i = x_i/X_j = x_j)$$

For each X_j that is a parent of X_i.

The variable of the place of work is based on a set of variables from the model of BBN that we proposed as an example: Type of activity of the individual, professional status of each individual.

4 Experimental Evaluation

A number of structures, both residential and industrial, as well as a roadway network, are included in our simulation model. The study area is represented by all of these components.

Figure 2 depicts the simulation of people's d in the Rabat region daily mobility, which was carried out using the multi-agent system Gama, which is based on the Dijkstra algorithm, which is a method for finding the shortest path in a network, in this instance the network of streets in Rabat. We simulated people's travel between their place of residence and their place of employment, using data from HCP's national survey of 2014 [6]. The black point represents the location of residence, while the blue point represents the place of work. In our simulation, we used the Chain Markov model, where the initial state, i.e. the

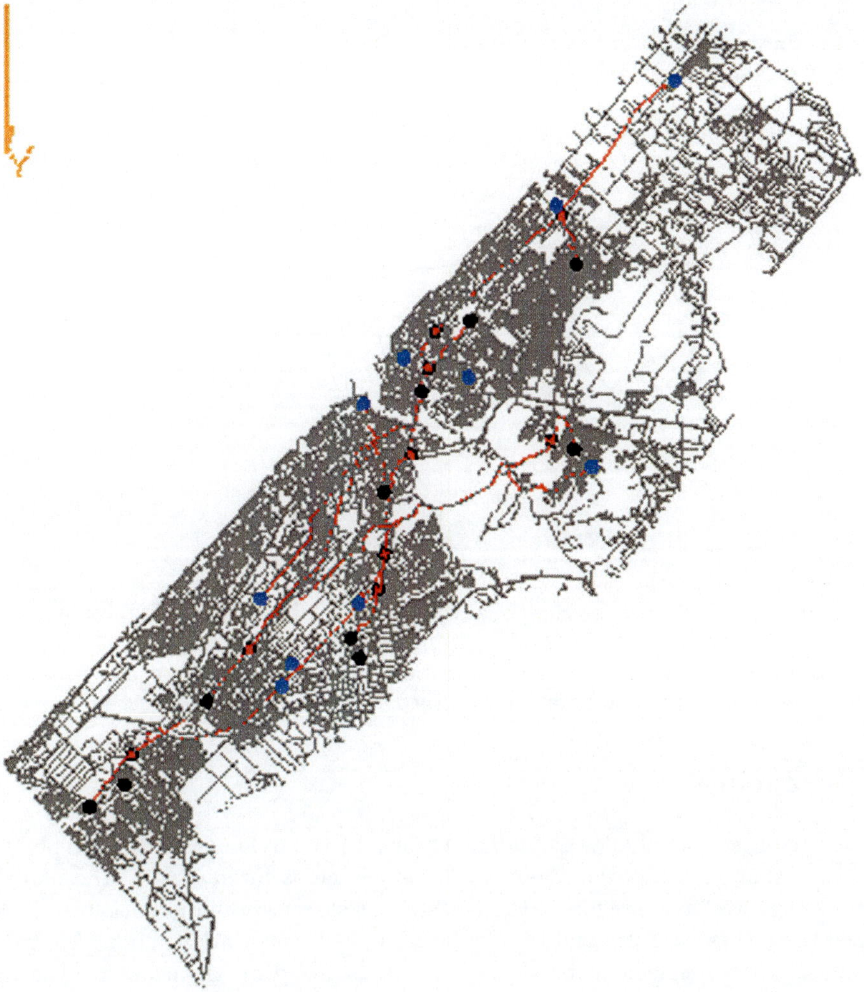

Fig. 2. Simulation of the daily mobility of people in the Rabat region using the Gama-platform.

location of departure, was the place of residence, and the next state was the site of employment, with a time factor of between 7 a.m. and 9 a.m. as the time to enter. Figure 3 depicts a chart made up of the distribution of the number of individuals who have joined their workplace and the number of people who have not yet been reached due to traffic issues such as traffic jams.

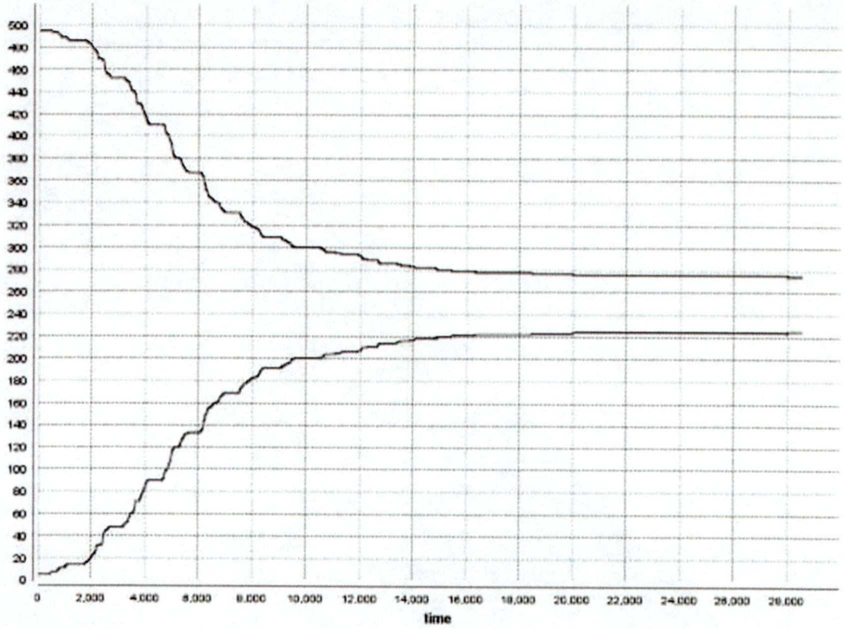

Fig. 3. The diagram represents the distribution of the people who have joined the work places and the people who have not yet joined.

5 Conclusion

The presented model allows building a vision of the dynamics of daily mobility in the urban space in the region of Rabat. What is interesting in the modeling is that we have integrated the multi-agent system in our situation we have used the Gama platform and the stochastic models the Markov model. We have simulated the displacements of the individuals based on the characteristics of each individual 'place of residence, place of work, professional status ...' are the information provided by HCP [6]. We may test and simulate many scenarios of a certain region using the model we've built. For instance, we might mimic people's mobility based on indications like 'time', allowing us to see who has arrived at work and who has not a cause of congestion or other reasons. This information enables decision-makers and urban planners to assess whether the city's architecture meets mobility needs or not. The validation of the model will be provided in the future study, which will be validated by comparing the everyday movement in the Rabat region utilizing validation methodologies.

References

1. Gambs, S., Killijian, M.-O., del Prado Cortez, M.N.: Next place prediction using mobility Markov chains. In: Proceedings of the 1st Workshop on Measurement, Privacy, and Mobility, MPM 2012 (2012). https://doi.org/10.1145/2181196.2181199

2. Kocabas, V., Dragicevic, S.: Bayesian networks and agent-based modeling approach for urban land-use and population density change: a BNAS model. J. Geogr. Syst. **15**(4), 403–426 (2012). https://doi.org/10.1007/s10109-012-0171-2
3. Ersin, A., Öncü, M., Bayar, R., Yılmaz, M.: Eskişehir Kentsel Büyüme Alanın Hücresel Otomat ve CA-Markov Zincirleri ile Analizi (1984–2056) (2020)
4. Agbinya, J.: Markov Chain and Its Applications an Introduction (2020)
5. Qbouche, K., Rhoulami, K.: Simulation daily mobility in Rabat region (2021). https://doi.org/10.1145/3454127.3454128
6. Institutional site of the High Commission for Planning of the Kingdom of Morocco. https://www.hcp.ma/
7. Hasb, R., Syed Ariffin, S.H., Fisal, N.: Mobility prediction method for vehicular network using Markov chain. Jurnal Teknol. **78** (2016). https://doi.org/10.11113/jt.v78.8885
8. Giret, A., Botti, V.: Multi agent systems of multi agent systems (2021)
9. Monti, S., Cooper, G.F.: Learning Bayesian belief networks with neural network estimators. In: Advances in Neural Information Processing Systems, pp. 578–584 (1997)
10. An official website Gama Platform. https://gama-platform.github.io/
11. Nguyen, T., Bouju, A., Estraillier, P.: Multi-agent architecture with space-time components for the simulation of urban transportation systems. Proc. - Soc. Behav. Sci. **54**, 365–374 (2012). https://doi.org/10.1016/j.sbspro.2012.09.756

Check for updates

Scenes Segmentation in Self-driving Car Perception System Based U-Net and FCN Models

Aimad Lahbas[1(✉)], Azhar Hadmi[2], and Amina Radgui[1]

[1] National Institute of Posts and Telecommunications (INPT), STRS Lab - INPT, Rabat, Morocco
{lahbas.aimad,radgui}@inpt.ac.ma

[2] Higher Institute of Audiovisual and Film Professions (ISMAC), STRS Lab - INPT, Rabat, Morocco

Abstract. Visual perception for mobile robots and self-driving cars performs the task of semantic segmentation in order to comprehend the road scenes. Most of researchers in computer vision focuses to improve the accuracy of their image segmentation models. In autonomous driving application, the accuracy is not the only important metric but also it requires a model that can achieve a good Intersection Over Union (IoU). We propose a U-Net architecture in this work that allows and maintain a decent balance in terms of accuracy and (IoU) compared with the FCN architecture evaluated on CamVid dataset [4].

Keywords: Deep learning · Semantic segmentation · U-Net · Scene understanding · Fully Convolutional Networks (FCN)

1 Introduction

Visual perception is one of the fundamental component of self-driving systems. It allows autonomous vehicles making decisions and understanding their surrounding in order to achieve a full autonomy. Visual perception system has become the fundamental task in autonomous driving cars also for mobile robots. This is especially true with the rise of deep learning methods, as demonstrated by the fact that a convolutional neural network (CNN) named "AlexNet" [10] easily defeated the best old techniques in the ImageNet large-scale picture classification competition. Deep learning based computer vision methods have successfully overcome a number of difficult perception issues that autonomous robotics and self-driving cars have a several challenges to overcome like object detection and tracking, lane detection, semantic segmentation [19]. The visual perception especially the semantic segmentation task in autonomous driving field, tend to

R. Saidi et al. (Eds.): ICATH 2021, LNDECT 110, pp. 10–18, 2022.
https://doi.org/10.1007/978-3-030-94188-8_2

perform an object segmentation methods to classify and label each pixel in image to understand and have the ability to analyze the road scene to get information that can help the ego vehicle to behave during different road scenarios.

2 Previous Works

Computer vision and especially the semantic segmentation task uses Convolutional neural networks that successfully perform this task by classifying each pixel in an image, which serves the goal of scene understanding. Scene parsing/understanding has important benefits in the mobile robots field [3,11,18]. The most notable benefit is in the autonomous driving area [5,16,20], and there is also the use of semantic segmentation in the medical application [15,21]. The most notable CNNs architectures used in semantic segmentation is FCN [12], and U-Net [15] that is the top of the list architecture used in medical applications [17]. However, deploying an autonomous driving system not only requires an architecture that only can afford the best performance in term of accuracy, but also we should focus on various aspects for semantic segmentation, like mean Intersection-Over-Union (mIOU) and pixel-wise accuracy. These aspects ensures the information about the architecture performance, in term of predicting the boundaries of each semantic label that appears in the scene, which is a critical aspect in the autonomous driving area. In this paper, we are using U-Net architecture [15] that is achieved significant results in medical application and apply it to perform scene understanding in autonomous driving application. Our contribution in this paper gives a good achievement in term of accuracy and mean intersection over union (mioU) compared to the common used architecture in scene understanding FCN [12].

3 Deep Learning

Deep learning models can learn and represent data at multiple levels of abstraction by simulating how the brain sees and analyzes information, thanks to the use of multiple processing layers. As a result, the complex structures of large-scale data are implicitly captured. The deep learning family of approaches includes neural networks, hierarchical probabilistic models, and a variety of unsupervised and supervised feature learning algorithms. Deep learning methods have recently attracted attention due to their ability to outperform previous state-of-the-art techniques in a variety of tasks, as well as the large amount of complex data from various sources (visual, audio, sensors, etc.). The desire to create a system that could simulate human brain sparked the development of neural networks. Walter Pitts and Warren McCulloch [13] developed a computer model based on human brain neural networks in 1943, which gave birth to deep learning. The McCulloch and Pitts (MCP) model is used to simulate the mental process. Since then, a number of significant contributions to the field have emerged, as shown in Fig. 1, including CNN and Long Short-Term Memory (LSTM) [7], which serve as the basis for computer vision applications like object detection [14], semantic segmentation [19], and motion tracking [6].

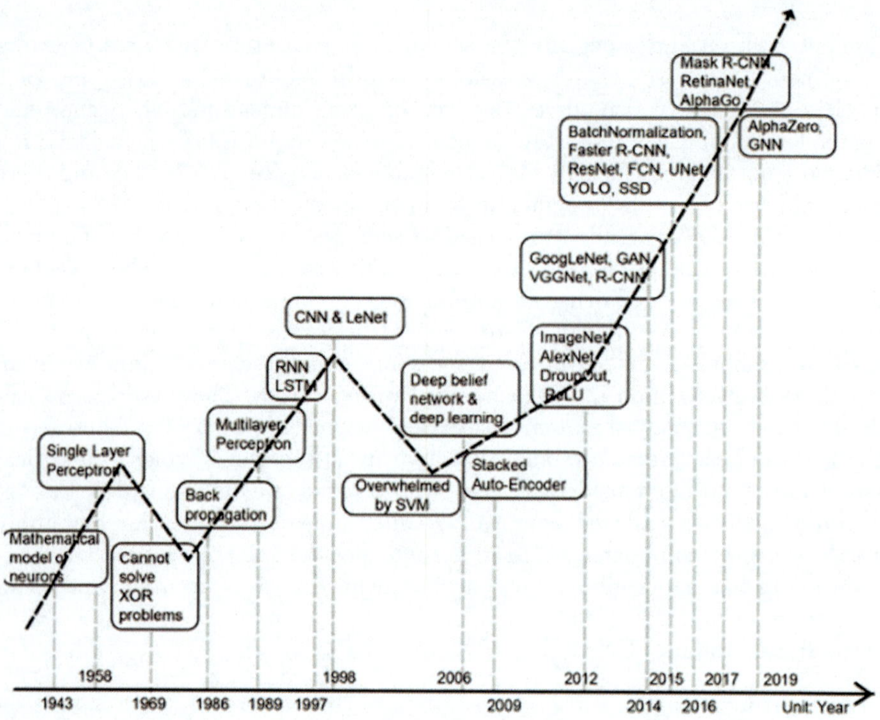

Fig. 1. A significant turning point in the historical memory of neural networks, opening the way for deep learning.

4 Methods

In our paper, we are comparing the U-Net architecture and FCN model to perform the semantic segmentation task in the field of self-driving cars based on CamVid dataset [4].

4.1 FCN Architecture

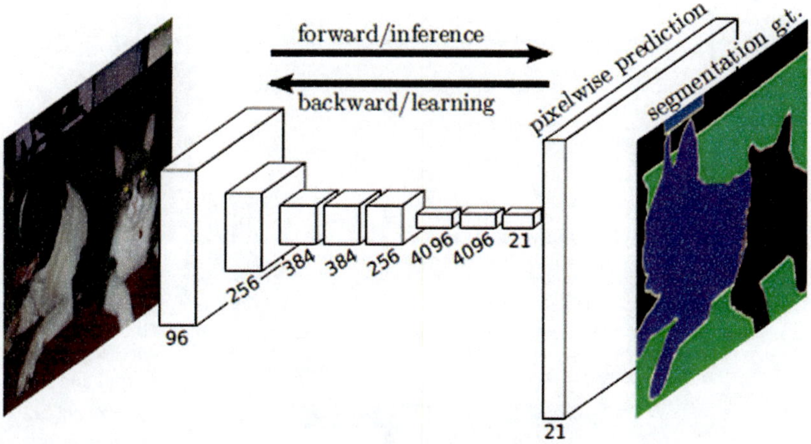

Fig. 2. FCN architecture representation.

As illustrated in Fig. 2, in a ConvNet, each data layer is a representation of a three-dimensional array that has a size of h * w * d. d is the feature or channel dimension, whereas h and w are spatial dimensions. The layer at the start is the input image with a pixel size of h * w, the color channels represented by d. Higher-level locations correspond to the image locations to which they are path-connected, which are referred to their receptive fields. ConvNets are based on the concept of translation invariance. Their fundamental components that contain convolution blocs, pooling and activation function [12] perform only on relative spatial coordinates and only on limited input regions.

4.2 U-Net Architecture

Figure 3 illustrates the U-net architecture. The multi-channel feature map is represented by the blue blocks. White blocks represent copied feature maps, with arrows indicating different operations [9]. The U-Net network architecture is a neural network that consists of encoders and decoders, left side called (encoder) and for right side (decoder) The encoder block is built using the standard CNN architecture. It's made up of two 3 * 3 convolutions that are used over and over again called (unpadded convolutions). For downsampling a Rectified Linear Unit (ReLU) function and a max pooling operation of 2 * 2 with stride 2 added to each operation [8]. With every step of downsampling, we add an extra number of feature channel. The decoder block starts with an upsampling of the feature map, followed by a 2 * 2 convolution called (up-convolution), which split the number of feature channels by two. Two 3 * 3 convolutions, each followed by a ReLU, and a sequences with a cropped encoder block feature map. Because of

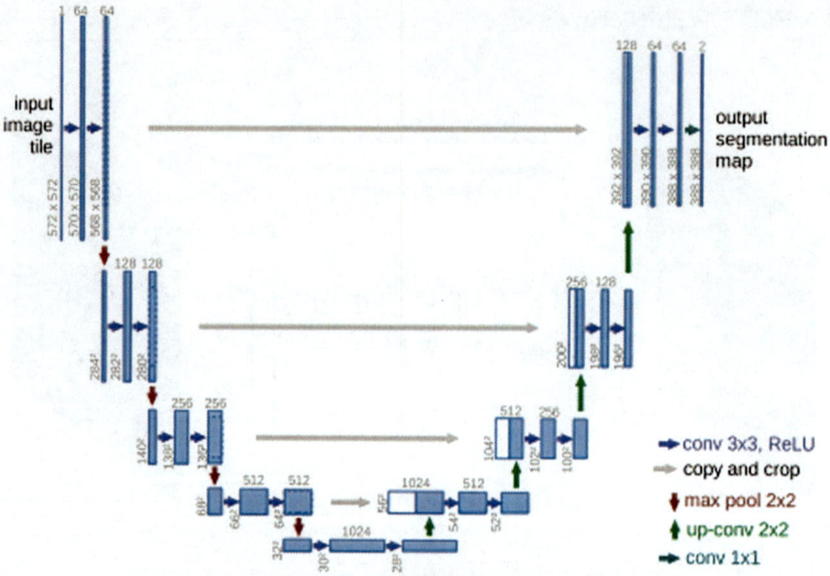

Fig. 3. U-net architecture. The blue rectangular blocks corresponds to multi-channel feature map, white rectangular blocks is copied feature maps representation, and arrows denote different operation [5].

the loss of boundary pixels within every convolution, cropping is necessary. At the final layer, a 1 * 1 convolution is used to map each component feature vector to the preferred number of classes. The whole network has 23 convolution layers.

4.3 Categorical Crossentropy

In our work, we are facing a semantic segmentation problem which is a multi class segmentation in our case. For this task we are adopting a categorical crossentropy [2] as the loss function used to train our models. This loss function is a perfect solution for multi-class classification tasks, because one example can be classified as belonging to one class with probability of 1 and another with 0. the loss function is defined as:

$$loss = -\sum_{i=1}^{N_{class}} y_i * \log y_{pred}$$

where y_i is the true label, N_{class} is the number of classes that our model should predict, and y_{pred} is the probability of each class.

5 Experiment

For the experiment section, we will discuss the experiment phase and show the segmentation results using our proposed architectures based on U-Net and FCN. This results allow to tackle the scene understanding applied to autonomous driving area by getting contextual information required to analyze road scenes scenarios.

5.1 Experiment Setup

We implement our architecture on top of tensorflow framework [1]. During implementation we are using Categorical cross entropy [2] for both used architectures to get the labeled class output. The optimizer adam [9] is used with learning rate of $1e^{-5}$ for U-Net and Stochastic Gradient Descent (SGD) [9] with learning rate of $1e^{-2}$ and momentum of 0.9 is used for FCN. Also a batch normalization [8] is included for both models. Our proposed architectures has trained using end-to-end training. We have tested our architectures on (CamVid) dataset [11] whos comprised of 700 frames (960×720) specified manually of 32 labels captured by a car mounted camera at daytime. The 700 images are splited up as follow : 367 with annotation for training, 100 validation images, and 233 for test. Also the experiment was conducted with (224×224) images to reduce time of training phase.

5.2 Results

To evaluate our architectures as semantic segmentation model, we are using two metrics that is significant in the autonomous driving area, which are the accuracy and the mean IOU (mIoU). In addition, to compare our result we are testing the FCN [12] and U-Net with Camvid dataset [4].

Experiment results on CamVid dataset are showed in Table 1 describing the segmentation results from our used models. Thus, U-Net model provides best results compared with FCN in term of accuracy also in term of (mIoU). Figure 4 shows qualitative segmentation results on CamVid dataset using our proposed architecture U-Net, also Fig. 5 shows predicted results of two scenes where we can see some sort of similarity in term of class prediction between both models, otherwise U-Net shows better performance in term of segmentation mask prediction. The obtained predicted images show that the U-Net architecture can produce accurate segmentation results (Table 1).

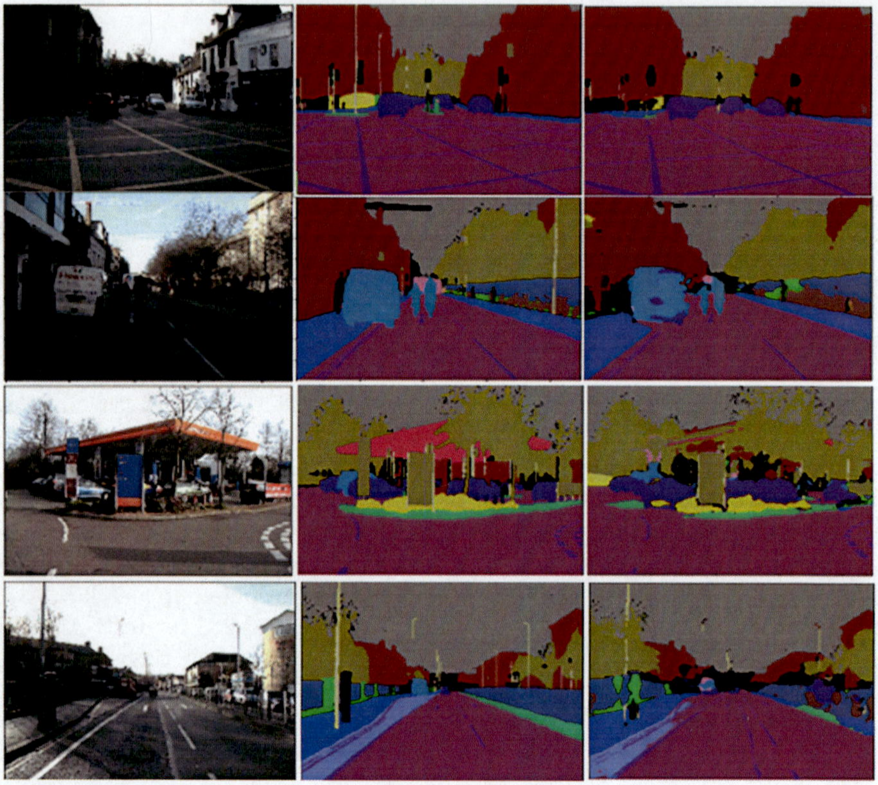

Fig. 4. Qualitative results on CamVid dataset, (left) the original images, (middle) ground truth, and (right) the predicted images.

Table 1. Comparison between results of our proposed architectures U-Net and FCN

Model	*mIoU %*	*Accuracy %*
U-Net	**52.60**	**89.07**
FCN	37.55	87.73

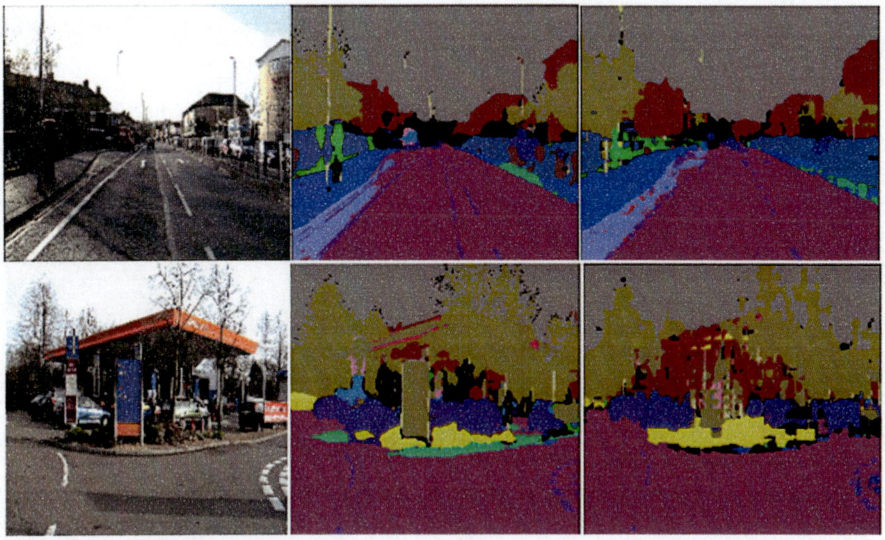

Fig. 5. Comparison results between models, (left) the original images, (middle) U-Net output, and (right) FCN output.

6 Conclusion

In our presented work, we have implemented a semantic segmentation architecture based encoder-decoder U-Net that was generally used in the medical application to perform semantic segmentation on the autonomous driving application field compared with the FCN architecture. We believe that the optimal balance of (mIoU) and accuracy, which are critical metrics in the autonomous driving application, make the U-Net architecture important to perform semantic segmentation for self-driving cars.

References

1. Abadi, M., et al.: TensorFlow: large-scale machine learning on heterogeneous systems (2015). https://www.tensorflow.org/. Software available from tensorflow.org
2. Bisong, E.: Introduction to Scikit-Learn, pp. 215–229 (2019)
3. Bonanni, T.M., Pennisi, A., Bloisi, D., Iocchi, L., Nardi, D.: Human-robot collaboration for semantic labeling of the environment, vol. 7 (2013)
4. Brostow, G.J., Fauqueur, J., Cipolla, R.: Semantic object classes in video: a high-definition ground truth database. Pattern Recogn. Lett. **30**(2), 88–97 (2009)
5. Cordts, M., et al.: The cityscapes dataset for semantic urban scene understanding, pp. 3213–3223 (2016)
6. Doulamis, N., Voulodimos, A.: FAST-MDL: fast adaptive supervised training of multi-layered deep learning models for consistent object tracking and classification, pp. 318–323 (2016)

7. Hochreiter, S., Schmidhuber, J.: Long short-term memory. Neural Comput. **9**(8), 1735–1780 (1997)
8. Ioffe, S., Szegedy, C.: Batch normalization: accelerating deep network training by reducing internal covariate shift, pp. 448–456 (2015)
9. Kingma, D.P., Ba, J.: Adam: a method for stochastic optimization. arXiv preprint arXiv:1412.6980 (2014)
10. Krizhevsky, A., Sutskever, I., Hinton, G.E.: ImageNet classification with deep convolutional neural networks. Adv. Neural. Inf. Process. Syst. **25**, 1097–1105 (2012)
11. Kundu, A., Li, Y., Dellaert, F., Li, F., Rehg, J.M.: Joint semantic segmentation and 3D reconstruction from monocular video. In: Fleet, D., Pajdla, T., Schiele, B., Tuytelaars, T. (eds.) ECCV 2014. LNCS, vol. 8694, pp. 703–718. Springer, Cham (2014). https://doi.org/10.1007/978-3-319-10599-4_45
12. Long, J., Shelhamer, E., Darrell, T.: Fully convolutional networks for semantic segmentation, pp. 3431–3440 (2015)
13. McCulloch, W.S., Pitts, W.: A logical calculus of the ideas immanent in nervous activity. Bull. Math. Biophys. **5**(4), 115–133 (1943)
14. Ouyang, W., et al.: DeepID-Net: object detection with deformable part based convolutional neural networks. IEEE Trans. Pattern Anal. Mach. Intell. **39**(7), 1320–1334 (2016)
15. Ronneberger, O., Fischer, P., Brox, T.: U-net: convolutional networks for biomedical image segmentation. In: Navab, N., Hornegger, J., Wells, W.M., Frangi, A.F. (eds.) MICCAI 2015. LNCS, vol. 9351, pp. 234–241. Springer, Cham (2015). https://doi.org/10.1007/978-3-319-24574-4_28
16. Ros, G., Sellart, L., Materzynska, J., Vazquez, D., Lopez, A.M.: The synthia dataset: a large collection of synthetic images for semantic segmentation of urban scenes, pp. 3234–3243 (2016)
17. Siddique, N., Paheding, S., Elkin, C.P., Devabhaktuni, V.: U-net and its variants for medical image segmentation: a review of theory and applications. IEEE Access **9**, 82031–82057 (2021). https://doi.org/10.1109/ACCESS.2021.3086020
18. Valada, A., Oliveira, G.L., Brox, T., Burgard, W.: Deep multispectral semantic scene understanding of forested environments using multimodal fusion. In: Kulić, D., Nakamura, Y., Khatib, O., Venture, G. (eds.) ISER 2016. SPAR, vol. 1, pp. 465–477. Springer, Cham (2017). https://doi.org/10.1007/978-3-319-50115-4_41
19. Wang, P., Chen, P., Yuan, Y., Liu, D., Huang, Z., Hou, X., Cottrell, G.: Understanding convolution for semantic segmentation, pp. 1451–1460 (2018)
20. Zhang, H., Geiger, A., Urtasun, R.: Understanding high-level semantics by modeling traffic patterns, pp. 3056–3063 (2013)
21. Zhu, W., Xiang, X., Tran, T.D., Xie, X.: Adversarial deep structural networks for mammographic mass segmentation. arXiv preprint arXiv:1612.05970 (2016)

Autonomous Parallel Parking Using Raspberry-PI

Mahmoud Wagdy, Ahmed Roushdy, Ahmed Sabry, Abdallah Mostafa, Mahmoud Magdy, and Amal S. Mehanna(✉)

Department of Digital media Technology, Future University in Egypt, New Cairo, Egypt
{20172368,20171380,20172312,20173233,mahmoud.abdo,amal.safwa}@fue.edu.eg

Abstract. Due to the observed growth in the number of vehicles in recent years, a parking spaces availability problem has occurred. In this paper, a proposed system is implemented to, achieve a low complexity autonomous parallel parking system, to decrease traffic congestion, achieve less fuel consumption, and prevent vehicle damage like scratches and other damages due to miscalculations from drivers. The target is to minimize the parking time using an autonomous system that leads to an increase in the efficiency of traffic flow. The algorithm will be implemented using a Raspberry-PI microcomputer. A hybrid hardware and software system will be used with a low error percentage.

Keywords: Parallel parking technique · Autonomous parking · Microcontroller · Mobile application

1 Introduction

In recent years, the availability of parking spaces became a concern after the increase in the number of vehicles, especially in big cities [3]. In the last decade, many intelligent driver assistance systems I- DAS, have been implemented by car manufactures such as driver drowsiness detectors, proximity indicators, and parallel parking systems. Different types of sensors are used to gather information from the surrounding environment to prevent any accident or damage [1,2]. In labs and industries, many researchers and automotive manufactures are working on developing automatic parking systems to achieve high security and increase safety for smart transportation [4,5] Information communication technologies traffic management, and infotainment. Nowadays, cameras and sensors are implemented on the roadside have gained increasing attention and importance in modern transportation systems in different areas including safety, infrastructures to collect data about environmental and traffic conditions and other sensors are attached to the vehicle.

Guerrero et al. [a1] discusses some of the challenges, like the applicability of the cooperative intelligent transportation systems and also how the integration between the sensor technology attached to the car and the transportation

R. Saidi et al. (Eds.): ICATH 2021, LNDECT 110, pp. 19–28, 2022.
https://doi.org/10.1007/978-3-030-94188-8_3

infrastructure can achieve a sustainable Intelligent Transportation System (ITS), Finally the authors proposed some of the challenges that need to be considered to implement a fully operational and ITS environment.

In this paper, a parking system has been constructed to overcome the challenges. The proposed algorithm is integrating the smartphone application and the embedded system technology into autonomous parallel parking by pressing one button in the smartphone. The paper first explains different types of parking systems in 2, then in Sect. 3, different autonomous parallel parking techniques are illustrated. The methodology and the proposed technique are discussed in Sect. 4. Section 5, provides the experimental results and the performance of the proposed technique. Finally, Sect. 6 represents the conclusion.

2 Types of Parking Techniques

In Fig. 1, three parking types have been listed and explained in Table 1. The proposed algorithm is applied to parallel parking.

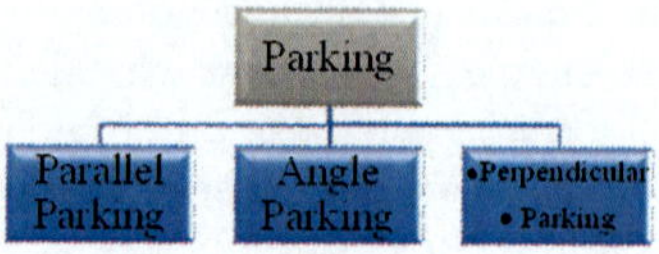

Fig. 1. Parking types

In the Parallel parking technique, all the vehicles are parking on the same line, parallel to the pavement and the target is to park in between two cars, one at the front and one at the back.

Table 1. Different parking types

Parking Type	Image
Parallel Parking	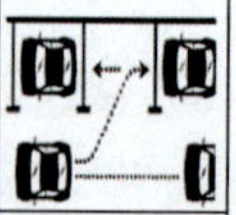
Angle Parking	
Perpendicular Parking	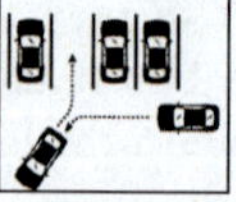

In the Angle Parking technique, all the vehicles are aligned in an angle which is much easier than the parallel way. The most difficult one is when The Vehicles are aligned perpendicular to the pavement, beside each other, that increases the number of sensors.

3 Parallel Parking Techniques

In 2020, S. Das et al. [1] proposed a fuzzy and conventional lateral automated parallel parking control algorithm using ultrasonic sensors to assist the driver based on a path planning algorithm and longitudinal aid for parking. The Experimental results showed that the performance of the proposed algorithm is accurate compared to other parallel parking techniques. The proposed model is integrated with a real vehicle to validate the parking efficiency.

J. Zhang et al. [2] described an autonomous parallel parking methodology for a front- wheel steering vehicle to solve the problem of longitudinal velocity planning and path planning problems. Vehicle path control strategy is based on longitudinal velocity, Lyapunov stability theory, and smooth handoff method. The proposed methodology performance is verified based on the model-in-the-loop simulation system.

In [3], a design and implementation of an algorithm for automatic parallel parking, applied to four-wheeled scaled vehicle mobile robots is presented to avoid a collision.

The PID controller is designed to handle the error generated between the previous objective and the real position.

The simulation results show that the proposed algorithm considers as a solution for the tracking problems.

4 Design and Implementation

4.1 System Description

In Fig. 3, the system architecture of the proposed model is shown. The system consists of three main components, the Raspberry-pi, the H-Bridge and the Ultrasonic sensor. The Raspberry-pi is the brain of the system, that uses the H-Bridge motor controller to control the speed, direction, and as a brake for the autonomous car.

The Ultrasonic sensor, is used to calculate the distances that are needed to park the vehicle in an accurate way, and also used as an environmental data collector.

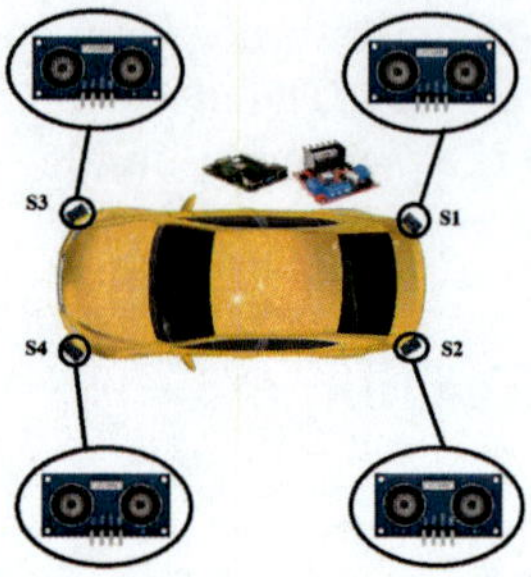

Fig. 2. Four sensors architecture

4.2 Hardware

A description of the components are listed in Table 2. The proposed system consists of four ultrasonic sensors attached to the car body, S1 on the back right, S2 on the back left, S3 on the front right, and S4 front left as shown in Fig. 2. All sensors are connected to the car raspberry Pi, the brain of the system, to send and receive data. Also, the car motor drivers are attached to the kit to receive the commands of motion and steering, the circuit diagram of the proposed model is illustrated in Fig. 4 and Fig. 5.

In Fig. 3, Raspberry pi is shown, it is a Linux-based with small size, which provides all functionalities of a desktop computer. Raspberry-pi is used for various applications such as parking systems, robots, and home automation. This kit has USB, Ethernet port, and various input and output pins to connect all available sensors. Here in our project, we have used Raspberry pi kit due to its processing power, small size, and variety of pins.

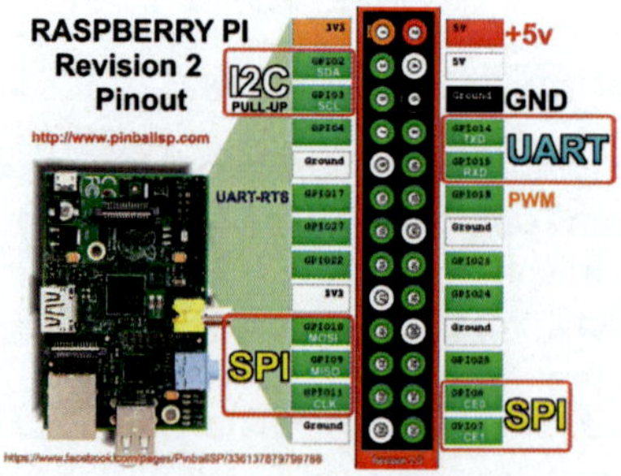

Fig. 3. H-bridge with Raspberry-PI circuit diagram

In the proposed system, HC-SR04 ultrasonic sensor has been used with raspberry pi to find the distance between the car and the object as shown in Fig. 4, HC-SR04 is a good choice because it provides high range accuracy with more transmitters and also has a stable performance. It has been used in various applications in Vehicle parking, and Robotics.

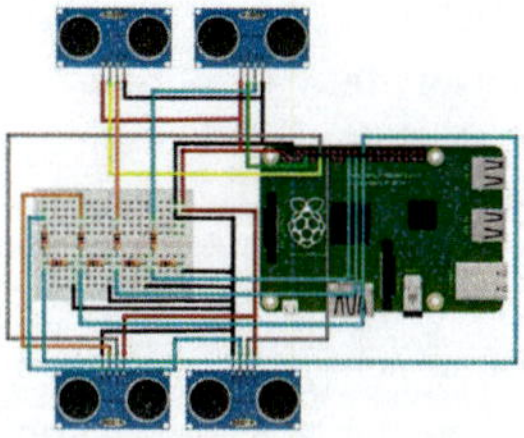

Fig. 4. Ultrasonic sensors with Raspberry-PI circuit diagram

We used geared DC motors to move and steer the car. With high torque. The direction of rotation would also be reversed. And the speed of the motor is proportional to the force.

An H-bridge is illustrated in Fig. 5 Often, it is used in robotics and we used it here in our proposed project to control speed and direction of a motor by applying voltage across a load for both directions.

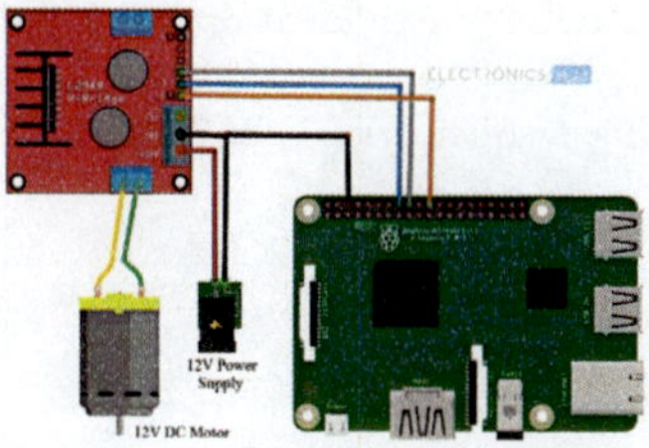

Fig. 5. H-bridge with Raspberry-PI circuit diagram

The GT-U7 module is a stand-alone GPS receiver, this module provides a high performance location detection even in the most challenging environments.

Table 2. Hardware components

Component	Description	Image
Raspberry- PI (PI-3 model B)	– Raspberry pi is a pocket-sized microprocessor with a Linux operating system. – The capability to connect sensors through GPIO pins – Analyze sensor data to detect obstacles and calculate the distance and control H-bridge ...	
Ultrasonic Sensors (HC-SR04)	– An ultrasound sensor is used to measure the distance and detect obstacles. – HC-SR04 has good performance with(0.3 cm resolution, from 2 to 450 cm detection range).	
H-Bridge (Dual Driver)	– H-bridge is a simple motor driver to control motor direction and speed – Peak current Io: 2A, Maximum power consumption: 20W	
GT-7U GPS module	– A low power consumption with high sensitivity. – It has extremely high location tracking sensitivity. – Operating frequency: L1 (1575.42 +/- 10MHz).	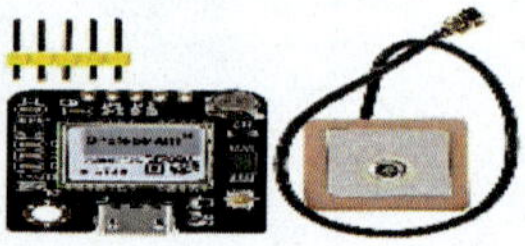
DS18B20 Temperature sensor	– Sensing Temperature : – Sensor type: Local • 55 °C ∼ 125 °C – Accuracy : • ±0.5°C (±2°C)	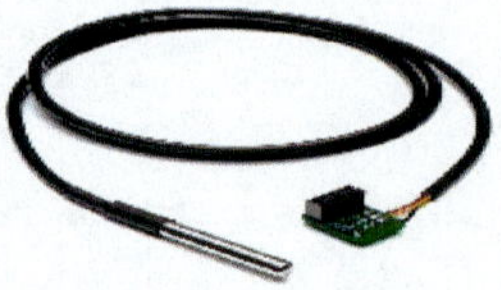
JGY370-3000 Dc Motor	– 12v DC Worm Gear Motor – 3000 rpm 6v -24v Reversed High Torque	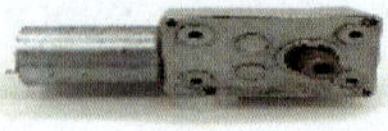

4.3 Methodology

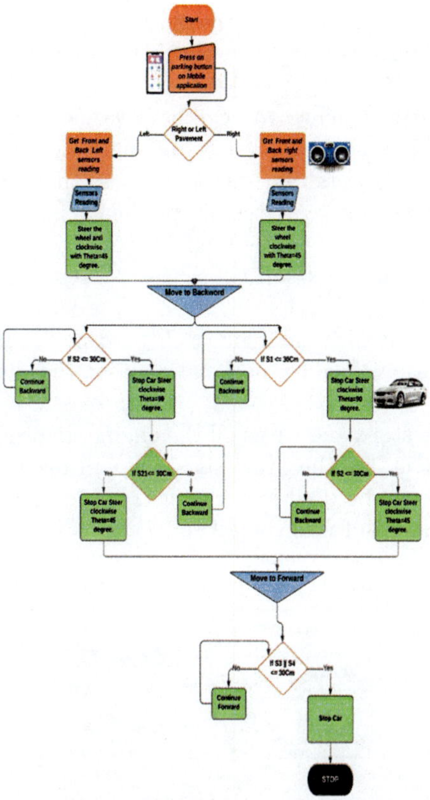

Fig. 6. Flow chart of parking steps

In this paper, the parallel technique is chosen, so the initial position of the vehicle is a major issue that the whole technique will depend on. The system starts by sending to the driver signals that contain the distance between his vehicle and the first front car to assure him that the initial position is proper.

The proposed system will collect information about the parking area using four ultrasonic sensors located on each corner of the vehicle to measure the distance between the two obstacles, the front car and the back one.

There are two scenarios for the proposed parallel parking, the right and left scenarios. For the right scenario, the driver must receive a reading from sensor 4 and sensor 1 that the distance between his vehicle and the adjacent one is 50 cm and for the left scenario, sensor 3 and sensor 2 are used.

The vehicle starts controlling the speed and directions by the control system according to the continuous sensor feedback, in order to place the vehicle in an appropriate way. Figure 8, will illustrate the parking steps. The flow chart given in Fig. 6, discusses the flow of both the right and the left parallel parking also discusses all the sensors location that are used in each scenario.

To place the vehicle on the right pavement:

- The controller requests reading from the front and back right ultrasonic sensor to collect information about the surrounding environment.
- According to the sensor reading, the controller starts to steer the wheel clockwise with Theta = 45°.
- The controller starts to move the car backwards with continuous feedback from two backward sensors on the left and right sides of the vehicle.
- When the S1 sensor detects that distance is less than or equal to the threshold1 = 30 cm then the brake system stops the car.
- The car starts to steer the wheel anticlockwise with an angle = 90° and keeps moving in a backwards direction till sensor 2's reading is less than or equal to 30 cm.
- Then steers the wheel with an angle = 45° and moves in a forward direction till sensor 3 or sensor 4 send a notification that the distance is less than or equal 30 cm.
- Finally, the vehicle stops and turns off the engine. In order to place the vehicle on the left pavement, all the previous steps will occur but instead of using sensor 1, sensor 2 will be used and vice versa, and instead of using sensor 4, sensor 3 will be used and vice versa (Fig. 7).

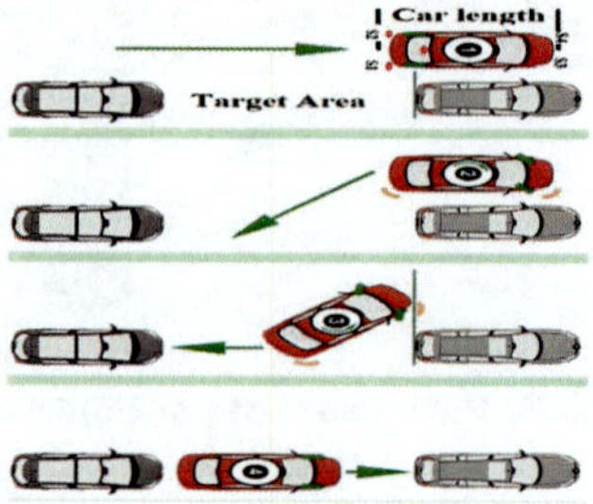

Fig. 7. Car parallel parking step

Mobile Application. Cross-platform application has been designed to track, control, and check the status of the car remotely. For the tracking part, the proposed application locates the car's location using a GPS module that is attached to the car Raspberry pi board. As a pre-processing phase, user can send a request to the microcontroller, via the proposed application, to investigate about battery level, car brake, and the status of motor drivers. An automatic notification will be sent from the system, to inform the driver that he placed the car in a proper location.

Some features are added to enhance the security of the vehicle against robbery. A high-definition cameras are added to capture frequent images for the surrounding environment in the current location, also the vehicle cannot start without the existence of an authorized person, some of the mobile application's tabs shown in Fig. 8.

Our application contains three main screens first, login page to access application functionality and connect the car, second screen to check all car sensors (temperature, location, obstacles around car) and return values to user before start engine, the last one for auto parking car by positioning the car at the parking location then select which direction to park then let car to handle.

Fig. 8. Mobile application screens

5 Results and Discussion

The proposed system is tested on different target areas. The average of the experimental results are listed in the both Tables 3, one standard car is used with different targeted areas.

The results show that that the system could not park the car in small areas and the car can only park when the distance is proper.

Table 3. Results 1

Category		Standard car 1	Standard car 2	Two-wheeler
Car dimension	Length (m)	3.72	3.72	1.97
	Width (m)	1.44	1.44	0.74
Target length (m)		5	3.90	2.70
Back sensor (cm)	S1	29.5	29.5	29.2
	S2	–	–	–
Front Sensor (cm)	S3	29.3	12	30
	S4	29	15	29.8
Pavement		Right	Right	Right
Error (cm)	Back	0.5	Cannot	0.8
	Front	0.85	Park	0.1

All the threshold values chosen to handle car motion are based on try and error for the sensors reading, sensor operating angles, and also the car height of the ground. Also the error rate and safety of the system are considered. In Table 3, two different types of vehicle were used. The first one was a tiny two wheeler car and the second was a car. The results show that the error raised in the car and was a bit small in the tiny two wheeler car.

6 Conclusion

Using the proposed autonomous parallel technique allows the driver to park his vehicle in a stress-free. The algorithm has been applied to different sizes of Vehicle. Tables 3 show a comparison between a normal size of a car, a big one, a tiny two wheeler and a bus. It can be observed from the results that there is a proportional relation between the size of the vehicle and the error percentage, adding more sensors can be used in large vehicles to reduce errors. In Table 3, the proposed system was applied to the same standard car but in a different parking space. The system didn't allow the user to park the car in a small area.

The proposed method is preferred. Firstly, it is easy to implement and it is economical as well. Ultrasonic sensors provides exact distance with a few error decimal places which can be negligible.

References

1. Ghita, N., Kloetzer, M.: Trajectory planning for a car-like robot by environment abstraction. Robot. Auton. Syst. **60**, 609–619 (2012)
2. Keall, M.D., Fildes, B., Newstad, S.: Real-world evaluation of the effectiveness of reversing camera and parking sensor technologies in preventing backover pedestrian injuries. Accid. Anal. Prev. **99**, 39–43 (2017)
3. Shin, J., Jun, H.: A study on smart parking guidance algorithms. Transp. Res. Part C **44**, 299–317 (2014)
4. Filatov, D.M., Serykh, E.V., Kopichev, M.M., Weinmesiter, A.V.: Autonomous parking control system of four-wheeled vehicle, pp 102–107. IEEE (2016)
5. Gupta, A., Divekar, R.: Autonomous parallel parking methodology for Ackerman configured vehicles. ACEEE Int. J. Commun. **1**(2), 22–27 (2010)
6. Chirca, M., Guillaume, M., Roland, C., Debain, C., Lenain, R.: Autonomous valet parking system architecture. In: IEEE International Conference of Intelligent Transportation Systems, vol. 18, pp. 2619–2624 (2015)
7. Das, S., Reshma Sheerin, M., Nair, S.R.P., Vora, P.B., Dey, R., Sheta, M.A.: Path tracking and control for parallel parking. In: 2020 International Conference on Image Processing and Robotics (ICIP), pp. 1–6 (2020). https://doi.org/10.1109/ICIP48927.2020.9367343
8. Zhang, J., Shi, Z., Yang, X., Zhao, J.: Trajectory planning and tracking control for autonomous parallel parking of a non-holonomic vehicle. Meas. Control (U.K.) **53**(9–10), 1800–1816 (2020). https://doi.org/10.1177/0020294020944961
9. Ballinas, E., Montiel, O., Castillo, O., Rubio, Y., Aguilar, L.T.: Automatic parallel parking algorithm for a car-like robot using fuzzy PD+I control. Eng. Lett. **26**(4), 447–454 (2018)

A Survey on the Application of the Internet of Things in the Diagnosis of Autism Spectrum Disorder

Fatima Ez Zahra El Arbaoui(✉), Kaoutar El Hari, and Rajaa Saidi

SI2M Laboratory, National Institute of Statistics and Applied Economics (INSEA), Rabat, Morocco
{f.elarbaoui,k.elhari,r.saidi}@insea.ac.ma

Abstract. In this paper, we provide a survey that explores the application of the Internet of Things (IoT) to diagnose Autism Spectrum Disorder (ASD). This disorder represents several limitations in social interaction, communication, etc. Research has shown that early detection has tremendous importance to reduce the negative impacts of autism. IoT innovations in the field of autism detection have proven to be a great potential to improve the clinical process. And yet, we survey studies on IoT for ASD detection and we classify them into: Eye-tracking technology, social interaction and behaviors monitoring, play behavior analysis, vocalizations analysis and facial expression recognition.

Keywords: Autism · IoT · Diagnosis of autism · Sensing technology · Autism Spectrum Disorder (ASD)

1 Introduction

Autism is a difficult condition that requires high accessibly to support and a strong social commitment. According to the Non-Governmental Organization "Vaincre l'autisme", autism affects 680,000 people in Morocco, including 12,800 births per year. It is necessary to sustain a treatment over the long-haul for the affected person since there is no known medicinal cure. Autism is variable; it has numerous symptoms and severity levels. Some people with autism may need special support, others are able to respond best to the appropriate treatment plans and therapies. Therefore, they can learn, work, and live independently.

Several methods have been used for screening ASD. Before the revision of the third version of Diagnostic and Statistical Manual of Mental Disorders (DSM) [51] or DSM-IV [53] and The International Classification of Diseases-10 (ICD-10) criteria [52], the Childhood Autism Rating Scale (CARS) [54] was invented to provide ratings related to autism and it was used by clinicians, teachers and in a parent interview. Afterwards, DSM-IV and ICD-10 were developed, and they consider the social deficits. The E-1 and E-2 scales [55] were used for direct observation of autism behaviors. Since the diagnosis of autism in school-age is more reliable, two methods where used to measure autism

R. Saidi et al. (Eds.): ICATH 2021, LNDECT 110, pp. 29–41, 2022.
https://doi.org/10.1007/978-3-030-94188-8_4

severity at this stage: the Autism Diagnostic Interview -Revised (ADI-R) [56] and the Autism Diagnostic Observation Schedule -Generic (ADOS-G), and its previous versions: the PL-ADOS and ADOS [57]. These frameworks provide standardized diagnose [1]. However, the principal limitation of these methods is the fact that they are extensive and time-consuming.

Meanwhile, IoT has become a trending topic in health care. Many IoT-driven sensors and devices are already used in hospitals for data collection and analysis, their growth is noticeable. The key advantages of IoT systems are efficiency and real-time data access, most of them are body- related, while the rest measure environmental parameters [19]. And yet, recent advances in IoT have emerged a powerful platform to develop solutions for autism detection. In this work, we explore several studies that demonstrate how connected devices can be used to diagnose autism. The remaining sections of this survey are organized as follows: Sect. 2 provides the proposed categories for the classification of selected studies, Sect. 3 presents a discussion based on surveyed research studies. Finally, in the Sect. 4, we present the conclusion.

2 The Taxonomy of IoT Approaches for Autism Detection

After surveying and analyzing the existing studies that involve the usage of IoT-based solutions for autism detection, we considered that IoT can be implemented through different methods, therefore, we propose to classify studies according to the following categories that illustrate those methods:

2.1 Eye-Tracking

Oculometry has been widely used by specialists to characterize autism. A Long Short-Term Memory (LSTM) neural network was implemented for eye movement monitoring to identify the autism status. The system detects whether the individual is affected or not [7]. 17 autistic children and 15 Typically Developing (TD) ones participated in this study. Visual activities of these children was recorded and was divided into testing data and training data for the LSTM network. Regarding the low amount of data, the network still faces uncertainty and over-fitting. Thus, it cannot replace the medical diagnose. A research by [22] proposed the application of eye-tracking techniques to detect autism at early stages. In this approach a camera tracks the visual preference of the child when watching video scenes. A cascade algorithm was implemented to analyze the gaze direction and the entire system was compared with the method M-Chat for validation. The results suggested that the visual preference can contribute in the detection of autism risk. Further analyses were applied using eye-tracking data and suggested objective indicators for the qualification of ASD risk and symptoms [23–25]. In [23], an Autism Risk Index (ARI) indicator was created to demonstrate how remote eye tracking can be useful to identify the likelihood of autism. This ARI was based on the visual preference to social and nonsocial information and it had a significant accuracy. Furthermore, the authors suggested that the ARI may be combined with other clinical measures. Similarly, eye tracking-based meters have shown consistent effectiveness for detecting ASD risk and assessing symptoms severity [25]. In this study, the authors were based on large realistic eye-tracking data to create an ARI and an Autism Symptom Index (ASI) indicators. ARI and ASI have provided quantitative evaluation of ASD's symptoms and risk. They showed a good diagnostic accuracy and a correlation with clinical observation; Strong Cross-validated correlations were noted between ASI and ADOS-2 total and sub-scale severity scores while ARI had a good classification accuracy for ASD. These results indicated that eye tracking meters may accelerate the progress of the diagnose field.

Deep Neural Networks (DNN) have been applied to analyze eye movements and fixation data to detect possible signs of ASD [24]. This work proposed a DNN for saliency prediction. It is a proof-of-concept that reveals the possibility of improving the clinical process by machine learning. The authors identified the difference in eye movement patterns between neurotypical and ASD individuals when they look at natural image stimuli to extract discriminative features. They have reported a rate of 92% of accuracy in classifying ASD. Using DNN and Random Forest (RF) classification, a research by [26] suggested learning from eye tracking data features that identify hallmark characteristics of ASD. However, this study considers facial emotion recognition as well. The authors conducted an eye tracking experiment using a modified Dynamic Affect Recognition Evaluation (DARE) method as a facial emotion recognition task. 23 individuals with ASD and 35 Typically Developing (TD) ones participated and statistical analyses were applied to detect the differences between the two groups. A notable difference in the visual attention and eye-movement patterns was observed. Based on observations and data analysis, the authors proposed then a RF classifier that identifies individuals with autism.

In addition, a study by [27] presented an approach for visualizing eye tracking data and the classification of ASD and TD children. According to this method, children with ASD may have troubles to engage and interact with other children since they don't respond immediately to the stimuli as TD children. A small delay may lead to a loss of important information. Based on findings from experiments and observations, the authors developed a new measure named Distance to Reference (D2R) point. The D2R takes into account both spatial and temporal aspects to identify a simple visual presentation of eye-tracking data. Moreover, a proof of concept study was conducted by [28]. The authors of this study examined the fixation time on areas-of-interest (AOI) represented in a female face. 37 ASD and 37 TD children aged between 4 and 6 year-old participated in this study. They watched a video of a girl speaking for 10 s. The results indicated that fixation time for the body and the mouth can differentiate between ASD and TD children with a classification accuracy of 85.1%. The SMI RED250 system was used as an eye-tracker and an SVM algorithm was applied for classification. In a similar study, a machine learning framework was proposed to detect children with autism using the face scanning eye movement patterns [29]. The machine learning model showed a good performance for classification. The rate of accuracy is 88.51%, 86.21% for specificity, and a sensitivity of 93.10%. The authors used a SVM algorithm to classify ASD and TD children and a data driven feature extraction method. During experiments, they found that ASD children looked longer below the left eye and briefly in the right eye and above the month compared with TD children.

Eye contact is essential for nonverbal communication. However, people with autism have an atypical pattern of eye-contact and they need special sessions of therapy. Wearable smart glasses were designed to assess and measure the eye-contact behavior of children with ASD [5]. The smart glasses were equipped with mechatronic sensors and controllers. During a session, the instructor and the child worn the smart device, a smartphone application called WatchMe was connected to the instructor's glasses to record data and monitor the progress. WatchMe allows the instructor to personalize the child reword based on their enhancement in eye contact behavior.

EyeXplain Autism was presented by [33] as an interactive eye-tracking system that aid in screening ASD. The system allows the visualization and the interpretation of eye tracking data so physicians can analyze them to diagnose ASD cases. Eye movements were collected by the Tobii X2–60 eye tracker. Then, a DNN model was developed for ASD classification. To improve the medical decision-making for physicians, an interface was implemented for visualization and statistical graphs. Advancement in fields of eye tracking and machine learning techniques, has embarked into creation of other methods of ASD classification. In [34], 64 children were recruited for this eye-tracking experiment (32 ASD-Diagnosed and 32 non-ASD). The authors applied lasso, group lasso, and Sparsely Grouped Input Variables for Neural Networks (SGIN) algorithms to select eye-tracking stimuli and predict ADOS, IQ, (SRS) and adaptive functioning in children. An SR Eyelink 1000 Plus sampling binocularly at 500 Hz was utilized for gathering data and Samsung SyncMaster 2233RZ 22-inch monitor for experiments display. 9647 eye-tracking features were obtained from 109 stimuli using region-of-interest ROI analyses. Then, SGIN was implemented and compared with other popular machine learning methods. The authors suggested that SGIN achieves the best accuracy in ASD classification.

In [38], an approach based on eye-tracking, visualization and ML was proposed to diagnose ASD at early stages. A group of 59 children were invited to watch some videos. The SMI remote eye-tracker was utilized to record eye movements including fixation, saccade, and blink, it provides the (x, y) coordinates of gaze direction that aid in drawing the virtual path to recognize the gaze behavior of autistic children. Experiments were conducted using various ML models: Logistic Regression, SVM, Naive Bayes and Random Forests. However, the authors demonstrated that neural network models are more accurate than other approaches.

A study reported in [45] presented a system that aims to identify if the visual behavior is associated with autism or not. The system compromises eye movement statistics, image saliency and facial expressions. Features that are extracted from those elements were used by classifiers based on random forests to detect the scanpaths of individuals with autism. The model achieves a rate of 75% of accuracy.

2.2 Social Interaction and Behaviors Monitoring

Through years of research and efforts, many robots were developed to support autism. Thus, a parrot-like robot was developed as a tool for screening ASD [12]. This approach is based on the child's behaviors and his social interaction. Data were collected by a toy equipped by sensors and classified into the three main groups specified by the

DSM-Fourth Edition-Text Revision (DSM-IV-TR). Using extracted features, the Random Forest tree in Weka software was applied to identify ASD and TD children. A similar approach was adopted by [13]. A robot with the form of a bullet was developed. It was equipped by sensors and connected to an iPad interface for data analysis. The robot is used to diagnose autism by observing children's interaction with the robot.

In [1], authors were inspired by funding from NASA and they developed a wearable system for qualifying the interaction and the behavior of children with autism (face-to-face time, proximity, and the physical movements). They suggested that the collected data are useful for autism recognition, therapy, involvement, and the improvement of monitoring of autistic children. The system detects abnormality in children interaction and provides teachers and parents an accurate report for intervention. On the other hand, the authors of [36] proposed using musical instruments to measure autism sensory disorders. They designed an Augmented Musical Instrument (AMI) as sensors that perceives and detects the hypo/hypersensitivity behaviors among children. The method used Bogdashina as a psychological method to evaluate sensitivity and produce sensory profiles.

The atypical motor development is not a core element for ASD diagnosis, still, some proofs demonstrate that motor problems can be an earlier predictor for autism [47]. Some researchers have investigated the use of machine learning algorithms to detect motor stereotypy [40]. In this study, autistic individuals aged between 13 and 20 years were participated. They wore three wireless accelerometers during experiments. Accelerometers data were transmitted to a receiver. A pattern recognition algorithm was implemented to recognize, measure and classify stereotypical motor movements. Results suggested that the proposed method is more efficient than the direct behavioral observation and rating scales. Another automated movement recognition technique was proposed in [42]. This research investigated the use of the Microsoft Kinect v.2 to detect automatically measure autistic body movements. 12 TD participants took part in this study, they performed some stereotypical movements (hand flapping, body rocking and spinning). The Microsoft Kinect V2 recordings were processed by two methods: Visual Gesture Builder (VGB) to build models using Random Forest Regression (RFR) and MatLab. Results demonstrated that VGB approach outperformed the MatLab one.

Overall, to assess problem behaviors among individuals with developmental disabilities, specialists observe directly the behavior to provide the proper intervention. A promising alternative for the clinical human observation was introduced in [41]. The study aimed to enhance this process by designing on-body accelerometers and developing activity recognition techniques that measure and classify problem behaviors automatically. Accelerometers have been attached to the limbs to collect tri-axial acceleration data then a lightweight segmentation where applied to these data in order to identify behavior episodes before classification. The proposed system allows clinicians to record data of problem behavior in natural environments. In [43], authors examined only the upper limb movements to correctly classify High-risk (HR) and Low-Risk (LR) children for ASD. 17 children at high risk for autism (having older members of family with autism) and 15 low-risk children (having TD older sibling) participated in this study. They wear three sensors that record data and transmit it via Bluetooth to a (personal computer) PC. Two machine learning techniques where applied for the classification: SVM and Extreme Learning Machine (ELM), the estimation of their accuracy was %75.0, and % 81.67, respectively.

To predict ASD and Attention Deficit Hyperactivity Disorder (ADHD), a study conducted by [46] provided a system that use of facial expression analysis and behavior 3D analysis. Rather than the development of an automatic diagnostic approach for ADHD and ASD, the authors of this work showed the relationship between ASD/ADHD and facial expressions and motions. They elaborated a new database of video recordings to qualify algorithms that are based on computer vision. The Kinect 2.0 device was used to record videos of 57 adults participated in the work. The participants read and listened to 12 short stories during recordings. Statistical machine learning models were applied for classification. Furthermore, motor development was examined to identify early signs of ASD risk [16]. This study proposed working with Opal wearable sensors. Sensors were connected directly with the infant and record his behaviors and activities. These data were used to create a model based on theory that establish the link between lower motion complexity to motor disability, and ASD. An important sign of autism is repetitive motor behaviors that can be represented by less complex behaviors. This study demonstrated that children at high risk for autism and with a diagnosis of ASD have lower motion complexity then others.

Robots can be used in autism as a new screening tool. This was demonstrated in [32]. In this research a Child-Robot Interaction (CRI) system based on a robotic platform and a computer vision framework for visual attention analysis was designed and developed in order to supervise the child-robot interaction and examine nonverbal indications. The computer vision framework is built upon a network of RGBD sensors. Some patterns were extracted from the analysis of the child's behavior. The authors suggested that these patterns are promising to improve the traditional methods of diagnosing ASD.

However, autism has various subtypes. Environmental, genetic, and biological causes can induce this diversity [17]. Depending on the autism type, some patients may need important help. To deal with this problematic, a Belief Rule Base (BRB) system based on IoT was proposed [31]. A BRB system can be an expert system that manipulates qualitative and quantitative information. It is enhanced from IF–THEN rule base to build an accurate knowledge representation schema. The BRB system is different from the classic rule-based system because a belief degree is related with each possible result of a rule [20]. The proposed BRB system of [31] collects symptoms thought sensing nodes and classifies the autistic children into groups of different types of autism to provide them the appropriate treatment plan. The system is based on three layer: a presentation layer to ensure the interaction between the system and the user, an application layer used for the data transfer between the system and the computer and a data processing layer that handles the belief rule base and the sensor data storage.

2.3 Play Behavior Analysis

As noted earlier, learning early signs of autism is fundamental for timely intervention strategies. A toy car was invented to diagnose autism in its early stages [4]. The toy car contains accelerators that collect data. An SVM algorithm was applied to discern between children with autism and neurotypical ones. Experiments show that the autistic child plays with car's wheels much more compared with a neurotypical child. The system is a part of a screening room and results must be proved by an expert. Another approach to diagnose autism based on activity recognition and movement classification was designed by MoVeas (Monitoring and Visualization of Early Autism Signs) project [9]. Relevant data was extracted during play sessions where children were observed when playing with a sensorized toy. The system examines two solutions for movement classification: time delay neural network (TDNN) and recurrent neural network (RNN). The first neural network was chosen for its simplicity and the second one because it can identify short sequences. Experimental results show that TDNN provides higher accuracy and small amount of data was lost.

Observing motions and behaviors is crucial in screening autism, it's a common way performed by specialists for an early detection. Therefore, a prototype for early detecting autism warnings was developed. It is based on accelerometers, magnetometers, and gyroscope sensors embedded in toys (truck and an airplane). A data fusion algorithm was used to measure movements and identify autism signs [6]. This prototype can be a part of the diagnosis clinical process. An improved version of this system was reported in [35]. Based on [6], the authors of [35] presented a better IoT-based approach that allows the remote observation and analysis of children movements when playing in a natural context, to identify how children behave with the toys. They added a new layer to the data fusion algorithm that implements neural networks. A motion capture device was developed to measure movements and accelerations of the toys and sent it as JavaScript Object Notation (JSON) record to a remote server where data is processed by a neural network to classify typical movements. The resulting neural network models were deployed by TensorFlow.js library.

2.4 Vocalizations Analysis

In DSM-5, Some voice characteristics are often observed and examined to identify ASD symptoms (annotation, rhythm, Abnormal volume, etc.). The vocal behavior was examined by [8]. Vocalizations were used to build an ASD detecting solution that can be implemented on smart devices. This solution is based on studies that shown that some features extracted from children's vocalizations can be exploited to reveal some ASD markers. The smart devices record vocalizations, and a platform is designed to process those vocalizations and extracts results. The software records two types of vocalizations: vocalization when interaction with the device and everyday communication. Besides, in an earlier study, an acoustic-prosodic analysis was applied on pre-verbal vocalizations of 18-month old toddlers [30]. The authors extracted several acoustic-prosodic parameters from recorded videos and audios using the VoiceSauce software and MATALAB. Data were used to train an SVM and Probabilistic Neural Network (PNN) classifiers. Statistics were calculated for each vocalization sample for all parameters across all children and employed to train the classifiers. The authors compared between the two classifiers, they found that the PNN is more efficient for classification. The accuracy, the sensitivity, and the specificity of the PNN classifier exceed 95%. The presented work may be beneficial for clinicians to identify autism in early stage.

In order to examine abnormal prosody, the authors of [44] used a lavalier microphone, audio capture and a personal computer to record and analyze the pitch of 30 children with ASD and 51 TD children older than 2 years-old. An SVM was used for voice classification. The proposed method is suggested to be important in the clinical assessment since it is more accurate compared with human hearing judgments as shown by experiments. A similar approach was introduced in [48]. A system has been elaborated to detect autism among children. It compromises wearable accelerometer, audio and a video camera sensor. Machine Learning algorithms were used for classification and identifying vocal stereotypy in four autistic children who have a lack in verbal communication.

2.5 Facial Expression Recognition

According to [15] facial expressions are a sign for health conditions. Sleep disorder and excessive crying are common indicators for irritability, social interaction difficulties, learning troubles, and ASD. The authors of [15] proposed then a system based on deep learning and IoT edge computing to exploit baby's facial expressions for an early intervention. They extracted 128 features from each picture, then, they developed a multi-headed 1D (Dimension) Convolutional Neural Network (CNN) to classify it into one of the three proposed categories (Happy, Crying, or Sleeping). The deep learning model is deployed on an edge device and it works as a web service that predict the category of an image sent via a REST API.

In [36], a new protocol, Response to Name (RTN) was proposed for ASD diagnosis. This protocol evaluates the ability of the child to response after calling him by name. It involves the participation of the child, a physician, and the parent. The parent sit in front of the child. The task it relies on three steps. First, the distraction of the child by some toys, then the physician starts to call the child, generally four times depending on the reaction of the child. If the child didn't respond, the parent takes the place of the physician and start calling the child. A multi-sensor system using vision-based algorithms was developed to evaluate the performance of the child and acquire data during the session. The system includes the pedestrian detection and skeleton extraction, head pose estimation, speech recognition, facial expression recognition and gaze estimation.

3 Discussion

The using of IoT to diagnose ASD is demonstrated to be a powerful tool ameliorate the clinical process. In this survey, we selected 35 papers that show method of applying IoT in order to diagnose autism. Based on the number of the surveyed articles for each technique, we calculated the percent of their usage as demonstrated in Fig. 1.

Eye-tracking techniques were the most applied to detect autism signs by tracking eye-movement and analyzing the visual pattern of ASD individuals. On the contrary, less articles committed to applying facial expression recognition to qualify ASD.

Machine learning and deep learning techniques were the most used for classification, prediction, decision-making and data visualization (SVM, DNN, CNN, etc.). We observed that those techniques were widely used to process data gathered from sensors and devices. Usually, indicators such as specificity, accuracy, precision, and sensitivity of the proposed technique were calculated to demonstrated its performance. We observed

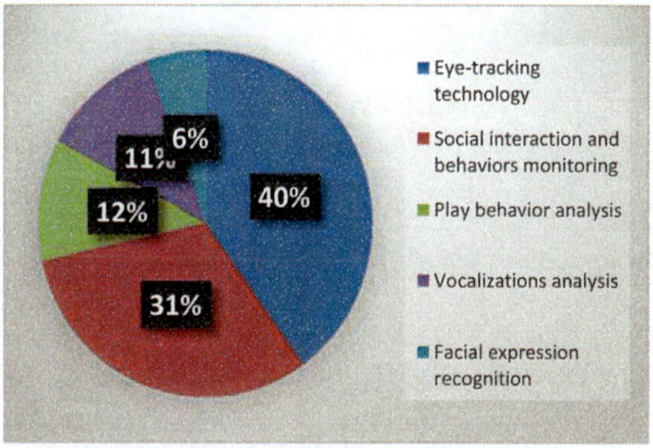

Fig.1. IoT-based approaches for autism diagnosis

that researchers sometimes applied more than one ML/DL method to compare between them and propose the most accurate one.

However, developing IoT-based systems for screening autism is hard and challenging. Fundamental issues were noted due to the small simples of data for methods validating and during experiments. Usually, few datasets can't demonstrate the efficiency of the method. On the other hand, sensors need to be strong, comfortable, invasive, and well implemented to avoid the loss of data and the distraction of children with ASD when they wear the device. ASD children are sensitive, sometimes they can't tolerate a wearable device. It is possible to establish an IoT system for screening ASD that can integrate non-wearable devices such as camera, microphones, etc. We calculated the percent of the use of wearable and non-wearable devices based on their presence in the surveyed studies as showed in Fig. 2.

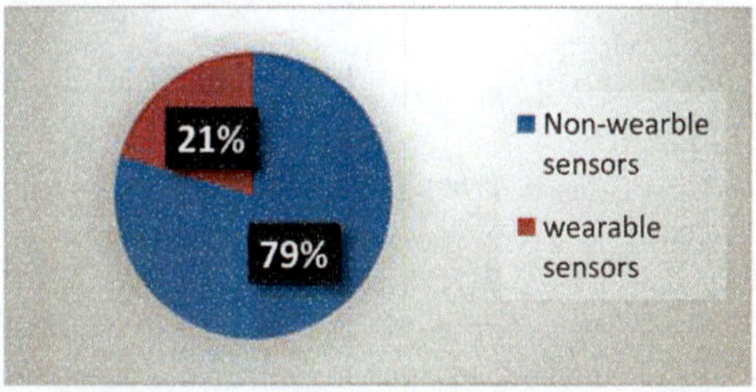

Fig.2. The percent of non-wearable and wearable sensors

The presented studies proved the feasibility of developing an IoT approach to diagnose ASD that can substitute the clinical process. However, improvements are required for a medical validation and adoption.

4 Conclusion

The application of IoT has a significant impact on improving the diagnosis mechanism of autism. There are researchers that suggest an early detection of autism signs to intervene rapidly. Most of studies exploit eye-tracking and behavior analysis to extract ASD markers. Different sensors were used, wearable and non-wearable ones. Each type of sensors is designed to acquire measures that can be processed and analyzed to detect autism signals.

Most of studies proposed a classification between children with autism and TD children by a ML or DL classifier algorithm based on the collected data. However, several challenges may face the algorithms such as accuracy and efficiency.

References

1. Shi, Y., et al.: An experimental wearable IoT for data-driven management of autism. In: 9th International Conference on Communication Systems and Networks (COMSNETS). IEEE (2017)
2. Cabibihan, J.J., et al.: Sensing technologies for autism spectrum disorder screening and intervention. Sensors (Basel) (2016)
3. Kowallik, A.E., Schweinberger, S.R.: Sensor-based technology for social information processing in autism: a review. Sensors (Basel) (2019)
4. Moradi, H., Amiri, S.E., Ghanavi, R., Aarabi, B.N., Pouretemad, H.-R.: Autism screening using an intelligent Toy car. In: Ochoa, S.F., Singh, P., Bravo, J. (eds.) UCAmI 2017. LNCS, vol. 10586, pp. 817–827. Springer, Cham (2017). https://doi.org/10.1007/978-3-319-67585-5_79
5. RajKumar, A., et al.: Wearable smart glasses for assessment of eye-contact behavior in children with autism. In: Proceedings of the 2019 Design of Medical Devices Conference DMD2019 (2019)
6. Lanini, M., Bondioli, M., Narzisi, A., Pelagatti, S., Chessa, S.: Sensorized Toys to identify the early 'red flags' of autistic spectrum disorders in preschoolers. In: Novais, P., et al. (eds.) ISAmI2018 2018. AISC, vol. 806, pp. 190–198. Springer, Cham (2019). https://doi.org/10.1007/978-3-030-01746-0_22
7. Carette, R., Cilia, F., Dequen, G., Bosche, J., Guerin, J.-L., Vandromme, L.: Automatic autism spectrum disorder detection thanks to Eye-tracking and neural network-based approach. In: Ahmed, M.U., Begum, S., Bastel, J.-B. (eds.) HealthyIoT 2017. LNICSSITE, vol. 225, pp. 75–81. Springer, Cham (2018). https://doi.org/10.1007/978-3-319-76213-5_11
8. Gong, Y., et al.: Automatic autism spectrum disorder detection using everyday vocalizations captured by smart devices. In: The 9th ACM Conference on Bioinformatics, Computational Biology, and Health Informatics (ACM BCB), pp. 456–473 (2018)
9. Bondioli, M., et al.: Analyzing the sensor data stream for monitoring and visualization of early autism signs (MoVEAS) (2021)
10. Yaneva, A., et al.: Review of early detection of autism based on eye-tracking and sensing technology. In: Conference: The Internet of Accessible Things (2018)
11. Moradi, H., Mohammad-Rezazadeh, I.: Recent advances in mechatronics devices: screening and rehabilitation devices for autism spectrum disorder. In: Zhang, D., Wei, B. (eds.) Advanced Mechatronics and MEMS Devices II. MN, pp. 283–296. Springer, Cham (2017). https://doi.org/10.1007/978-3-319-32180-6_13

12. Dehkordi, P.S., Moradi, H., Mahmoudi, M., Pouretemad, H.R.: The design, development, and deployment of roboparrot for screening autistic children. Int. J. Soc. Robot. **7**(4), 513–522 (2015). https://doi.org/10.1007/s12369-015-0309-8
13. Golliot, J., et al.: A tool to diagnose autism in children aged between two to five old: an exploratory study with the robot QueBall. In: The 10th Annual ACM/IEEE International Conference on Human-Robot Interaction Extended Abstracts, pp. 61–62 (2015)
14. Kaur, N., et al.: A systematic analysis of detection of autism spectrum disorder: IOT perspective. Int. J. Innovative Sci. Mod. Eng. (IJISME) **6**, 10–13 (2020)
15. Pathak, R., Singh, Y.: Real time baby facial expression recognition using deep learning and IoT edge computing. In: The 5th International Conference on Computing, Communication and Security (ICCCS) (2020)
16. Wilson, R.B., et al.: Using wearable sensor technology to measure motion complexity in infants at high familial risk for autism spectrum disorder. Sensors (Basel) (2017)
17. Grandin, T.: How people with autism think. In: Schopler, E., Mesibov, G.B. (eds.) Learning and Cognition in Autism, pp. 137–156. Plenum Press, New York (1995)
18. Grandin, T.: Visual abilities and sensory differences in a person with autism. Biol. Psychiat. **65**(1), 15–16 (2009)
19. Jayatilleka, I., Halgamuge, M.N.: Internet of things in healthcare: smart devices, sensors, and systems related to diseases and health conditions. In: Real-Time Data Analytics for Large Scale Sensor Data, pp. 1–35. Elsevier (2020)
20. Chang, L.L., et al.: Structure learning for belief rule base expert system: a comparative study. Knowl. Based Syst. **39**, 159–172 (2013)
21. Lord, C., Risi, S.: Frameworks and methods in diagnosing autism spectrum disorders. Ment. Retard. Dev. Disabil. Res. Rev. **4**, 90–96 (1998)
22. Vargas-Cuentas, N.I., et al.: Diagnosis of autism using an eye tracking system. IEEE 2016 Global Humanitarian Technology Conference (GHTC 2016) (2016)
23. Frazier, T.W., et al.: Development of an objective autism risk index using remote eye tracking. J. Am. Acad. Child Adolesc. Psychiatry **55**, 301–309 (2016)
24. Jiang, M., Zhao, Q.: Learning visual attention to identify people with autism spectrum disorder. In: Proceedings of the IEEE Conference on Computer Vision and Pattern Recognition, pp. 3267–3276 (2017)
25. Frazier, T.W., et al.: Development and validation of objective and quantitative eye tracking based measures of autism risk and symptom levels. J. Am. Acad. Child Adolesc. Psychiatry **57**(11), 858–866 (2018)
26. Ming, J., et al.: Classifying individuals with ASD through facial emotion recognition and eye-tracking. In: Annual International Conference of the IEEE Engineering in Medicine and Biology Society, pp. 6063–6068 (2019)
27. Falck-Ytter, T., et al.: Visualization and analysis of eye movement data from children with typical and atypical development. J. Autism Dev. Disord. **43**(10), 2249–2258 (2013)
28. Wan, G., et al.: Applying eye tracking to identify autism spectrum disorder in children. J. Autism Dev. Disord. **49**(1), 209–215 (2019)
29. Liu, W., et al.: Identifying children with autism spectrum disorder based on their face processing abnormality: a machine learning framework. Autism Res. **9**(8), 888–898 (2016)
30. Santos, J., et al.: Very early detection of autism spectrum disorders based on acoustic analysis of pre-verbal vocalizations of 18-month old toddlers. In: 2013 IEEE International Conference on Acoustics, Speech and Signal Processing, Vancouver, Canada, May 2013, pp. 7567–7571 (2013)
31. Alam, M.E., et al.: An IoT-belief rule base smart system to assess autism. In: The 4th International Conference on Electrical Engineering and Information and Communication Technology (iCEE-iCT). IEEE, New York (2018)

32. Ramirez-Duque, A.A.: Robotassisted diagnosis for children with autism spectrum disorder based on automated analysis of nonverbal cues. In: 7th IEEE International Conference on Biomedical Robotics and Biomechatronics (Biorob), pp. 456–461. IEEE (2018)
33. de Belen, J.R.A., et al.: EyeXplain autism: interactive system for eye tracking data analysis and deep neural network interpretation for autism spectrum disorder diagnosis. In: Conference: CHI 2021: CHI Conference on Human Factors in Computing Systems (2021)
34. Beibin, L., et al.: Selection of eye-tracking stimuli for prediction by sparsely grouped input variables for neural networks: towards biomarker refinement for autism. In: ACM Symposium on Eye Tracking Research and Applications, pp. 1–8 (2020)
35. Bondioli, M., Chessa, S., Narzisi, A., Pelagatti, S., Piotrowicz, D.: Capturing play activities of young children to detect autism red flags. In: Novais, P., Lloret, J., Chamoso, P., Carneiro, D., Navarro, E., Omatu, S. (eds.) ISAmI 2019. AISC, vol. 1006, pp. 71–79. Springer, Cham (2020). https://doi.org/10.1007/978-3-030-24097-4_9
36. Wang, Z., et al.: Early children with autism spectrum disorder via response-to-name protocol. IEEE Trans. Ind. Inform. **17**, 587–595 (2021)
37. Pandkar, C., et al.: Automations in the screening of autism spectrum disorder. Technium Rom. J. Appl. Sci. Technol. **2**(5), 123–131 (2020)
38. Carette, R., et al.: Learning to predict autism spectrum disorder based on the visual patterns of eye-tracking scanpaths. In: Healthinf, pp. 103–112 (2019)
39. Benoit, E., et al.: Musical instruments for the measurement of autism sensory disorders. J. Phys. Conf. Ser. **1379**, 012035 (2019)
40. Goodwin, M.S., Intille, S.S., Albinali, F., et al.: Automated detection of stereotypical motor movements. J. Autism Dev. Disord. **41**, 770–782 (2011). https://doi.org/10.1007/s10803-010-1102-z
41. Plötz, T., et al.: Automatic assessment of problem behavior in individuals with developmental disabilities. In: Proceedings of the 2012 ACM Conference on Ubiquitous Computing, pp. 391–400 (2012)
42. Kang, J.Y., et al.: Automated Tracking and Quantification of Autistic Behavioral Symptoms Using Microsoft Kinect, pp. 167–170. IOS Press, Amsterdam (2016)
43. Wedyan, M., Al-Jumaily, A.: Early diagnosis autism based on upper limb motor coordination in high risk subjects for autism. In: Proceedings of the 2016 IEEE International Symposium on Robotics and Intelligent Sensors (IRIS), pp. 13–18 (2016)
44. Nakai, Y., et al.: Detecting abnormal word utterances in children with autism spectrum disorders: machine-learningbased voice analysis versus speech therapists. Percept. Mot. Skills **124**, 961–973 (2017)
45. Startsev, M., Dorr, M.: Classifying autism spectrum disorder based on scanpaths and saliency. In: Proceedings - 2019 IEEE International Conference on Multimedia and Expo Workshops, ICMEW 2019. Institute of Electrical and Electronics Engineers Inc., pp. 633–636 (2019)
46. Jaiswal, S., et al.: Automatic detecion of ADHD and ASD from expressive behaviour in RGBD data. CoRR, pp. 762–769 (2016)
47. Travers, B.G., et al.: Motor difficulties in autism spectrum disorder: linking symptom severity and postural stability. J. Autism Dev. Disord. **43**(7), 1568–1583 (2013)
48. Min, C.H., Fetzner, J.: Vocal stereotypy detection: an initial step to understanding emotions of children with autism spectrum disorder. In: Annual International Conference of the IEEE Engineering in Medicine and Biology Society, pp. 3306–3309 (2018)
49. Smith, B.A., Trujillo-Priego, I.A., Lane, C.J., Finley, J.M., Horak, F.B.: Daily quantity of infant leg movement: wearable sensor algorithm and relationship to walking onset. Sensors **15**, 19006–19020 (2015)
50. Wang, S., et al.: Atypical visual saliency in autism spectrum disorder quantified through model-based eye tracking. Neuron **88**(3), 604–616 (2015)

51. American Psychiatric Association. Diagnostic and Statistical Manual of Mental Disorders, 3rd (ed.) (1987)
52. World Health Organization. International statistical classification of diseases and related health problems 10th edn (2016)
53. American Psychiatric Association. Diagnostic and Statistical Manual of Mental Disorders, 4th edn (1994)
54. Schopler, E., Van Bourgondien, M.E., Wellman, G.J., Love, S.R.: Childhood Autism Rating Scale, 2nd edn. Western Psychological Services, Los Angeles (2010)
55. Butler, S., Lord, C.: Rimland diagnostic form for behavior-disturbed children (E-2). In: Volkmar, F.R. (eds.) Encyclopedia of Autism Spectrum Disorders. Springer, New York (2013). https://doi.org/10.1007/978-1-4419-1698-3_914
56. de Bildt, A., et al.: Autism Diagnostic Interview-Revised (ADI-R) algorithms for toddlers and young preschoolers: application in a Non-US sample of 1,104 children. J. Autism Dev. Disord. **45**(7), 2076–2091 (2015). https://doi.org/10.1007/s10803-015-2372-2
57. Bastiaansen, J.A., Meffert, H., Hein, S., et al.: Diagnosing autism spectrum disorders in adults: the use of Autism Diagnostic Observation Schedule (ADOS) module 4. J Autism Dev Disord **41**, 1256–1266 (2011). https://doi.org/10.1007/s10803-010-1157-x

Cost Reduction in Smart Grid Considering Greenhouse Gas Emissions Using Genetic Algorithm

F. Z. Zahraoui[1,2](✉), H. E. Chakir[3], and H. Ouadi[1]

[1] Smartilab, ERERA, ENSAM, Mohammed V University, Rabat, Morocco
[2] EMSI Rabat, Morocco 3 EEIS-Lab, Rabat, Morocco
[3] ENSET Mohammedia, Hassan II University, Casablanca, Morocco

Abstract. The indiscriminate nature of renewable energy sources does not help the stability of the Micro Grid (MG), especially the power balance: equilibrium of produced and consumed power. MG uses one or multiple sources of renewable energy, even if it is connected to the electrical main grid the need to add auxiliary sources is explained due to some unexpected main grid power outages. In this setting, this paper proposes the management of the energy production in the MG considering: (i) consumption and weather predictions, and the main grid fees (ii) charging mode strategy of the energy storage system (ESS); (iii) energy buying or selling decision from/to the main grid. The objectives of this paper are: (i) reducing the daily energy bill of the MG; (ii) optimizing CO_2 emissions. This management is done using a Genetic Algorithm which is an evolutionary computation, and the obtained results are compared to a conventional management system.

Keywords: Optimal power flow · Micro-grid · Operational planning · Photovoltaic system · Hybrid system · Energy storage system · Genetic Algorithm

1 Introduction

Protecting our environment is becoming a major issue in the electricity grid. The introduction of different renewable energy sources in islanded or distributed grids, responds to the need to protect our planet by reducing CO_2 emissions and as an alternative to fossil energies. The structure of the grid has then changed [1,2,14].

Smart grids (SG) provide accurate data on-demand profiles, distributed sources will optimize their generated power according to the stated demand to improve efficiency and safety [3,4]. Toward consumers, SG maintains trustworthy and safe services and helps to minimize electricity expenses. For the grid operator, the SG furnishes information on the ideal power dispatch to avoid

R. Saidi et al. (Eds.): ICATH 2021, LNDECT 110, pp. 42–53, 2022.
https://doi.org/10.1007/978-3-030-94188-8_5

electricity outages: the supply and demand equilibrium can be easily made and updated.

There are several methods to calculate the optimized distribution of energies in an MG:

- Reduce the energy cost in the SG [5,10];
- Ensure the power balance: power generated and power of the loads [13,26];
- Ensure the stability of the grid [27,28].

Several algorithms have been used as optimization strategies in SG to reach these goals. Different combinations of hybrid electric generators have been used in these SGs. Below, a review of some related papers:

The authors of [5], used the Bellman algorithm in a hybrid grid to optimize the production and the sale costs to reduce the CO_2 emission in a grid containing PV cells, batteries, and a gas turbine. In [6], a genetic algorithm based on dynamic pricing is used to minimize the daily electricity cost. It enhances bidirectional interactions and reduces power imbalance between the thermal power plants, the renewable energy systems and, the 33 buildings connected to the grid. G. Lei et al., in [7], worked on a system made of wind turbines, photovoltaic panels as renewables. A hydrogen tank, an electrolyzer, and a fuel cell are used as a storage system. Employing a modified seagull optimization technique, the target of this paper is to obtain the lowest power generation price and the components' best size. In [8], a Multi-Objective Particle Swarm Optimization (MOPSO) is used to reduce: the outages of Power Supply Probability and the Energy price; taking into account the Greenhouse gasses. In this SG, solar PVs, Wind Turbines, and ESS are connected. A QuasiOppositional Swine Influenza Model-based Optimization with Quarantine was applied in [9], to minimize operational cost in an MG taking into consideration ESS size optimization. The MG is grid-connected and combined with Photo-Voltaic cells, Wind Turbines (WT), Fuel Cells, and Micro-Turbines. L. Wei in [10], proposes reinforcement learning based on energy control in SG including the intermittent behavior of the electric vehicles and wind turbines connected to the main grid. In [11], ϵ-constraint approach is applied in a day-ahead scheduling problem to reduce three objectives: the operation cost and the CO_2 emission, the cost of load curtailment, and the coordination of shiftable Loads from the WTs output power. This smart microgrid can trade energy with the main grid and can also receive energy from diesel generators and wind turbines. The authors of [12] aim to satisfy power reduction of the energy costs in real-time (ECRT) by considering the latency factor (FoL). The study was done considering a SG that generates electricity from Photovoltaic (PV), hydro and thermal power; to efficiently manage and control the power operations. In [13], a Linear programming method is used for optimal energy management to minimize the power price and maximize the electricity trade with the power grid. The system has a wind turbine, photovoltaic cells, and energy storage system, the results were compared with the Genetic Algorithm method.

Green House Gas emissions are not considered in [6,9,10,12,13]. In [5,8] the grid cost has a fixed price, and it's independent of the market energy prices during the day (on-peak/off-peak periods). The energy storage system has not been used in papers [6,10,13] even if these systems have proved their benefits in time and peak shifting and energy generation cost reduction [29]. ESS enables the exchange of the generated energy from the PV system from off-peak periods to on-peak ones. Authors of [8,13], have neglected battery cost and its maintenance. [11,12] deal with problems from demand-side management, unlike our study where we will optimize the generated power of distributed sources.

The studied MG is connected to the main grid. It is composed of (see Fig. 1):

- Photovoltaic field of 10 kWp, considered as renewable energy;
- a Wind turbine system of 150 kWp, considered as renewable energy;
- Gas Turbine of rated power 30 kW, considered as an auxiliary source;
- a Lithium-ion battery of 125 kW, considered as a storage system;
- a main single-phase grid 220V/50 Hz.

The goal of this paper is to present a power management strategy over 24 h, to reduce greenhouse emissions and electricity power bills. This optimization depends on the weather forecast, load profile, the ESS state of charge SOC, the electricity price, and the energy cost of each source. We will establish a technical and economic model for the different energy sources and propose a Genetic Algorithm model (GA) to minimize the energy cost, respect the environment, and respond to the load.

This article is structured as follows: Sect. 2 details the considered MG architecture, the electric model, functional costs, and the technical constraints associated with each source. The proposed GA model management optimization program is also described in this Sect. 2. Section 3 is dedicated to the simulation results. They are discussed and compared with a conventional management system.

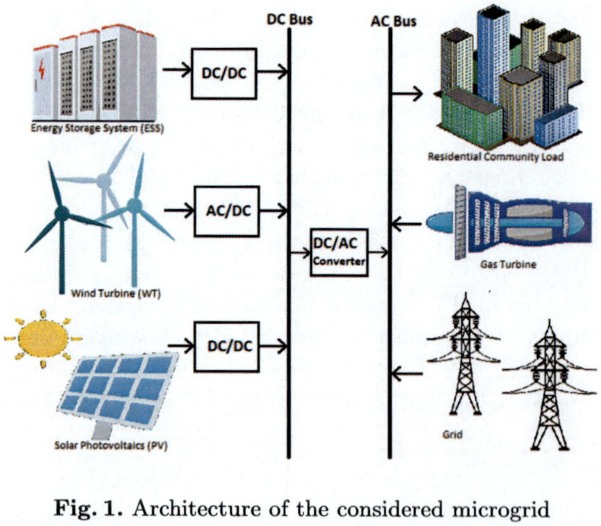

Fig. 1. Architecture of the considered microgrid

2 System Description

2.1 Distributed Sources Model

Photovoltaic Model: A photovoltaic (PV) cell is illustrated with a diode. This solar system generates an hourly power determined with [13,14]:

$$P_{PV\_GEN} = A_{PV} \times \eta_{PV} \times I_{PV} \tag{1}$$

$P_{PV\_GEN} = A_{PV}$ power generated in 1 h, A_{PV} is the surface of PV array (m^2), η_{PV} is the efficiency of PV cell and I_{PV} is the solar irradiation incident on the PV array in 1 h (kWh/m^2).

Wind Turbine Model: The maximum aerodynamic power P_{PV_GEN}, generated by these turbines, is calculated with [16]:

$$P_{WT\_GEN} = \begin{cases} 137.17V^2 & V_{cut_{in}} \leq V \leq V_r \\ 137.17V_R^3 & V_R \leq V \leq V_{cut\_out} \\ 0.0 < V_{cut\_in}\ orV > V\_cut\_out & \end{cases} \tag{2}$$

Where: V, V_R, V_{cut-in}, and $V_{cut-out}$ are respectively actual wind speed, rated speed, cut-in speed, and cut-out speed.

Gas Turbine Model: A Gas Turbine (GT) is used as an auxiliary source. The turbine has an instant response time, but this combustion turbine causes CO, CO_2, and NOx emissions [15]. Economical cost and emission characteristics are considered in the participation of this generator.

Grid Model: Due to the intermittency of the renewable sources, the grid must supply the load and charge the batteries and also be able to buy the excess power produced by PV cells and WT. The electricity cost varies throughout the day: high cost for peak periods, medium cost for standard periods, and low cost for off-peak periods. The National Office of Electricity and Water (ONEE) is the unique buyer and distributor of energy in Morocco. Electricity is produced by other independent companies. In 2015, the 58–15 law, allowed private producers to sell their surplus energy to the national grid, but not more than 20% of their annual production.

2.2 Optimization Problem

Objective Function (Cost Function)

$$Cp = \sum_{t=t_0}^{24} (C_{ESS} + C_{GT} + C_{Grid} + C_{CO_{2eq}}) \tag{3}$$

Photovoltaic and Wind Turbine Cost. In this work, the power produced by renewables: PV cells and Wind Turbines are considered with no cost to maximize the use of these generators.

Gas Turbine Cost. The cost of Gas Turbine (GT) power is the sum of the used natural gas, the ON/OFF GT operation Cost, and the CO_2 equivalent operation Cost:

$$C_{GT} = Cost\,(P_{GT\_GEN}) + \delta\, C_{MGT_{ON/OFF}} + C_{CO_{2eq}} \tag{4}$$

For a period of Δt, the cost of natural gas is calculated by [17]:

$$Cost\,(P_{GT_GEN}) = M_{gas} \times Cg \times \Delta t \tag{5}$$

whit

$$M_{gas} = \frac{E_{elec}}{d_g \times \eta_G} \tag{6}$$

Where M_{gas} is the consumed gas mass, C_g is the cost of 1 Kg of natural gas, d_g is the natural gas energy density ($d_g = 13.5$ Kwh/kg), and η_G is the GT efficiency (Fig. 2).

$$C_{CO_{2eq}}\,(k) = M_{CO2_{eq}}\,(k) \times Cost_{penal_CO2} \tag{7}$$

The estimate of greenhouse gas emissions is limited to the most polluting and harmful gases: NOx, CO, and CO_2 gas [15]. The equivalent CO_2 emission cost ($C_{CO_{2eq}}$) is calculated by considering the equivalent CO_2 mass [23,24] and the environmental penalty price. In this work, we use a penalty cost (Cost_{penal_CO2}) of €30 per ton of equivalent CO_2 [19].

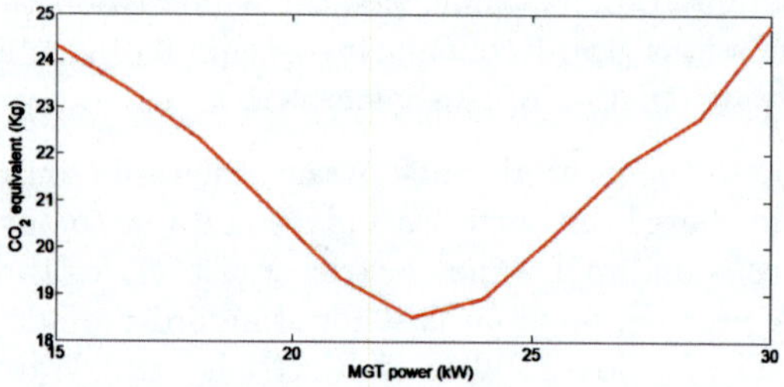

Fig. 2. Amount of CO_2 gas emissions Vs. Turbine power.

Figure 2 presents the variation of the equivalent CO_2 cost depending on the generated power. *Grid Cost*: The electricity tariff for the industry can be given as [13,22]. If during a day, it is found that the recorded power has exceeded the advocated power, the difference between these 2 values will be charged. This power is known as the Subscribed Power Exceeding Charge (SPEC) determined as follows [22]:

$$SPEC = \frac{1.5 \times fixed\_increase}{365} \times (P_A - P_S) \tag{8}$$

P_S is the subscribed power and P_A is the maximum power exported through 24 h. As a result, the cost of the power traded with the main grid is:

$$C_{Grid} = G_{cost\_buy} + SPEC - G_{cost\_sell} \tag{9}$$

Battery (ESS). Cost: Lithium-ion (Li-ion) batteries are used as a storage system. This technology has good efficiency performances and a life cycle [20]. The model used in this algorithm is performed by [20,21,30]. It is estimated by using Wöhler's model where the mechanical wear level involves the counting of charge and discharge cycles.

$$SOH\,(k) = SOH\,(k-1) \times (1 - \frac{N^{eq_{100}}_{cycles}(k)}{N^{100\%}_{cycles\_max}}) \tag{10}$$

At each planning step, battery aging is associated with a wear cost (battery replacement cost) defined with:

$$C_{ESS}\,(k) = \frac{Bic \times \Delta SOH(k)}{1 - SOH_{min}} \tag{11}$$

Where ΔSOH is:

$$SOH\,(k) = SOH\,(k-1) \times (1 - \frac{N^{eq_{100}}_{cycles}(k)}{N^{100\%}_{cycles\_max}}) \tag{12}$$

Bic is the initial investment price of ESS.

Constraints: The power balance (13) is the major requirement that the management system should respect between distributed generators, renewables, storage systems, and demand in the MG.

$$P_{PV} + P_{WT} + P_{Grid} + P_{GT} - P_{ESS} = P_{LD} \tag{13}$$

If available, a reserve of 10% of WT generators' power and PV cells' power is imposed by electrical energy standards. Considering this, the power reference of the different energy sources must respect these inequalities:

$$P_{PV} \leq 0.9 \times \; P_{PV\_GEN} \tag{14}$$

$$P_{WT} \leq 0.9 \times \; P_{WT\_GEN} \tag{15}$$

Also, 10% reserve of the available power in both battery and Gas Turbine, is imposed by electrical energy standards.

$$P_{ESS} \leq 0.9 \times P_{ESS\_MAX} \tag{16}$$

$$0.5 \times P_{GT\_gen} \leq P_{GT} \leq 0.9 \times P_{GT\_GEN} \tag{17}$$

It is advisable to use the turbine over fifty percent of its rated power. Above 50%, the efficiency increases, and the CO_2 emanation weakens [18,19].

The State of Charge SOC_{min} and SOC_{max} must satisfy (18), (19), and (20), to increase life expectancy [21] storage autonomy.

$$\mathrm{SOC}_{min} \leq SOC \leq \mathrm{SOC}_{max} \tag{18}$$

$$\Delta SOC(k) \leq \Delta SOC_{max} \tag{19}$$

$$SOH\,(k) \geq SOH_{min} \tag{20}$$

Genetic Algorithm Application. Genetic algorithms (GA) were developed by J. H. Holland in the 1970s. GA mimics "the theory of evolution" [25]. It is applied due to its random nature and ease of adaptability. The components of GA are:

- A fitness function (the objective function that will be optimized)
- A population of chromosomes (the first population is randomly chosen chromosomes)
- A selection that will reproduce the new generation chromosomes
- Crossover to produce the new generation chromosomes
- Mutation to ensure a good diversity between the offspring.

The generation and replacement cycles are repeated until a stopping criterion is met. This criterion can be a fixed number of generations, a calculation time threshold, or/and a satisfactory fitness. The best optimal solution is then given.

The steps of the Genetic Algorithm are presented in the flowchart in Fig. 3. In our smart MG, the GA is used for the optimization of the energy cost and CO_2 emanation.

We use the GA tool in MATLAB to find the minimum of a constrained nonlinear multivariable function. GA solves problems with the form:

$$\left.\begin{array}{l} A \times x \leq B \\ Aeq \times x = Beq \end{array}\right\} linear\ constraints \tag{21}$$

$$\left.\begin{array}{l} C(x) \leq 0 \\ Ceq(x) = 0 \end{array}\right\} linear\ constraints \tag{22}$$

$$LB \leq\ x \leq\ UB \rightarrow\ bounding\ of\ variables \tag{23}$$

Selection Strategy: Stochastic uniform. The parents of the next generations are selected using a stochastic uniform strategy. Each parent is allocated a section proportional to its scaled value. The selection process proceeds by advancing in equal-sized steps. At each step, a parent is chosen.

Mutation Strategy: Adaptive feasible. Offspring are generated depending on the success or the failure of the last generation. A mutation step length is chosen to satisfy Constraints and bounds.

Crossover Strategy: Scattered. First, scattered crossover produces a random binary vector to form a child. Then, chooses, from the first parent, the genes where there is a '1' in this vector. The genes where there is a '0' are chosen from the second parent.

Non-linear Constraints Strategy: Augmented Lagrangian. To satisfy linear, non-linear constraints and bounds, a new minimization problem is defined: a function is formulated using the fitness function and the non-linear constraints using the Lagrangian and the penalty parameters. The penalty factor is increased if the problem's constraints aren't accurate and satisfied, otherwise, the penalty factor decreases.

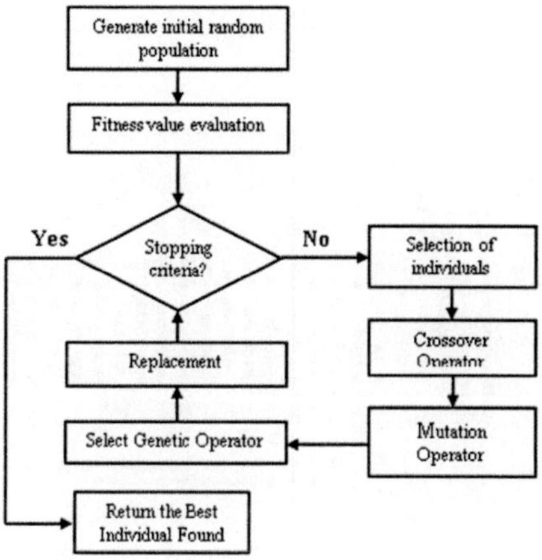

Fig. 3. How Genetic Algorithm works

3 Simulation and Results

3.1 Simulation Strategy

The considered day is discretized in twenty-four steps of 1 h. The resolution of the optimization problem, defined in section II chapter B, is done using MATLAB environment. Hereafter is the adopted strategy:

- The first generation offspring are created to respect the imposed constraints: (i)the power balance, (ii)the battery SOC in the acceptable limits, The population size is pre-defined.
- Then, mutation and crossover are applied to search the minimum cost of Cp (defined in (3)) among the population. If one constraint or more is not respected, a penalty cost is applied to Cp.
- The best offspring are kept to be in the next generation.
- These steps are repeated until the constraints' tolerance is obtained, or the maximum number of generations is reached.

3.2 Simulation Results Analysis

To assess the strength of the suggested management in this MG, we compared the obtained results to management based on common sense. The latter consists of choosing renewable sources first as they are considered with no cost to maximize the use of these generators while using the storage system as a backup energy source: the surplus of power will first charge the battery. The operator will then, distribute the remaining needed power among the remaining sources depending on the power cost factors [21,31]. The excess will be sold to the Grid (Fig. 4).

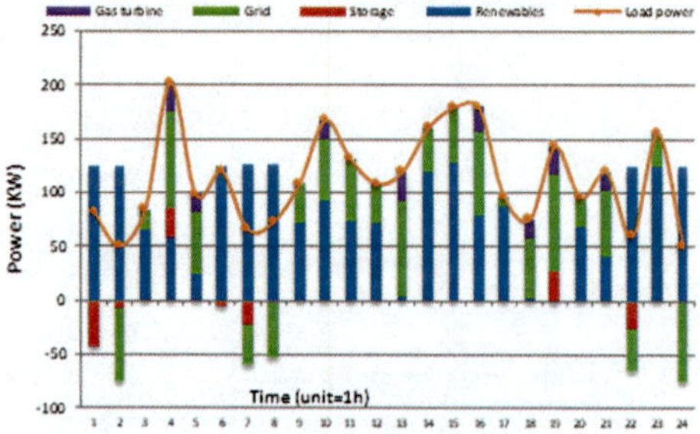

Fig. 4. Power restricted dispatching

The optimal distribution power proposed by GA is presented in Fig. 5. The excess power is sold to the grid to minimize energy costs and/or to charge/discharge the ESS (at midnight, the SOC must be at 50%). The GT power is requested to ensure the power equilibrium but its use is minimized to reduce CO_2 emissions. Over the studied day, the suggested management attains a cumulative price of €131,87 (see Fig. 6). Compared to the common-sense management where the energy cost is €180,87. That makes an economy of 27%.

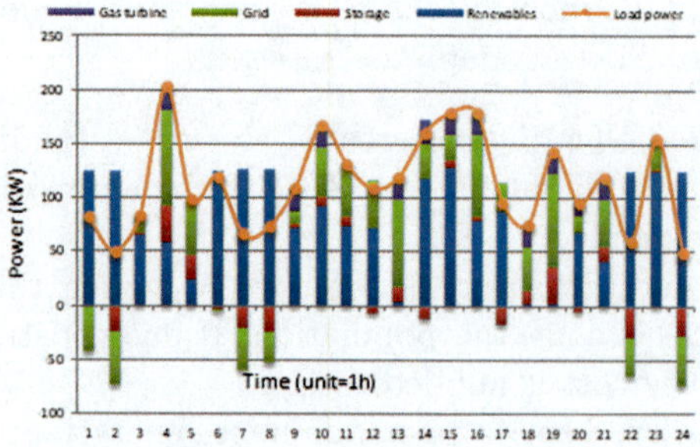

Fig. 5. Grid, Gas turbine, and Battery participation to the power balance in GA optimization

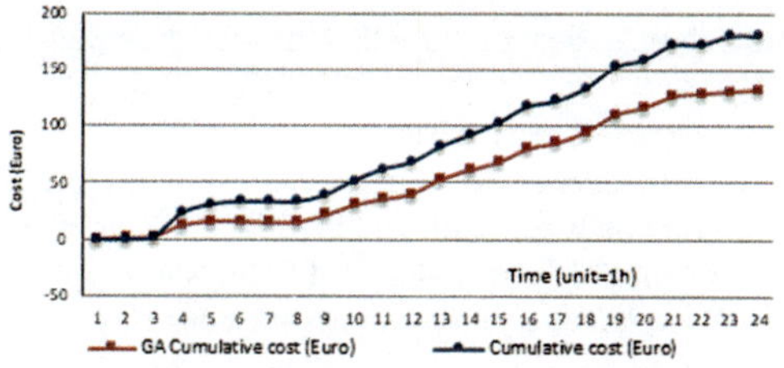

Fig. 6. Micro-Grid total cost comparison

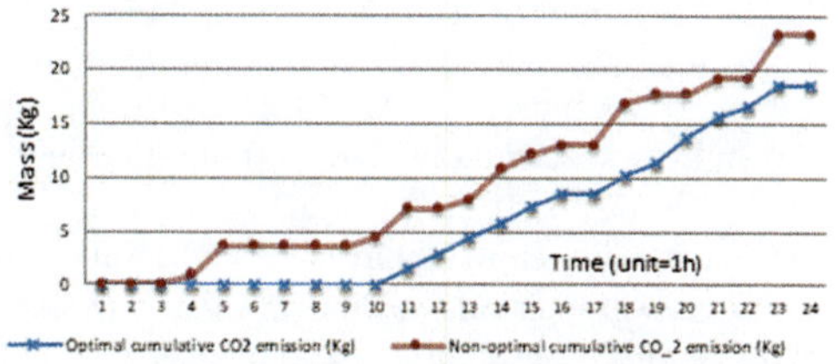

Fig. 7. GT equivalent CO_2 emanation

The contribution of the GT in the power mix has been limited. That implies less burning of fuel, and result though, in fewer greenhouse gases emanation. Air pollution is reduced by 20.4% (see Fig. 7).

4 Conclusion

This paper proposes a power management strategy over 24 h, to reduce greenhouse emissions and electricity power bills in a Smart Micro Grid. We applied genetic algorithm optimization in a grid-connected MG, composed of PV cells, wind turbines, a GT and, an ESS. The participation of the renewables in the power mix is important, these energy sources are supported by stable secondary sources: batteries and the gas turbine (auxiliary source) to also overcome unexpected main grid power outages. The proposed management delivers persistent, secure, and reduced-price electricity supplies. It results in a financial profit of 27% of electricity charges compared to the rules-based management. CO_2 emanation rate is well decreased by 5 Kg during the considered day (a reduction of 20.4%).

References

1. Hatziargyriou, N.: Microgrids: Architectures and Control, p. 3 (2014)
2. Fang, X., Misra, S., Xue, G., Yang, D.: Smart grid–the new and improved power grid: a survey. IEEE Commun. Surv. Tutor. **14**(4), 944–980 (2012)
3. Hledik, R.: How green is the smart grid? Electr. J. **22**(3), 29–41 (2009)

4. Momoh, J.A.: Smart grid design for efficient and flexible power grids operation and control. In: Proceedings of the IEEE/PES Power Systems Conference and Exposition (PSCE), pp. 1–8. IEEE (2009)
5. Boulal, A., Chakir, H.E., Drissi, M., Griguer, H., Ouadi, H.: Optimal management of energy flows in a multi-source grid. In: 2018 Renewable Energies, Power Systems & Green Inclusive Economy (REPS-GIE), Casablanca, pp. 1–6, April 2018
6. Huang, P., Xu, T., Sun, Y.: A genetic algorithm based dynamic pricing for improving bi-directional interactions with reduced power imbalance. Energy Build. **199**, 275–286 (2019)
7. Lei, G., Song, H., Rodriguez, D.: Power generation cost minimization of the grid-connected hybrid renewable energy system through optimal sizing using the modified seagull optimization technique. Energy Rep. **6**, 3365–3376 (2020)
8. Barakat, S., Ibrahim, H., Elbaset, A.A.: Multi-objective optimization of grid-connected PV-wind hybrid system considering reliability, cost, and environmental aspects. Sustain. Cities Soc. **60**, 102178 (2020)
9. Sharma, S., Bhattacharjee, S., Bhattacharya, A.: Operation cost minimization of a micro-grid using Quasi-oppositional swine influenza model based optimization with quarantine. Ain Shams Eng. J. **9**(1), 45–63 (2018)
10. Wei, L.: Energy drive and management of smart grids with high penetration of renewable sources of wind unit and solar panel, p. 7 (2021)
11. Chamandoust, H.: Day-ahead scheduling problem of smart micro-grid with high penetration of wind energy and demand side management strategies. Sustain. Energy Technol. Assess. 12 (2020)
12. Khalil, M.I., Jhanjhi, N.Z., Humayun, M., Sivanesan, S., Masud, M., Hossain, M.S.: Hybrid smart grid with sustainable energy efficient resources for smart cities. Sustain. Energy Technol. Assess. **46**, 101211 (2021)
13. Azaroual, M., Ouassaid, M., Maaroufi, M.: Optimal control for energy dispatch of a smart grid tied PV-wind-battery hybrid power system. In: 2019 Third International Conference on Intelligent Computing in Data Sciences (ICDS), pp. 1–7, October 2019
14. Ashok, S.: Optimised model for community-based hybrid energy system. Renew. Energy **32**, 1155–1164 (2007)
15. Canova, A., Chicco, G., Genon, G., Mancarella, P.: Emission characterization and evaluation of natural gas-fueled cogeneration microturbines and internal combustion engines. Energy Convers. Manag. **49**, 2900–2909 (2008)
16. Keshta, H.E., Malik, O.P., Saied, E.M., Bendary, F.M., Ali, A.A.: Energy management system for two islanded interconnected micro-grids using advanced evolutionary algorithms. Electr. Power Syst. Res. **192**, 106958 (2021)
17. Boulal, A., Chakir, H.E., Drissi, M., Ouadi, H.: Energy bill reduction by optimizing both active and reactive power in an electrical microgrid. IREE **15**(6), 456 (2020)
18. Boicea, A., Chicco, G., Mancarella, P.: Optimal operation of a microturbine cluster with partial-load efficiency and emission characterization. In: 2009 IEEE Bucharest PowerTech, pp. 1–8 (2009)
19. Kanchev, H.: Gestion des flux énergétiques dans un système hybride de sources d'énergie renouvelable: Optimisation de la planification opérationnelle et ajustement d'un micro réseau électrique urbain, Thesis 2014, Central School of Lille, Technical University of Sofia (2014)
20. Delaille, A.: Development of New State-of-Charge and State-of-Health Criteria for Batteries Used in Photovoltaic Systems, University Pierre et Marie Curie, Ph.D. Report (French) (2006)

21. Riffonneau, Y., Bacha, S., Barruel, F., Ploix, S.: Optimal power flow management for grid connected PV systems with batteries. IEEE Trans. Sustain. Energy **2**(3), 309–320 (2011)
22. http://www.one.org.ma/
23. Shine, K.P., Fuglestvedt, J.S., Hailemariam, K., Stuber, N.: Alternatives to the global warming potential for comparing climate impacts of emissions of greenhouse gases. Climatic Change **68**, 281–302 (2005)
24. International Panel on climate change. "Climate change 2001: Working group I: The scientific basis", Section 4, table 6.7, IPCC 2007
25. Holland, J.H.: Adaptation in Natural and Artificial Systems: An Introductory Analysis with Applications to Biology, Control, and Artificial Intelligence. MIT Press, Cambridge (1992)
26. Yan, C., Wang, F., Pan, Y., Shan, K., Kosonen, R.: A multi-timescale cold storage system within energy flexible buildings for power balance management of smart grids. Renew. Energy **161**, 626–634 (2020)
27. Shi, Z., et al.: Artificial intelligence techniques for stability analysis and control in smart grids: methodologies, applications, challenges and future directions. Appl. Energy **278**, 115733 (2020)
28. Tan, K.M., Babu, T.S., Ramachandaramurthy, V.K., Kasinathan, P., Solanki, S.G., Raveendran, S.K.: Empowering smart grid: a comprehensive review of energy storage technology and application with renewable energy integration. J. Energy Storage **39**, 102591 (2021)
29. Crespo Del Granado, P., Pang, Z., Wallace, S.W.: Synergy of smart grids and hybrid distributed generation on the value of energy storage. Appl. Energy **170**, 476–488 (2016)
30. Rigo-Mariani, R., Sareni, B., Roboam, X., Turpin, C.: Optimal power dispatching strategies in smart-microgrids with storage. Renew. Sustain. Energy Rev. **40**, 649–658 (2014)
31. Pazouki, S., Haghiafm, M.R.: Market based operation of a hybrid system including wind turbine, solar cells, storage device and interruptable load. In: 18th Electric Power Distribution Conference, pp. 1–7 (2013)

Check for updates

Modelling of Cavitation in Transient Flow in Pipe-Water Hammer

Ouafae Rkibi(✉), Nawal Achak, Bennasser Bahrar, and Kamal Gueraoui

Team of Modeling and Simulation of Mechanical and Energetic, Faculty of Sciences, Mohammed V/Rabat, Rabat, Morocco

Abstract. This study provides a theoretical and numerical modeling of transient vaporous cavitation in a horizontal pipeline, anchored to the upstream reservoir. The model approach is, essentially, based on that of the column separation model (CSM). The basic system of partial differential equations to solve is a hyperbolic type and adapts perfectly to the method of characteristics. This code, allows us, taking into account the unsteady part of the friction term, to determine at any point of the pipe, and at each instant, the average piezo metric head, the average discharge and the change in volume of the vapor cavity. This study illustrates the effect of the presence of air pockets, resulting in cavitation, on the amplitude of the pressure wave. The calculation results are in good agreement with those reported in the literature.

Keywords: Cavitation model · Column separation model of vapor · Method of characteristics · Transient flow · Unsteady Friction model · Vapor pressure

Ratings

C =	water hammer velocity [m/s].
D =	internal diameter of the pipe [m].
e =	thickness of the section [m].
E =	Young's modulus of the pipe [Pa].
f_g =	quasi-stationary part of the friction term [-].
f_u =	unsteady part of the friction term [-].
g =	acceleration of gravity [m/s^2].
K =	volume modulus of elasticity of water [Pa].
P =	pressure in a pipe section [Pa].
P_{vap} =	vapor pressure of water [Pa].
S =	cross section of the pipe [m^2].
V =	flow velocity at time t [m/s].
V_u =	flow velocity upstream of the air pocket [m/s].
ε =	roughness of the pipe wall [m]
μ =	Poisson's ratio [-].
Ψ =	weight factor [-]
∀ =	volume of the discrete steam cavity [m^3].
ρ =	water density [kg/m^3]

The original version of this chapter was revised: The second author's name has been changed to "Nawal Achak". The correction to this chapter is available at https://doi.org/10.1007/978-3-030-94188-8_49

R. Saidi et al. (Eds.): ICATH 2021, LNDECT 110, pp. 54–60, 2022.
https://doi.org/10.1007/978-3-030-94188-8_6

1 Introduction

In transient pipe flows, transient vaporous cavitation occurs when the pressure drops below the saturating vapor of the fluid. Air bubbles appear in the fluid. These cavities can cause implosion and damage to the various components of the hydraulic system [1, 2]. There are two types of vapor cavitation, localized vapor cavitation (high vacuum) and distributed vapor cavitation (low vacuum). In some regions where evaporation is produced by pressure drop, the continuous medium is ruptured by creating columns separated from the fluid, this is the Column Separation Model (CSM). Several studies have been made using this model, notably the work of Anton Bergant and al [3], Jian Junshu [4], Paquette [5]. In this article we will study the discrete vapor cavity model (DVCM) with consideration of the shear stress of Darcy-Weisbach that reflects the unsteady friction term introduced by Zielke Vardy-Brown [6]. We then examine the evolution of the pressure head and the effect of velocity of the flow at the valve and at the midpoint of the pipeline, and the changes in the volume of air bubbles resulting from cavitation.

2 Equation and Assumption

The flow is assumed unidirectional compressible barotropic and constant entropy. Assume also that the water hammer speed is very large compared to the average flow velocity, and it is not affected by its biphasic nature. The column separation model in systems (tank-horizontal-line valve) are described by the hyperbolic system formed by Eqs. (1) and (2) derived from the conservation of mass and momentum averaged over a cross section of the pipe., [1, 2]:

$$\frac{\partial P}{\partial t} + \rho C^2 \frac{\partial V}{\partial x} = 0 \tag{1}$$

$$\frac{1}{\rho}\frac{\partial P}{\partial t} + \frac{\partial V}{\partial t} = -(f_q + f_u)\frac{V|V|}{2D} \tag{2}$$

Water hammer Speed:

$$C = \sqrt{\frac{\frac{K}{\rho}}{1 + \left(\frac{1}{\mu^2}\right)\frac{kD}{eE}}}$$

With

$$f = (f_q + f_u)$$

Where:

f_q: is the part related to the quasi-stationary flow.
f_u: is the part related to the un-stationary flow.
The first term is the solution of the Coolbrok equation [1, 2].

$$\frac{1}{\sqrt{f_q}} = 2\log_{10}\left(\frac{2{,}51}{Re\sqrt{f_q}} + \frac{\frac{\varepsilon}{D}}{3{,}71}\right) \tag{3}$$

3 Numerical Solution

By introducing the water hammer velocity [1, 2]

$$C=\left(\frac{\partial\rho}{\partial P}+\frac{\rho}{S}\frac{\partial S}{\partial P}\right)^{\frac{1}{2}}$$

The system of Eqs. (1) and (2) transformed along the characteristic curves of slopes

$$\left.\frac{dx}{dt}\right|_{C^{\pm}}=\pm C$$

We obtain at each calculation node at each instant the algebraic system:

$$P_{i,t}-P_{i-1,t-\Delta t}+\rho C\big((V_u)_{i,t}-(V_d)_{i-1,t-\Delta t}\big)=\rho C\Delta t T_f \tag{4}$$

$$P_{i,t}-P_{i+1,t-\Delta t}+\rho C\big((V_d)_{i,t}-(V_u)_{i+1,t-\Delta t}\big)=-\rho C\Delta t T_f \tag{5}$$

The continuity equation for the volume of the discrete vapor cavity is described by:

$$(\forall_g)_{i,t}=(\forall_g)_{i,t-2\Delta t}+(\Psi((V_d)_{i,t}-(V_u)_{i,t})$$
$$+(1-\Psi\big((V_d)_{i,t-2\Delta t}-(V_u)_{i,t-2\Delta t}\big))S\,\mathrm{x}\,2\Delta t \tag{6}$$

The unsteady term of friction is given by the convolution product [6]:

$$f_u(t)=\frac{32\nu}{DV|V|}\int_0^t\frac{\partial V}{\partial\tau}W_0(t-\tau)d\tau \tag{7}$$

$$W_{app}(\tau)=\sum_{l=1}^{N}m_l e^{-n_l\tau} \tag{8}$$

$$\tau=\frac{4\nu}{D}t$$

Where coefficients m_l and n_l relative to the weighting function W are determined for laminar flow by Zielke [6] and Vardy and al [7], Vitkovsky and al [8] for turbulent flow.

From the function f_u a function y_l is defined by:

$$f_u(t)=\frac{32\nu}{DV|V|}\sum_{l=1}^{N}m_l e^{-n_l\tau}$$

$$y_l(t)=\int_0^t\frac{\partial V}{\partial t^*}m_l e^{\frac{-n_l 4\nu}{D^2}(t-t^*)}dt^*$$

The explicit expression of the y_l function is:

$$y_l(t+2\Delta t)=e^{\frac{-n_l 4\nu}{D^2}\Delta t}\left\{e^{-n_l K\Delta t}y_l(t)+m_l[V(t+2\Delta t)-V(t)]\right\} \tag{9}$$

4 Application and Results

In this application, we consider a turbulent flow, for two steady-state flow values $V_0 =$ 0.3 m/s and $V_0 =$ 1.4 m/s. The flow is in the horizontal copper pipe anchored to the upstream to a tank filled with water and of height H_0, ending at the downstream to a valve that closes abruptly. The parameters of the fluid and the pipe are summarized in Table 1:

Table 1. Installation data.

Tank height H_0 *(m)*	26
Internal diameter of the pipe (mm)	22.1
Pipe length (m)	37.2
Pipe thickness (mm)	1.63
Kinematic viscosity of water at 20 °C (m^2/s)	1.11E−6
Poisson Coefficient	0.34
Modulus of elasticity by volume of water (GPa)	2.2
Young's modulus of copper (GPa)	120
Vapor pressure at 20 °C (m)	0.23
Density of water at 20 °C (Kg/m^3)	1000

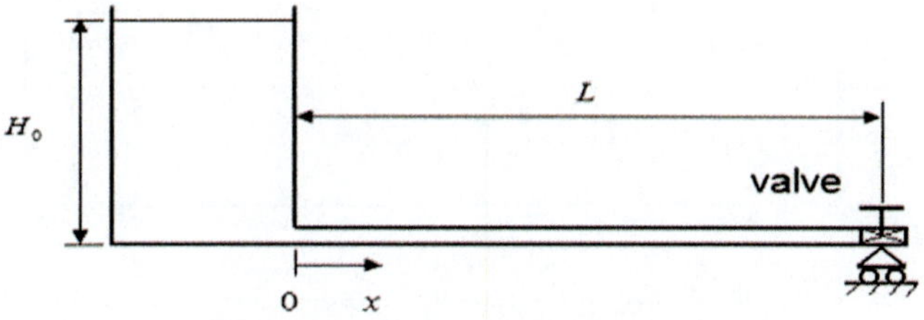

Fig. 1. Diagram of the system studied

The Fig. 1 and 2 show, respectively, the temporal variation of the water head pressure at the valve in the case of the classical water hammer without or with cavitation for the velocity value $V_0 = 0.3$ m/s corresponding to the steady-state flow.

The Figs. 3 and 4 show, in the same conditions, for a steady-state velocity $V_0 =$ 1.4 m/s, the time evolution of the pressure at the valve without or with cavitation. In the results section, we illustrate an attenuation of the pressure wave due to the friction of the fluid on the pipe wall. This study shows that, for low velocity of the quasi-stationary flow, the phenomenon of the classical water-hammer is predominant, whereas for high speeds, the first over pressure; the most dangerous for the hydraulic system, is caused by the instantaneous closing of the valve, and the other peaks are due to the collapse of the air pockets present in the fluid. In the case of cavitation, the presence of air pockets leads to greater attenuation of the pressure wave amplitude compared to the case of a conventional water-hammer. During cavitation, the formation of the air pocket, its volume increases, and by collapse, an over pressure takes place that can affect the internal surface of the pipe wall causing its wear (Fig. 5).

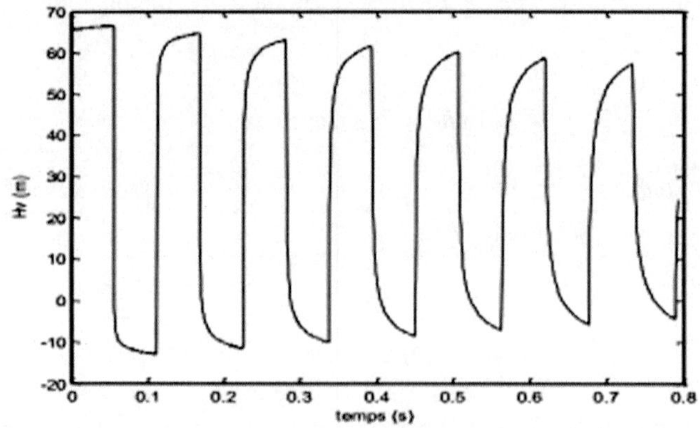

Fig. 2. Pressure in the middle of the pipe in the case of the water hammer for $V_0 = 0.3$ m/s

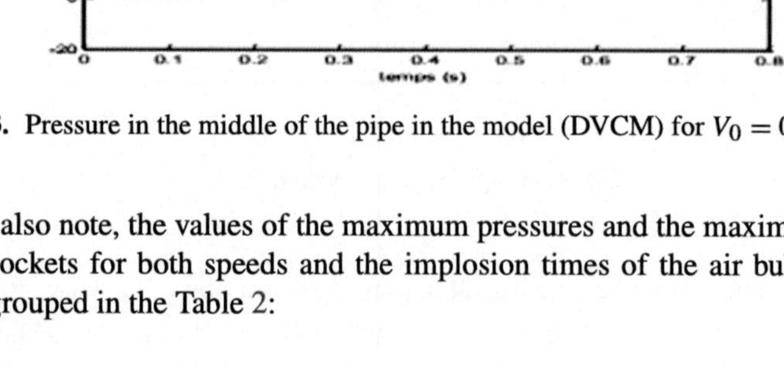

Fig. 3. Pressure in the middle of the pipe in the model (DVCM) for $V_0 = 0.3$ m/s

We can also note, the values of the maximum pressures and the maximum volumes of the air pockets for both speeds and the implosion times of the air bubbles. These values are grouped in the Table 2:

5 Conclusion

This paper presents the study of transient cavitation flow in a horizontal straight copper pipe, and an initial data set under turbulent conditions was collected during transient events caused by rapid valve closure. This study, clearly, shows the possibility of a simple numerical treatment of cavitation. It has highlighted, in addition to the local nature of cavitation, the importance of the turbulent flow regime, the value of the velocity of the permanent regime of the flow, and the air pockets resulting of cavitation. In view of a

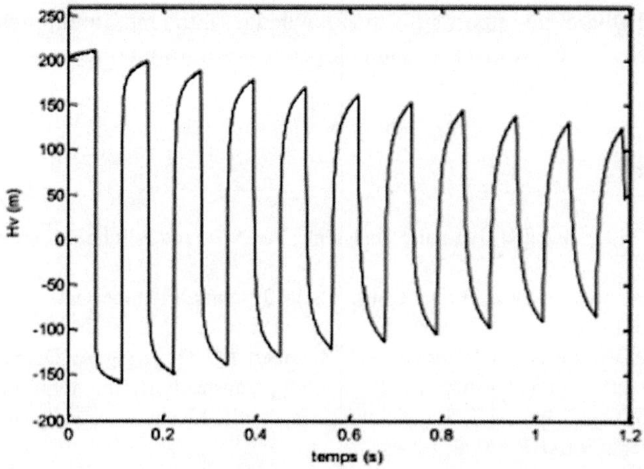

Fig. 4. Pressure in the middle of the pipe in the case of water hammer for $V_0 = 1.4$ m/s

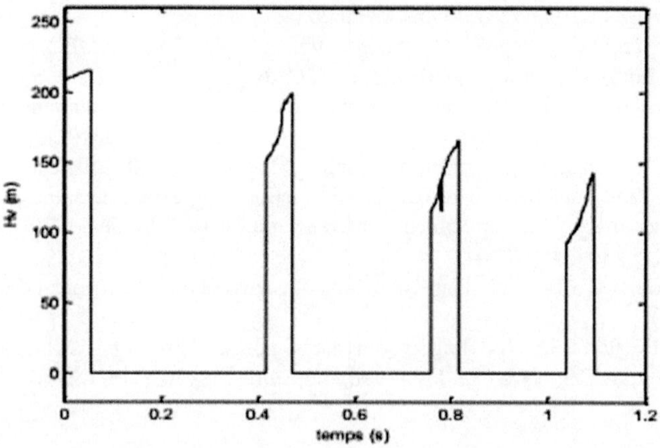

Fig. 5. Pressure in the middle of the pipe in the model (DVCM) for $V_0 = 1.4$ m/s

Table 2. Values of the maximum pressures and maximum volumes of the air pockets for both speeds and the implosion times of the air bubbles.

V₀ =0.3m/s	$\forall_{max}$ (m^3)	1st collapse (s)	$H_{\forall_{max}}$ (m)	t($H_{\forall_{max}}$) (s)
N=32	7.8105 E-7	0.1261	87.9842	0.1817

V₀ =1.4m/s	$\forall_{max}$ (m^3)	1st collapse (s)	$H_{\forall_{max}}$ (m)	t($H_{\forall_{max}}$) (s)
N=32	4.093 E-5	0.4172	214.7657	0.0565=2L/C

practical level where the important is the knowledge of the maximum pressure likely to occur, it is logical to think that this computer code is a relevant understanding of the tool in transient cavitation flow in pipes.

References

1. Chaudhry, M.H.: Applied Hydraulic Transients. Van Nostrand Reinhold Company. New York, USA (1987)
2. Wylie, E.B., Streeter, V.L.: Fluid Transients in Systems. Prentice Hall, Englewood Cliffs, USA (1993)
3. Bergant, A., V'itkovsk'y, J. P., Simpson, A., Lambert, M., Tijsseling, A.: Discrete vapor cavity model with efficient and accurate convolution type unsteady friction term. In: Kurokawa, J. (ed.) Proceedings 23rd IAHR Symposium on Hydraulic Machinery and Systems, Yokohama, Japan, Paper 109 IAHR (2006)
4. Shu, J.J.: Modeling vaporous cavitation on fluid transients. Int. J. Press. Vessel. Pip. **80**, 187–195 (2003). https://doi.org/10.1016/S0308-0161(03)00025-5
5. Paquette, Y.: Fluid-Structure Interaction and Cavitation Erosion. PhD thesis. Universit'e Grenoble Alpes (2017). HAL Id: hal-02066203
6. Zielke, W.: Frequency dependent friction in transient pipe flow. ASME J. Basic Eng. **90**(1), 109–115 (1968). https://doi.org/10.1115/1.2926516
7. Vardy, A.E., Brown, J.M.B.: Transient turbulent friction in fully rough pipe flows. J. Sound Vib. **270**(1–2), 233–257 (2004). https://doi.org/10.1016/S0022-460X(03)00492-9
8. Vítkovský J., Stephens M., Bergant A., Lambert M., Simpson A.R. Efficient and accurate calculation of Zielke and Vardy-Brown unsteady friction in pipe transients. In: Murray, S.J. (ed.) Proceedings of the 9th International Conference on Pressure Surges, BHR Group, Chester, UK, vol. 2, pp 405–419 (2004)
9. Chen, Y., Heister, S.D.: Two-phase modelling of cavitated flows. Comput. Fluids **24**(7), 799–806 (1995)
10. Chen, Y., Heister, S.D. Modeling hydrodynamic nonequilibrium in cavitating flows. Trans. ASME J. Fluids Eng. **118**, 172–178 (1996). https://doi.org/10.1115/1.2817497

The Contribution of GIS for the Modeling of Water Erosion, Using Two Spatial Approaches: The Wischmeier Model and SWAT in the High Oum Er-Rbia Watershed (Middle Atlas Region)

Younes Oularbi[1(✉)], Jamila Dahmani[1], and Fouad Mounir[2]

[1] Laboratory of Plant, Animal and Agro-Industry Productions, Department of Biology, Faculty of Sciences, Ibn Tofail University, Kénitra, Morocco
younes.oularbi@uit.ac.ma

[2] National Forestry School of Engineers, 511, Salé, Morocco

Abstract. Soil erosion is a multifaceted phenomena that is influenced by a variety of natural processes, resulting in a decline in soil fertility and agricultural production. The purpose of this study is to quantify the impact of this phenomena by employing the Geographic Information Systems and Remote Sensing in order to assess soil losses in the high Oum Er-Rbia watershed in the Middle Atlas area. To this end, two models were used, namely the Wischmeier model and the Soil and Water Assessment Tool (SWAT) model for estimating the amounts of sediment generated in the study area. The data to be used for the implementation of the models are: a digital terrain model, maps of erosivity, erodibility and topographic factors, a land cover map, a pedological map and climatic data from rainfall climate stations. An extension called QSWAT is used at the QGIS level for loss calculation using the SWAT model. The results obtained showed that the entire study area is affected by the phenomena of water erosion and generate land losses ranging from 20 to 100 t/ha/year for the two models used.

Keywords: GIS · Watershed · Water erosion · Wischmeier model · SWAT · QSWAT

1 Introduction

Soil erosion is a multifaceted phenomena that is influenced by a variety of natural processes, resulting in a decline in soil fertility and agricultural production. Each year. In many environments, water is one of the major causes of erosion [6]. Water erosion can be defined as the process of detachment and removal of soil material by water it's a natural process that can be enhanced by human action. Depending on the soil, the regional environment, and weather conditions, erosion can range from extremely slow to very fast [4]. It's a real danger for the sustainability

R. Saidi et al. (Eds.): ICATH 2021, LNDECT 110, pp. 61–72, 2022.
https://doi.org/10.1007/978-3-030-94188-8_7

of natural resources, the mobilization of surface water and the socio-economic development of rural areas. Around the World, 75 billion tons of soil are removed of agricultural land because of erosion, and approximately 20 million hectares of land were already gone [8]. In Morocco according to a report done by the Ministry of Agriculture and Agricultural Development (MAMVA, 1996), water erosion affects 2.1 million hectares of agricultural land [3]. Water erosion is a major cause of soil degradation around the world, as well as the primary cause of desertification. According to the United Nations Convention on Desertification, the term "desertification" means "land degradation in arid, semi-arid and sub-humid areas" [11]. To fight against desertification, it is necessary to: - Know the state of degradation of the watershed; - Ranking and prioritization of sub-watersheds and/or areas, potentially and actually, most affected by erosion and constituting sources of sediment production; - Identification of sub-watersheds or priority areas for intervention - Establishment of an adequate erosion control program to remedy soil degradation. The quantification of water erosion use purpose-built models. This study applied Geographic Information Systems and Remote Sensing to assess soil losses in the High Oum Er-Rbia watershed in the Middle Atlas region by using two spatial approaches: Wischmeier model and SWAT. The results are expected to provide theoretical support for adjusting and optimizing the Establishment of an adequate erosion control program to remedy soil degradation at the High Oum Er-Rbia watershed. The study area is located in the Middle Atlas, which is one of the three great mountain ranges which constitute the framework of Moroccan geographical space and which are: The Rif in the North, the Middle and High Atlas in the center and the Anti-Atlas in the South. This chain is also characterized by its diverse natural resources and location as a transit center. In this mountainous region, altitudes vary between 600 and 2400 m; it contains the most beautiful cedar groves in Morocco and the Mediterranean basin. This World ecological heritage is of major economic, social, cultural and tourist interest, hence the need for its conservation and development.

2 Materials and Methods

2.1 Study Area

Located southwest of the central Middle Atlas in the Beni mellal-Khenifra region (Fig. 1), the studied basin is delimited by the Hercynian Central Massif lies to the west, the Causse of Ajdir lies to the north, and the plain of the High Moulouya lies to the east. The high-basin of Oum Er Rbia is part of the Middle Atlas Mountain and Central Massif. According to the 1:500000 scale geological maps of Rabat (1976), the region's geological formations are dominated by Cretaceous sub-tabular limestone formations, Liassic dolomitic limestone, Triassic doleritic basalts and red clays, as well as Paleozoic flyschs and quartzite. The longitude ranges from 5°03′W to 5°55′W, and the latitude ranges from 32°35′N to 33°12′N, With an altitude ranging from 700 to 2400 m, the terrain presents a diversity of landforms, including structural forms, closed depressions, ravines, and accumulation forms represented by alluvial terraces. The study area covers approximately 371560 hectare. It belongs to the upper basin of Oum Errabiaa which is made up of liassic limestone soils belonging to the folded massif of the Middle Atlas and resting on an impermeable substratum of salt clays of the Triassic. In addition, the study area has a Mediterranean climate with hot, dry summers and mild, wet winters. The annual average precipitation is between 400 and 700 mm, and the minimum and maximum temperatures usual values are respectively 5 and 50 °C. The forest vegetation consists mainly of holm oaks and cedars which are mainly located in the upper and middle part of the watershed. The studied watershed is drained by the Oum Er Rbia River and its tributaries Srou and Chbouka, which cross the Atlas Mountain chain.

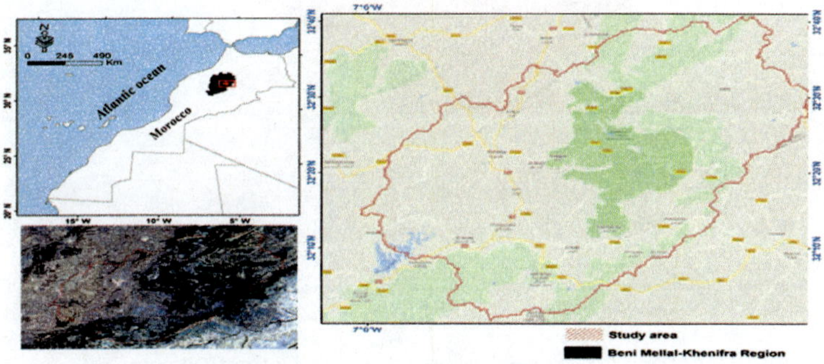

Fig. 1. Location of the study area.

2.2 Revised Universal Soil Loss Equation (RUSLE)

The Revised Universal Soil Loss Equation [12] is an improved version of the USLE. Many of the factor estimation procedures have been improved, including a new procedure for calculating the cover factor, slope length, and steepness factors. The RUSLE model now includes climatic factors based on an expanded database of rainfall-runoff in the Western United States. This model, like USLE, is based on the principle of overlaying various thematic maps that represent the main erosion factors. Climate aggressiveness, soil erodibility, inclination and length of slope, land use, and anti-erosion practices [12], are among these factors. As defined by the formula:

$$A = RxKxLSxCxP \tag{1}$$

Where A: Computed annual soil loss, R: Rainfall-runoff erosivity factor, K: Soil erodibility factor, LS: Topographic factor combining slope length and slope steepness (dimensionless), C: Cover-management (vegetation cover type), P: Supporting practices factor.

The RUSLE model was implemented in Quantum QGIS software to calculate the Geo-spatial rates of soil loss in the High Oum Rabia watershed. To do this the digital elevation model (DEM), average annual precipitation, soil type map, and NDVI map were used to produce RUSLE factors.

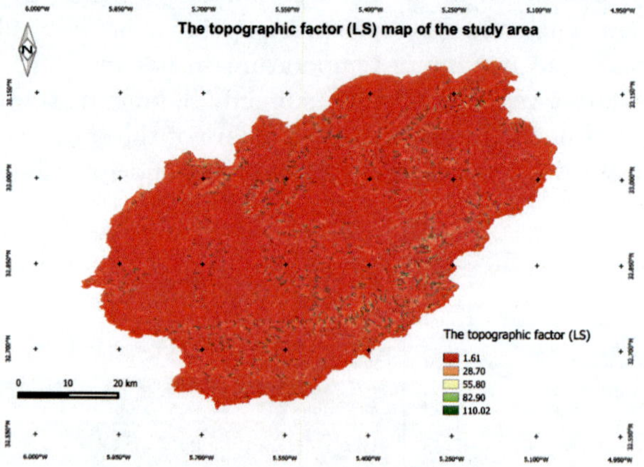

Fig. 2. The topographic factor (LS) map

- Topographic factor (LS) reflects the combined effects of slope length (L) and slope (S) on erosion [12]. The implementation of the LS factor in QGIS software produced a map which reflect the combined effect of degree and slope (Fig. 2).
- Land cover factor (C) is the ratio of soil loss from cultivated land to the comparable loss from clean-tilled, continuous fallow land under particular conditions [12]. This factor is strongly intertwined to land use types [9], a soil with highest vegetation cover will be more preserved. To calculate this factor we used the normalized NDVI vegetation index wish provides information on vegetation cover and health [2]

$$C = exp[-(\alpha NDVI/(\beta - NDVI))] \tag{2}$$

by applying the values 2 and 1 respectively to alpha and beta (Fig. 3).
- Supporting practices factor (factor P) was set the to 1 because we didn't take into account this factor in the studied area.
- The soil erodibility factor K is an indicator of inherent soil erodibility determined experimentally under standard conditions [12], The FAO soil map and the equation of Wischmeier and Smith [12] were used to determine this factor (Fig. 4).
- The rainfall aggressiveness index R is the product of rainfall kinetic energy and maximum rainfall intensity over a 30-minute duration [12]. To estimate this component (Fig. 5), the Rango and Arnoldus equation [10] was used to

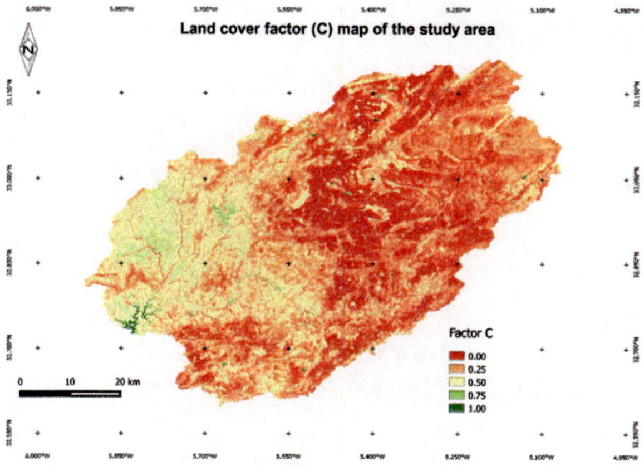

Fig. 3. The land cover factor (C) map

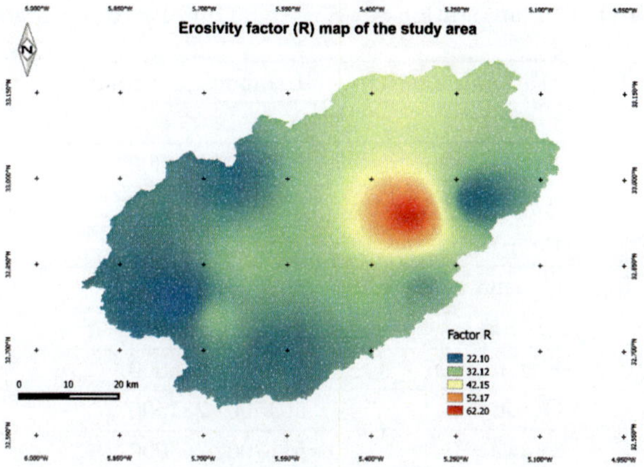

Fig. 4. The rainfall aggressiveness index R map

10 sites in the research region during a 20-year period (1980–2002) to obtain this climate index:

$$LnR = 1.74log(Pi^2/P) + 1.29 \quad (3)$$

The global methodological process used for the characterisation of losses in soil by the RUSLE model on the study area is presented in (Fig. 6).

2.3 Soil and Water Assessment Tool (SWAT)

The Agricultural Research Service (ARS) of the United States Department of Agriculture (USDA-ARS) and the Agricultural Experiment Station in Temple, Texas, collaborated on the SWAT (Soil and Water Assessment Tool) distributed model. It's a well-known model for calculating sediment loss [7] and Water Balance Estimation [1]. Input data used by the model are (Table 1):

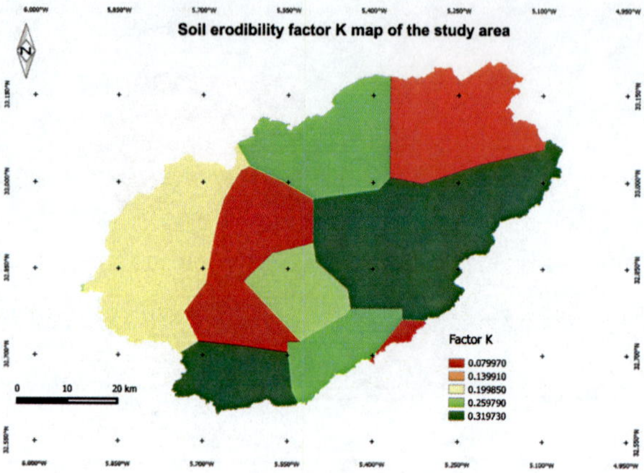

Fig. 5. Soil erodibility factor K map

Table 1. Climate stations in the High Oum Er-Rbia watershed

Station name	Lambert coordinate	
	X	*Y*
Arhbalou n Ikhwane	506000	259800
Azrou n Ait Lahcen	484000	233500
Beni Khlil	474200	240000
Chacha n Amellah	467500	243250
El Heri	478500	251125
Ke1rnuchen	508500	246000
Ouiouane	504500	281500
Sanwal	517100	263000
Taghat	476220	266940
Taghzout	461850	235350

- Digital Terrain Model for the definition of the hydrographic network. To delimit the sub-watersheds, outlets can be chosen.; in this study we used a digital terrain model of 30 m (SRTM) resolution (Fig. 7).
- Information on land use and soil type is cross-referenced to define the different HRU. The land cover map was produced using a recent Landsat satellite image from the OLI 8 collection from the year 2018 with a spatial resolution of 30 m (Operational Land Imager). An open source software is used first to download and prepare the images (atmospheric correction, geometric correction, merging and extraction of the study area), then a supervised classification with the SVM (Support Vector Machine) algorithm for the elaboration of land use maps of with an overall accuracy exceeding 90%. The class of the land use map obtained present: Dense Forest 16%, Clear Forest 25%,

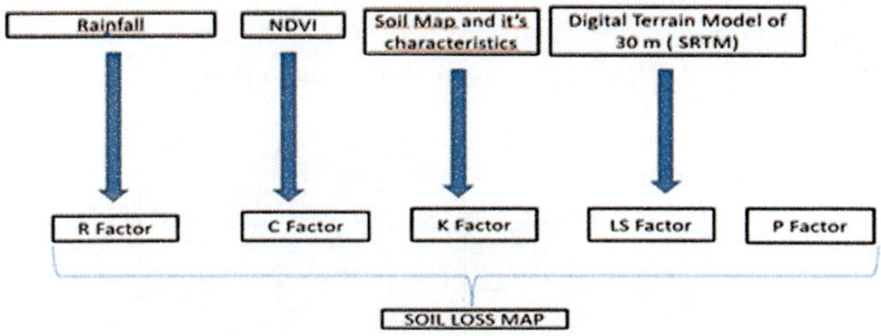

Fig. 6. Methodology used for the characterization of Loss of soil in land cover on the study area using RUSLE

Culture 26%, Shrubs 12%, Shrubbery 8%, Rock or bare soil 9%, Water 0.65%, establishment 0.60% (Fig. 8).
- The food and agriculture organization of the United States (FAO) published the Harmonized World Soil Database (HWSD), which was used to create the soil map (Fig. 9).
- real data or simulation data from the SWAT database wish taking into account climatic data: daily precipitation, temperatures (min, max, daily average), wind speed, relative humidity and solar radiation.

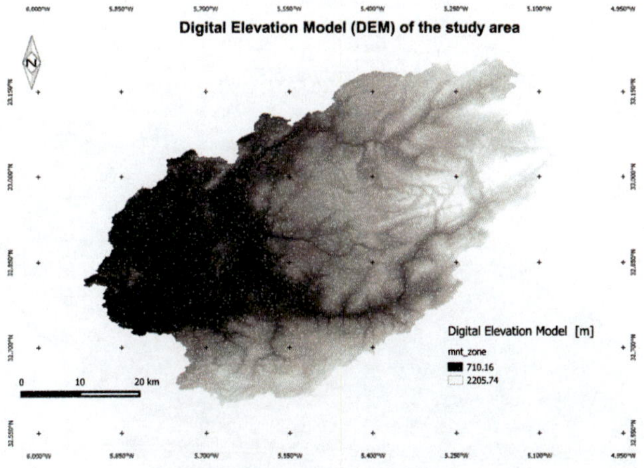

Fig. 7. Digital Terrain Model map

Three thematic maps are required by the SWAT model: soil raster map, Digital Elevation Model (DEM), slope map, and land use raster, all of which were created with Quantum GIS software. The model implementation was done in QSWAT, which is a Quantum GIS interface to set up and run the SWAT model. The global methodological process used for the characterisation of losses in soil by SWAT model on the study area is presented in Fig. 10.

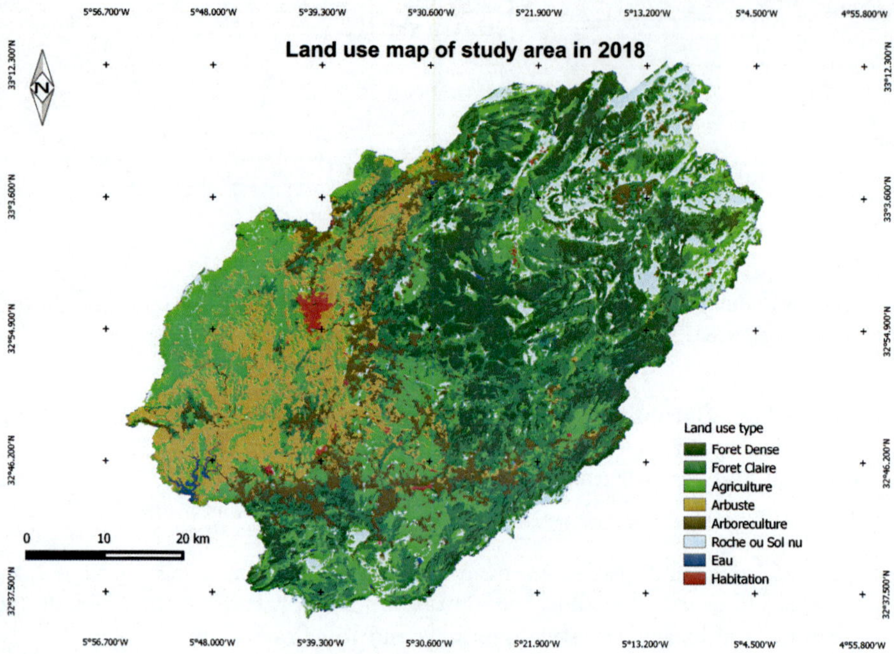

Fig. 8. Land use map

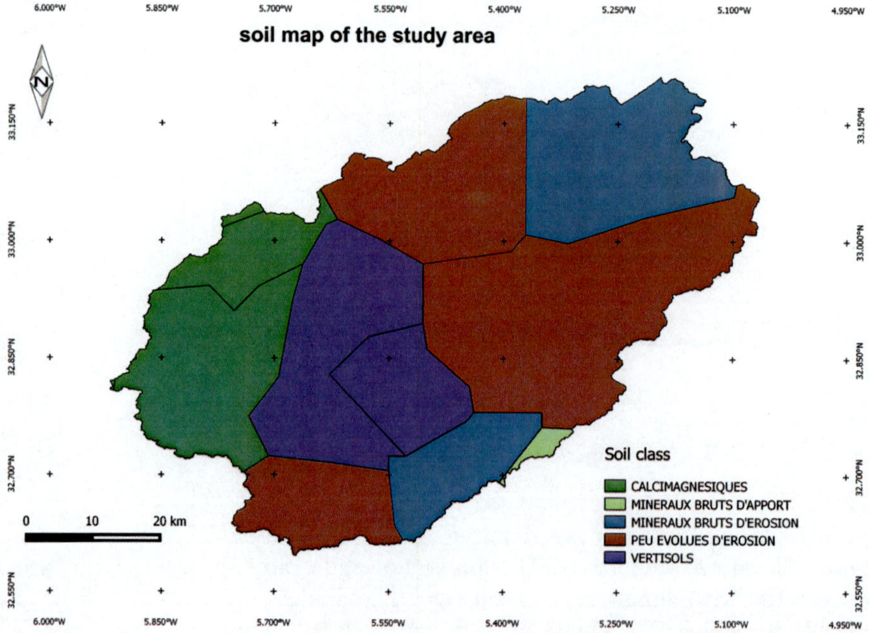

Fig. 9. Soil map of th study area

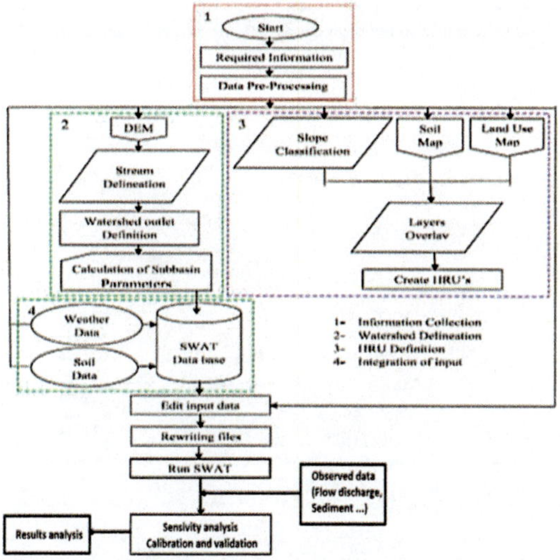

Fig. 10. Methodology used for the characterization of Loss of soil in land cover on the study area using SWAT model

3 Results and Discussion

The average soil losses resulting from the application of two models, rusle and swat, are between 0 and 20 t/ha/year, which shows that the area does not present a great variability of the water erosion phenomenon in the basin. The same results found by Yjjou et al. (2014) who reported soil losses in the Oum Rbiaa basin in the Middle Atlas ranging from 50 and 400 t/ha/year [13] and Mazouzi K. et al. who found that 64% of Oued Mikkès watershed upstream of the Sidi Chahed dam (Meknes region, Morocco) had soil loss values less than 20 t/h/year [5].

3.1 Revised Universal Soil Loss Equation (RUSLE)

The soil losses map (Fig. 11) which was generated using the Revised Universal Soil Loss Equation (RUSLE) shows that almost 74% of the studied area has soil losses values that do not exceed 20 t/h/year, which are relatively reliable losses. Soil losses (t/ha/year) in the basin were grouped into 6 classes of values see table number (Table 2).

The first class concerns areas with a soil loss of less than 20 t/ha/year. It represents almost 74% of the surface area of the basin, spread over the whole study area, the areas that are downstream of the watershed. The second class concerns areas with soil loss between 20 and 210 t/ha/year. It represents almost 25% of the watershed area. The third class consists of areas with soil loss between 210 and 290 t/ha/yr. It represents 0.5% of the basin area. The fourth and fifth classes include areas with soil loss between 290 and 350 and 350 and 400 t/ha/yr

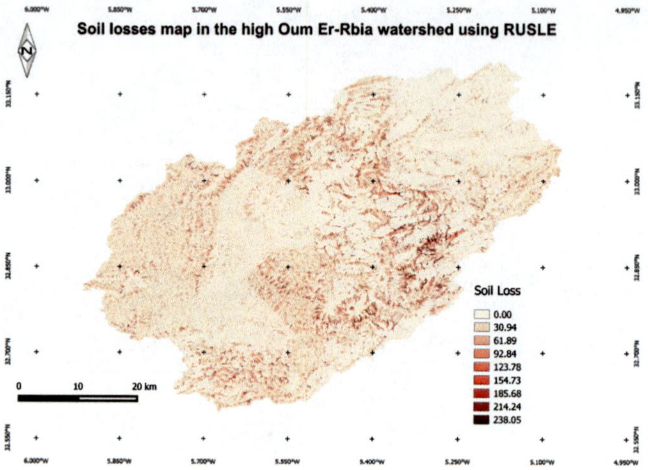

Fig. 11. Soil losses map in the high Oum Er-Rbia watershed using RUSLE

Table 2. Soil losses map of the High Oum Er-Rbia watershed using RUSLE model

Soil loss class(t/ha/yr)	RUSLE model	
	Percentage [%]	*Area [ha]*
Less than 20	73.907	229400
20–210	24.942	77418
210–290	0.519	1611
290–350	0.198	614
350–400	0.105	324
Higher than 400	0.329	1315

respectively. They constitute less than 1% of the watershed area. While the sixth class which concerns the zones whose soil loss exceeds 440 t/ha/year represents 0.30% of the global surface of the study area 20 and 210 t/ha/year. These last two classes concern the mountainous zones and the zones with friable substrate.

3.2 SWAT Model Application

The soil losses map (Fig. 11) created using the Soil and Water Assessment Tool (SWAT) shows the annual soil losses per tonne and per hectare for each sub-basin, as well as the most eroded and water-erosion zones. It is to highlight that the SWAT model was successfully calibrated using real data from the Oum Er Rbia station, with a determination coefficient of 0.72. The produced map shows that almost 79% of the study area has soil loss values that do not exceed 20 t/h/year, which are relatively reliable losses (Fig. 12).

Soil losses (t/ha/year) in the basin were grouped into 6 classes of values see table number (Table 3). The first class concerns the areas with a soil loss of less

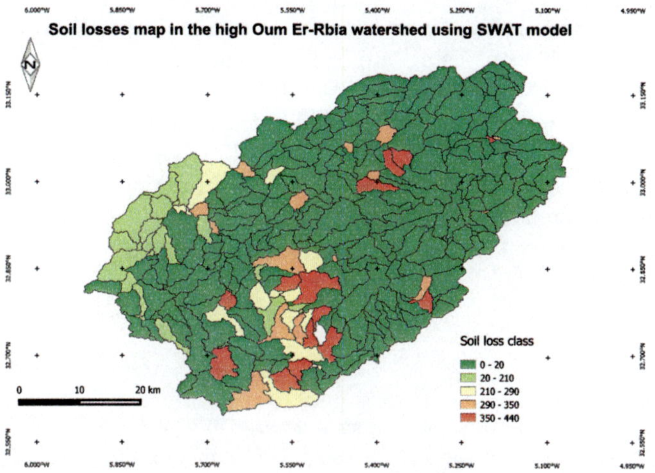

Fig. 12. Soil losses map in the high Oum Er-Rbia watershed using SWAT model

Table 3. Soil losses map of the High Oum Er-Rbia watershed using SWAT model

Soil loss class(t/ha/yr)	SWAT model	
	Percentage [%]	*Area [ha]*
Less than 20	79.038	243200
20–210	6.012	18500
210–290	5.78	17800
290–350	3.867	11900
350–400	4.16	12800
Higher than 400	1.138	3500

than 20 t/ha/year. It represents almost 79% of the surface area of the basin, spread over the whole study area. The second class concerns areas with soil loss between 20 and 210 t/ha/year. It represents almost 6% of the watershed area. The third class consists of areas with soil loss between 210 and 290 t/ha/yr. It represents 5.78% of the basin area. The fourth class include areas with soil loss between 290 and 350 represents 3.86% and fifth which include areas with soil loss varying from 350 to 400 t/ha/yr represents 4.16% respectively. They constitute less than 1% of the watershed area. While the sixth class which concerns the zones whose soil loss exceeds 440 t/ha/year represents 1.13% of the global surface of the study area. These last two classes concern areas with a friable substrate and mountainous regions.

4 Conclusion

Monitoring soil losses using different models can support land management and decision-making by detecting the most important actions and locations to fight against water erosion risks. The study of erosion risk in the High Oum Er-Rbia watershed was carried out using two soil loss quantification models: 1) Revised Universal Soil Loss Equation (RUSLE) and 2) Soil and Water Assessment Tool (SWAT). The different input data involved in the process of applying the models were determined and their combination in a GIS environment made it possible to obtain the soil losses in the basin. The results obtained show that the High Oum Er-Rbia watershed is subject to a soil loss of which almost more than 70% of the basin area presents an annual loss that does not exceed 20 t/ ha using the two quantification models.

References

1. Bouslihim, Y., Rochdi, A., El Amrani Pazza, N.: Water balance estimation in semiarid Mediterranean watersheds using SWAT model. In: Ksibi, M., et al. (eds.) EMCEI 2019. ESE, pp. 1537–1543. Springer, Cham (2021). https://doi.org/10.1007/978-3-030-51210-1_245
2. Jensen, J.R.: Remote Sensing of the Environment an Earth Resource Perspective. Prentice Hall, Upper Saddle River (2000)
3. MAMVA: Résultats de l'enquête nati onale sur les terres agricoles soumises á l'érosion, Worck accomplished by Bouhouch. S. Ministry of Agriculture and Agricultural Development (1996)
4. Maximillian, J., Brusseau, M., Glenn, E., Matthias, A.: Pollution and environmental perturbations in the global system. In: Environmental and Pollution Science, pp. 457–476. Elsevier (2019)
5. Mazouzi, K., El Hmaidi, A., Bouabid, R.: Quantification de l'érosion hydrique, par la méthode rusle, au niveau du bassin versant de l'oued mikkès en amont du barrage sidi chahed (région de meknès, maroc) (2021)
6. Mishra, A.K., Placzek, C., Jones, R.: Coupled influence of precipitation and vegetation on millennial-scale erosion rates derived from 10be. PLoS ONE **14**(1), e0211325 (2019)
7. Neitsch, S., Arnold, J., Kiniry, J., Williams, J., King, K.: Soil and water assessment tool theoretical documentation: Version 2005. Grassland soil and water research laboratory. Agricultural Research Service: Temple, Texas, USA (2005)
8. Pandey, A., Mathur, A., Mishra, S., Mal, B.: Soil erosion modeling of a Himalayan watershed using RS and GIS. Environ. Earth Sci. **59**(2), 399–410 (2009)
9. Prasannakumar, V., Vijith, H., Abinod, S., Geetha, N.: Estimation of soil erosion risk within a small mountainous sub-watershed in Kerala, India, using revised universal soil loss equation (RUSLE) and geo-information technology. Geosci. Front. **3**(2), 209–215 (2012)
10. Rango, A., Arnoldus, H.: Aménagement des bassins versants. Cahiers techniques de la FAO **36** (1987)
11. Stiles, D.: Desertification as dryland degradation (1994)
12. Wischmeier, W.H., Smith, D.D.: Predicting rainfall erosion losses: a guide to conservation planning. No. 537, Department of Agriculture, Science and Education Administration (1978)
13. Yjjou, M., Bouabid, R., El Hmaidi, A., Essahlaoui, A., El Abassi, M.: Modélisation de l'érosion hydrique via les sig et l'équation universelle des pertes en sol au niveau du bassin versant de l'oum er-rbia. In. J. Eng. Sci. (IJES) **3**(8), 83–91 (2014)

Communication Systems, Signal and Image Processing for Humanity

A Compact Frequency Re-configurable Antenna for Many Wireless Applications

Younes Karfa Bekali[1](✉), Asmaa Zugari[2], Brahim El Bhiri[3], Mohammed Ali Ennasar[2], and Mohsine Khalladi[2]

[1] LCS, Faculty of Sciences, Mohammed V University in Rabat, Rabat, Morocco
bekali@um5r.ac.ma
[2] ITSL, Faculty of Sciences, Adelmalek Essaadi University, Tetouan, Morocco
[3] SMARTiLab EMSI-Rabat, Rabat, Morocco
b.elbhiri@emsi.ma

Abstract. To answer the upcoming wireless communication system objectives and functionalities an important number of antennas on the same equipment should be integrated. This situation involves many problems related to the size, weight, volume of the board. Reconfigurable antennas are the latest developed concept by improving the antenna design. This new antenna design transforms the conventional fixed-form antenna on a dynamic one by changing its radiation and frequency properties in a permanent way. The simulations result and experimental measurements show the performance of our design approaches.

Keywords: Reconfigurable antenna · Compact frequency · Antenna design · Multi-standard

1 Introduction

The fast progress of mobile communication required the use of new model of antennas. Wireless communication system requires low profile, light weight, important gain, and simple design antennas to assure mobility, reliability, and high efficiency characteristics [1]. In addition, the important demand for new services and applications and this very important number of users, a genius techniques must be developed for the future radio systems. On the other hand, the recent advances in wireless communication systems, such as GSM (880 MHz–960 MHz) and DCS (1710–1880 MHz) in Europe, PCS (1850–1990 MHz) in USA, WLAN, WLL, 3G etc., have encouraged a important interest in micro-strip antennas [2,3]. Indeed, due to their capability to answer the particular limitations of size, weight and specifically cost imposed by the new mobile systems, the field of micro-strip antennas have grown vastly in the last years. Nevertheless, the antennas inherent narrow bandwidth and low gain is one of their major weakness. This problem have attracted the attention of the researchers. The patch antenna is used in various fields like aircrafts, missiles, GPS system, and broadcasting etc. [4]. The patch antennas can be prepared into a various shapes as circular,

R. Saidi et al. (Eds.): ICATH 2021, LNDECT 110, pp. 75–83, 2022.
https://doi.org/10.1007/978-3-030-94188-8_8

triangular, rectangular, square etc. and are light in weight, small size, low cost, simplicity of manufacture and easy integration to circuits. In other hand, the interest considered to multi-band antennas is growing. The reason is especially involved to reduce the number of the embedded antennas combining multiple applications. Many techniques are proposed and developed, we note Frequency Selective Surface [5,6], the icker profile for folded shorted patch antennas [7], using slots [8], use of the thicker substrate [9], etc. The feeding patch size and the dielectric should be taken with more caution. This combination with the use of the method of the microstrip line [10] gives a very important overview of the physical antenna. Those antennas will be able to cover the GSM, the DCS, PCS, UMTS and WLAN applications namely in ISM band used by systems BLUETOOTH and Wi-Fi. In other hand, the reconfigurable antennas are discussed and many models are developed according to the antenna parameter. A reconfigurable antenna is capable of modifying dynamically its frequency and radiation characteristics in a reversible manner [11]. In addition, reconfigurable antennas present a latest improvement in antenna design that changes the conventional fixed-form antenna [12]. There are many sorts of reconfiguration methods for antennas. Mostly they are electrical [13] (RF-MEMS, diodes, or varactors), optical, physical, and using materials [14,15]. In this paper and to response this objective, an internal, low profile new design for reconfigurable antenna will be presented adapting several techniques as the ones cited above. Three geometrics design from the same model are demonstrated. This new design and its geometric characteristics will be presented in this paper. Ansoft HFSS and CST are used to simulate results with Far Field Antenna. In addition, this paper presents the radiation patterns over the whole frequency bands which are acceptable.

2 Antenna Design and Reconfiguration

2.1 Geometry of the Proposed Antenna

The configuration and geometry of the proposed frequency reconfigurable antenna for WLAN, Wi-Fi, WiMAX, UMTS band, X band and Radio Altimeter is presented in Fig. 1. The preliminary antenna shape is taken from the basic design mentioned in [1]. The designed antenna uses FR4 low cost substrate (1.6 mm thickness and 0.02 tangent loss) with a compact size of $28 \times 16\,\text{mm}^2$ and dielectric constant $\epsilon_r = 4.4$ backed by the truncated metallic ground surface. In order to modify the current distribution, two metallic strips with 1mm width are integrated in the radiating element. This allows changes the resonance frequency under different states (modes) of switches. These strips provide a similar working of the two PIN diodes, thereby making it a frequency reconfigurable antenna. The dimensional details of the design are tabulated in Table 1.

2.2 Geometry of the Proposed Antenna

In order to take a closer look at the impact of the presence or the absence of two metal strips which was inserted into the patch, the simulation outcomes in

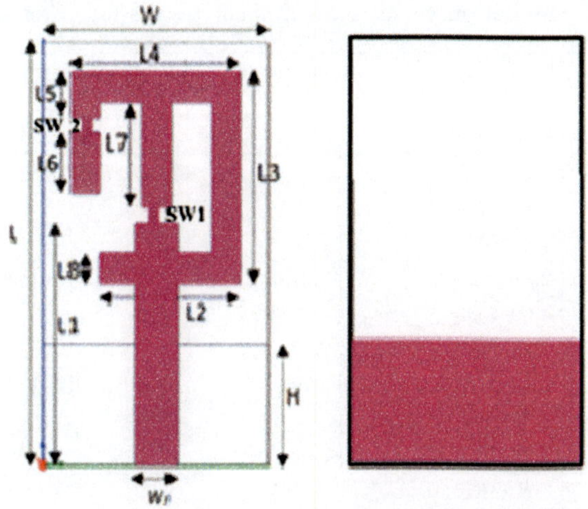

Fig. 1. The geometry of frequency reconfigurable planner antenna: Front view and Rear view.

Table 1. Detailed dimensions of the proposed antenna.

Parameter	Value (mm)	Parameter	Value (mm)
L	28	L_3	14
W	16	L_4	12
W_f	3	L_5	3
H	8	L_6	4
L_1	16	L_7	7
L_2	10	L_8	2

terms of the reflection coefficient for the different cases are shown in Fig. 2. In the first case (ON-ON), it can be observed that the antenna has a three different operating band when the two switches are ON as shown in Fig. 2. In the second case (ON-OFF), the first switch (SW1) is ON (represented by the presence of the metal strip) while the second switch (SW2) is OFF (represented by the absence of the metal strip), the switches will act as an open circuit and as a short circuit respectively. In this situation, as observed in Fig. 2, the antenna work as a dual-band antenna. The antenna in this case resonates at 3.1 and 7.1 GHz. In the last case (OFF-ON), when SW1 is OFF while SW2 is ON, the antenna is also working as a dual-band antenna as shown in Fig. 2. The antenna resonates at 5 and 7.72 GHz in this case. Therefore, controlling the operating bands is realised by changing switches states.

Table 2. Simulation versus measurement results for different states

	Impedance bandwidth (GHz)	Frequency (GHz)	Applications
	ON-ON state		
CST simulation	2.89–3.27	3	Maritime Radiolocation Service Wifi
	4.8–5.57	5	
	6.5–8.03	7.4	
HFSS simulation	2.89–3.27	3.07	Maritime Radiolocation Service Wifi
	4.69–5.62	4.97	
	6.22–8.36	7.38	
Measurement	2.44–3.15	2.74	Maritime Radiolocation Service Wifi
	4.75–5.69	5	
	6.24–8.65	7.93	
	ON-OFF state		
CST simulation	3.1–3.7	3.34	WiMAX, fixed satellite communication
	6.48–8.11	7.35	
HFSS simulation	2.89–3.27	3.07	WiMAX, fixed satellite communication
	4.69–5.62	4.97	
Measurement	2.44–3.15	2.74	WiMAX, fixed satellite communication
	4.75–5.69	5	
	OFF-ON state		
CST simulation	2.16–2.33	2.25	WLAN, WiMAX, UMTS, X-band Satellite communication
	4.59–5.72	5	
	7–8.34	7.72	
HFSS simulation	2.07–2.23	2.17	WLAN, WiMAX, UMTS, X-band Satellite communication
	4.15–5.29	4.53	
	65.86–9.22	7.72	
Measurement	1.79–2.08	21.97	WLAN, WiMAX, UMTS, X-band Satellite communication
	4.47–6.05	5.18	
	7–9.22	8.25	

2.3 Analyzing the Current Distribution

To offer a clear understanding of the resonance behaviour of the designed antenna, the current distribution of different frequencies was simulated using computer simulation technology (CST) microwave studio. Figure 3 displays the current distribution of the antenna at various frequencies considering different states of the switches. The high-density current at 3 GHz when both switches are ON is observed along the edge of the middle strip while the current density becomes less along the other parts as shown in Fig. 3a. At 5 GHz, the simulated current distribution is shown in Fig. 3 b when SW1 is OFF and SW2 is ON. As observed, the high-density current distributed along the middle strip as well as along the edge of the upper horizontal strip. From Fig. 3c, which shows the current at 7.35 GHz when SW1 is ON and SW2 is OFF; the high intensity current

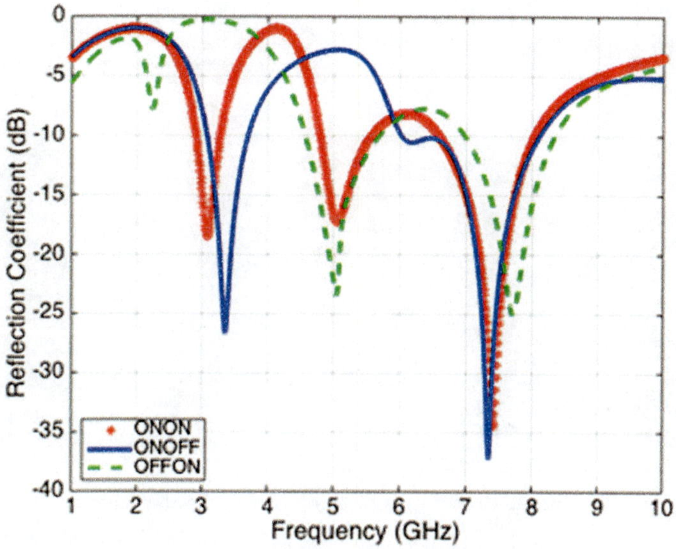

Fig. 2. Comparison of the reflection coefficient of the frequency reconfigurable antenna.

can be observed along the upper horizontal strip as well as along the lower right edge. It can be concluded that by changing switches states, different discontinuities in current distribution were observed along antenna parts, which helped to achieve different multi-band modes.

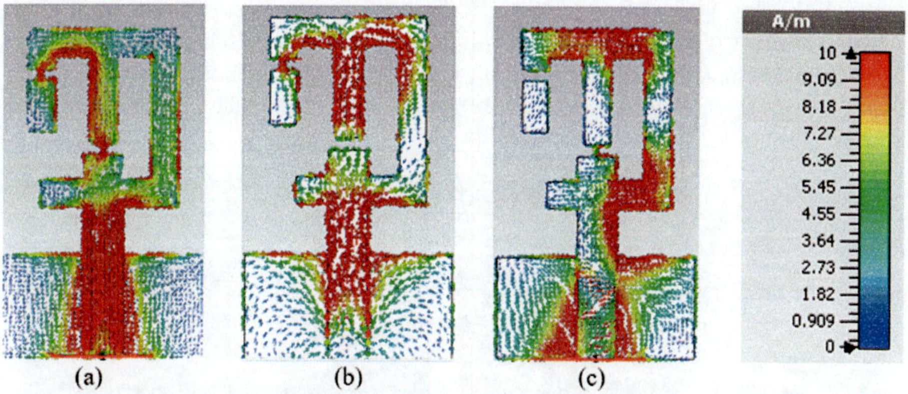

Fig. 3. The surface current distributions of the proposed antenna at: (a) 3 GHz, (b) 5 GHz, (c) 7.35 GHz

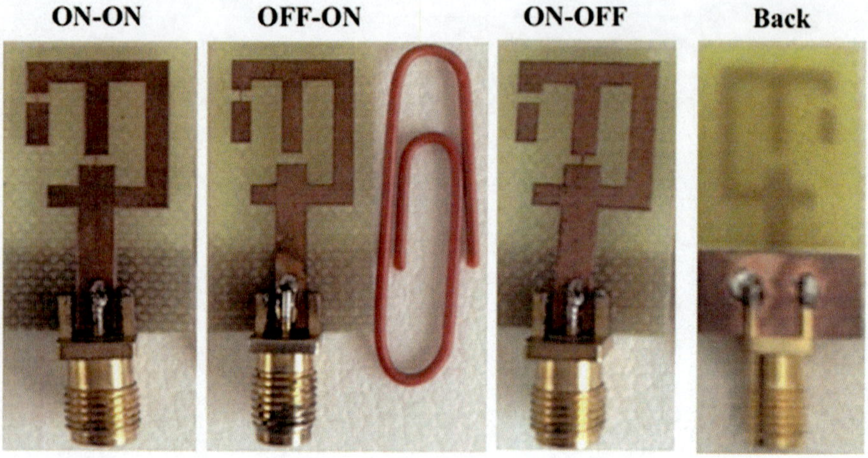

Fig. 4. Prototypes of the fabricated antennas

3 Results and Discussions

Three prototypes of the designed antenna have been fabricated to validate the design concept. The fabrication models depicted in Fig. 4 were made using the LPKF Proto Mat E33 machine. The measurement of the reflection coefficient results was carried out using the Rohde and Schwarz ZVB 20 vector network analyzer. Then, as shown in Fig. 5, the far field radiation patterns were measured by Far fair Antenna Measurement Systems of Geozondas.

A comparison between simulated reflection coefficient curves, which were simulated using CST software and HFSS solver, and measured reflection coefficient curves are shown in Fig. 6 of different states of switches. The proposed antenna presents an excellent matching with a 50 Ω source. The measured results are in accordance with the simulation results, especially with CST results. The

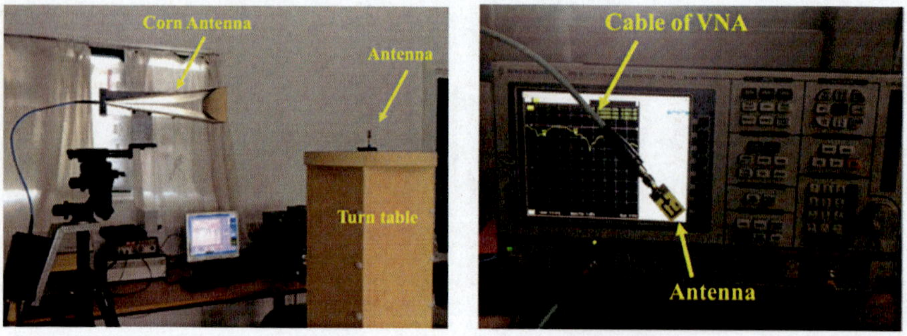

Fig. 5. Far Field Antenna Measurement System's principle and measurement the return loss S11

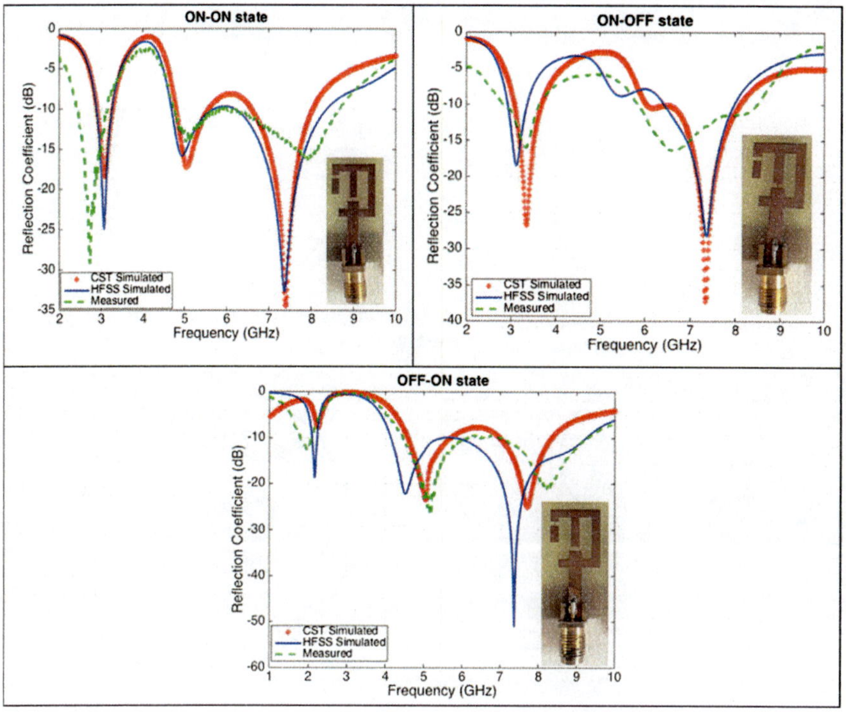

Fig. 6. Simulated and measured reflection coefficient for different states of switches

slight difference between simulated and measured curves, however, is due to the imprecision of fabrication and port loss, but the measured curves still cover the desired frequency bands. The measured reflection coefficient curves show that the designed antenna has a triple operating band at 2.74 GHz, 5 GHz and 7.93 GHz for ON-ON state, while a dual operating band at 3.34 GHz and 6.65 GHz for ON-OFF state were achieved. Furthermore, a triple operating band at 1.97 GHz, 5.18 GHz and 8.25 GHz for OFF-ON were achieved. For commercial applications, the metal strips can be replaced with PIN diodes. A comparison between simulation and measurement results for the proposed antenna are summarized in Table 2.

The simulated and measured radiation patterns of the proposed antenna at 3 GHz, 5 GHz and 7.35 GHz are plotted in Fig. 7. As can be seen from this Figure, the simulated and measured radiation patterns in H-plane have an omnidirectional style, whereas the E-plane has almost an "8" shape radiation. This shows that the antenna is apposite for integration with portable devices.

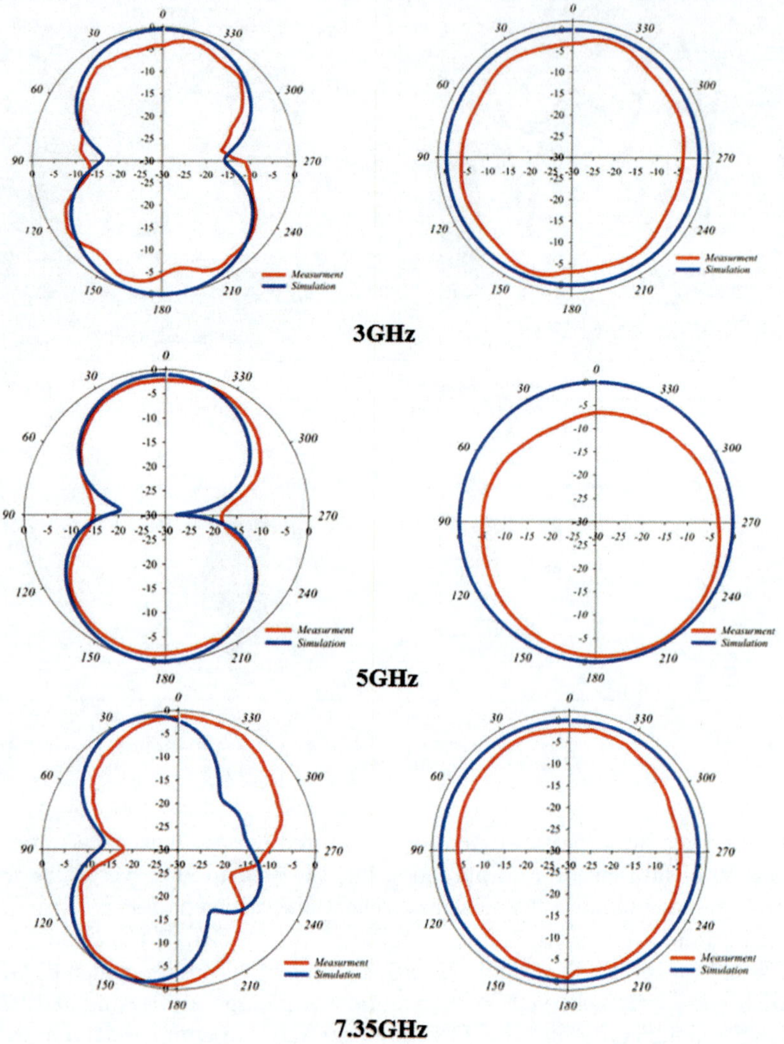

Fig. 7. Simulated and measured reflection coefficient for different states of switches

4 Conclusion

Reconfigurable antennas attract a massive attention of the forthcoming wireless applications. Many techniques have been used to answers those objectives. This paper work shows a model for reconfigurable antennas where the simulation results using 3 GHz, 5 GHz and 7.35 GHz bands are correlated with the realistic ones. The impact of the antenna parameters using the two metal strips which was inserted into the patch (ON-ON, ON-OFF or OFF-ON) is shown and visible. Results confirm that our antenna design is very compact, very easy to fabricate and apposite for integration with portable devices. In the upcoming work we will develop a solution to control dynamically the two metal strips.

Acknowledgment. The authors would like to thank LCS-FSR UM5 and SMARTi-Lab/EMSI for support and infrastructure.

References

1. Bekali, Y.K., Zugari, A., Essaaidi, M., Khalladi, M.: A novel design of multi-band microstrip patch antenna for wireless communications. In: Mediterranean Microwave Symposium, MMS 2018, Istanbul, Turkey, 31 October 2018–2 November 2018, Added to IEEE Xplore, 17 January 2019
2. James, J.R., Hall, P.S.: Handbook of Microstrip Antennas. I.E.E. Electromagnetic Waves Series 28. Peter Peregrinus LTD (1989)
3. Best, S.R.: Electrically small multiband antennas. In: Sanchez-Hernandez, D.A. (ed.) Multiband Integrated Antennas for 4G Terminals, pp. 1–32, Artech House, Boston (2008). ISBN 978-1-59693-331-6
4. Govardhani, I., Swetha, K., Venkata Narayana, M., Sowmya, M., Ranjana, R.: Design of microstrip patch antenna for WLAN applications using Back to Back connection of Two E-Shapes. Int. J. Eng. Res. Appl. **2**(3), 319–323 (2012)
5. Chen, J.-S.: Multi-frequency characteristics of annularring slot antennas. Microw. Opt. Technol. Lett. **38**(6), 506–511 (2003)
6. Liu, Y.-T., Su, S.-W., Tang, C.-L., Chen, H.-T., Wong, K.-L.: On-vehicle low-profile metalplate antenna for AMPS/GDM/DCS/PCS/UMTS multiband operations. Microw. Opt. Technol. Lett. **41**(2), 144–146 (2004)
7. Liu, Y.-S., Sun, J.-S., Lu, R.-H., Lee, Y.-J.: New multiband printed meander antenna for wireless applications. Microw. Opt. Technol. Lett. **47**(6), 539–543 (2005)
8. Eratuuli, P., Haapala, P., Vainikainen, P.: Dual frequency wire antennas. Electron. Lett. **32**(12), 1051–1052 (1996)
9. Salonen, P., Keskilammi, M., Kivikoski, M.: New slot configurations for dual-band planar inverted antenna. Microw. Opt. Technol. Lett. **28**(5), 293–298 (2001)
10. Yang, F., Zhang, X.X., Ye, X., Rahmat-Samii, Y.: Wide-band Eshaped patch antennas for wireless communications. IEEE Trans. Antennas Propag. **49**(7), 1094–1100 (2001)
11. Bernhard, J.T.: Reconfigurable antennas. Synth. Lect. Antennas **2**, 1–66 (2007). https://doi.org/10.2200/S00067ED1V01Y200707ANT004
12. Deshmukh, V.Y., Chorage, S.S.: Review of reconfigurable antennas for future wireless communication. In: 2020 International Conference on Emerging Smart Computing and Informatics (ESCI), pp. 28–33 (2020). https://doi.org/10.1109/ESCI48226.2020.9167528
13. Panagamuwa, C.J., Chauraya, A., Vardaxoglou, J.C.: Frequency and beam reconfigurable antenna using photoconducting switches. IEEE Trans. Antennas Propag. **54**(2), 449 (2006)
14. Chiao, J.C., Fu, Y., Chio, I.M., DeLisio, M., Li, L.Y.: MEMS reconfigurable Vee antenna. In: IEEE MTT-S International Microwave Symposium, vol. 4, pp. 1515–1518 (1999)
15. Rodrigo, D., Jofre, L., Cetiner, B.A.: Circular beam-steering reconfigurable antenna with liquid metal parasitics. IEEE Trans. Antennas Propag. **60**(4), 1796 (2012)

A Review on Visible Light Communication System for 5G

Mohamed El Jbari(✉), Mohamed Moussaoui, and Noha Chahboun

Information and Communication Technologies Laboratory (LabTIC), National School of Applied Sciences of Tangier, Abdelmalek Essaadi University, Tetuan, Morocco
mohamed.eljbari2@etu.uae.ac.ma, m.moussaoui@uae.ac.ma

Abstract. Visible-Light Communications (VLC) is an emerging wireless communication technology, which appears as a promising solution for very high speed 5G wireless networks in short-range communications. Is based on Intensity Modulation (IM). The implementation of VLC is based on Intensity Modulation/ Direct Detection (IM/DD) where the signal is directly modulated onto the instantaneous optical carrier power and direct detection-based photo-detector is used to generate an electrical current that is directly proportional to incident light power. The incident light, on the other hand carries no just the data signal but is also noise. In this paper, the proposed VLC-based signal transmission system model using IM/DD is presented. The injection current is modulated by a NRZ binary pseudo-random bit sequence. At the receiver, the photodiode produces a noisy electrical current which is proportional to the input optical intensity.

Keywords: 5G · VLC · Intensity Modulation Detection Direct (IM/DD) · LED · Laser Diode LD · On-off Keying (OOK) · PhotoDiode PD

1 Introduction

The most of the valuable RF bands have already been used in existing wireless systems. So RF spectrum is becoming rapidly increasingly congested.

In addition, as it's proportional to the channel bandwidth, the most effective way to rapidly expand channel capacity is to increase the channel bandwidth and expand the available spectrum. In this context, it is inevitable to switch to the VL band. Visible-Light Communication (VLC) has emerged as a promising complementary technology offering the highest data transmission capacity for 5G wireless networks in both indoor and outdoor environments [3].

Visible Light communication (VLC) involves the electromagnetic spectrum domain in the range of 428 THz to 750 THz with corresponding wavelength range of 380 nm at 750 nm [4], when compared with microwave frequencies band of 3 kHz to 300 GHz, this utilizes a very larger frequency range [1, 2]. This is therefore an environmentally-friendly green communication technology [5]. The implementation of VLC is based on Intensity Modulation/ Direct Detection (IM/DD) where the signal is directly modulated onto the instantaneous optical carrier power and direct detection-based photo-detector is

R. Saidi et al. (Eds.): ICATH 2021, LNDECT 110, pp. 84–95, 2022.
https://doi.org/10.1007/978-3-030-94188-8_9

used to generate an electrical current that is directly proportional to incident light power. In this paper, we present and study a VLC-based signal transmission system model using IM/DD and provide a comprehensive study on advances in this technology, focusing on technical challenges for practical implementation.

2 Comparison Between VLC and RF Technologies

LED lighting offers many advantages such as low energy consumption, much longer lifespan, high efficiency, specific spectrum and environmental friendliness. Those advantages made the LEDs a ubiquitous replacement for the classical illumination sources and more suitable for the high speed communications due principally to its wider modulation bandwidth. In Table1, we have brief comparison between VLC and RF technologies, where VLC has many advantages over RF.

VLC is considered as an alternative solution for RF spectrum where the light is already present. The Light-Emitting Diode (LED) driver is a main part of Visible Light Communication (VLC) [6, 7].

Transfer data from mobile device to a different device can be done only by aiming and aligning a light the light beam per one device to another, making the data much more difficult to intercept from outside [8]. Indeed, the interference would be negligible for indoor applications compared to RF signals.

VLC communication technologies can be further divided into two, real-time and non-real-time systems [10]. when real-time data processing of data received is needed, Real-time VLC systems are utilized. However, Offline VLC systems do not require instant processing [11–13],

More complex modulation schemes can be used in non-real-time systems to provide high spectral efficiency and throughput such as orthogonal frequency division multiplexing (OFDM), discrete multi-tone (DMT) modulation technology. However, On-Off keying (OOK) is the most suitable modulation scheme for real-time VLC systems. Therefore, VLC systems can use different modulation schemes for different applications to improve the overall performance. The deployment of high spectral efficiency modulation techniques that take into account the constraints of IM/DD is a challenge in the design of high data-rates VLC systems. Very high data transmission can only be achieved by IM/DD, despite the narrow bandwidth of the LEDs which limits VLC technology for fifth-generation (5G), its frequency bandwidth is around 300 THz [14].

3 VLC System Architecture

VLC system principally is composed of a VLC emitter that are modulates the light emitted from LEDs or the LDs and of a VLC receiver which a photoreceptor element, located in the light field (VLC channel) of the LED (PIN photodiode or avalanche photodiode (APD)) [16–18], detects these variations (the modulated signal of light) and translates them into electrical pulses. The transmitter and receiver are separate from one

Table 1. Comparison of VLC, RF communication technologies [9].

Property	VLC	RF
Bandwidth of the carrier	400–700 nm (very large and unregulated)	300 GHz (saturated and regulated)
Electromagnetic interferences and hazard	No	Yes
Range	Short to Medium	Short to long (outdoor)
Security	Good	Poor
Service	Illumination and Communication	Communication
Noise sources	Sunlight and other ambient lights	All electrical and electronic appliances
Power consumption	Relatively low	Medium
Mobility	Limited	Good
Coverage	Narrow	Mostly wide

other, that are connected by the VLC channel [15]. In VLC systems, line-of-sight is an obligatory condition. The diagram of a VLC system is shown in Fig. 1.

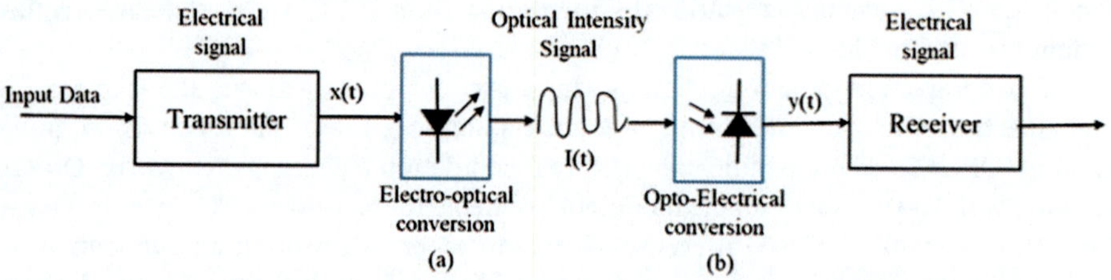

Fig. 1. Block diagram of VLC system: (a). Transmitter; (b). Receiver

3.1 Transmitter

A VLC transmitter is the device that converts data into the optical media messages spreading in free space using visible light. VLC transmitter must the emission of light and transmission the data at the same time.

3.2 LED

The spontaneous emission of a forward-biased PN semiconductor are important from a LED. However, the P-type region, electrons are the minority carriers and holes are the majority carriers (see Fig. 2.), the majority carriers in the N-type region are electrons and holes as minority carriers [19].

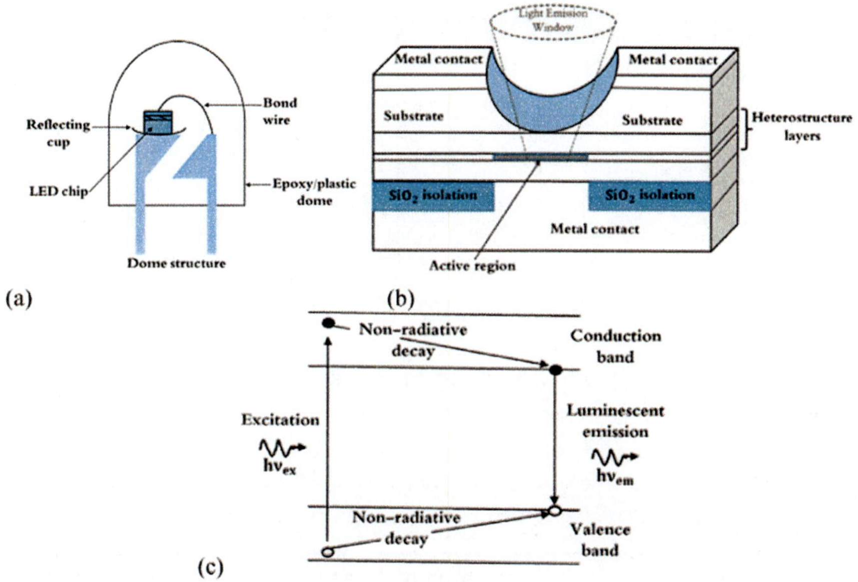

Fig. 2. Dome LED (a) LED structures: planar LED and Illustration of surface emission (b), and White light production (c).

The optical transmit power can be expressed by:

$$P(\alpha) = P_{Opt}.\cos(\alpha) \tag{1}$$

Where α is the angle between the directions of emission is the perpendicular to the emitter surface, and P_{Opt} is the output optical power from the normal direction.

- *The characteristic curve P-I*

As shown in Fig. 3 (a), the optical power output by a LED is proportional to the injected electric current. The energy yield is the ratio into P_{opt} the output power optical and the injection current I then $\frac{dP_{opt}}{dI}$, that we express:

$$\frac{dP_{opt}}{dI} = \frac{hv}{q} = \frac{hc}{\lambda q} \tag{2}$$

Where $h = 6,63.10^{-34} J.s$ is the planks constant, ν is the optical frequency, q is the electron charge, and C is the speed of light.

The optical power emitted by an LED has a linear relationship with the injection current I, in the form:

$$P_{opt} = \eta_q \eta_{ext} \frac{hc}{\lambda q}.I \tag{3}$$

Where η_q is external efficiency, and η_{ext} is internal quantum efficiency.

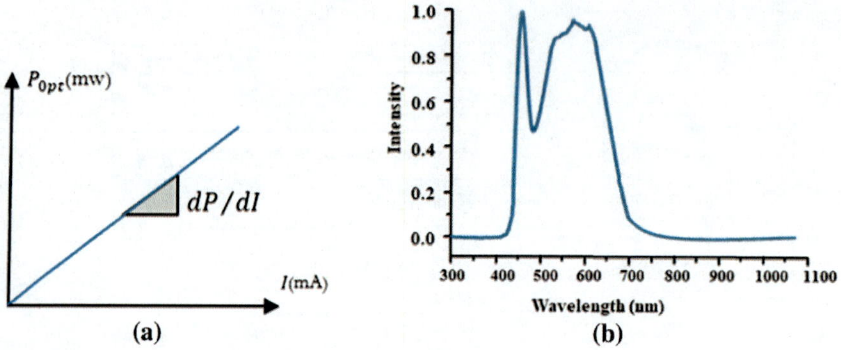

Fig. 3. (a) The characteristic curve of the emission power of the LEDs-Injection current. (b) White LED light spectrum in VLC system.

VLC transmitter has an essential component is the encoder that transforms the data throughout modulated message. The encoder controls the communication of LEDs based on binary data and certain data rate. The binary data is converted in an amplitude-modulated light beam. Generally, the light emitted by the LEDs are the current modulated by amplitude modulation using the On-Off Keying (OOK) type [20].

3.3 Laser Diodes

PIN-diodes as they are positive-negative having undoped intrinsic (I) region, which are used in laser diode. Charge injection is used in some LDs to power the diode laser (injection LDs). Some use optical pumping and are called optically pumped semiconductor lasers. GaAs, indium phosphide, gallium antimonide, or gallium nitride are the most common LD materials. All of them are made up of compound semiconductors. LDs semiconductors are based on the stimulated emission from the PN junction of forwarding biased semiconductors. LDs have higher efficiency spectrum purity compared to LEDs [21], due to the spatial and spectral consistency of the stimulated emission [22]. Among direct modulation lasers as a vertical cavity surface emitting lasers (VCSELs),its surface emission, in which light emitted perpendicular to the wafer plane. In Fig. 4, as showing an laser optical cavity in length L that the semiconductor material of the cavity furnishes by the optical gain g, and its loss α according to length, where n is refractive index for a material cavity, R_1 and R_2 are the reflectivity by tow facets [22].

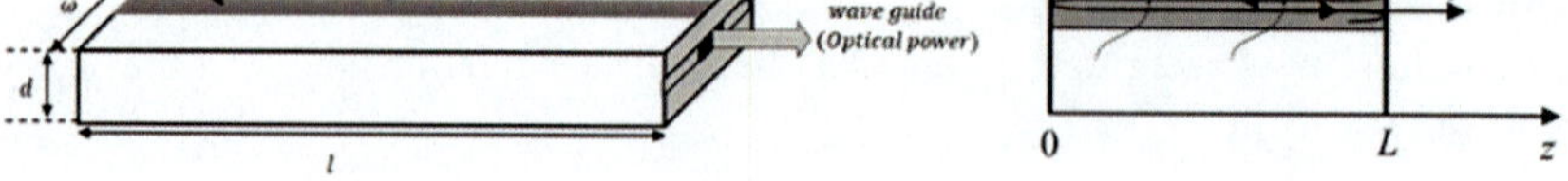

Fig. 4. The representation of the dimension of a laser cavity and a laser cavity with faceted reflectivity R_1 and R_2.

3.3.1 Rate Equations

The temporal dynamics of a single-mode semiconductor laser diode is modeled utilizing a pair of a coupling of differential equations characterizing the relationship into the density of photons and that of the charge carriers of the active region of the diode, in relation to the injection current $I(t) = J.lw$. Widely used in simulation and in dynamic analysis [24–28], this temporal formalism called rate Eqs. (3) and (4) is defined as:

$$\frac{dN(t)}{dt} = \frac{J}{qd} - \frac{N(t)}{\tau} - 2\Gamma v_g a(N - N_0)P(t) \tag{4}$$

$$\frac{dP(t)}{dt} = 2\Gamma v_g a(N - N_0)P(t) - \frac{P(t)}{\tau_{ph}} + R_{sp} \tag{5}$$

Where N(t) and P(t) are the photon density and the carrier density, respectively, about the laser cavity, that are expressed in units $[cm^3]$. J is the density of injected current in $[C/cm^2]$, and v_g is a group velocity from the light wave in $[cm/s]$, d is the depth of active layer. τ and τ_{ph} are the lifetimes of electron and photon, respectively. R_{sp} is the spontaneous emission rate, which indicates the spontaneously generated photons density of second which are coupled in LD mode, so the R_{sp} is in $[cm^3.s^{-1}]$.

The integration of these rate equations, accomplished by an adaptive Runge-Kutta algorithm of order 4, solution of the P(t) and N(t) concentrations of photons and charge carriers of the active region [23]. From these 2 temporal evolutions are deduced the intensity P(t) of laser emission.

The optical power in output LD is the flux of photons over the outer surface that is expressed by:

$$P_{opt}(t) = P.(lwd)hv2\alpha_m v_g \tag{6}$$

Finally, the injection current has a relation with the current density defined by $I = J.wl$ [22], then we give:

$$P_{opt} = \frac{(I - I_{th})h\nu}{q} \cdot \frac{\alpha_m}{\alpha_m + \alpha} \tag{7}$$

However, the optical power is totally exiting by two laser extremity facets. Where α is the rate of material absorption, and α_m is the photon leakage rate per the facet mirrors, $\frac{\alpha_m}{\alpha_m+\alpha}$ is coefficient by the external efficiency [22]. Where:

$$\alpha_m = \frac{-\ln(R_1 R_2)}{4L} \tag{8}$$

3.3.2 Modulation Type

Electro-optical modulation is a one-step very important in the optical transmitter, and transform electrical signals on optical signals. The injection current determines the optical power of LD, and direct OOK modulation is viable solution for converting the electrical signal (i.e. the injection current) into an optical signal then this linear modulation is achieved. However, the LD characteristic with direct modulation is a practical question for only applications in an optical communication systems. The curve of Fig. 5 represents the principle of the direct modulation of the intensity of the semiconductor laser. DC bias current I_B is sometimes necessary to guarantee that LD operates well above the threshold. To modulate the LD, combine the electric current signal with I_B. The efficiency of the modulation is dictated by the slope of the characteristic curve LD (i.e. Power-Current I). The optical power in output is linearly proportional to modulation of current, if this curve is absolutely linear, such as:

$$P_{opt}(t) \approx R_c(I_B - I_{th}) + R_c I(t) \tag{9}$$

Where $R_c = \frac{\Delta P_{opt}}{\Delta I}$ is the slop of LD, and I_{th} is its threshold current.
Curve characteristic P-I.
In this case, the optical power is neglected level at the threshold.

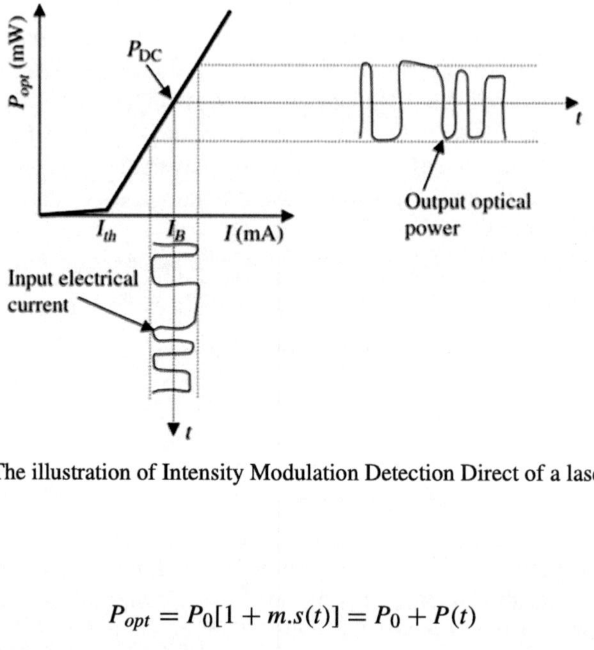

Fig. 5. The illustration of Intensity Modulation Detection Direct of a laser diode.

Where:

$$P_{opt} = P_0[1 + m.s(t)] = P_0 + P(t) \tag{10}$$

P_0 is the average optical power.
The electric current signal is given by:

$$I = I_B[1 + m.s(t)] = I_B + i(t) \tag{11}$$

Where s(t) is the signal component, and m represents the modulation index, where:

$$m = \frac{\mathrm{I_{max}}}{\mathrm{I_B}} \tag{12}$$

3.3.3 Laser Noises

At constant bias current, an LD will produce fluctuating optical power due to the random nature of the phenomena of spontaneous generation and recombination of electron-hole pairs. However, a practical LD has both a noise intensity which is troublesome in the case of direct detection when the modulation rate of the laser is low as shown in Fig. 6. The performance like an optical communication systems that utilizes LD as an optical element can be harmed by intensity noise. Even though stimulated emission dominates the emission process for a semiconductor laser operating beyond the threshold, spontaneous emission nevertheless generates a less percentage of photons. The irregular nature of spontaneous photons causes optical noisy intensity in an LD. When a steady injection current is used to bias an LD, the spontaneous emission are affected cause by the density carriers and photons to vary about their equilibrium value (Agrawal and Dutta 1986) [26].

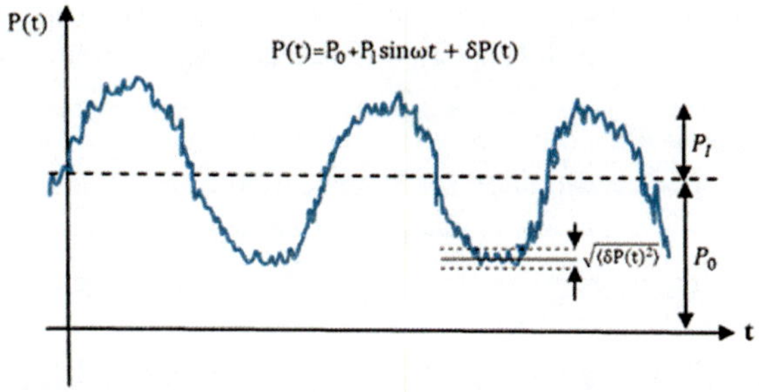

Fig. 6. Illustrative curve of the optical signal envelope with amplitude noise

The relative intensity noise (RIN) is the absolute ratio value of the intensity noise power to the average optical power, which is a regularly used LD specification parameter, that is giving as:

$$RIN(f) = \frac{\langle \delta P(t)^2 \rangle}{{P_0}^2} = \frac{\langle (P(t) - P_0)^2 \rangle}{{P_0}^2} \tag{13}$$

Where P_0 is the average optical power, $\langle (P(t) - P_0)^2 \rangle$ is the mean square intensity fluctuation (variance). RIN is expressed in [Hz^{-1}], [dB/Hz].

4 The VLC Channel

In VLC communication, the channel represents the free-space optical between the transmitter LED or LD and the receiver PD. IM/DD is a technique that characterizes the operational principle of a VLC system. It is characterized by its capacity to transmit the carrier signal, and, it is influenced by many factors such as attenuation, interference, and noise. It is mathematically represented by its transfer function h, it is written as:

$$y(t) = h(t) \otimes (t) + v(t) \tag{14}$$

Where $x(t)$ is the waveform that transportes data and modulates the instantaneous transmitted power (see Fig. 1.), and, y(t) is the received signal (Optical Intensity I(t)). $v(t)$ Is the background noise, ⊗ is the convolutional operator.

5 Receiver: Photodetectors

As a result of propagation in the optical channel, the high speed modulation emerging from the channel is first detected by a photodetector. As such, the small dimensions, the high sensitivity, and the relatively short response time of photosensitive semiconductor compounds justify their use with regard to IM / DD optical communications links. This class of detectors is based on the design of photosensitive diodes of the positive-intrinsic-negative (p-i-n) or avalanche effect (APD) type, the latter having an internal gain ensuring the amplification of the detected signal. Light is detected by a regions layer of n and p-type semiconductors (sometimes, a lightly doped P-type layer) into two terminals N (electrons) and P (holes) and of the device. This layer serves as the device's optically active area [27]. The external light source is detected by a semiconductor, while the impinging photons arrived at semiconductor may provide electrons in the valence band with the proper quantity of energy to go into conduction, by a phenomenon is called the photon absorption. Because these photo-generated of electrons are free to move in the semiconductor, they could act as charge carriers when an electrical field is applied externally. Figure 7 illustrates this process.

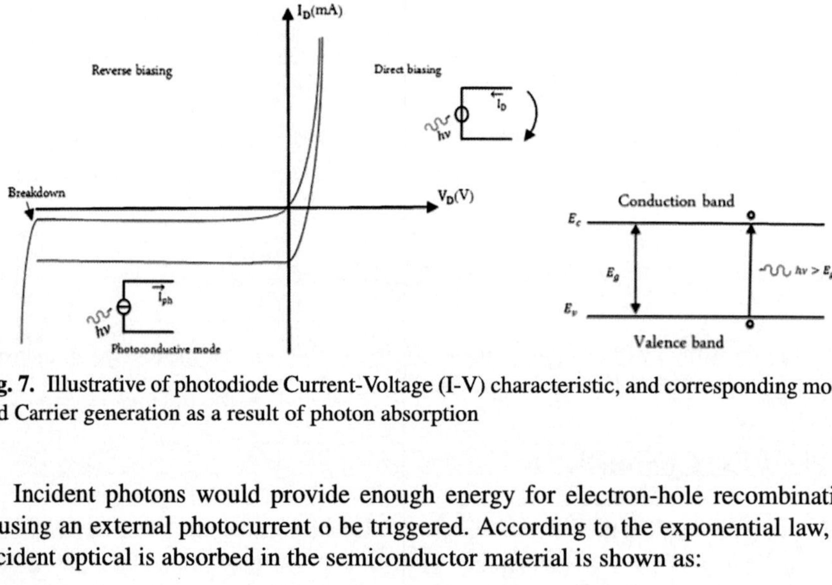

Fig. 7. Illustrative of photodiode Current-Voltage (I-V) characteristic, and corresponding modes, and Carrier generation as a result of photon absorption

Incident photons would provide enough energy for electron-hole recombination, causing an external photocurrent o be triggered. According to the exponential law, the incident optical is absorbed in the semiconductor material is shown as:

$$P(x) = P_0\left(1 - e^{-\alpha_s(\lambda)x}\right)(1 - r) \tag{15}$$

Where: $R(\lambda)$ is the spectral responsivity [mA/mW], and M is the Gain of the APD ($M = \frac{I_M}{I_P}$).

The shot noise variance is under these conditions defined by:

$$\sigma_{sh}{}^{2} = 2qI_{P}BM^{2}F(M) \tag{16}$$

Where: B is the bandwidth and F(M) is the noise figure. This Equation is valid for an APD; for a PIN photodiode, the term $M^2F(M)$ is replaced by 1. The photodiode has a noise source that are related to the incident optical power. It should also be noted that the noise performance in an APD is worse than that of PIN photodiodes.

5.1 Bit Error Rate

We can measure the visible light communication performance, using the SNR, which is expressed by a relation between the received power and the ambient noise, in the form:

$$SNR = \frac{R^2 P_0{}^2}{\sigma_{sh}{}^2 + \sigma_{Th}{}^2} \tag{17}$$

Where: R denotes the photodiode's photoresponsivity at the receiver, $\sigma_{sh}{}^2$ (giving in Eq. (16)), and $\sigma_{Th}{}^2$ is the noisy thermal variance [29].

$$\sigma_{Thermal} = \frac{8\pi kT_k}{G}\eta AI_2B^2 + \frac{16\pi^2 kT_k\Gamma}{g_m}\eta^2A^2I_3B^3 \tag{18}$$

Where k is the Boltzmann constant, T_k is the absolute temperature, η is the capacity per unit area, G is the open-loop voltage gain, A is the receiver side area, Γ is the Field-Effect Transistor (FET) channel noise factor, g_m is the FET transconductance, and I_2 is the noisy bandwidth current, I_3 is the noise's bandwidth coefficient.

For OOK optical intensity modulation, the Bit Error Rate (BER) can be expressed according to this relation:

$$BER = Q\left(\sqrt{SNR}\right) \tag{19}$$

Where $Q(x) = \frac{1}{2}\int_{x}^{\infty} e^{(-y^2)/2}dy$: is Average BER.

6 Conclusion and Future Directions

In this paper, we have discussed the new VLC wireless communication technology which has attracted special attention as a promising solution for ultra-high-speed 5G wireless networks in two main transmission environments, to the intra-building (indoor) and outdoors (Outdoor VLC), their advantages and challenges. As result, VLC will be critical technologies in meeting the high-demanding requirements of the 5G network and beyond communication systems. We've discussed semiconductor-based light sources, such as LEDs and LDs, which are commonly used in optical communications. Appliqued the Intensity Modulation/Direct Detection to ensure very high data transmission rates despite limiting the bandwidth of the LEDs appropriate to provide the length desired emission wave. The orientation of the internet of people to the internet of things (IoT), to the industrial internet of things needs VLC which can act as a catalyst to accelerate their revolution.

Acknowledgment. This work is part of the research activities conducted at Abdelmalek Essaadi Tetuan University, National School of Applied Sciences Tangier, Information and Communication Technologies Laboratory (LabTIC), "Design and optimization of a multi-user MIMO VLC visible light communication system".

References

1. Karunatilaka, D., Zafar, F., Kalavally, V., Parthiban, R.: LED based indoor visible light communications: state of the art. IEEE Commun. Survey Tutorial **17**(3), 1649–1678 (2015)
2. Ibhaze, A.E., Orukpe, P.E., Edeko, F.O.: Li-Fi prospect in internet of things network. In: Kacprzyk, J. (ed.) Advances in Information and Communication. Advances in Intelligent Systems and Computing, pp. 272–280. Springer, Cham (2020). https://doi.org/10.1007/978-3-030-39445-5_21
3. Wu, S., Wang, H., Youn, C.H.: Visible light communications for 5G wireless networking systems: from fixed to mobile communications. IEEE Netw. **28**, 41–45 (2014)
4. Li, S., Pandharipande, A., Willems, F.M.J.: Two-way visible light communication and illumination with LEDs. IEEE Trans. Commun. **65**(2), 740–750 (2017)
5. Tsiatmas, A., Baggen, C.P.M.J., Willems, F.M.J., Linnartz, J.P.M.G., Bergmans, J.W.M.: An illumination perspective on visible light communications. IEEE Commun. Mag. **52**(7), 64–71 (2014)
6. Loose, F., Duarte, R.R., Barriquello, C.H., Dalla Costa, M.A., Teixeira, L., Campos, A.: Ripple-based visible light communication technique for switched LED drivers. In: 2017 IEEE Industry Applications Society Annual Meeting, IAS 2017, pp. 1–6. IEEE Industry Applications Society, Janua (2017)
7. Modepalli, K., Parsa, L.: Dual-purpose offline LED driver for illumination and visible light communication. IEEE Trans. Ind. Appl. **51**(1), 406–419 (2015)
8. Teixeira, L., Loose, F., Brum, J.P., Barriquello, C.H., Reguera, V.A., Dalla Costa, M.A.: On the LED illumination and communication design space for visible light communication. IEEE Trans. Ind. Appl. **55**(3), 3264–3273 (2019)
9. Ghassemlooy, Z., Popoola, W., Rajbhandari, S.: Optical wireless communications: system and channel modelling with MATLAB, ISBN 978-4398-5188-3 (2012). https://doi.org/10.1201/b12687
10. Shin, H., Park, S., Lee, K., et al.: Investigation of visible light communication transceiver performance for short-range wireless data interfaces. In: Proceedings of the 7th International Conference on Networking and Services, Venice/Mestre, pp. 213–216 (2011)
11. Han, C.-X., Sun, X.-Z., Cui, S.-G.: Design of 100 Mbps white light LED based visible light communication. In: Proceedings of IEEE the 4th International Conference on Systems and Informatics, Hangzhou, pp. 1035–1039 (2017)
12. Wang, Y.-Q., Wang, Y.-G., Chi, N., Yu, J.-J., Shang, H.-L.: Demonstration of 575-Mb/s downlink and 225-Mb/s uplink bidirectional SCM-WDM visible light communication using RGB LED and phosphor-based LED. Opt. Exp. **21**(1), 1203–1208 (2013)
13. Wu, F.M., Lin, C.T., Wei, C.C., Chen, C.W., Chen, Z.Y., Huang, H.T.: 3.22-Gb/s WDM visible light communication of a single RGB LED employing carrier-less amplitude and phase modulation. In: Proceedings of Optical Fiber Communication Conference and Exposition and the National Fiber Optic Engineers Conference, Anaheim, p. 13 (2013)
14. Wang, Y.-Q., Huang, X.-X., Zhang, J.-W., Wang, Y.-G., Chi, N.: Enhanced performance of visible light communication employing 512 QAM NSC-FDE and DD-LMS. Opt. Exp. **22**(13), 15328–15334 (2014)

15. Jovicic, A., Li, J., Richardson, T.: Visible light communication: opportunities, challenges and the path to market. IEEE Commun. Mag. **51**(12), 26–32 (2013)
16. VLCC, Visible light communications consotium (vlcc). http://www.vlcc.net/
17. I. Nakagawa Laboratories, Nakagawa laboratories creates the next generation ubiquitous society using visible light communication. http://www.naka-lab.jpl
18. Nakagawa, M.: Illuminative light communication device, pat. US7583901B2 (2003). http://www.google.com/patents/US7583901
19. IEEE standard for local and metropolitan area networks–part 15.7: Short-range wireless optical communication using visible light. IEEE Std 802.15.7-2011, pp. 1–309 (2011). https://doi.org/10.1109/IEEESTD.2011.6016195
20. Introduction to Fiber-Optic Communications. 2020 Elsevier Inc. Chapter 3: Light sources for optical communications. https://doi.org/10.1016/B978-0-12-805345-4.00003-2
21. Ali, A., Li, Q., Fu, H., Mehdi, S.R.: Blue laser diode-based visible light communication and solid-state lighting. In: Antenna Systems (2021)
22. Corvini, P.J., Koch, T.L.: Computer simulation of hia-bit-rate optical fiber transmission using single-frequency lasers. J. Lightw. Technol. **LT-5**(11), 1591–1595 (19870
23. Cartledge, J.C., Burley, G.S.: The effect of laser chirping on lightwave system performance. J. Lightw. Technol. **7**(3), 568–573 (1989)
24. Cimini, L.J., Greenstein, L.J., Saleh, A.A.M.: Optical equalization to combat the effects of laser chirp and fiber dispersion. J. Lightw. Technol. **8**(5), 649–659 (1990)
25. Hinton, K., Stephen, T.: Modeling high-speed optical transmission system. IEEE J. Sel. Areas Commun. **11**(3), 380–392 (1993)
26. Bowers, J., Hemenway, B., Gnauck, A., WiltAlduraibi, D., et al.: High-speed InGaAsP constricted-mesa lasers. IEEE J. Quant. Electron. **22**(6), 833–844 (1986)
27. Agrawal, G., Dutta, N.: Long-wavelength semiconductor lasers. In: Long Wavelength Semiconductor Lasers, Van Nostrand Reinhold, New York (1986)
28. Al-Khaffaf, D.A.J., Alsahlany, A.M.: A cloud VLC access point design for 5G and beyond. Opt. Quant. Electron. **53**, 472 (2021). https://doi.org/10.1007/s11082-021-03132-2
29. Côté, S., Mémoire, à la Faculté des études supérieures de l'Université Laval: Elaboration d'un simulateur de liaison de communication optique à intensité modulée et détection direct, longue portée un grand débit, p. 62 (1999)

Sleep Stages Detection Based BCI: A Novel Single-Channel EEG Classification Based on Optimized Bandpass Filter

Said Abenna(✉), Mohammed Nahid, and Hamid Bouyghf

Faculty of Science and Technology, Hassan II University, Casablanca, Morocco

Abstract. The automatic recognition of the sleep stages from a single-channel EEG allows opening a new trick on the knowledge of the approaches of building ideas from dreams and imagination in the brain and then to specify the human thought originality. In this paper, optimization algorithms were applied like a genetic algorithm (GA) to improve the prediction accuracy and searching for the best frequency bandwidth-sensitive to each sleep-stage, this process shows that the prediction accuracy has been increased from values of 70% to values of 95%, these results indicate a great possibility to separate between EEG data of sleep-stages.

Keywords: Brain-computer interface (BCI) · Electroencephalogram (EEG) · Data analysis · Signal processing · Feature extraction · Parameters optimization

1 Introduction

Automatic sleep staging systems might be the best option for obtaining consistent and reliable sleep stages classification data. The EEG signal is the most important polysomnography (PSG) signal for determining the sleep stages. Sleep phases categorization approaches based on single-channel EEG signals are now widely explored because, unlike complex PSG devices, where related bipolar devices have less sleep interference and are more convenient to use. Traditional approaches [7, 8, 11, 13], as well as the newest deep and machine learning technology to automatically produce features [11, 16] have been utilized to identify sleep stages in these investigations. REM (rapid eye movement) and non-REM sleep are separated into three stages. Sleep usually starts in the first stage, which non-REM. The person sleeps well and wakes up easily at this point, the eyelids move slowly, the muscles relax, and the heart and respiration rate begin to settle down at this point [15]. Patients frequently move on the second stage of non-REM after the initial stage. Brain waves slow down, rapid waves erupt intermittently, and eye movements come to a halt, it takes a person half of the night at this point [15]. The third stage, non-REM, is much slower in the human brain, which produces extremely slow triangle waves. Further-more, the heart and respirator stages are both extremely sluggish [15]. It's called deep sleep because it's difficult to wake up when a person is in this stage. The eyes move rapidly in different directions during REM sleep. Dreams usually take place at a specific point in time. Hands and feet were temporarily immobilized to prevent

R. Saidi et al. (Eds.): ICATH 2021, LNDECT 110, pp. 96–105, 2022.
https://doi.org/10.1007/978-3-030-94188-8_10

dream interpretation. With elevated heart rate and blood pressure, breathing becomes erratic, rapid, and shallow [1, 15]. The majority of EEG research by sleep stages includes two phases: extracting functionality from EEG data and classifying derived pieces using classification algorithms. This research is crucial in the development of a sleep assessment system. Two of the problems of sleep categorization are improving classification accuracy [8, 17]. In this study, optimization algorithms are proposed to improve the prediction of EEG signals in general and the recognition of sleep-stages precisely [4], knowing that the optimizers were applied to improve the light gradient boosting machine (LGBM) parameters [5, 6] and the band-pass filter to determine the best bandwidth that characterizes each stage as illustrated in Fig. 1, and then we notice that the accuracy of predicting EEG signals has been increased from the order of 70% to values higher than 88%. The following is how the rest of the article is structured: The dataset used, data pre-processing, classification, and optimization techniques proposed for the prediction system are presented in Sect. 2, the results are presented and analyzed in Sect. 3, and the study finishes with a synopsis of future work in Sect. 4.

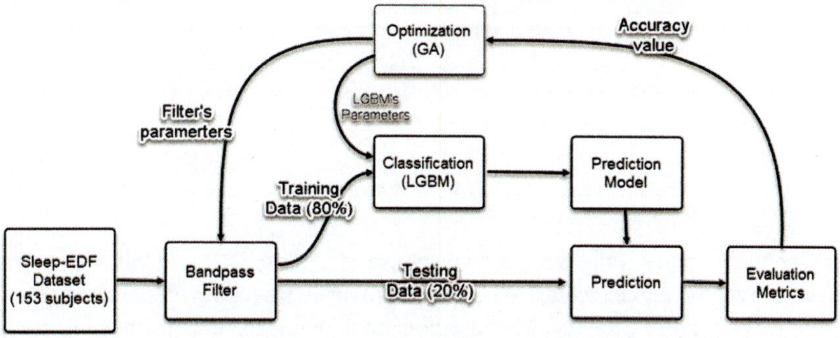

Fig. 1. Illustrate EEG data pre-processing and classification.

Table 1. The number of samples used for each task during training.

All samples	Tasks				
	Wake	Stage 1	Stage 2	Stage 3/4	REM
10100	6900	700	1800	100	600

2 Materials and Methods

Table 1 depicts the distribution of EEG data used for task training and testing, with the understanding that all classifications are binary, with the first-class containing only the data from the main task and the second class containing the rest of the data from all other classes, as shown in Fig. 2a.

2.1 Dataset

PhysioNet contributed the single-channel EEG data for this investigation, which came from the Advanced Sleep-EDF databases provided by PhysioNet [7, 11, 16]. The electrode EEG signal on Fpz-Cz and Pz-Oz placed as follow the international system 10-20, horizontal EOG signal, and EMG signal are all included in the Sleep-EDF data-base. The number of records is then increased to 61, resulting in a sleep-EDF database with more records. The Sleep Box Group (SC: Healthy Subjects) and the Sleep Telemetry Group are two different groupings of subjects in the database (ST: Objects with Sleep Difficulties). All EEG data are shown at 100 Hz and then manually labeled by professionals to different stages, which are referred as "W", "1", "2", "3/4", and "R" in this work.

2.2 Data Pre-processing

In this work the data from the different 153 subjects were used, knowing that all the data is collected and that an algorithm prefers to select only 10100 samples under a cross-subject of all data, then this data is filtered with a bandpass was fixed by the GA optimizer, after we extracted 8 characteristics from every 20 samples (at every 0.2 second), ultimately this data is randomized and subdivided into 80% of data for training and the remaining 20% is used for testing.

2.3 Power Spectral Density (PSD)

Power spectral density is the frequency response of a periodic or random signal (PSD). PSD represents the signal intensity distribution that is frequency-dependent. The data from the EEG is saved as a discrete time-domain signal. Eq. (1) [3] is used to compute the PSD of all frequencies:

$$PSD = \frac{1}{m}\left|\sum\nolimits_{i=1}^{m} X[i] * e^{-j2\pi fi}\right|^2 \tag{1}$$

With 'm' is the number of all samples in raw EEG ('X') and 'f' is the data sampling frequency.

2.4 Feature Extraction

For each EEG segment, the group of 136 functions was recovered primarily from the same frequency band [2]. The self-flow coefficient with time, frequency, waveform, cepstral, and nonlinear properties of auto-regressive are separated into six main groups, which can improve classification performance by extracting data [9]. The hubris of functional spaces, on the other hand, might lead to a measurement curse. Unrelated traits or substantial redundancy have a significant impact on classification accuracy. You can eliminate irrelevant and redundant functionality using correlation and redundancy analysis [9]. When employing a single-channel EEG, we must apply the Kurtosis, Median, Skew, IQR, Spec, Mean, Min, and Max features to increase the accuracy value.

2.5 Gradient Boosting Model (GBM)

The classification stage has a signal processing final step to show the performance of a system using the accuracy value. We Chose the LGBM algorithm because of its great resiliency in comparison to other approaches employed in the research. LGBM is dedicated to reducing challenges through accelerated GBRT training [5, 6]. With the same precision, Jiajun et al. [10] cite four new technologies existed only in LGBM:

1 - The intergenerational leaf technique is to take it all at once, rather than in stages. This strategy minimizes folio node separation loss and keeps the tree's depth to a minimum to avoid installation.
2 - To reduce the learning time during the division phase, histogram methods are employed to discover the corresponding division values.
3 - Gradient-based sampling is recommended to speed up iterations (GOSS). This policy gives a precise estimation of the information acquired from a smaller data set.
4 - Exclusive feature combination (EFB) groups should have mutually exclusive functionality.

2.6 Genetic Algorithm (GA)

The method for choosing the genetic algorithm's characteristics encodes the fitness function (LGBM) parameters and determines the initial number of designs [6]. Every particle in the crowd is a good fit for the potential answer. After that, each population's function value is calculated, calculates the likelihood of a mutation. The following generation is formed using appropriate genetic procedures like intersection, mutation, and choice until overall performance reaches a certain index or a set number of rounds is accomplished [12].

3 Results and Discussion

The tests were carried out on a 2.4 GHz desktop with 6 GB of RAM, four Intel®Core (TM) i5 CPUs, and Windows 10 64-bit operating system. Figure 2 illustrates the EEG signals of the cross-subject extracted from 153 subjects, where Fig. 2a shows the signals of the different tasks during acquisition in training stage (Awake: in blue, Stage 1: in red, Stage 2: in green, Stage 3/4: in viola, REM: in brown), we notice that the total of signals is colored in blue (Awake stage) because this task is the most dominant. Figure 2b illustrates the PSD of the raw EEG, knowing that there is a randomization of signals on a large bandwidth of frequencies, except in five frequencies with high power: $f = \{1, 25, 36, 43, 49\}$, also the Fig. 2.c shows that all tasks data are very mixed between them, which means that the data is non-separated and therefore low prediction values. Figure 3 illustrates the spectral and spatial distribution of EEG data after each pre-processing step, such as Fig. 3a shows the PSD of EEG signals after the application of optimized band-pass filters during the 'W' task recognition, Fig. 3b shows distortion on the spatial distribution of data after employing of the optimized bandpass filter, which improves the prediction accuracy Table 2 compared to the single state of Fig. 2c. Moreover, after the extraction of all features, Fig. 3c shows an improvement in the spatial distribution of EEG data for well-separated between tasks, which improves the classification of the data.

3.1 Performance Evaluation

After training, [5, 6, 14] analyze the effectiveness of several functional options, including classifiers, using the following metrics: True positive results (TP) are samples that are correctly classified and correctly predicted. In the proper classification and prediction task, the true-negative task (TN) is a type of meditation. The false-negative (FN) system is misclassified as a concentration sample, whereas the false-positive (FP) system is misclassified as a meditation sample. The performance of the classifier is often determined using the following parameters based on the above measurements: The most widely used empirical metric is accuracy (AC), which does not distinguish between the number of correct classifications in different categories:

$$AC = \frac{TP + TN}{TP + TN + FP + FN} \tag{2}$$

F-Measure (F1): This parameter is harmonic average weight precision with recall:

$$F1 = \frac{2 * TP}{2 * TP + FP + FN} \tag{3}$$

Zero One Loss (ZOL) is the number of false markers predicted:

$$ZOL = FP + FN \tag{4}$$

3.2 Classification Results

Table 2 shows optimization of band-pass filter settings using a set of optimization algorithms in parallel to the goal of finding the right combination of parameters with a high accuracy value, knowing that in this table, we notice that the accuracy value increases from 74.26% to 89.11%, so that the bandwidth found remains stable for that Fb = 21.6158 and Fh = 43.8278, this frequency combination can be observed in the PSD graph of Fig. 3b. Table 3 illustrates a parameters optimization of the LGBM classifier, aimed at finding good values of n-estimator (N-est) and learning-rate (L-rate) corresponding with a high accuracy value, such as this table shows that the accuracy value increases from 83.17% to 90.10% and that the parameters of classification stabilize on a N-est = 153 and L-rate = 0.09639 in case of the prediction of the awake task ('W'). Table 3 shows that the use of the filter parameters found for case of task 'W' can be also carried out for task '1', such that it remains only sought a good combination of the classifier parameters, knowing that this table shows that the accuracy value increases from 97.03% to 98.02% with values of N-est = 50 and L-rate = 0.09639.

Table 2 illustrates the optimization of filter parameters to reconstitute task '2' easily with a good accuracy value, such that this table shows that the accuracy value has been

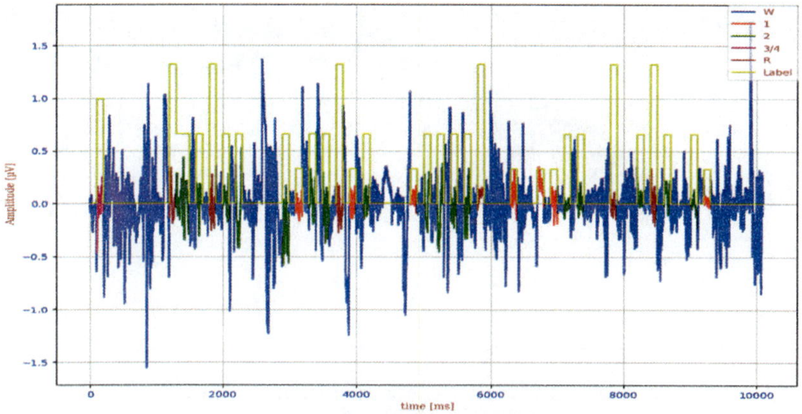

a) Electrodes position

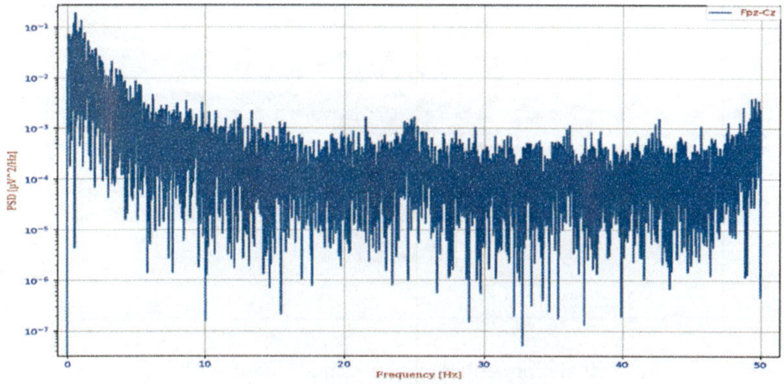

b) PSD

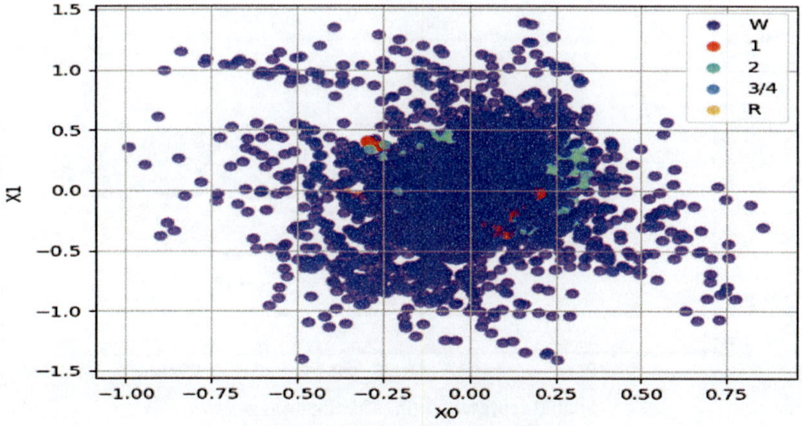

c) Spatial representation

Fig. 2. Example of EEG signals plotting for sleep-EDF dataset (Fpz-Cz channel).

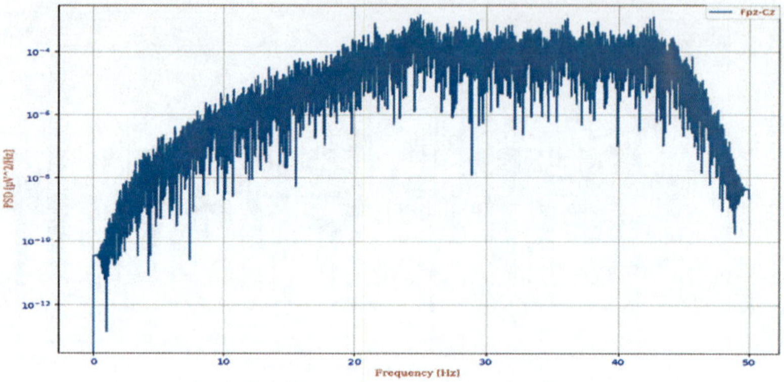

a) PSD, after using the bandpass filter

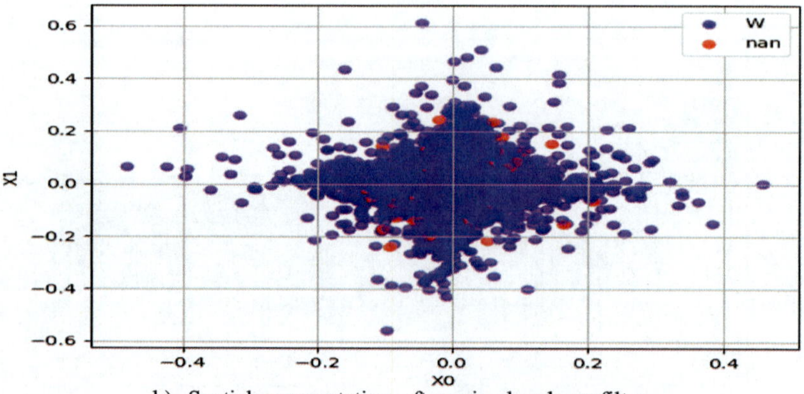

b) Spatial representation, after using bandpass filter

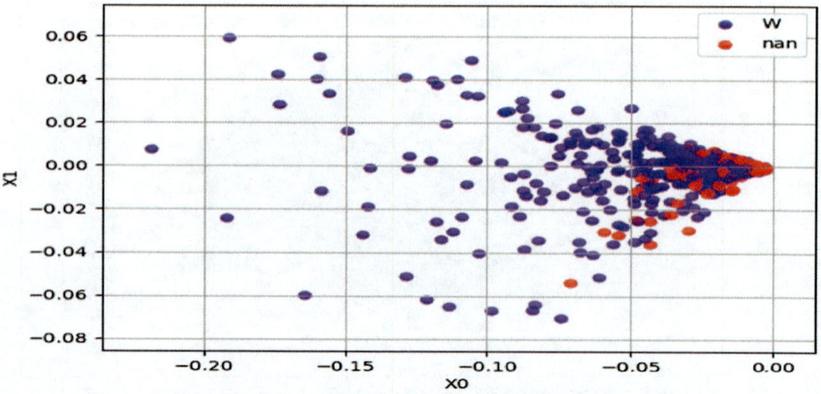

c) Spatial representation, after using features extraction

Fig. 3. EEG signals plotting after the employing of the bandpass filter and the features extraction in case of 'W'-task signals.

Table 2. Filter parameters optimization using GA of single-channel ('Fpz-Cz') and for each task

Tasks	Fb	Fh	Classification results		
			AC (%)	F1 (%)	ZOL
W	24.56	30.47	74.26	70.74	26
	30.40	44.73	83.17	80.04	17
	21.62	43.83	89.11	87.31	11
2	27.75	38.21	70.30	53.24	30
	29.99	47.51	77.23	58.35	23
	34.02	43.92	84.16	75.06	16
	34.07	43.94	87.13	77.54	13
R	32.80	46.48	91.09	56.73	9
	23.95	25.37	94.06	60.95	6
	21.9	26.4	97.03	82.56	3

Table 3. LGBM optimization using GA for single-channel ('Fpz-Cz') and each task.

Tasks	N-est	L-rate	Classification results		
			AC (%)	F1 (%)	ZOL
W	35	0.424	83.17	80.04	17
	206	0.981	86.14	83.40	14
	13	0.096	90.10	88.36	10
1	46	0.096	97.03	69.24	3
	47	0.096	98.02	82.82	2
	50	0.096	98.02	82.82	2
2	85	0.588	81.19	67.17	19
	28	0.187	85.15	74.08	15
	87	0.162	88.12	79.72	12
R	49	0.543	96.04	73.97	4
	105	0.745	97.03	82.56	3
	206	0.826	97.03	82.56	3

increased from 70.30% to 87.13% for a value of Fb = 34.067 and Fh = 43.9358, this bandwidth can be shown also in the PSD graph of Fig. 2b. Table 3 shows a process for setting the parameters to be classified and increase the accuracy value up to 88.12% in the case of task recognition '2', knowing that the best parameters found are for a value of N-est = 87 and L-rate = 0.1618. Table 2 illustrates an optimization of the band-pass filter and well predicted task '3/4', such as the accuracy value increased from 91.9%

Table 4. All subject classification results for sleep-EDF extended using LGBM.

Tasks	Classification results		
	AC (%)	F1 (%)	ZOL
W	90.10	88.36	10
1	98.02	82.82	2
2	88.12	79.72	12
¾	100.0	100.0	0
R	97.03	82.56	3

to 97.03%, and the filter parameters stabilize on a value of Fb = 21.9 and Fh = 26.4. Table 3 shows the new LGBM settings the stable pus for great accuracy values have been increased to 97.03%, such that the best parameters to classify the data are N-est = 206 and L-rate = 0.8262. Table 4 shows the final results of the classification of the data for each sleep stage, such that it is noted that the totality of the prediction values is 88% more than that shown by the possibility of manufacturing sleep stage analysis devices based on EEG signals only, which makes it possible to detect sleep disorders. On the other hand, the most important stage in this study is REM because, in this stage, the brain is totally isolated from the interactions of the body which implies that the acquisition signals are due to pure imagination, which allows giving the chance to analyze the tasks of motor imagery deep in the future.

4 Conclusion

In conclusion, the resulting system shows great performance to develop the recognition of sleep stages from bass to high precision values, by applying the GA algorithm to optimize the bandpass filter parameters during preprocessing step and the classifier parameters (LGBM) to increase the accuracy values of the sleep stages prediction from 70% to values over of 88% for each stage, which shows the possibility of using this system to diagnosing of sleep disorders and also to analysis the EEG signals for REM stage using the bandwidth between 21.9 and 26.4 Hz. Therefore, to fully understand the mechanisms of imagination in the brains. We hope that by sharing our findings, other researchers will be able to achieve similar success in the future. And our future work will be focused on real-time applications.

References

1. Your Guide to Healthy Sleep. U.S. Department of Health and Human Services and National Heart Lung and Blood Institute (2011). https://www.nhlbi.nih.gov/health-topics/all-publications-and-resources/your-guide-healthy-s
2. Aarabi, A., Grebe, R., Wallois, F.: A multistage knowledge-based system for EEG seizure detection in newborn infants. Clin. Neurophysiol. **118**, 2781–2797 (2007). https://doi.org/10.1016/j.clinph.2007.08.012

3. Abenna, S., Nahid, M., Belbachir, A.K.: Brain-computer interface: rhythm alpha analysis for eyes signals. In: The Fourth International Conference on Intelligent Computing in Data Sciences. IEEE (2020). https://doi.org/10.1109/ICDS50568.2020.9268719
4. Abenna, S., Nahid, M., Bajit, A.: BCI: classifiers optimization for EEG signals acquiring in real-time. In: 2020 6th IEEE Congress on Information Science and Technology (CiSt) (2021). https://doi.org/10.1109/CiSt49399.2021.9357209
5. Abenna, S., Nahid, M., Bajit, A.: Brain-computer interface: A novel EEG classification for baseline eye states using LGBM algorithm. In: Motahhir, S., Bossoufi, B. (eds.) Digital Technologies and Applications. ICDTA 2021. Lecture Notes in Networks and Systems, vol. 211, pp. 189–198. Springer, Cham (2021). https://doi.org/10.1007/978-3-030-73882-2_18
6. Abenna, S., Nahid, M., Bajit, A.: Motor imagery based brain-computer interface: improving the EEG classification using delta rhythm and lightGBM algorithm. Biomed. Signal Process. Control **71**(103102) (2022). https://doi.org/10.1016/j.bspc.2021.103102
7. Alickovic, E., Subasi, A.: Ensemble SVM method for automatic sleep stage classification. IEEE Trans. Instr. Meas. 1–8 (2018). https://doi.org/10.1109/TIM.2018.2799059
8. Diykh, M., Li, Y.: Complex networks approach for EEG signal sleep stages classification. Expert Syst. Appl. **63**, 241–248 (2016). https://doi.org/10.1016/j.eswa.2016.07.004. ISSN 09574174
9. Ghimatgar, H., Kazemi, K., Sadegh, M., Aarabi, A.: An automatic single-channel EEG-based sleep stage scoring method based on hidden Markov model. J. Neurosci. Methods **324**(May), 108320 (2019). https://doi.org/10.1016/j.jneumeth.2019.108320. ISSN 0165-0270
10. Jiajun, H., Chuanjin, Y., Yongle, L., Huoyue, X.: Ultra-short term wind prediction with wavelet transform, deep belief network and ensemble learning. Energy Convers. Manag. (2020). https://doi.org/10.1016/j.enconman.2019.112418
11. Jiang, D., Lu, Y., Ma, Y., Wang, Y.: Robust sleep stage classification with single-channel EEG signals using multimodal decomposition and HMM-based refinement. Expert Syst. Appl. **121**, 188–203 (2019). https://doi.org/10.1016/j.eswa.2018.12.023. ISSN 09574174
12. Lin, J., Chen, H., Li, S., Liu, Y., Li, X., Yu, B.: Accurate prediction of potential druggable proteins based on genetic algorithm and bagging-SVM ensemble classifier. Artif. Intell. Med. **98**, 35–47 (2019). https://doi.org/10.1016/j.artmed.2019.07.005
13. Memar, P., Faradji, F.: A novel multi-class EEG-based sleep stage classification system. IEEE Trans. Neural Syst. Rehabil. Eng. **26**(1), 84–95 (2018). https://doi.org/10.1109/TNSRE.2017.2776149
14. Ripley, R.M., Tarassenko, L.: Neural network models for breast cancer prognosis. Neural Comput. Appl. **7**, 367–375 (1998)
15. Stuburic, K., Gaiduk, M., Seepold, R.: A deep learning approach to detect sleep stage. Proc. Comput. Sci. **176**, 2764–2772 (2020). https://doi.org/10.1016/j.procs.2020.09.280. ISSN 1877-0509
16. Tsinalis, O., Matthews, P.M., Guo, Y.: Automatic sleep stage scoring using time-frequency analysis and stacked sparse autoencoders. Ann. Biomed. Eng. **44**(5), 1587–1597 (2015). https://doi.org/10.1007/s10439-015-1444-y
17. Xiao, M., Yan, H., Song, J., Yang, Y., Yang, X.: Sleep stages classification based on heart rate variability and random forest. Biomed. Signal Process. Control **8**(6), 624–633 (2013). https://doi.org/10.1016/j.bspc.2013.06.001. ISSN 1746-8094

Check for updates

3D Numerical Study of Sound Waves Behavior in the Presence of Obstacles Using the D3Q15-Lattice Boltzmann Model

Jaouad Benhamou(✉), Salaheddine Channouf, and Mohammed Jami

Laboratory of Mechanics and Energetics, Department of Physics, Faculty of Sciences, Mohammed First University, 60000 Oujda, Morocco
benhamou_jaouad1718@ump.ac.ma

Abstract. The study of the propagation of sound waves in confined spaces with obstacles is a very important research topic. It has many applications in engineering, especially in the domain of civil engineering. The most important of these applications is the diffraction in rooms. In this paper, a three-dimensional (3D) numerical study of the wave propagation in a cavity filled with air and containing obstacles is studied. The numerical method chosen to perform the simulations is the lattice Boltzmann method (LBM) associated with the D3Q15 model. This method is checked by treating the flow caused by a lid-driven cavity. The acoustic waves are created by the vibration of the left wall of the 3D cavity. Firstly, the propagation of the waves through the 3D obstacle of the square section is investigated. Secondly, the behavior of the waves when they cross an opening of small and large dimensions is studied.

Keywords: Lattice Boltzmann method · Sound waves · Diffraction · Obstacle

1 Introduction

In physics, sound or acoustic waves correspond to the propagation of mechanical disturbances in an elastic medium. At certain frequencies, the human ear detects these disturbances and interprets them as sound. Unlike light waves, acoustic waves, do not propagate in a vacuum. They necessarily travel in a material environment (solid, liquid, gas ...). They can interact in different physical phenomena such as reflection, diffraction, absorption, refraction, and interference [1]. These phenomena are very important and represent many applications in different fields. This work focuses on the study of sound diffraction and its interest in the field of civil engineering, in particular, in the acoustic insulation of buildings.

The phenomenon of diffraction is the modification of the characteristics of a wave (direction, shape ...) when it encounters an obstacle or passes through a small aperture. It is generally observed with all waves: sound, radio waves, X-rays... The study of sound diffraction has been known in the literature for many years. For example, Spence [2] studied theoretically the diffraction of plane sound waves by circular apertures and disks.

R. Saidi et al. (Eds.): ICATH 2021, LNDECT 110, pp. 106–115, 2022.
https://doi.org/10.1007/978-3-030-94188-8_11

Pierce [3] investigated the diffraction of sound waves in corners and over large obstacles. Seznec [4] used the boundary element technique to address the diffraction of acoustic waves around barriers. Jin et al. [5] presented a new analytical technique to study the sound diffraction of a partially inclined barrier. Piechowicz [6] exploited the Huygens-Fresnel theory to describe the diffraction of sound waves at the edge of a sound barrier. Recently, Rabisse et al. [7] developed a numerical method to simulate the propagation of sound in a confined space, taking into account the scattering and diffraction of sound in a chamber. Huang et al. [8] studied the impact of top edge impedance on acoustic barrier diffraction.

In this work, the diffraction of sound waves around barriers in a 3D cavity filled with air is studied numerically employing the lattice Boltzmann method. This method is defined as a statistical approach. It is based on lattice gases and the kinetic theory developed by Boltzmann. It is widely used in the fields of fluid dynamics. It allows the modeling of different physical phenomena on a small scale with high accuracy [9].

2 Numerical Method and Boundary Conditions

2.1 D3Q15-Lattice Boltzmann Model

The LB method is used to perform the simulations of the wave propagation in the cavity. For accuracy reasons, the multiple relaxation time (MRT) model is used. In this work, this model is based on the D3Q15 scheme (Fig. 1) and its implementation can be carried out using the following linear Boltzmann equation [9–11]:

$$f_i(x_i + c_i\Delta t, t + \Delta t) - f_i(x_i, t) = \Omega_i \tag{1}$$

where the parameters $f_i, t, x_i, c_i, \Delta t$ and Ω_i are the distribution function, time, node of the LBM lattice, D3Q15-LBM velocities, time step, and collision operator, respectively.

For the MRT schema, the operator Ω_i is expressed in terms of the inverse matrix (M^{-1}) of the M transformation matrix, the relaxation matrix (S) and the equilibrium (m_i^{eq}) and non-equilibrium (m_i) moments:

$$\Omega_i = -M^{-1}S[m_i - m_i^{eq}] \tag{2}$$

The diagonal matrix S can be defined in terms of 15 relaxation times s_i:

$$S = diag(s_0, s_1, s_2, s_3, s_4, s_5, s_6, s_7, s_8, s_9, s_{10}, s_{11}, s_{12}, s_{13}, s_{14}) \tag{3}$$

The relaxation times s_i, and the vectors m_i and m_i^{eq} used in this work are those given in the reference [12].

The two matrices M and M^{-1} are composed of 15 columns and 15 rows (15 × 15). They allow to link the moments to the distribution functions as follows:

$$m = Mf \text{ and } f = M^{-1}m \tag{4}$$

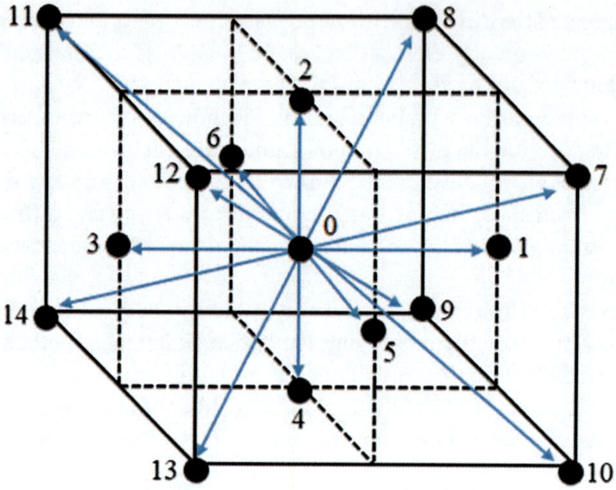

Fig. 1. The D3Q15-LBM model.

The mathematical expression of M is given as [13]:

$$M = \begin{pmatrix} 1 & 1 & 1 & 1 & 1 & 1 & 1 & 1 & 1 & 1 & 1 & 1 & 1 & 1 & 1 \\ -2 & -1 & -1 & -1 & -1 & -1 & -1 & 1 & 1 & 1 & 1 & 1 & 1 & 1 & 1 \\ 16 & -4 & -4 & -4 & -4 & -4 & -4 & 1 & 1 & 1 & 1 & 1 & 1 & 1 & 1 \\ 0 & 1 & -1 & 0 & 0 & 0 & 0 & 1 & -1 & 1 & -1 & 1 & -1 & 1 & -1 \\ 0 & -4 & 4 & 0 & 0 & 0 & 0 & 1 & -1 & 1 & -1 & 1 & -1 & 1 & -1 \\ 0 & 0 & 0 & 1 & -1 & 0 & 0 & 1 & 1 & -1 & -1 & 1 & 1 & -1 & -1 \\ 0 & 0 & 0 & -4 & 4 & 0 & 0 & 1 & 1 & -1 & -1 & 1 & 1 & -1 & -1 \\ 0 & 0 & 0 & 0 & 0 & 1 & -1 & 1 & 1 & 1 & 1 & -1 & -1 & -1 & -1 \\ 0 & 0 & 0 & 0 & 0 & -4 & 4 & 1 & 1 & 1 & 1 & -1 & -1 & -1 & -1 \\ 0 & 2 & 2 & -1 & -1 & -1 & -1 & 0 & 0 & 0 & 0 & 0 & 0 & 0 & 0 \\ 0 & 0 & 0 & 1 & 1 & -1 & -1 & 0 & 0 & 0 & 0 & 0 & 0 & 0 & 0 \\ 0 & 0 & 0 & 0 & 0 & 0 & 0 & 1 & -1 & -1 & 1 & 1 & -1 & -1 & 1 \\ 0 & 0 & 0 & 0 & 0 & 0 & 0 & 1 & 1 & -1 & -1 & -1 & -1 & 1 & 1 \\ 0 & 0 & 0 & 0 & 0 & 0 & 0 & 1 & -1 & 1 & -1 & -1 & 1 & -1 & 1 \\ 0 & 0 & 0 & 0 & 0 & 0 & 0 & 1 & -1 & -1 & 1 & -1 & 1 & 1 & -1 \end{pmatrix} \quad (5)$$

2.2 Boundary Conditions

The boundary conditions used in this paper are the usual Bounce-Back conditions. The principle of these conditions is to define the unknown functions f_i that come from outside to the studied domain from the known functions. They are defined in the opposite directions of f_i [9, 11].

After calculating the functions f_i and implementing the boundary conditions, the macroscopic density (ρ) and velocities (u, v, w) can be calculated as:

$$\rho = \sum\nolimits_{i=0}^{14} f_i, \ (u, v, w)\rho = \sum\nolimits_{i=0}^{14} f_i c_i \quad (6)$$

where u, v and w are the velocities along the x, y, and z axes, respectively.

3 D3Q15-LBM Model Validation

The lid-driven cavity is a popular benchmark problem in the bibliography for validating different types of numerical methods. Thus, this physical problem is chosen to validate our numerical model. Indeed, it concerns a cubic cavity filled with a viscous fluid. The upper wall moves with a constant velocity ($u_0 = 0.1$). However, the remaining surfaces are characterized by a no-slip condition, i.e., the velocity is equal to 0 (see Fig. 2).

Our validation is based on the comparison of the velocity profiles with the results reported in the bibliography. For this purpose, a Reynolds number (*Re*) set at 400 is selected to confront our numerical results with those obtained by Ku et al. [14], Jiang et al. [15], and Ding et al. [16].

Figure 3 illustrates the profile of the x–component of the dimensionless velocity (U) plotted along the vertical centerline (left), and the y–component of the velocity (V) along the horizontal centerline (right). From this figure, we can note that the simulation results closely match the literature results. Then, we can say that our numerical model is well validated and can be used to simulate different physical problems.

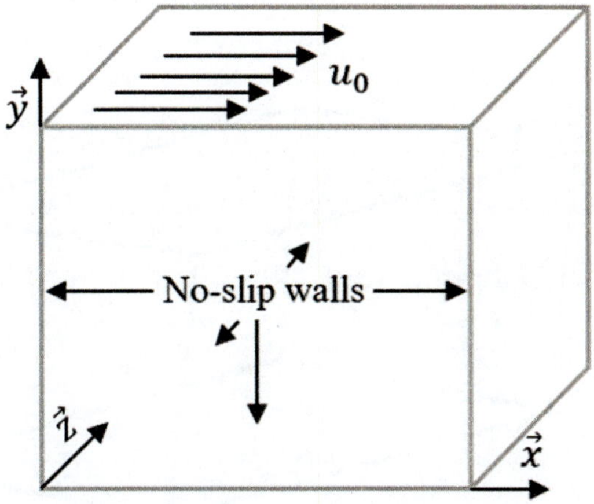

Fig. 2. The lid-driven cavity configuration.

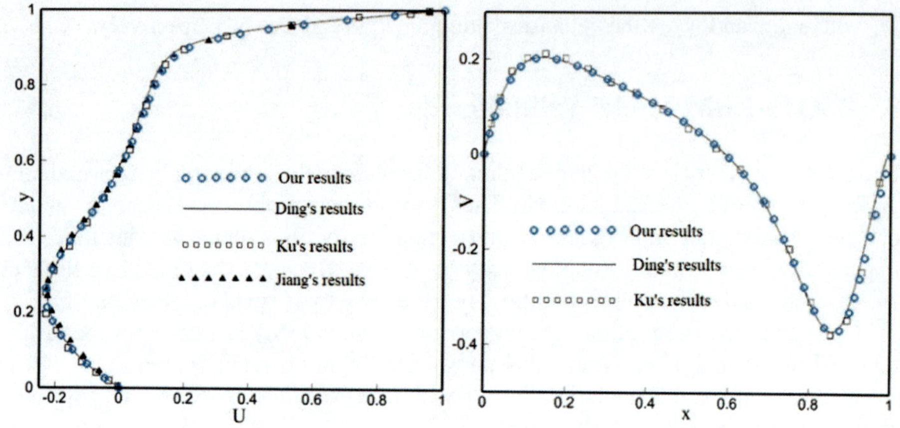

Fig. 3. Comparison of U and V velocities for $Re = 400$.

4 Results and Discussion

The acoustic waves are produced by the vibration of the whole left vertical wall of the cavity (Fig. 4). This cavity is identified by a length Lx, a width Ly, and a height Lz. These dimensions define the number of points of the 3D lattice used ($Lx = 300$, $Ly = 220$, and $Lz = Ly$).

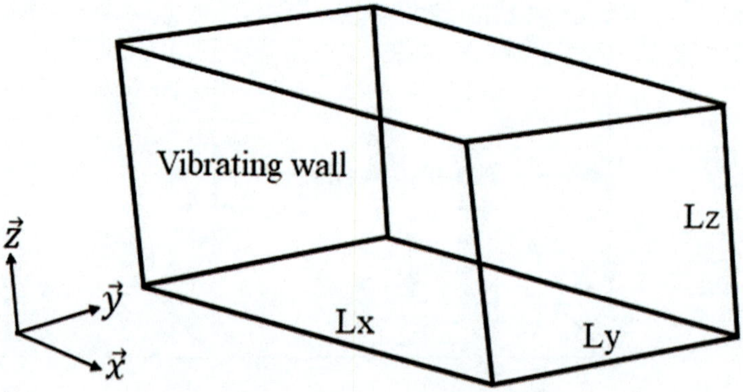

Fig. 4. Illustration of the geometry studied.

The technique used to model the waves is the acoustic point source. The principle of this technique is to emit the sound equally in all directions, i. e. with a "spherical symmetry". Its mathematical model can be described by the following sinusoidal function [11, 17, 18]:

$$\rho = 1 + \rho_a \, sin(\frac{2\pi t}{T}) \quad (7)$$

where ρ_a, t, and T are the amplitude, the time, and the LBM period. The values of these parameters used in this work are those used in references [11, 17, 18].

In the 2D case, this acoustic model is well studied in our previously published work [11, 18] on the study of acoustic waves. In the 3D case, our previous work started by applying the D3Q19-LBM model to the study of acoustic waves generated by a square source [19]. To show the performance of LBM simulations, in this work, another lattice Boltzmann model is chosen to simulate the waves: it concerns the D3Q15-LBM model. The first simulation starts with the study of the diffraction of sound around a 3D obstacle of a square section. This problem presents an important application of sound diffraction in a room in the presence of a solid body. Figure 5 illustrates the diffraction of sound around an obstacle with a square section located at the distance $Lx/3$ from the vibrating wall. The waves emitted by the vibration of the wall are plane. When they meet this obstacle, their shape changes. An acoustic shadow appears behind the obstacle, i.e. there is a region in which the waves do not propagate. The vertical section of the 3D cavity at the position $y = Ly/2$ illustrates clearly this phenomenon (Fig. 6).

The wavelength (λ_{lbm}) utilized in this work can be calculated from the speed of sound (c_s) and period T. In LBM units, c_s is fixed at $1/\sqrt{3}$ for the D3Q15 lattice. For this value of c_s and a period $T = 40$, λ_{lbm} is defined by almost 23 points ($\lambda_{lbm} = 23.09$). The square section of the obstacle is formed by $Ly/5 \times Lz/5$ nodes. This indicates that the obstacle has a dimension greater than the wavelength. The tests carried out have shown that the larger the size of the obstacle compared to λ_{lbm}, the longer and sharper the acoustic shadow behind it.

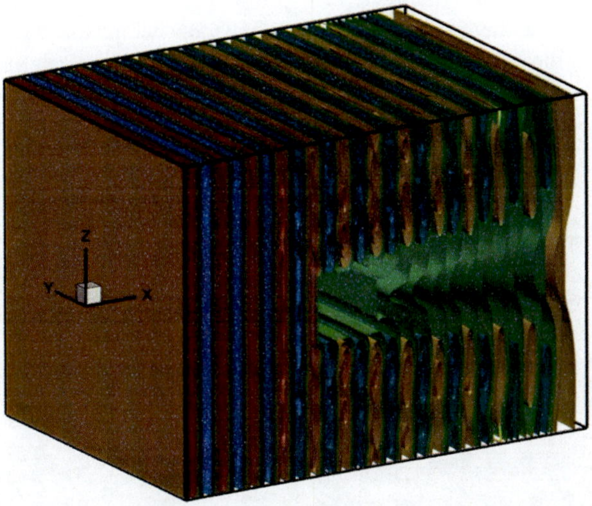

Fig. 5. Diffraction of sound around 3D obstacle at time $t = 500$.

The second simulation studies the behavior of plane waves generated by the vibration of the cavity wall as they pass through an opening. Figures 7 and 9 show the 3D diffraction of the sound caused by two vertical planes located at the position $x = Lx/3$. These vertical planes are spaced by a distance d. This distance characterizes the opening through which the waves diffract.

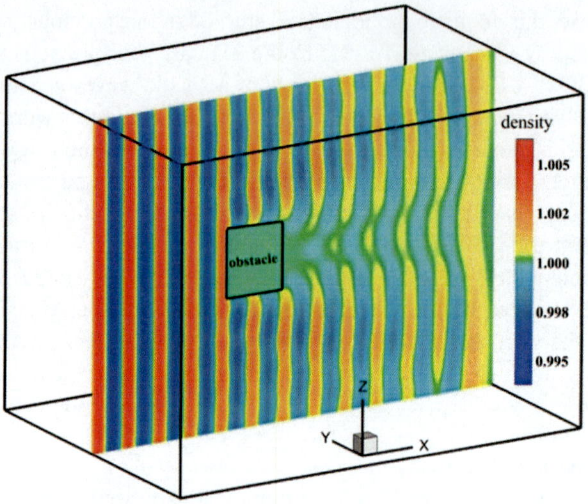

Fig. 6. Vertical section at $y = Ly/2$ of the 3D cavity depicted in Fig. 5.

The diffraction through an aperture also depends on the ratio between the parameters d and λ_{lbm}. We start with the case where the aperture is much larger than the wavelength ($d = 5\lambda_{lbm}$). Figure 7 shows that there is no diffraction. The wave passes freely through the aperture. However, there are areas of acoustic shadow behind the barriers. In addition, small changes in the waveforms are noted near the edges of the aperture as shown in Fig. 8.

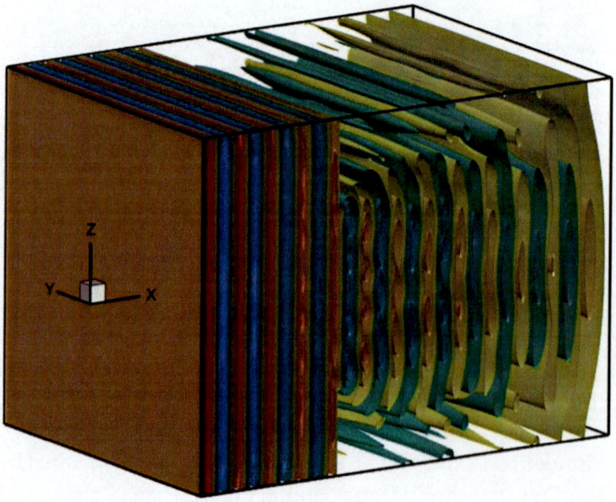

Fig. 7. 3D diffraction of sound through an opening larger than the wavelength at $t = 500$.

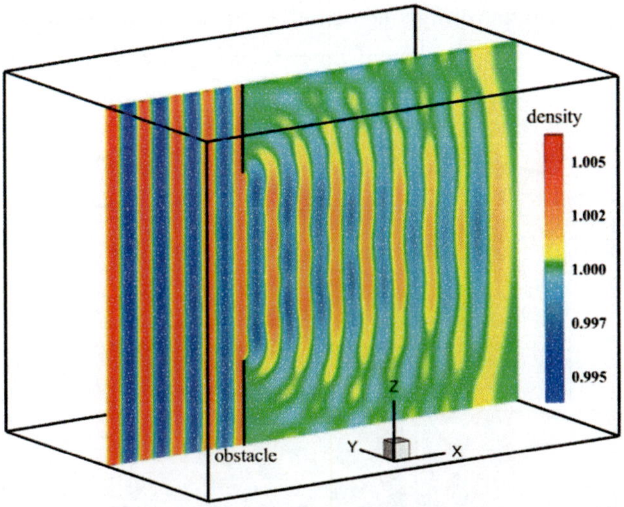

Fig. 8. Vertical section at $y = Ly/2$ of the 3D cavity depicted in Fig. 7.

Figure 9 shows the case where the aperture is smaller than the wavelength ($d = 18 < \lambda_{lbm}$). In this figure, the diffraction process is clearly observed. The plane waves have become half spheres. In this case, the aperture starts to react as a new spherical radiation point that propagates sound as a wave. Thus, the acoustic shadows are eliminated and the sound ceases to be considered as emitted from its initial location but is moved from the point of secondary spherical radiation, i.e. from the aperture. The vertical section of the 3D cavity shown in Fig. 9 clearly shows the diffraction phenomenon (Fig. 10). This vertical section illustrates that the direction of propagation of the wave is modified

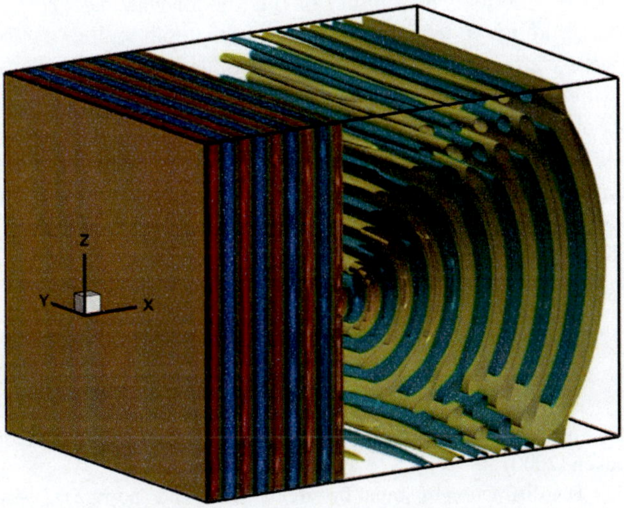

Fig. 9. 3D diffraction of sound through an opening smaller than the wavelength at time $t = 500$.

after its encounter with the diffracting barrier: it propagates in different directions with a maximum angular deviation from the initial direction of propagation.

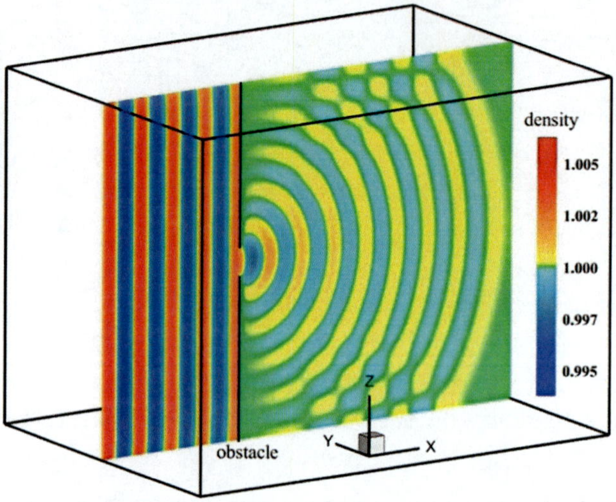

Fig. 10. Vertical section at $y = Ly/2$ of the 3D cavity depicted in Fig. 9.

5 Conclusion

This paper presented a 3D numerical study of the phenomenon of diffraction around obstacles. Numerical simulations were carried out using the lattice Boltzmann approach based on the D3Q15 model. The validity of this model was checked by studying the flow of a viscous fluid driven by the movement of the upper wall of a cubic cavity. The results found were well validated with those of the literature. Then, the focus was put on the study of the behavior of sound waves when they encounter an obstacle and pass through two openings. The simulations carried out showed that the waves keep the same properties when they pass through an opening five times larger than the wavelength. However, when the aperture is smaller than the wavelength, the plane waves produced by the vibration of the left wall have become half-spheres. These studies represent very important applications of the waves in the civil engineering sector, in particular when it concerns the acoustic insulation of a room.

References

1. Kinsler, L.E., Frey, A.R., Coppens, A.B., Sanders, J.V.: Fundamentals of Acoustics, 4th edn. Wiley, Hoboken (2000)
2. Spence, R.D.: The diffraction of sound by circular disks and apertures. J. Acoust. Soc. Am. **20**, 380–386 (1948). https://doi.org/10.1121/1.1906389

3. Pierce, A.D.: Diffraction of sound around corners and over wide barriers. J. Acoust. Soc. Am. **55**, 941–955 (1974). https://doi.org/10.1121/1.1914668
4. Seznec, R.: Diffraction of sound around barriers: use of the boundary elements technique. J. Sound Vib. **73**, 195–209 (1980). https://doi.org/10.1016/0022-460X(80)90689-6
5. Jin, B.J., Kim, H.S., Kang, H.J., Kim, J.S.: Sound diffraction by a partially inclined noise barrier. Appl. Acoust. **62**, 1107–1121 (2001). https://doi.org/10.1016/S0003-682X(00)00094-3
6. Piechowicz, J.: Sound wave diffraction at the edge of a sound barrier. Acta Phys. Pol. A (2011). https://doi.org/10.12693/APhysPolA.119.1040
7. Rabisse, K., Ducourneau, J., Faiz, A., Trompette, N.: Numerical modelling of sound propagation in rooms bounded by walls with rectangular-shaped irregularities and frequency-dependent impedance. J. Sound Vib. **440**, 291–314 (2019). https://doi.org/10.1016/j.jsv.2018.08.059
8. Huang, X., Zou, H., Qiu, X.: Effects of the top edge impedance on sound barrier diffraction. Appl. Sci. **10**, 6042 (2020). https://doi.org/10.3390/app10176042
9. Mohamad, A.A.: Lattice Boltzmann Method: Fundamentals and Engineering Applications with Computer Codes. Springer, London (2011). https://doi.org/10.1007/978-0-85729-455-5
10. Jami, M., Moufekkir, F., Mezrhab, A., et al.: New thermal MRT lattice Boltzmann method for simulations of convective flows. Int. J. Therm. Sci. **100**, 98–107 (2016). https://doi.org/10.1016/j.ijthermalsci.2015.09.011
11. Benhamou, J., Jami, M., Mezrhab, A., et al.: Numerical study of natural convection and acoustic waves using the lattice Boltzmann method. Heat Transf. **49**, 3779–3796 (2020). https://doi.org/10.1002/htj.21800
12. Premnath, K.N., Abraham, J.: Three-dimensional multi-relaxation time (MRT) lattice-Boltzmann models for multiphase flow. J. Comput. Phys. **224**, 539–559 (2007). https://doi.org/10.1016/j.jcp.2006.10.023
13. D'Humières, D., Ginzburg, I., Krafczyk, M., et al.: Multiple-relaxation-time lattice Boltzmann models in three dimensions. Philos. Tran. Roy. Soc. A: Math. Phys. Eng. Sci. **360**, 437–451 (2002). https://doi.org/10.1098/rsta.2001.0955
14. Ku, H.C., Hirsh, R.S., Taylor, T.D.: A pseudospectral method for solution of the three-dimensional incompressible Navier-Stokes equations. J. Comput. Phys. **70**, 439–462 (1987). https://doi.org/10.1016/0021-9991(87)90190-2
15. Jiang, B.N., Lin, T.L., Povinelli, L.A.: Large-scale computation of incompressible viscous flow by least-squares finite element method. Comput. Methods Appl. Mech. Eng. **114**, 213–231 (1994). https://doi.org/10.1016/0045-7825(94)90172-4
16. Ding, H., Shu, C., Yeo, K.S., Xu, D.: Numerical computation of three-dimensional incompressible viscous flows in the primitive variable form by local multiquadric differential quadrature method. Comput. Methods Appl. Mech. Eng. **195**, 516–533 (2006). https://doi.org/10.1016/j.cma.2005.02.006
17. Salomons, E.M., Lohman, W.J.A., Zhou, H.: Simulation of sound waves using the lattice Boltzmann method for fluid flow: benchmark cases for outdoor sound propagation. PLoS One **11**, e0147206 (2016). https://doi.org/10.1371/journal.pone.0147206
18. Benhamou, J., Jami, M., Mezrhab, A.: Application of the lattice Boltzmann method to the acoustic wave in a rectangular enclosure. In: Proceedings of 2nd International Conference on Advanced Technologies for Humanity - ICATH (2021). https://doi.org/10.5220/0010427200420047
19. Benhamou, J., Channouf, S., Jami, M., et al.: Three-dimensional Lattice Boltzmann model for acoustic waves emitted by a source. Int. J. Comput. Fluid Dyn., 1–22 (2021). https://doi.org/10.1080/10618562.2021.2019226

Deep Learning for Building Extraction from High-Resolution Remote Sensing Images

Abderrahim Norelyaqine[1(✉)] and Abderrahim Saadane[2]

[1] Department of Mineral Engineering, Mohammedia School of Engineers, Rabat, Morocco
[2] Department of Geology, Faculty of Sciences of Rabat, Rabat, Morocco

Abstract. The extraction of buildings from satellite images of very high spatial resolution is an important issue for many applications mainly related to the implementation of public policies, urban development, land use planning, and the updating of geographic databases, and during the last two decades, it has been the subject of much research. Several existing classical techniques have been proposed in remote sensing images, but they have several limitations that prevent segmenting buildings with high accuracy. We propose, in this paper, a U-net architecture with a ResNet50 encoder for the extraction of buildings from Massachusetts building datasets, in order to automate the production chain of urban three-dimensional models. The results obtained in this study show a very promising segmentation with huge accuracy, it outperforms many presented models with 82.63% of intersection over union (IoU).

Keywords: Deep learning · Building extraction · Remote sensing · Very high resolution

1 Introduction

Feature extraction from remote sensing images is a crucial and important topic in the field of remote sensing. How to accurately identify and draw features from remotely sensed images is extremely important for change detection, disaster assessment, land cover detection and other areas. However, due to the complex and diverse features contained in real scenes, the extraction of specific features from remotely sensed images is sensitive to interference from background factors.

In remote sensing images of the urban area, the building area is more than 80%, so the extraction of buildings plays an irreplaceable role in various human activities. In modern society, buildings are also important identification objects in maps and geographic information systems. With the construction of geographic information systems, the technology of automatic building extraction appears and continues to develop. An urban area information system can play an essential role in many areas such as urban development, urban change monitoring, urbain planning, population estimation, and topographic map production. The renewal of human observation methods and the rapid growth of urban construction put forward higher requirements for automatic building extraction technology. On the one hand, today's human construction activities are rapidly

R. Saidi et al. (Eds.): ICATH 2021, LNDECT 110, pp. 116–128, 2022.
https://doi.org/10.1007/978-3-030-94188-8_12

changing the information about the city. On the other hand, humans have more abundant means to observe the Earth, and the demand for digital maps is also very high.

With the construction of geographic information systems, the technology of automatic building extraction appears and continues to develop. An urban area information system can play an essential role in many areas such as urban development, urban change monitoring, urbain planning, population estimation, and topographic map production. The renewal of human observation methods and the rapid growth of urban construction put forward higher requirements for automatic building extraction technology. On the one hand, today's human construction activities are rapidly changing the information about the city. On the other hand, humans have more abundant means to observe the Earth, and the demand for digital maps is also very high.

The application of high-resolution satellites, aerial drones, radar, and other equipment has led to a massive increase in available data. In this case, how to use massive amounts of data to quickly and thoroughly extract urban information to update. Geographic information systems have become a focal point of research in this area.

Traditionally, the main work of extracting buildings from aeronautical, aerospace, and remote sensing images have focused on empirically designing an appropriate feature to express building and build a set of features to better understand and automatically detect buildings, with indicators that are commonly used such as pixel [1], spectrum [2], the texture [3], and shadow [4]. However, these indicators will change considerably with the season, light, atmospheric conditions, quality of sensors, style of building and environment. Therefore, this feature method can often only process specific data, but cannot be genuinely automated. And using traditional methods to analyze and extract building features from remote sensing imagery has some complexity and does not effectively extract spatial structure.

Artificial intelligence and specifically modern deep learning computer vision technology has made significant progress over the past ten years. It is used in several applications such as video analysis, image classification, image processing for robots and autonomous vehicles, and object recognition. Several computer vision applications need intelligent segmentation to understand features images content and facilitate extraction of each part.

Segmentation of geospatial objects plays an important role in the object extraction task. It can provide location and semantic information for objects of interest. It is part of a special semantic segmentation process. The objective is to separate the pixels in the image into two subsets of background regions and foreground objects. At the same time, it must additionally attribute a unified semantic label to each pixel of the foreground object region.

Segmentation of geospatial objects in HSR images is more difficult compared to natural scenes, for three main reasons:

- In HSR remote sensing images, objects always have large-scale changes, which causes problems at multiple scales, making it difficult to locate and identify objects [5].
- The foreground ratio is much smaller than the natural image, causing the problem of foreground to background imbalance.
- The background of HSR remote sensing image is more complicated, and due to the large intra-class difference, it is easy to cause serious false alarms [6].

The target object segmentation task for natural images is directly considered as a semantic segmentation task in computer vision, and due to the multiscale problem its performances are mainly limited. Consequently, the latest current general semantic segmentation methods focus on multi-scale modeling and scale-sensitive [7].

Today's image segmentation technology uses computer vision deep learning models to analyze each pixel of the image that represents the real object. Deep learning can learn patterns from specific inputs to predict object classes, which was unimaginable. The task of extracting buildings from remote sensing images can be studied as a subset of the semantic segmentation task of images, i.e. developing a segmentation algorithm to separate buildings of the background.

In recent years, deep learning (DL) technology has developed rapidly. Semantic image segmentation based on deep learning methods is also evolving every day, and more and more researchers are trying to solve feature extraction problems through deep learning. Many studies have proven that the use of deep learning methods can significantly improve the accuracy of soil feature extraction. [8] proposed a deep neural network (DNN)-based neural dynamic tracking framework for road network extraction. Study results show that this method is more efficient and accurate than traditional methods.

Since early 2015, deep learning convolution neural networks (CNN) show excellent potential for image classification and target detection, it has been gradually introduced in the fields of remote sensing [9]. The success of CNN is that it can automatically learn a multilayer feature expression that maps the original input to a continuous vector (a regression problem) or a binary label (a classification problem). This self-learning feature-capability gradually outperforms and replaces the traditional manual feature method; in particular, it provides a more automated and robust solution for automatic building extraction on which this paper focuses. A series of general CNN architectures have been gradually developed on this basis, such as VGGNet [10], ResNet [11], etc. Among them, ImageNet is composed of 10 million natural images covering 1,000 categories, which indirectly supports the explosion of deep learning methods.

However, building extraction is a semantic segmentation task at the pixel level. Using CNN will result in a large increase in memory overhead, low computational efficiency, and limited perceptual areas. According to this, Fully Convolutional Network (FCN) [12] removed the fully connected layer in the traditional CNN and deconvoluted the feature map at the end to generate a segmentation result consistent with the resolution of the input image; it has been used in the field of remote sensing target extraction [13]. FCN is specially developed and widely used for semantic segmentation: all pixels in the image are given category labels, including SegNet [14], U-Net [15] and many other variants. Among them, U-net is based on FCN, adopts symmetric structure design, and achieves high extraction accuracy in medical image segmentation. Many neural network architectures have adopted a jump connection structure similar to U-Net and achieved good classification results in practice [16]. Abderrahim et al. [17] mounted the use of the U-net model for road segmentation of Massachusetts roads dataset.

This paper uses the U-net deep learning method to perform automatic intelligent building detection in Massachusetts aerial images.

2 Related Work

Automation of map production based on aerial image analysis was first studied in [18] when they used neural networks to facilitate a manual tracing process. The authors [18, 19] improved it by using another neural network to classify texture patterns. Both networks are fed by features extracted from satellite images, as CNN's are not yet used. Ahmadi et al. [20] propose an active contour model to extract buildings. They test their method on a single image. And to capture the desired features of the image, the model involves dynamic surfaces that move in an image domain. Vakalopoulou et al. [21] designed models trying to extract the outlines of buildings from satellite images. In the first case, they used a CNN to classify the patches as containing buildings or not; then, they performed patch segmentation with an SVM. They extract the contours after segmentation and evaluate their model on Quickbird and Worldview 2 images for Attica, Greece. They do not specify what a true positive is, i.e. how to consider that an extracted contour corresponds to the manually extracted contour (ground truth). This lack of clarification prevents their model from being compared to others. Their model is evaluated on three images, containing a total of only 1438 buildings. Wu Guangming et al. [22] proposed an improved U-net with dual constraints, which optimized the parameter update process that improves the accuracy of building extraction. In [23] authors used an innovative non-local residual U-shaped method (ENRU-Net) of image processing that uses an encoder-decoder structure. It presents a very remarkable improvement when compared with other semantic segmentation methods in the extraction of buildings on the Massachusetts data with an overall accuracy of (94.12%). Cai et al. [24] have proposed a fully convolutional MHA-Net network wich consists of multipath HDC, the encoding network and DUC operation. To detect and extract building using high-resolution Aerial images. In [25] authors introduced a lightweight and efficient memory model, RCA-Net, that can accurately capture inter-channel connections using pre-trained layers of ResNet-50 and the ECA module, for extracting building Footprints. They showed satisfactory results on high resolution aerial images (Massachusetts, Inria).

3 Methodology

3.1 Data Preprocessing

Since the original remote sensing image is too large with a resolution of 1500×1500 pixels, the direct input of the original image into the neural network requires a large amount of memory and video memory, so it must be preprocessed as an input layer is 256×256 pixels, to simplify the training process.

3.2 U-net Architecture Overview

In this paper, the U-net architecture [15] was used. In the field of semantic segmentation, this architecture has proven to be a state-of-the-art solution. The input belongs to the encoder/decoder architecture, where the encoder is responsible of extracting features from the image, while the decoder is in charge of reconstructing the segmentation map from the features obtained from the original image. The decoder and encoder can be composed of any number of blocks. A decoder block composed of two convolutional layers and a transposed convolution layer, while an encoder block composed of two convolutional layers and a max pool layer. Number of filters can vary in the convolution layers, but it is usually divided by two. Also, the corresponding blocks of encoders and decoders are connected by skip links, which transmit some information provided by the encoder to the decoder in the image reconstruction phase.

Convolution Layer
Convolution layer allows to drag a matrix over an image, and for each pixel it uses the sum of the multiplication of that pixel by the value of the matrix. This technique allows us to find parts of the image that could be interesting to us. and each convolution kernel has particular parameters that are learned by the algorithm from the training data. In particular, through the gradient backpropagation technique, which allows the adjustment of the parameters according to the gradient value of the loss function (Fig. 1).

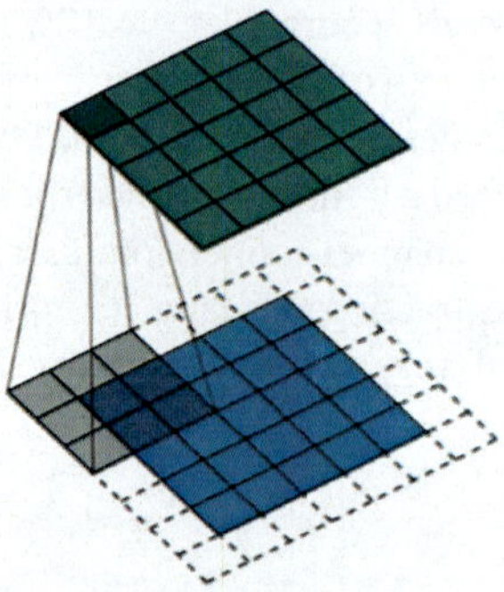

Fig. 1. Convolution with a kernel of size 3 * 3 and a step of 2

Max Pooling
Similar to the convolution layer, the sampling layer is responsible for reducing the spatial size of the feature maps, but it retains the most important information. There are different types of sampling including Max Pooling, Average Pooling, etc. Sampling consists in applying a kernel of size n × n on the activation map by dragging it with a previously defined step (the step is usually equal to the size of the kernel n to avoid the overlapping effect). Max Pooling returns the maximum value of the part of the image covered by the kernel, he allows us to remove the noise (Fig. 2).

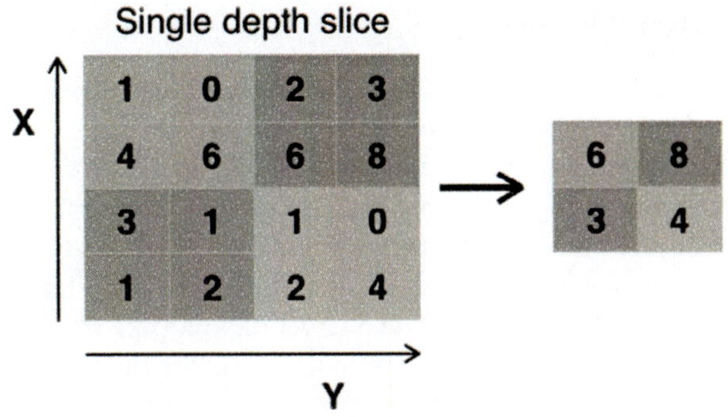

Fig. 2. Pooling operation with size 2 * 2 and a step of 2

The most common form is a pooling layer with 2 × 2 size kernels applied with a step of 2 thus reducing the input dimension by 2, eliminating 75% of the activations and leaving the depth dimension unchanged. In Fig. 5, the number of activations is reduced from 16 to 4 by applying pooling.

3.3 ResNet

The residual neural network, better known by the name ResNet [6], appears as a result of the problem of backpropagation of the gradient and the increase of the learning error. Indeed, when the neural network is too deep, the gradient reduces to zero and becomes very low to update the weights of the different layers, hence the difficulty of learning.

With ResNet as encoder, the optimization of deep networks is ensured by using residual connections. This allows gradients to pass through two convolution layers, but also to pass directly through a jump to the next layers by connecting the input of the nth layer to the output of the (n + a) th layer (Fig. 3).

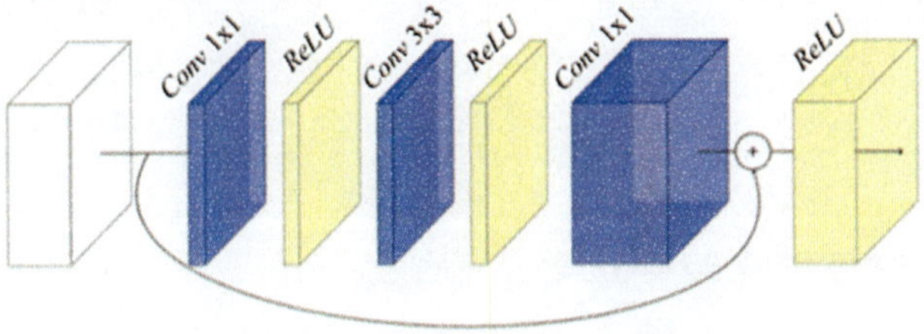

Fig. 3. Residual convolutional block

This residual block can be modeled by the following equation:

$$x_l = H_l(x_l - 1) + x_l - 1$$

With x_l the output of the residual block and H_l a compound function which represents all the layers present in the residual block.

3.4 U-net with Resnet Encoder

Since the encoder is part of an architecture that is structurally identical to the architectures used to classify images, it can be replaced by a classification model by removing the last layers used to make the output to the classification model. This results in a model that generates a feature vector, which is connected to the decoder part. The corresponding blocks of this model are connected to the corresponding blocks in the decoder to obtain the U-net architecture. The interest of using other models as encoders in U-net is to use a model that has already been searched on a larger dataset (like imagenet), to get better feature vectors at the output of the encoder. In this case, the model weights can be used without modification, or these weights can be used as a starting point for new training. In this project, we used a model of different classification architectures called Resnet50 (Fig. 4).

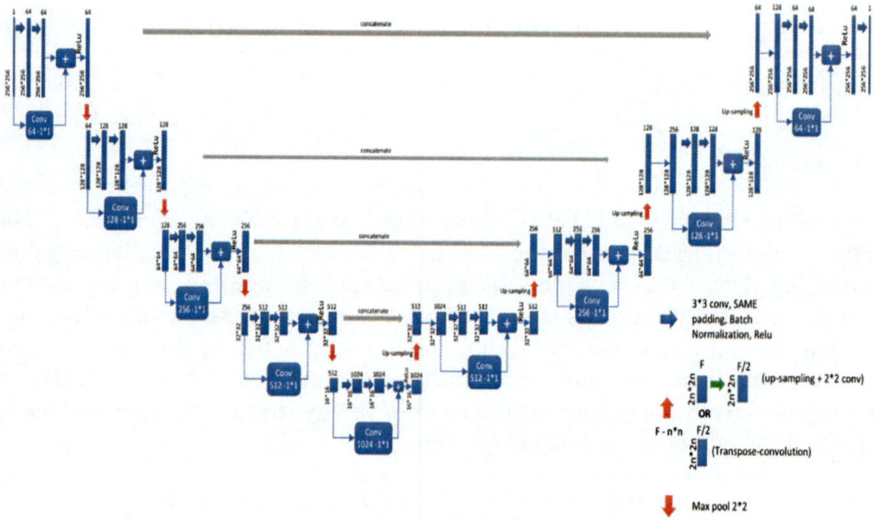

Fig. 4. U-net architecture with Resnet encoder

4 Experiments and Analysis

4.1 Dataset Description

In this paper, we have used aerial imagery dataset of Massachusetts buildings [26] especially the City of Boston. The collection is available online by the University of Toronto, based on OpenStreetMap data. Each image has a pixel size of 1500 × 1500. The data covers rural and cities areas, covering an area of over 2600 square kilometres; the dataset was divided into a training set (1108 images), a test set (49 images) and a validation set (14 images). As shown in Fig. 5, each image contains a corresponding binary map on which background are marked in black and the buildings in white.

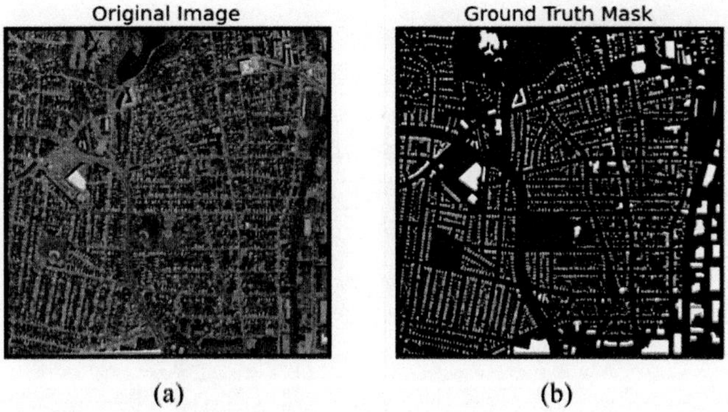

Fig. 5. (a) Original RGB image, (b) Ground Truth Mask of Massachusetts buildings dataset

4.2 Data Augmentation

To train our models, we need huge amounts of data. Indeed, the quantity and especially the quality of the dataset will have a major role in the elaboration of a good model. It is logical to have data that are comparable between them, and that have the same format, the same size and length, etc. And it is from these constraints that the problems begin. Having specific data according to our problem with the above mentioned points can often be impossible. This is where data augmentation will be very useful.

The principle of data augmentation is based on the principle of artificially increasing our data, which is a solution to the problem of limited data by applying transformations. We will be able to increase the diversity and therefore the learning field of our model, which will be able to better adapt to predict new data.

There are a series of simple and effective methods for data enhancement among which we have chosen Horizontal Flip, Vertical Flip and Random Rotate as shown in Fig. 6.

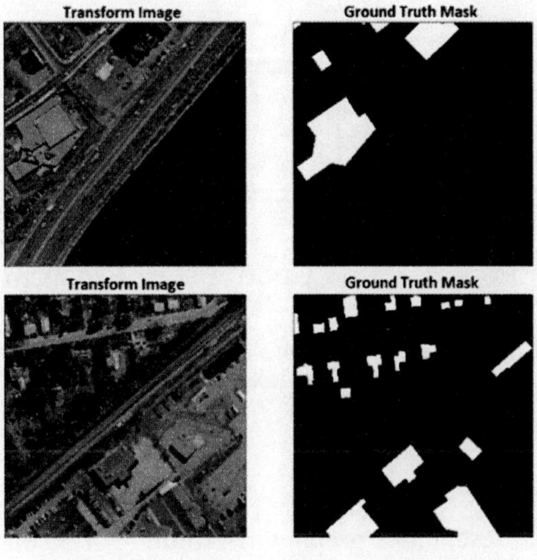

Fig. 6. Example of data augmentation

4.3 Evaluation Metrics

To evolve our model we used the IoU score - or Intersection over Union - is a way to measure the quality of object detection by comparing, in a training dataset, the known position of the detected object in the image with the prediction made by the model. The IoU is the ratio between the area of the union of the sets and the area of the intersection of the considered bounding boxes (Fig. 7).

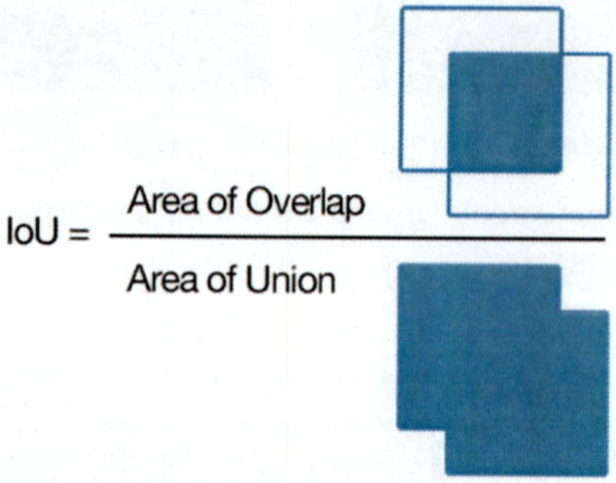

Fig. 7. Computing intersection over union

From this measurement, an acceptable tolerance threshold will be defined to determine if the prediction is correct or not. It is therefore the IoU that will determine if a prediction is:

- True positive: The object is correctly detected.
- False positive: The object is detected when it is not present in the ground truth.
- False negative: the model does not detect an object when one is present in the ground truth (Fig. 8).

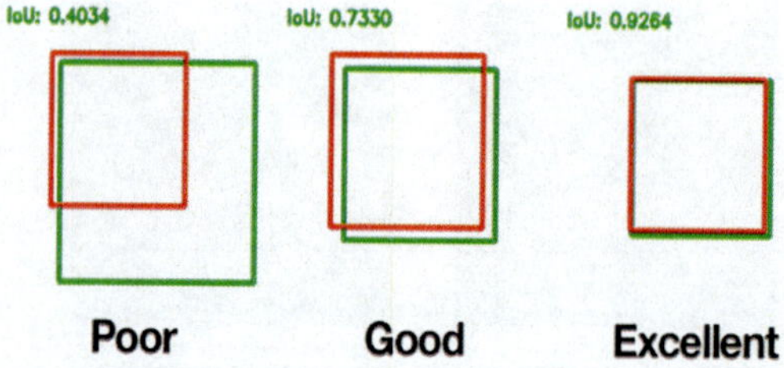

Fig. 8. Eexample of computing Intersection over Unions

4.4 Comparison and Analysis

Figure 9 represents a representative graph when training at 60 epochs the U-net model on the Massachusetts remote sensing images, the IoU score on the training set is about

82.63%, obtaining good results. And it represents the loss function curve, and the model trains at about 60 epochs and tends to be stable and close to 0.09 (Fig. 10).

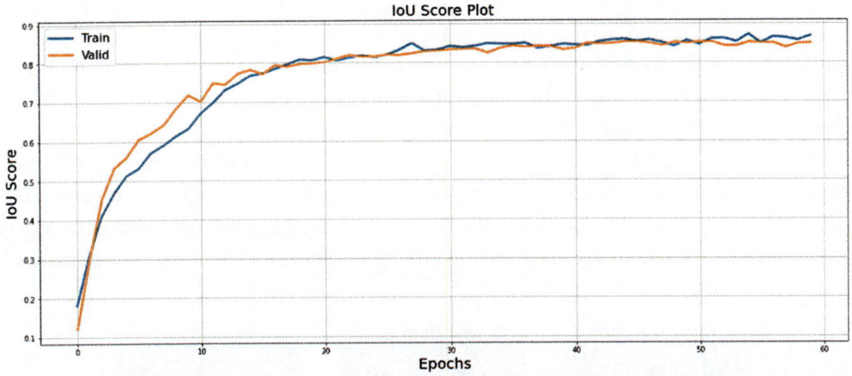

Fig. 9. IoU score plot

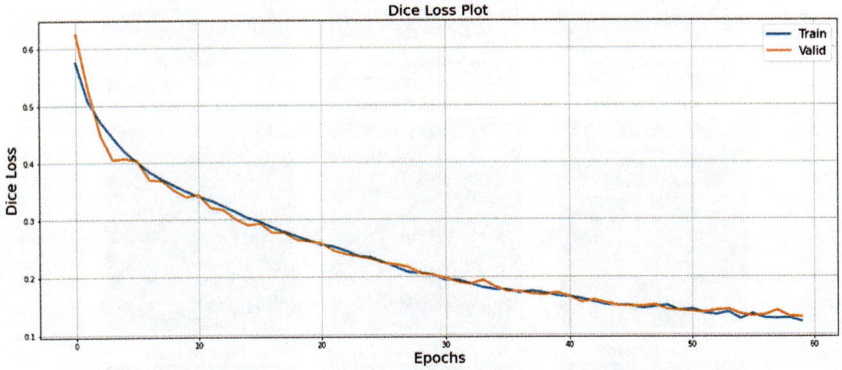

Fig. 10. Dice-coefficient loss function plot

Table 1. Comparison of our results (%) with four other methods in testing data.

Method	IoU
SegNet [14]	67.98
U-Net [15]	70.99
ENRU-Net [23]	73.02
MHA-Net [24]	74.46
Ours	**82.63**

Almost all the most advanced building extraction methods are based on FCN structures. Table 1 shows that our model proposed in this paper on the testing data for the

Massachusetts dataset obtained its best results, better than the other basic models Segnet, U-net and 11% better than the last results obtained in 2021 [24]. 11% can be considered a significant improvement.

At the same time, these results also reflect that compared to traditional methods (which are often difficult to exceed 50% accuracy), deep learning-based methods have pushed building extraction to a new level of automation. And as shown in Fig. 11 our model fully reflects the IoU measure obtained in the test, it can better distinguish buildings and backgrounds with high accuracy.

We can also notice that in some places the buildings are not well segmented. Namely, in the dataset, there are images in which the buildings are not marked in output as. All this affects the fact that the model can not converge in the learning process to a given objective, which translates into such results in the validation and learning set. It can also

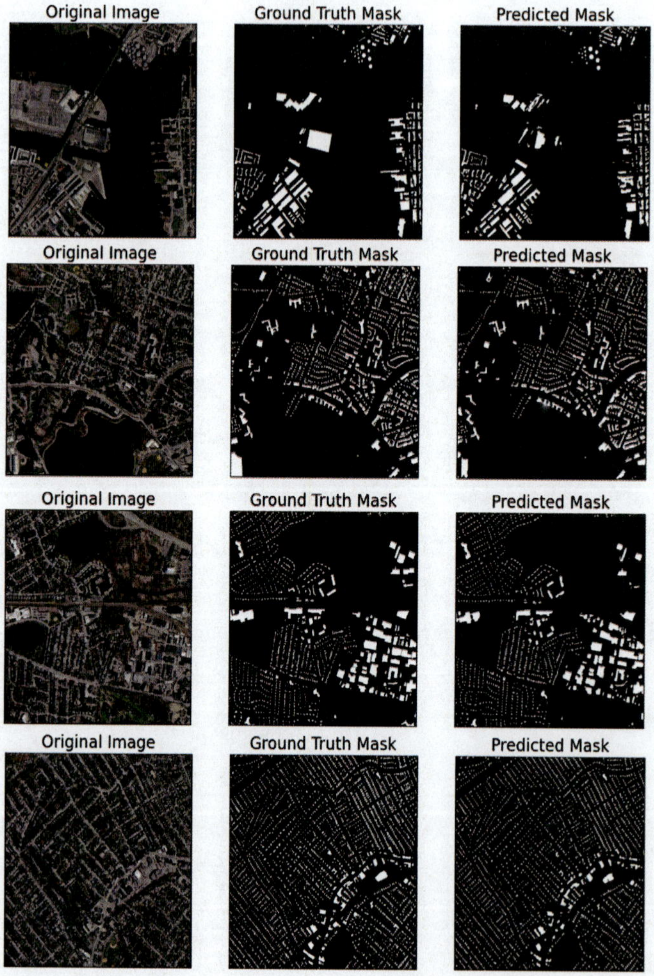

Fig. 11. Some samples of predictions in massachusetts building dataset.

be noticed that the resolution so high for the original images is not sufficient to make some buildings clearly visible.

5 Conclusion

U-net with a ResNet50 encoder is applied to building extraction from remote sensing images to accurately extract buildings in the target area. Moreover, when buildings are extracted in urban areas in a complex environment, environmental information and building information are easily confused, making the extraction poor. In response to the above problems, an improved U-net neural network is proposed, improving the detailed information in the network transmission process and improving the model's ability to obtain details in the 2600 km^2 coverage area. After training and testing on Massachusetts remote sensing image datasets and comparing with other methods, the results show that the average values of IoU coefficients of the U-net model proposed in this paper will reach 82.60, respectively, which is better than the other models compared.

The solar shading and the difference in the characteristics of the building itself will impact the integrity of the building extraction. In future work, it will be necessary to study the shadow and more and more building characteristics in the image to improve the building extraction effect.

References

1. Sirmacek, B., Unsalan, C.: Building detection from aerial images using invariant color features and shadow information. In: 2008 23rd International Symposium on Computer and Information Sciences, pp. 1–5. IEEE (2008)
2. Zhong, S.-H., Huang, J.-J., Xie, W.-X.: A new method of building detection from a single aerial photograph. In: 2008 9th International Conference on Signal Processing, pp. 1219–1222. IEEE (2008)
3. Zhang, Y.: Optimisation of building detection in satellite images by combining multispectral classification and texture filtering. ISPRS J. Photogr. Remote Sens. **54**(1), 50–60 (1999)
4. Chen, D., Shang, S., Wu, C.: Shadow-based building detection and segmentation in high-resolution remote sensing image. J. Multim. **9**(1), 181–188 (2014)
5. Deng, Z., Sun, H., Zhou, S., Zhao, J., Lei, L., Zou, H.: Multi-scale object detection in remote sensing imagery with convolutional neural networks. ISPRS J. Photogr. Remote Sens. **145**, 3–22 (2018)
6. Deng, Z., Sun, H., Zhou, S., Zhao, J.: Learning deep ship detector in SAR images from scratch. IEEE Trans. Geosci. Remote Sens. **57**(6), 4021–4039 (2019)
7. Chen, L.-C., Zhu, Y., Papandreou, G., Schroff, F., Adam, H.: Encoder-decoder with atrous separable convolution for semantic image segmentation. In: Proceedings of the European Conference on Computer Vision (ECCV), pp. 801–818 (2018)
8. Wang, J., Song, J., Chen, M., Yang, Z.: Road network extraction: a neural-dynamic framework based on deep learning and a finite state machine. Int. J. Remote Sens. **36**(12), 3144–3169 (2015)
9. Guo, J., Pan, Z., Lei, B., Ding, C.: Automatic color correction for multisource remote sensing images with Wasserstein CNN. Remote Sens. **9**(5), 483 (2017)
10. Simonyan, K., Zisserman, A: Very deep convolutional networks for large-scale image recognition (2014)

11. He, K., Zhang, X., Ren, S., Sun, J.: Deep residual learning for image recognition. In: Proceedings of the IEEE Conference on Computer Vision and Pattern Recognition, pp. 770–778 (2016)
12. Long, J., Shelhamer, E., Darrell, T.: Fully convolutional networks for semantic segmentation. In: Proceedings of the IEEE Conference on Computer Vision and Pattern Recognition, pp. 3431–3440 (2015)
13. Qingsong, S., Chao, Z., Yu, C., Xingli, W., Xiaojun, Y.: Road segmentation using full convolutional neural networks with conditional random fields. J. Tsinghua Univ. **58**(8), 725–731 (2018)
14. Badrinarayanan, V., Kendall, A., Cipolla, R.: SegNet: a deep convolutional encoder-decoder architecture for image segmentation. IEEE Trans. Pattern Anal. Mach. Intell. **39**(12), 2481–2495 (2017)
15. Ronneberger, O., Fischer, P., Brox, T.: U-net: Convolutional networks for biomedical image segmentation. In: Navab, N., Hornegger, J., Wells, W., Frangi, A. (eds.) MICCAI 2015. LNCS, vol. 9351, pp. 234–241. Springer, Cham (2015). https://doi.org/10.1007/978-3-319-24574-4_28
16. Rakhlin, A., Davydow, A., Nikolenko, S.: Land cover classification from satellite imagery with U-net and lovász-softmax loss. In: Proceedings of the IEEE Conference on Computer Vision and Pattern Recognition Workshops, pp. 262–266 (2018)
17. Abderrahim, N.Y.Q., Abderrahim, S., Rida, A.: Road segmentation using u-net architecture. In: 2020 IEEE International conference of Moroccan Geomatics (Morgeo), pp. 1–4. IEEE (2020)
18. Hunt, B.R., Ryan, T.W., Sementilli, P.J., DeKruger, D.: Interactive tools for assisting the extraction of cartographic features. In: Image Understanding and the Man-Machine Interface III, pp. 208–218. International Society for Optics and Photonics (1991)
19. DeKruger, D., Hunt, B.R.: Image processing and neural networks for recognition of cartographic area features. Pattern Recogn. **27**(4), 461–483 (1994)
20. Ahmadi, S., Zoej, M.V., Ebadi, H., Moghaddam, H.A., Mohammadzadeh, A.: automatic urban building boundary extraction from high resolution aerial images using an innovative model of active contours. Int. J. Appl. Earth Observ. Geoinf. **12**(3), 150–157 (2010)
21. Vakalopoulou, M., Karantzalos, K., Komodakis, N., Paragios, N.: Building detection in very high resolution multispectral data with deep learning features. In: 2015 IEEE International Geoscience and Remote Sensing Symposium (IGARSS), pp. 1873–1876. IEEE (2015)
22. Guangming, W., Qi, C., Shibasaki, R., Zhiling, G., Xiaowei, S., Yongwei, X.: High precision building detection from aerial imagery using a U-Net like convolutional architecture. Acta Geodaetica Cartogr. Sinica **47**(6), 864 (2018)
23. Wang, S., Hou, X., Zhao, X.: Automatic building extraction from high-resolution aerial imagery via fully convolutional encoder-decoder network with non-local block. IEEE Access **8**, 7313–7322 (2020)
24. Cai, J., Chen, Y.: MHA-net: multipath hybrid attention network for building footprint extraction from high-resolution remote sensing imagery. IEEE J. Sel. Top. Appl. Earth Observ. Remote Sens. (2021)
25. Das, P., Chand, S.: Extracting building footprints from high-resolution aerial imagery using refined cross AttentionNet. IETE Tech. Rev. 1–12 (2021)
26. Mnih, V.: Machine learning for aerial image labeling. University of Toronto (Canada) (2013)

Automatic Searching of Deep Neural Networks for Medical Imaging Diagnostic

Zakaria Rguibi(✉), Abdelmajid Hajami, and Zitouni Dya

Faculty of Science and Technology, Hassan First University of Settat, Settat, Morocco
{rguibi.fst,abdelmajid.hajami,zitouni.dya}@uhp.ac.ma

Abstract. Medical imaging diagnosis is the most assisted method to help physicians diagnose patient diseases using different imaging test modalities. But Imbalanced data is one of the biggest challenges in the field of medical imaging. To advance this field, this work proposes a framework that can be used to find the optimal DNN architectures for a database with the challenge of imbalanced datasets. In our paper, we present a framework for automatic deep neural network search for medical imaging diagnosis.

Keywords: Architecture growth · Deep neural networks (DNNs) · Medical imaging · Imbalanced data problem

1 Introduction

Neural architecture search (NAS) is the research area that has emerged from various efforts to automate the process of designing architectures. that can be trained efficiently and effectively to perform a given task well. designing neural architectures that can be trained efficiently and effectively to perform a given task well. The process of finding the optimal DNN model is strongly dependent on human analysts, Therefore, our next step as a researchers is to automate the process of designing the networks.

Finding this optimal DNNs is by packing computational nodes capable of performing simple mathematical operations, such as summations, multiplications, and convolutions. Layering multiple convolutional layers is ideal for processing two-dimensional data, such as that found in medical imaging, although their exact architecture depends on the problem. The use of CNNs in medical imaging diagnostics is not exclusive to the present work. There are many studies in the literature that have applied them to aid in the diagnosis of many diseases. CNN models have been rapidly adapted by the medical imaging research community. Indeed, CNNs have the ability to be parallelized with GPUs and have demonstrated exceptional performance in computer vision. The reason that CNNs in medical imaging have shown promising results in the segmentation of

---

Watch Laboratory for Emerging Technologies (LAVETE).

R. Saidi et al. (Eds.): ICATH 2021, LNDECT 110, pp. 129–140, 2022.
https://doi.org/10.1007/978-3-030-94188-8_13

brain pathologies and in an editorial of deep learning techniques in computer-aided detection, segmentation, and pattern analysis.

Neural architecture search (NAS) is the procedure of discovering better neural architectures than human designs automatically. In [1] and [2] we show that the use of Optimization to perform DNN architecture search can be accomplished by developing models using only EAs.

This work proposes an algorithm for the automatic design of DNNs for medical imaging diagnostics with the Imbalanced dataset problem. The algorithm grows a DNN architecture using residual layer blocks until it surfs the data with good control of whether the datasets had the Imbalanced problem. Therefore, our work presents a framework to search optimal DNNs for use with medical imaging diagnostic datasets in the presence of Imbalanced dataset.

The rest of the paper is organized as follows: literature review and the backround covering related work is discussed in Sect. 2 and 3. Then, Sect. 4 and 5 covers thethe proposed algorithm and experimental design. In Sect. 6, we discuss the analysis the expermental results then goal orientations and comparisons used in Sect 7. Finally, the conclusion of this work is presented in Sect. 8.

2 Previous Works

In this section, we will give the readers a brief summary of some previous work as you can see in Fig. 1.

In this kind of research we have many works that try to automate the processes of finding the optimal architecture, all of this work is close to each other in the deep learning context but the most important thing is that they do not explain their approaches in medical view. More details can be found in the section GOAL ORIENTATIONS AND COMPARISONS. The major drawback of standard DNN architecture finding algorithms is that all of them try to find the global optimal architecture through trial and error of many DNN settings without any control of whether the data is unbalanced or not, so that the proposed configuration is based solely in terms of loss and trial errors.

The SCANN [6], STEERAGE [5], NeST [3], and AutoGrow [4] Certain important approaches use the grow-and-prune paradigm. While AutoGrow and NeST can be employed with DNNs, SCANN and STEERAGE have employed the grow-and-prune paradigm to design classical ANNs. **see** Fig. 1.

3 Backround

In this paragraph, we present the conceptual background of the suggested algorithm. the suggested algorithm, consisting of the residual networks and metric analysis. Specifically, the residual networks are the constructive blocks used during the growth of the DNN architecture.

Paper Title	reference
"NeST: A neural network synthesis tool based on a Grow-and-Prune paradigm,"	[3]
"AutoGrow: Automatic layer growing in deep convolutional networks,"	[4]
"STEERAGE: Synthesis of neural networks using architecture search and Grow-and-Prune methods,"	[5]
"SCANN: Synthesis of compact and accurate neural networks,"	[6]
"Automatic Searching and Pruning of Deep Neural Networks for Medical Imaging Diagnostic"	[7]
"NAS-Unet: Neural Architecture Search for Medical Image Segmentation"	[10]
"MS-NAS: Multi-Scale Neural Architecture Search for Medical Image Segmentation "	[9]

Fig. 1. Previous works

3.1 Residual Networks

Residual network is a particular class of neural network which was first introduced in the paper "Deep Residual Learning for Image Recognition" by Kaiming He, Xiangyu Zhang, Shaoqing Ren and Jian Sun . The models are built by layering operations on top of each other. In most cases, each layer performs two operations, a weight operation and an activation function operation. First, the weight operation processes the input of the layer by multiplying its inputs by its inputs by its related weights. Next, the result of the weight process enters an activation function, usually a nonlinear function. Finally, the input to the next layer is the output of the activation function of the current layer.

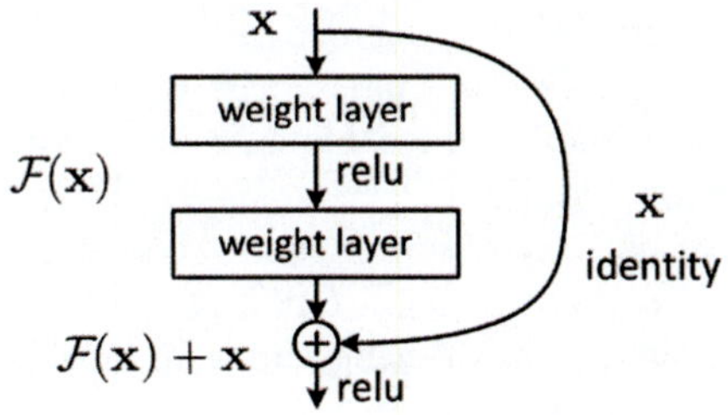

Fig. 2. Single residual block

The employment of residual blocks allows DNNs to have additional layers without facing gradient vanishing problems. In this way The suggested algorithm generates a model using residual blocks, that is, convolutional layers with a short connection across two layers developed by He et al. this building block, and the DNNs that use them are called the ResNets.

3.2 Multi-metrics Analysis

One good approach to dealing with the data imbalance in binary classification is The resampling techniques to optimize performance metrics that are built to handle data imbalance. In addition, using inappropriate measures of performance, such as accuracy or RUC, leads to weak generalization outcomes since classifiers are prone to predict greater class size. That is why The Matthews correlation coefficient (MCC) is extensively employed in medicine as a metric of performance. We were interested in developing a new classifier that uses the MCC metric to handle unbalanced data.

That is why in our paper we will use ROc and MCC as a two objective function for giving the optimal architecture.

3.3 the Matthews Correlation Coefficient for Imbalanced Data

The Matthews correlation coefficient was first proposed by B.W. Matthews to evaluate the prediction performance of the secondary structure of proteins. It has since become a popular performance measure in biomedical research. MCC and area under the ROC curve (AUC) were selected as optional measures in the US FDA-led CQM-II initiative, aimed at reaching consensus on best practices for the development and validation of predictive models for personalized medicine [8].

Therefore, both the MCC and the AUC are robust to data imbalance. The one limitation of using the AUC is that there is no explicit formula for computing the AUC. On the other hand, The MCC has a similar form and is very well suitable to be used to construct the optimal classifier for unbalanced data.

4 Proposed Algorithm

This paragraph provides extensive information about the suggested algorithm. The summary of the proposed algorithm can be viewed in Fig. 3. Models are developed by creating iterative residual blocks on top of the existing model, without ever deleting a formerly added block. After finding a proper and over-designed model structure.

In our algorithm we have two inputs: the images data sets and the number of new block G trial. and the algorithm outputs is a trained model . the main advantage of our approach is that during the searching process we control the optimal configuration not based only in the erea under the curve(AUC) of the receiver operating charachteristic (ROC) but mostly in the MCC factor. On the other hand, the simple measure of accuracy uses only one decision threshold, which is not very complete [14].

Algorithm 1: Proposed algorithm

```
Input: image classification dataset (X), Gtrial .
Output: our model.
dnn ← CreatesanemptyDNNobject;
dnn ← AddRandomResidualBlock(dnn);
previous_fitness ← Evaluate(dnn, X);
Blocks Without Improvement = 0;
while Blocks Without Improvement ≤ 1 do
    dnn_temp ← dnn;
    for j=1 to Gtrial do
        dnn_test ← dnn;
        dnn_temp ← AddRandomResidualBlock(dnn_temp);
        test_fitness ← Evaluate(dnn_temp, X);
        if j=1 then
            dnn ← dnn_temp;
            best_fitness ← test_fitness;
        end
        else if j>1 and test_fitnessisbetterthanbest_fitness then
            dnn ← dnn_temp;
            best_fitness ← test_fitness;
        end
    end
end
if best_fitness_is_better_than_previous_fitness;
 then
    Blocks Without Improvement = 0;
end
else
    Blocks Without Improvement = Blocks Without Improvement + 1;
end
dnn
```

Fig. 3. Proposed algorithm

5 Experimental Design

This paragraph discusses the specifics of the experimental design that was used to evaluate the proposed algorithm, this includes the datasets chosen to test the algorithm, the selected metrics for evaluation, and the algorithm parameters used for evaluation.

5.1 Medical Imaging Data Sets

Medical imaging diagnosis is the most assisted method to help physicians diagnose patients' diseases using different imaging test modalities. Thus, our algorithm is developed to be used with binary classification problems where the task is to classify whether the input data is belonging to a specific class or not. This kind of task is prevalent in the field of medical diagnosis, where a doctor has to use imaging and decide if the patient has a specific disease. In addition, due to the use of convolutional layers and residual blocks, the suggested algorithm is ideal to be used with imaging data. Thus, two publicly available medical imaging datasets with healthy and diseased patient data are chosen to evaluate the proposed algorithm.

The first chest X-ray data set Fig. 4 [13] is composed of color images from patients without and with COVID-19. All images have different 224 pixels. No additional image from other sources is used resolutions. Thus, they are all resized to a resolution of 224× during the algorithm evaluation, and no data augmentation is used during the training time.

training images	Normal chest patients	covid images	test images
13808	10192	3616	289

Fig. 4. The first data

The second chest X-rays data set Fig. 5 is composed images without and with pneumonia from patients , Similar to the COVID-19 data set, all images are resized to 224 × 224.

training images	Normal chest patients	pneumonia images	test images
5856	1349	3883	624

Fig. 5. The Second data

5.2 Evaluation Metrics

The performance of the composed model is evaluated with multiple parameters in each of the above datasets. In addition, learning is performed using stochastic gradient descent (SGD) with a learning rate equal to 0.001 (for COVID-19 data) and 0.0008 (for pneumonia data) and a Nesterov momentum equal to 0.9 for a total of 150 epochs.

The outcome of all models evaluated is the likelihood that the given data (X) belongs to a disease case [p(y = 1—X)]. Therefore, the expected classification (y^) of a model is a function of a decision threshold (t) and is given as follows:

$$\mathrm{y}\hat{} = \begin{cases} 0, & if\, p(y=1|X) < t \\ 1, & if\, p(y=1|X) >= t \end{cases}$$

In addition, AUC calculates the sensitivity (Se) and specificity (Sp) of a model below several thresholds (T). The threshold that maximizes sensitivity and specificity simultaneously is chosen to be the decision threshold, t, as described in the following:

$$t = \text{argmax}\ (\text{Se}(x) \cdot (1 - \text{Sp}(x)))$$

Se(x) and Sp(x) is the model's sensitivity, specificity they are defined as follows:

$$Sensitivity = Recall = \frac{TP}{TP + FN}$$

$$Specificity = \frac{TN}{FP + TN}$$

In this section, I would like to clarify many aspects firstly the TP, TN, FP, and FN when we talking about True positives are the number of times the model correctly predicted and true negatives are the number of times the model correctly predicted false negatives are the number of instances where the model incorrectly predicted and finally false positives are the number of instances where the model incorrectly predicted the classification accuracy (Acc) obtained using a decision threshold of 0.5, Therefore, four metrics is used to evaluate all models presented in Section V: Acc, AUC, MCC, and F1-score with a specified threshold. In addition, all results are obtained when assessing the DNN models with the test sets of the previous data sets.

5.3 Algorithm Parameters

At this level, we will describe different parameters used in our experiment Fig. 6. Firstly our main objective function is to find two important parameters the first one is the best validation AUC during the model evaluation then finally check the model with the MCC factor as the important parameter. Many hyperparameters are fixed firstly the learning rate is 0.0008 and 0.001 for the first and the second data respectively as at any iteration of the process two-block selected randomly to be tested. On the other side, we use a minimum four number of layers in a block and maximum of 12 layers which is proportional to the number of layers per block from ResNet18 to ResNet50. With a number of convolutional filters in a block is randomly chosen between 8 and 64 which is also proportional to the other ResNets.

parameter	value
number of trials (Gtrial)	2
stop DNN evaluation after x epochs without improvement	2
Learning rate to evaluate a candidate solution	0.0008
minimum number of layers in a residual block	4
maximum number of layers in a residual block	12
minimum number of convolutional filters in a residual block	8
maximum number of convolutional filters in a residual block	64

Fig. 6. Algorithm parameters

6 Experimental Results and Discussion

This section shows the experimental results obtained with the suggested approach. The results of every phase of the proposed algorithm are reported separately with their explanations.

The complete results of the proposed algorithm are presented in Figs. 7, 8, 9, 10, 11 and 12, where the best and average results of five unrelated runs are pointed out. The best result is selected according to the run that produced the best AUC with the best MCC on the test set, while the average results are the means of all evaluation criteria.

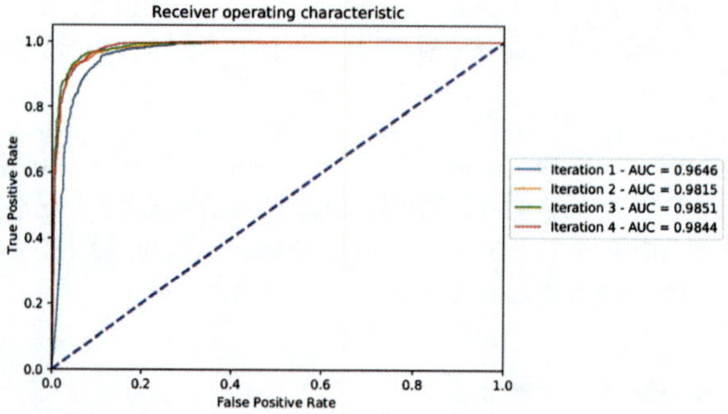

Fig. 7. ROC-AUC-Curves for pneumonia datasets

The most striking accomplishment of the algorithm is its ability to find such competitive models by simply increasing their computational complexity with each iteration while respecting the MCC factor.

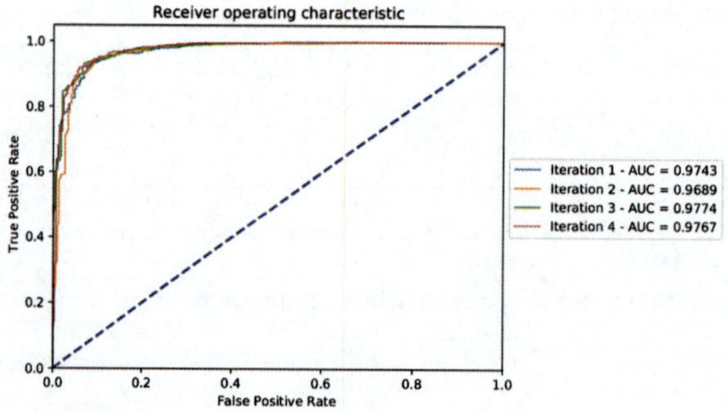

Fig. 8. ROC-AUC-Curves for Covid-19 datasets

This process of producing DNN models has higher accuracy and AUC and also F1 scores. But Figs. 11 and 12 demonstrate that the best DNN model generated by the suggested algorithm has a good MCC value. Thus, the MCC factor integration can be considered as an integral part of the suggested algorithm.

Test Loss	Test Acc	Test AUC	Test Sensitivity	Test Specificity
0.72168	0.788462	0.937081	0.905128	0.84188

Fig. 9. Algorithm results on X-RAY data sets

Test Loss	Test Acc	Test AUC	Test Sensitivity	Test Specificity
0.65168	0.8656	0.9631	0.8891	0.9195

Fig. 10. Algorithm results on COVID-19 data sets

We considered the following simulation, to illustrate the adequacy of the MCC for imbalanced data, : First, we use X-ray data as clean and well-structured data and then the covid dataset as imbalanced data in the first phase, all data give us good metrics (acc test, AUC, sensitivity, specificity) in the second phase, we want to demonstrate the question of how we can detect imbalanced data in the medical context as an important thing.

As we can see ideal evaluation on data collected from a multi-centric prospective enrollment of patients with eligibility criteria matching the characteristics of the Target population for whom the test will be applied in the real world.

Test AUC	f1-score	MCC
0.937081	0.9704	0.8851

Fig. 11. Algorithm results on COVID-19 data sets

Test AUC	f1-score	MCC
0.9631	0.983128	0.4097

Fig. 12. Algorithm results on COVID-19 data sets

7 Goal Orientations and Comparisons

The rapid growth in the supply of radiological diagnostic software based on Artificial Intelligence (AI) is opening up new opportunities to improve the quality and productivity of medical imaging.

Beyond technical robustness, the performance of AI software must be evaluated from a clinical point of view in order not to overestimate its capabilities in a real environment. This presentation aims to clarify the methods for evaluating the clinical performance of AI-based classification systems.

In this section we will provide a comparative study with some works in the same objective and provide our contributions.

Firstly the major comparison is with this papers [4–7], where try to automatic it the process of finding DNN models for medical imaging diagnosis. In their study generally they use the same approach when we talking i the deep learning context, But this is not the most important in healthcare context.

In our propose the accuracy and AUC is not really means that the model is highly recommended in medical imaging context. in fact, high classification performance does not mean better patients outcomes. because high accuracy doesn't necessarily mean good performance and accuracy doesn't inform about false prediction so is not reliable on imbalanced datasets. In [7] they give us a good accuracy test but with the same data used in our approach we see that the MCC factor is close to 0 and that mean that their framework can not give us an idea about imbalanced data-sets.

When we talking about AUC we talk indirectly to PPV(or Precision) and NPV. PPv is the share of correctly classified positive tests among all positive tests and informs only about TP and FP . In this way a test that classifies as abnormal(positive)only 1 case (the "most abnormal" case) and misses the rest of abnormal cases will have a 100% PPv and overlook numerous FN. NPV is the share of correctly classified negative tests among all negative tests and informs only about TN and FN . In this way a test that classifies as normal (negative)only 1 case (the "most normal" case) and misses the rest of normal cases will have a 100% NPV and overlook numerous FP. So if prevalence increases, PPV increases and NPV decreases (and vice versa).

As a results Accuracy and AUC do not capture effectively the performance of a test in imbalanced data-sets. our contribution overcome all precedent works because in healthcare context it is very important to know about the true negative samples. In our approach the use of MCC mathematical properties is very important to taking into account class imbalance data-sets . Good MCC score means correct prediction in both classes independently of their balance.

8 Conclusion and Future Work

In this paper, We investigated two major directions for finding the optimal DNNs model: have a the high accuracy and AUC and controlling the model by the MCC factor if we have imbalanced datasets .the important thing in this paper is that we unveiled certain orientations to improve the quality and productivity of medical imaging in real context. in fact that , high classification performance does not mean better patients outcomes. because high accuracy doesn't necessarily mean good performance and accuracy doesn't inform about false prediction so is not reliable on imbalanced datasets.

This is why, being able to explain DNNs can help the researchers of these models to diagnose and further improve the system, So Our future work will really focus on the black box of the deep learning model to understand the logic behind the decision side of deep learning in the context of medical imaging.

MCC Formule:

$$MCC = \frac{(TP * TN) - (FP * FN)}{\sqrt{[(TP + FP) * (FN + TN) * (FP + TN) * (TP + FN)]}}$$

References

1. Junior, F.E.F., Yen, G.G.: Particle swarm optimization of deep neural networks architectures for image classification. Swarm Evol. Comput. **49**, 62–74 (2019). https://linkinghub.elsevier.com/retrieve/pii/S2210650218309246
2. Sun, Y., Xue, B., Zhang, M., Yen, G.G.: A particle swarm optimization-based flexible convolutional autoencoder for image classification. IEEE Trans. Neural Netw. Learn. Syst. **30**(8), 2295–2309 (2019). https://ieeexplore.ieee.org/document/8571181/
3. Dai, X., Yin, H., Jha, N.K.: NeST: a neural network synthesis tool based on a Grow-and-Prune paradigm. IEEE Trans. Comput. **68**(10), pp. 1487–1497 (2019). https://ieeexplore.ieee.org/document/8704878/
4. Wen, W., Yan, F., Chen, Y., Li, H.: AutoGrow: Automatic layer growing in deep convolutional networks (2019). arXiv:1906.02909
5. Hassantabar, S., Wang, Z., Jha, N.K.: SCANN: Synthesis of compact and accurate neural networks (2019). arXiv:1904.09090
6. Hassantabar, S., Dai, X., Jha, N.K.: STEERAGE: Synthesis of neural networks using architecture search and Grow-and-Prune methods (2019) arXiv:1912.05831
7. Fernandes, F.E., Yen, G.G.: Automatic searching and pruning of deep neural networks for medical imaging diagnostic. In: IEEE Transactions on Neural Networks and Learning Systems. https://doi.org/10.1109/TNNLS.2020.3027308
8. Boughorbel, S., Jarray, F., El-Anbari, M.: Optimal classifier for imbalanced data using Matthews correlation coefficient metric. PLoS ONE **12**(6) (2017). https://doi.org/10.1371/journal.pone.0177678
9. Yan, X., Jiang, W., Shi, Y., Zhuo, C.: MS-NAS: multi-scale neural architecture search for medical image segmentation. In: Martel, A.L., et al. (eds.) MICCAI 2020. LNCS, vol. 12261, pp. 388–397. Springer, Cham (2020). https://doi.org/10.1007/978-3-030-59710-8_38
10. Weng, Y., Zhou, T., Li, Y., Qiu, X.: NAS-Unet: neural architecture search for medical image segmentation. IEEE Access 1 (2019). PP, https://doi.org/10.1109/ACCESS.2019.2908991
11. Kermany, D.S., et al.: Identifying medical diagnoses and treatable diseases by image-based deep learning. Cell **172**(5), 1122–1131 (2018). https://linkinghub.elsevier.com/retrieve/pii/S0092867418301545

12. He, K., Zhang, X., Ren, S., Sun, J.: Deep Residual learning for image recognition. In: IEEE Conference on Computer Vision and Pattern Recognition (CVPR), 2016, pp. 770–778 (2016). https://doi.org/10.1109/CVPR.2016.90
13. Kermany, D.S., et al.: Identifying medical diagnoses and treatable diseases by image-based deep learning. Cell **172**(5), 1122–1131 (2018). https://linkinghub.elsevier.com/retrieve/pii/S0, 92867418301545
14. Huang, J., Ling, C.X.: Using AUC and accuracy in evaluating learning algorithms. IEEE Trans. Knowl. Data Eng. **17**(3), 299–310 (2005). http://ieeexplore.ieee.org/document/1388242/

PCA SVM and Xgboost Algorithms for Covid-19 Recognition in Chest X-Ray Images

R. Assawab[(✉)], Abdellah Elzaar, Abderrahim El Allati, Nabil Benaya, and B. Benyacoub

Laboratory of R&D in Engineering Sciences, Faculty of Sciences and Techniques Al-Hoceima, Abdelmalek Essaadi University, Tetouan, Morocco
{Assawabrachida,abdellah.elzaar}@etu.uae.ac.ma

Abstract. Covid-19 is a life-threatening epidemic, which makes it an active research topic. Our work aims to use the power of Machine Learning algorithms to develop an intelligent system for recognition of covid-19 in chest X-ray images. In this paper, we proposed a hybrid model based Principal Component Analysis (PCA) technique, Support Vector Machine (SVM) and Xgboost algorithms for Covid-19 recognition in chest X-ray images. Our method merges the best properties of PCA and SVM to perform the recognition process. The PCA algorithm used to extract features from X-ray images, SVM implemented as a binary classifier and finally Xgboost used to boost the effectiveness of our model and to avoid the overfitting. Our model shows a satisfactory result with less complex model architecture.

Keywords: Machine learning · Support Vector Machine (SVM) · Principal Component Analysis (PCA) · Covid-19 · Xgboost algorithm · X-ray chest images

1 Introduction

Coronavirus disease (COVID-19) can cause severe pneumonia and would have a major impact on the healthcare system. There have been more than 179 million confirmed cases worldwide, Including 3million deaths, Reported to World Health Organization (WHO). Early diagnosis is essential for proper treatment to ultimately reduce the strain on the healthcare system. On the other side, The doctors, Biologists and the researchers work together in fact to discover more information about this field, Produce more data and take a decision that will service public health and the rapid development of effective vaccines. The gold-standard method for diagnosing COVID-19 is the polymerase chain reaction (PCR) test, Which detects the coronavirus's genetic material in a nose or throat swab [10]. However, People with COVID-19 symptoms who test negative should be retested, Especially in areas where the virus is widespread. Studies

R. Saidi et al. (Eds.): ICATH 2021, LNDECT 110, pp. 141–148, 2022.
https://doi.org/10.1007/978-3-030-94188-8_14

have shown that abnormalities in the chest are caused due to COVID-19, Which are visible in Chest X-rays [11].

In order to overcome the limitations of diagnostic process of COVID-19 within constrained accessible time and achieving good results, We formulated this problem for object recognition tasks. Many researchers' works on object recognition filed, They have proposed different Machine Learning models to detect COVID-19 using radiographs or CT images. For example Ghoshal et al. [8] a Bayesian Convolutional Neural network to estimate model uncertainty in COVID-19 prediction. Wang et al. [15] evaluates the diagnostic performance of a deep learning algorithm to detect COVID-19 during the influenza season the external testing dataset showed a total accuracy of 79.3%. Another model proposed by Wang et al. [14] where public dataset of chest X-ray images achieved an accuracy of 92.4%. X-ray images is even more challenging.

In this study, We aim to use the power of machine learning algorithms to provide an intelligent system for recognition of covid19 in chest X-ray images. To support that, We have mainly focused on 2 challenges: The first challenge is to involve simple and reliable architecture in this case, We used PCA Algorithm to extract features from X-ray images than we implemented SVM as a binary algorithm, Because of its better performance and we feed it by this features. The second challenge is to accelerate our system and predict the best accuracy we have succeed by applying xgboost.

The rest of this paper is organized as follows. Section 2 presents some related work. Then, Sect. 3 details our proposed methodology. Section 4 presents the details of our experimental setup and the obtained results. Finally, Sect. 5 concludes the current study and describes some possibilities for future work.

2 Related Work

In this part, We will explore different techniques and methods that are done in the field of prediction and diagnosis of COVID-19. Several works have been conducted on this area. Among the best studies proposed, That gives a high recognition rate using Support Vector Machine, We mention:

Dixit et al. [7] developed a framework for the detection of SARS-COV-2 without manual intervention. This framework consisted of 3-step. The first step is data preprocessing, where the author use K-means clustering to extract features. The second step is to optimize the selected features by using a hybrid differential evolution algorithm. Then, the optimized feature is fed to the SVM classifier.

Zhang et al. [16] proposed machine learning algorithms to create a detection model for COVID-19 gravity. After detecting 32 features significantly related to COVID-19 gravity, The support vector machine (SVM) demonstrated promising detection accuracy. Further check whether there is any redundancy among these 32 features.

Hassanien et al. [12] tested three machine learning techniques to predict patient recovery. The support vector machine was tested on the data provided and the average absolute error was 0.2155. A thorough analysis of other machine learning algorithms was tested and compared to the SVM results.

Barstugan et al. [9] trained four datasets by taking patches of size 16×16, 32×32, 48×48, 64×64 from 150 CT images. The features were extracted by using: Local directional model (LDP), Gray level co- occurrence matrix (GLCM), Gray level size zone matrix (GLSZM), Gray level run length matrix (GLRLM), and discrete wavelet transform (DWT), the author classified the extracted features with Support vector machines (SVM). Two-time, five-time and ten-time cross- validation were implemented during the classification process.

Sethy et al. [13] developed a method based on depth functions and support vector machines (SVM) to detect coronavirus-infected patients using radiographic images. For classification, SVM is used instead of depth learning based classifiers, As the latter require a large number of datasets for training and verification. The deep features of the fully connected layer of the CNN model are extracted and fed to SVM for classification. SVM classifies the X-ray images affected by the crowns of the others. SVM was evaluated to detect COVID-19 using the deep features of 13 different CNN models.

These methods use a lot of features and that makes the classification more complicated, so to solve this problem we propose to reduce the number of features into tow component by using the PCA Algorithm.

Following this context, This article proposes to contribute for early diagnosis of COVID-19 using Support Vector Machine, Assisted with Principal Component Analysis (PCA) technique and Xgboost algorithm.

3 Proposed Approach

In our work, We combined PCA, SVM and Xgboost machine learning algorithms to perform the recognition process. We implemented Principal Component Analysis (PCA) as a feature extractor algorithm from covid-19 X-ray chest images, The extracted features are then transmitted to SVM as input data for classification and finally Xgboost is used to ameliorate the recognition process and to avoid the overfitting problem. An overview of the proposed system is presented in Fig. 1.

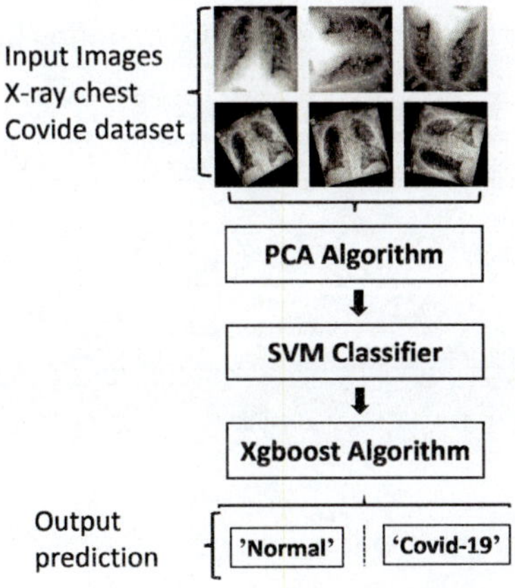

Fig. 1. Architecture of the proposed PCA-SVM-Xgboost model.

3.1 PCA Algorithm

Principal Component Analysis is a powerful algorithm for dimensionality reduction, The main idea of PCA is simplifying a data set by reducing the dimensionality of a dataset to lower dimensions for analysis.

In general, We can express PCA as a linear transformation of the image vector to the projection feature vector as given below:

$$Y = G^t X \tag{1}$$

where, G is the transformation matrix with dimension K x 1, Y is the K x N feature vector matrix and X is the higher dimension vector obtained by representing all the images into a single vector $X = \{x_1, x_2, x_3, ..., x_N\}$ Where, each x_i is a vector of dimension "n" obtained from the M x N dimension image. Huge dimensionality data can be time consuming and have negatively impact the recognition system. For this reason, PCA is used to reduce the dimensions of the input data by reducing the number of axes or components of the feature vector. In our case we used PCA to reduce the dimensions of chest X-ray images and also to extract the feature vector. The extracted features from the input images are feeded to SVM classifier for classification.

3.2 SVM Classifier

The SVM is a supervised ML, it can be treated as artificial neural network if the sigmoid function is used as the activation function to update the weights. SVMs tend to move features to higher dimensions so that they can be separated by a hyperplane, the goal of using SVM is to maximize this hyperplane as the more features are separated. Consider that:

- $D = [x_i, d_i]^n$ the series of data points;
- n is the size of data;
- d_i represents the target value;
- x_i represents the input.

The SVM estimates the function as given in the following two equations:

$$g(x) = w\Delta(x) + b. \tag{2}$$

$$R_{SVM}(C) = \frac{1}{2}v^2 + C(\frac{1}{n})\sum_{1}^{n} L(x_i, d_i) \tag{3}$$

Where:

- $\Delta(x)$ the high-dimensional space feature;
- b is a scalar;
- w is a vector;
- $C(\frac{1}{n})\sum_1^n L(x_i, d_i)$ the empirical error;
- b and w can be assessed by Eq. (3).

SVM algorithm works perfectly on binary classification problem and performs less with noisy and large amount of data. After the SVM, The Xgboost is applied to boost the speed and performances of our Model system.

3.3 Xgboost Algorithm

Xgboost is known as the classification and regression tree (CART). It has been widely employed in industry due to its high performance in problem-solving and minimal requirement for feature engineering. A detailed XGBoost algorithm is mentioned below:

Algorithm 1. XGboost Algorithm

Initialize $f_0(x)$
for $k = 1$ to M **do**
  compute $g_k = \frac{\partial(L(y,f))}{\partial f}$
  compute $h_k = \frac{\partial^2(L(y,f))}{\partial f^2}$
  Determine the structure by choosing splits with maximized gain
  $A = \frac{1}{2}[\frac{G_L^2}{H_L} + \frac{G_R^2}{H_R} + \frac{G^2}{H}]$
  Determine the leaf weights $w^* = -\frac{G}{H}$
  Determine the base learner $b(x) = \sum_j^T wI$
  Add trees $f_k(x) = f_{k-1}(x) + b(x)$
end for
Result: $f(x) = \sum_k^M f_k(x)$

Before feeding the data images to our Model, We first apply a preprocessing step, The preprocessing step consists of resizing the input images and make them all in one input dimension. We set all images to 500×500 input size. The PCA algorithm reduce the dimensionality of feature vector from 250000 axes to 2 axes. The extracted features are feeded to SVM classifier and then to Xgboost algorithm.

4 Experimental Results

4.1 Dataset

The dataset is organized into 2 folders (covid, normal) which represent the main classes of the dataset. The data is divided into train set and test set. Train data contains 2050 images of X-ray chest and Test data contains 1000 images (Table 1).

Table 1. Dataset class

No	Class label	Train	Test
1	Normal	1025	500
2	Covid	1025	500
Total		2050	1000

4.2 Results and Discussion

We implemented our architecture using Python programming language. The process ran under a core i5 processor of 3.6 GHz with 8 GB memory. To boost the speed of the training operation. We evaluated the performance of our proposed model by using test data which is not including in the training set. Table 2 shows the accuracy of our proposed method.

Table 2. Comparison of accuracy of PCA-SVM and PCA-SVM-Xgboost

Approach	Accuracy (%)
1st Approach: PCA-SVM based model	82.23%
2nd Approach: PCA-SVM-Xgboost based model	95.43%

The Table 2 shows that the PCA-SVM-Xgboost based approach shows a recognition score better than the method in which we use PCA-SVM.

To truly understand the accomplishment of our model we have to compare the results to other approaches (detailed mentioned in Table 3).

Table 3. Comparing our results with other state of art algorithms

Approach	Accuracy (%)
Wang et al. [15]	79.3
Xu et al. [4]	86.7
Li et al. [5]	88.9
Ismael et al. [6]	90.53
Wang et al. [14]	92.3
PCA-SVM-Xgboost based model	95.43%

In this section, We present some comparisons between our proposed approach and the recent state of the art. Most of the approaches mentioned in Table 3, Use the power of convolutional network to detect covid19 in X-ray images. Although they have obtained good results, They have some limitations, As we know deep neural networks has a complicated architecture and needs more memory capacity to generate complex real-time calculations. They also need a larger data set. As we know. The Corona virus is still new and the available data is very small, For this reason we did not use deep learning algorithms and thought of a new, Simple and inexpensive method.

Our proposed PCA-SVM model with Xgboost achieved higher results with 95.43% accuracy compared to all other approaches. This demonstrates the effectiveness of our method for working with difficult databases with character similarity.

5 Conclusion

In this work, We used two various machine learning models to get accurate features and classify x-ray chest images. The first method is based PCA-SVM, PCA is used to extract features from x-ray images and SVM is used for binary classification. The second method is based PCA-SVM-Xgboost. In this method Xgboost is used to avoid overfitting and accelerate the learning process. The second method achieves a recognition rate of 95.43% which is greater than the first one with a recognition rate of 82.23%. As a future work we plan to implement more challenging dataset with diversity of classes.

6 Data Availability

The dataset generated or analyzed during this study is publicly available at [17].

Acknowledgements. The authors would like to thank the referee for the important remarks, which improved our results. R. ASSAWAB acknowledges financial support for this research from the "Centre National pour la Recherche Scientifique et Technique" CNRST, Morocco.

References

1. Sun, J., et al.: A prospective observational study to investigate performance of a chest X-ray artificial intelligence diagnostic support tool across 12 US hospitals, arXiv preprint arXiv:2106.02118 (2021)
2. Baloch, S., Baloch, M.A., Zheng, T., Pei, X.: The coronavirus disease 2019 (COVID-19) pandemic. In: The Tohoku Journal of Experimental Medicine, pp. 271–278. Tohoku University Medical Press (2020)
3. Sonbhadra, S.K., Agarwal, S., Nagabhushan, P.: Target specific mining of COVID-19 scholarly articles using one-class approach. Elsevier (2020)
4. Xu, X., Jiang, X., Ma, C., et al.: Deep learning system to screen coronavirus disease 2019 pneumonia, pp. 1e29 (2020)
5. Li, X., Li, C., Zhu, D.: COVID-Xpert: an AI powered population screening of COVID-19 cases using chest radiography images (2020)

6. Ismael, A.M., Şengür, A.K.: Deep learning approaches for COVID-19 detection based on chest X-ray images. Expert Syst. Appl. **164**, 114054 (2021)
7. Dixit, A., Mani, A., Bansal, R.: CoV2-detect-net: design of COVID-19 prediction model based on hybrid DE-PSO with SVM using Chest X-ray images. Inf. Sci. 114054 (2021)
8. Ghoshal, B., Tucker, A.: Estimating uncertainty and interpretability in deep learning for coronavirus (COVID-19) detection, arXiv preprint arXiv:2003.10769 (2020)
9. Barstugan, M., Ozkaya, U., Ozturk, S.: Coronavirus (COVID-19) classification using CT images by machine learning methods. arXiv preprint arXiv:2003.09424 (2020)
10. Ornelas-Ricardo, D., Jaloma-Cruz, A.R.: Coronavirus disease 2019: hematological anomalies and antithrombotic therapy. Tohoku J. Exp. Med. **251**, 327–336 (2020)
11. Pham, T.D.: Classification of COVID-19 chest X-rays with deep learning: new models or fine tuning. Health Inf. Sci. Syst. Proc. **9**, 1–11 (2021)
12. Salama, A., Darwish, A., Hassanien, A.E.: Artificial intelligence approach to predict the COVID-19 patient's recovery (2021)
13. Sethy, P.K., Behera, S.K., Ratha, P.K., Biswas, P.: Detection of coronavirus disease (COVID-19) based on deep features and support vector machine (2020)
14. Wang, L., Lin, Z.Q., Wong, A.: Covid-net: a tailored deep convolutional neural network design for detection of COVID-19 cases from chest x-ray images. Sci. Rep. **10**, 1–12 (2020)
15. Wang, S., et al.: A deep learning algorithm using CT images to screen for Corona Virus Disease (COVID-19). Eur. Radiol. **31**, 1–9 (2021). https://doi.org/10.1007/s00330-021-07715-1
16. Yao, H., et al.: Severity detection for the coronavirus disease 2019 (COVID-19) patients using a machine learning model based on the blood and urine tests. Front. Cell Dev. Biol. **8**, 683 (2020)
17. Asraf, A., Islam, Z.: COVID19, pneumonia and normal chest X-ray PA dataset. Mendeley Data **V1** (2021). https://doi.org/10.17632/jctsfj2sfn.1

Machine Learning Algorithms for Forest Stand Delineation Using Yearly Sentinel 2MSI Time Series

Anass Legdou[1(✉)], Aouatif Amine[1], Said Lahssini[2], Hassan Chafik[3], and Mohamed Berada[3]

[1] Ibn Tofail University, ENSAK, Kenitra, Morocco
anass.legdou@uit.ac.ma
[2] National Forestry School of Engineers, Sale, Morocco
[3] University of Moulay Ismail, ENSAM, Meknes, Morocco
chafik.hassan@edu.umi.ac.ma

Abstract. Forest stand types maps is fundamental tools for sustainable forest management that needs to be regularly updated. This work aims at the use of satellite images time series and machine learning techniques to automate and improve the efficiency of forest stands maps production. A time series of "sentinel2MSI" satellite images has been classified using supervised classification using various machine learning (ML) algorithms whose effectiveness has been proven in several research studies. The produced stands map, when each type is defined according to conventional criteria, was assessed and gives satisfactory results with a high variability depending on the used classification algorithm. Temporal segmentation of the archive appears to be a feasible and a robust means of increasing information extraction from the Sentinel archive.

Keywords: Forest stand map · Sentinel2 MSI's time series · ML

1 Introduction

Forest management integrates changing knowledge, technology, and societal demands, to become a real tool for sustainable forest ecosystem management. It fulfills the fundamental objective of preserving wooded areas, maintaining environ-mental quality and biodiversity, and improving the ability to perform socio-economic functions (FAO, 2007).

Forest management as a strategic tool that does not have a universal model and needs to be periodically updated. It constitutes the set of analyses, then syntheses and choices which, periodically, organize the actions to be carried out on the man-aged area in order to make them coherent and effective. The map of forest stand types is considered to be one of the fundamental tools for planning interventions over time and space.

At the end of each management period, updating of forest stand types' map is required. Nevertheless, the methods used to update the map, despite their efficiency and accuracy, remain empirical and require consider-able time and human resources.

R. Saidi et al. (Eds.): ICATH 2021, LNDECT 110, pp. 149–158, 2022.
https://doi.org/10.1007/978-3-030-94188-8_15

The establishment of forest stands types map by remote sensing has attracted the attention of the scientific community. The first studies were focused on the use of Landsat images at medium spatial resolutions [1–3] in which limitations were noted especially in large areas and in the case of heterogeneous forest stands [4–6]. Other research has used hyper spectral imagery data, which has yielded better results than multi spectral imagery [7–9]. Nevertheless, most very high resolution (VHR) studies are limited to small areas due to cost [10].

Studies relating to the classification of trees in larger areas have shown that the spectral signatures of forest species are subject to stationary conditions and the state of development of the species [11]. This variability of the spectral signature can increase in mountain areas where slope, exposure, and altitude [12] combined with the phenological cycle [13, 14] affect it considerably. Thus, the use of time series 'sentinel 2 MSI has shown promising results [15–18] and has confirmed the potential of Sentinel-2 data for forest species.

The use of machine learning algorithms has become indispensable with the presence of big and complex data [19], particularly in the case of multi-temporal imageries and the processing of stationary variables. In this sense, several remote sensing studies have focused on the use of machine learning algorithms in data processing, notably Random Forest (RF) and Support Vector Machines (SVM), with promising results [10].

For this purpose, a time series of "sentinel2 MSI" type satellite images has been subject to supervised classification, from which a map of settlement types, by using conventional criteria, has been drawn up. This classification has been carried out using some machine learning algorithms whose performance has been demonstrated in several research studies.

2 Materials and Methods

2.1 Study Area

Seheb's canton is part of Sidi M'Guild is located in Ifran province, Fes-Meknes region, Morocco. It covers an area of about 2,212 hectares. From the orographic point of view, the said study area of the Middle Atlas plateau characterized by a tabular structure, faultier than folded. It is a medium mountain which culminates at about 2,200 m of altitude. One then observes at the level of this massif, a topographical diversity characterized by depressions, ravines, slopes more or less strong and a set of slopes. Climatically, the massif is located in a sub-humid bioclimatic environment with a cold variant with a WSAS (Winter Spring Autumn Summer)-type rainfall regime where the maximum rainfall is collected during the winter and the drought is accentuated during the summer season. Hydrologically, seheb 's canton is part of two sub-catchment areas: watershed Sebou and watershed Oum Rbia. For the vegetation component, we meet especially cedar atlas (Ca), and holm oak (Qr)species (Fig. 1).

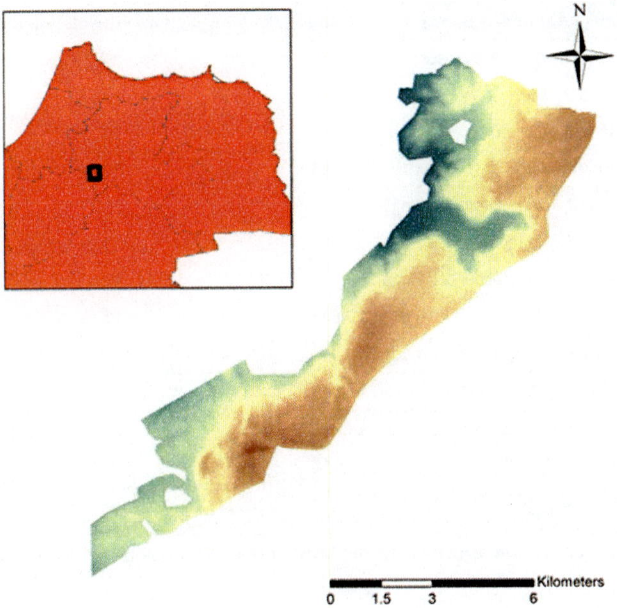

Fig. 1. Study area

2.2 Conventional Criterion for Establishing a Map of Settlement Types

Establishing a map of settlement types is based on the following criteria: a. species composition (pure or in a mixture); b. Density (Dense: cover > 45% rated 1; Moderately dense (normal): 25% < cover < 45% rated 2; Clear: 5% < cover < 25% rated 3; Sparse: Cover < 5% rated 4); c. Development status (young stand (j); mature stand (a); old stand (v)); d. Forest regime (Woodland (f); Coppice (t)).

2.3 Used Data

The paper used Sentinel 2 image for our study area for 4 periods (January, April, August, and November 2018) with cloud cover less than 10%. The images were downloaded from the USGS Earth explorer platform. The sentinel 2 images contain 13 bands (Fig. 2) in which We had used 5 bands (Red band (R), near infrared band (NIR), vegetation red edge band (B8A), two short-wave infrared bands (SWIR1 (Band 11), SWIR2 (Band 12))) of it for each mentioned date. Shuttle Radar Topography Mission (STRM) DEM was used to generate the Altitude and Aspect Maps.

2.4 Indices and Machine Learning Algorithms Used

The Normalized Difference Vegetation Index (NDVI). [21] defined the normalized difference vegetation index as:

$$NDVI = (Band\ Near\ Infrared - Band\ Red)/(Band\ Near\ Infrared + Band\ Red) \quad (1)$$

Sentinel-2 Bands	Central Wavelength (µm)	Resolution (m)
Band 1 - Coastal aerosol	0.443	60
Band 2 - Blue	0.490	10
Band 3 - Green	0.560	10
Band 4 - Red	0.665	10
Band 5 - Vegetation Red Edge	0.705	20
Band 6 - Vegetation Red Edge	0.740	20
Band 7 - Vegetation Red Edge	0.783	20
Band 8 - NIR	0.842	10
Band 8A - Vegetation Red Edge	0.865	20
Band 9 - Water vapour	0.945	60
Band 10 - SWIR - Cirrus	1.375	60
Band 11 - SWIR	1.610	20
Band 12 - SWIR	2.190	20

Fig. 2. Characterization of the bands composing sentinel 2MSI [20]

The Soil Adjusted Vegetation Index (SAVI). [22] defined the Soil-Adjusted Vegetation Index as:

$$SAVI = (1 + 0.5) * (BNIR - Bred)/(BNIR + Bred + 0.5) \quad (2)$$

Machine Learning Algorithms. In this study the following machine learning algorithms were used: Random Forest (RF), support vector machine (SVM), k-nearest neighbors (KNN), Linear Discriminant Analysis (LDA), and Adaboost (ADA).

2.5 Accuracy Report

The performance of the algorithms used has been tested through the following parameters: Accuracy, Precision, Recall, and F1 score.

2.6 Approach Used

The overall approach used in this study is summarized in Fig. 3.

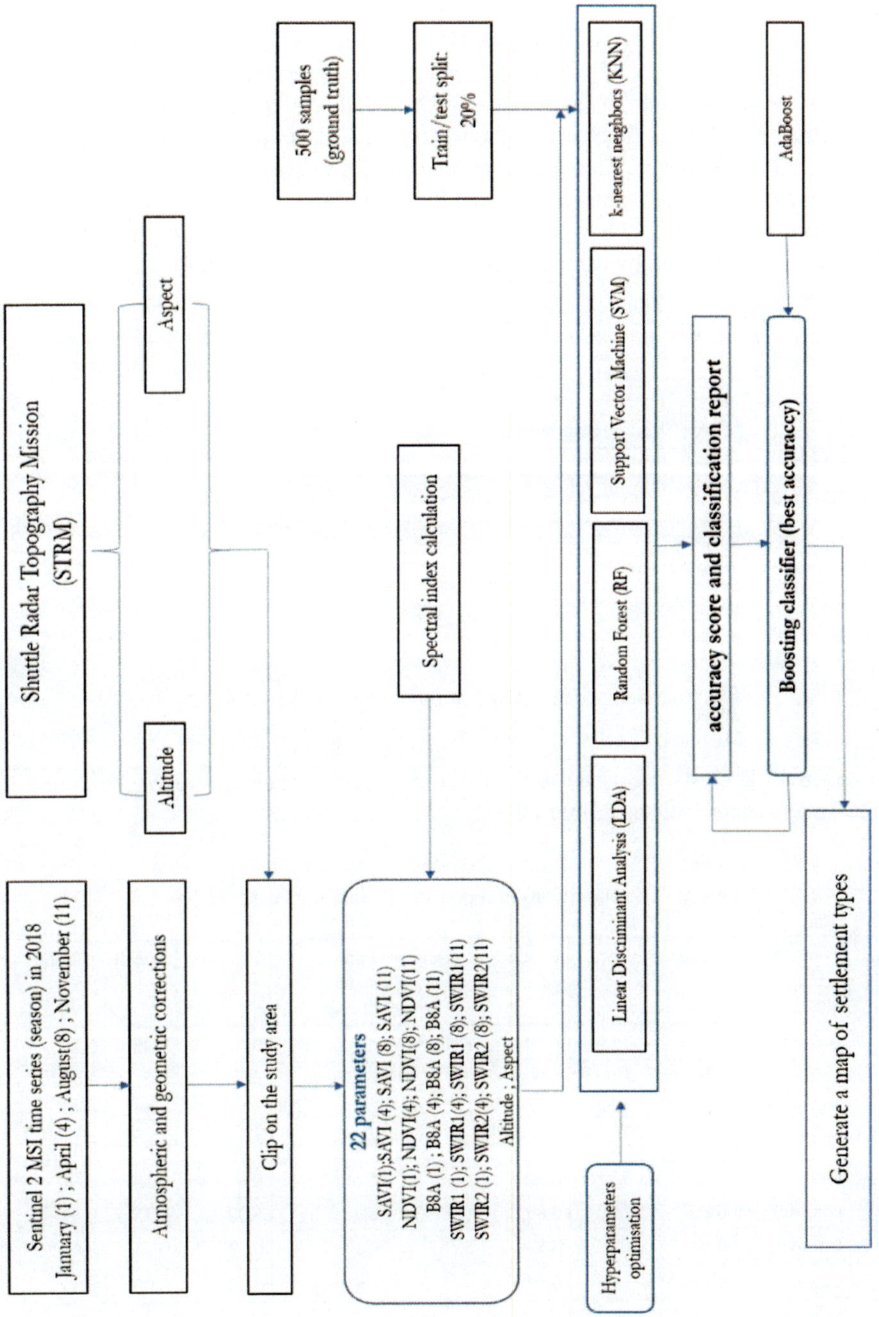

Fig. 3. Overall approach

3 Result and Discussion

Thus, the classes resulting from the classification operation using the machine learning algorithms are as follows: Ca2fj (Young cedar trees moderately dense), Ca3fv (Old cedar grove of light density), Ca2jfQr1ta (Young medium dense cedar grove mixed with mature holm oak thickets of high density), Ca3fvQr3ta (Old cedar grove of light density mixed with mature holm oak thickets of light density), Qr2tj (Young thickets of medium dense holm oak) and, bare soil. The classification result for each classifier used is related in Fig. 4. The highest classification is achieved by using Random Forest classifier with 95% of an accuracy, the second best score is achieved by using LDA classifier with 80% and lowest score is achieved by using respectively SVM and KNN with 37% and 39%.

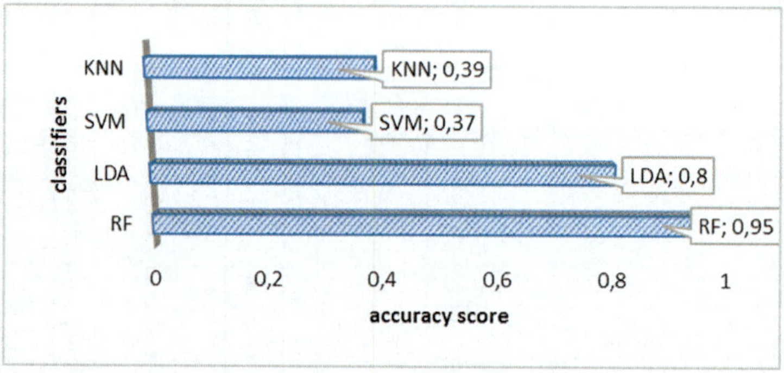

Fig. 4. Machine learning algorithm's accuracy score

To deepen the analysis, we used the classification report of each classifier according to the identified species classes (Table 1).

Table 1. The classification report of all classifiers used

Classifier	Forest stand classes	Precision	Recall	f1-score	Classifier	Forest stand classes	Precision	Recall	f1-score
RF	Ca2fj	1.00	1.00	1.00	KNN	Ca2fj	0.17	0.25	0.2
	Ca3fv	1.00	1.00	1.00		Ca3fv	0.5	0.4	0.44
	Ca2jfQr1ta	1.00	1.00	1.00		Ca2jfQr1ta	0.33	0.22	0.27
	Ca3fvQr3ta	1.00	0.88	0.93		Ca3fvQr3ta	0.33	0.5	0.4
	Qr2tj	0.92	0.92	0.92		Qr2tj	0.64	0.58	0.61
	bare soil	0.80	1.00	0.89		bare soil	1.00	0.25	0.4
LDA	Ca2fj	0.80	1.00	0.89	SVM	Ca2fj	0.29	0.75	0.41
	Ca3fv	0.71	1.00	0.83		Ca3fv	0.00	0.00	0.00

(continued)

Table 1. (*continued*)

Classifier	Forest stand classes	Precision	Recall	f1-score	Classifier	Forest stand classes	Precision	Recall	f1-score
	Ca2jfQr1ta	0.88	0.78	0.82		Ca2jfQr1ta	0.00	0.00	0.00
	Ca3fvQr3ta	1.00	0.25	0.40		Ca3fvQr3ta	0.00	0.00	0.00
	Qr2tj	0.79	0.92	0.85		Qr2tj	0.44	0.92	0.59
	bare soil	0.80	1.00	0.89		bare soil	0.00	0.00	0.00

It appears that the RF classifier has well performed in the first 3 classes with a little less in the last 3 classes.

The combination between RF and Adaboost (RF-ADA) classifiers allowed to reach 98% of accuracy and increasing the RF accuracy score about 3%. The classification report of this combination is related in Table 2.

Table 2. The classification report of RF-ADA classifier

Classifier	Forest stand classes	Precision	Recall	f1-score
RF-ADA	Ca2fj	1.00	1.00	1.00
	Ca3fv	1.00	1.00	1.00
	Ca2jfQr1ta	1.00	1.00	1.00
	Ca3fvQr3ta	1.00	1.00	1.00
	Qr2tj	1.00	0.92	0.96
	bare soil	0.80	1.00	0.89

The classification report shows that the RF-ADA classifiers in the first 4 classes with a little less in the last 2 classes.

Figure 5 presents an example of reflectance of a selected pixel for each forest species classes, and each parameter used. It appears that the reflectance of each forest stands species variates depending on the season, vegetation indices, and sentinel bands used.

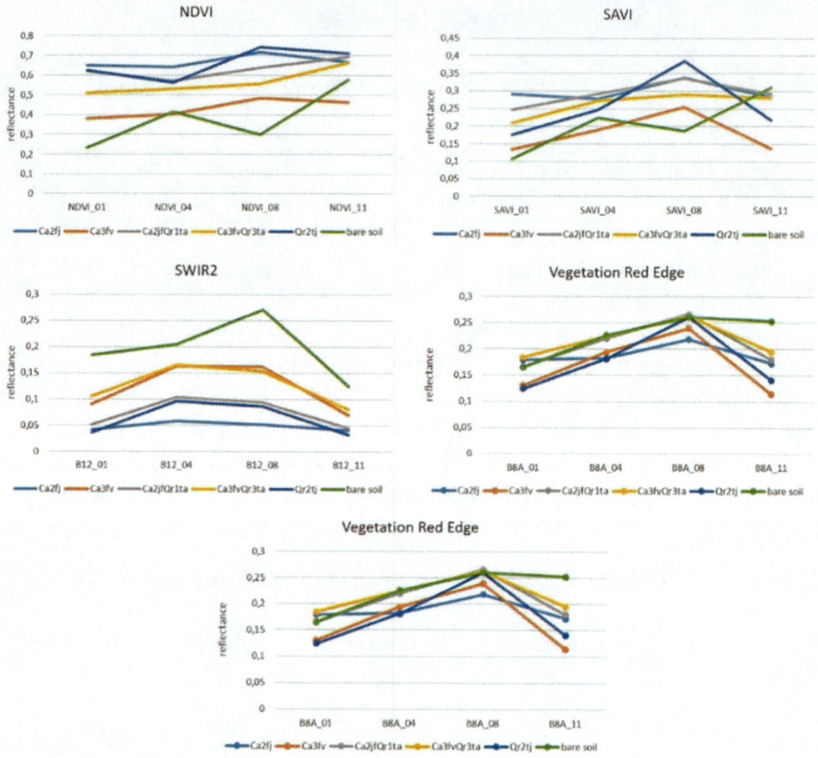

Fig. 5. An example of reflectance of a selected pixel for each forest species classes, and for each parameter used

The forest stand species Map generated after using RF-ADA classifiers is presented in Fig. 6.

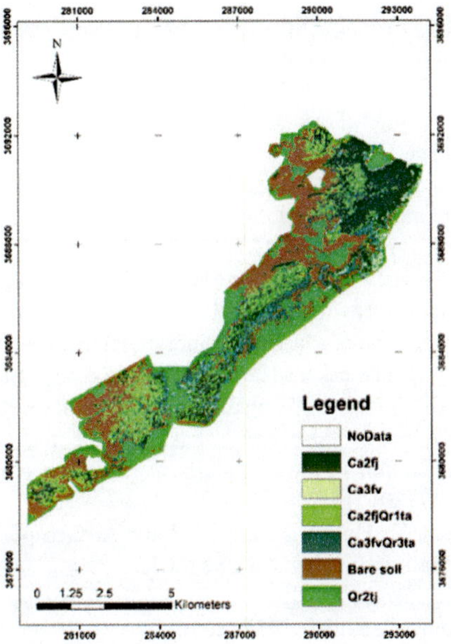

Fig. 6. Forest stand species' Map by using RF-ADA

4 Conclusion

The supervised classification using the different algorithms gave satisfactory results with a high variability. Temporal segmentation of the archive appears to be a feasible and robustness of increasing information extraction from the Sentinel archive.

According to Sentinel 2 MSI time series, we are going to push the analysis to study of other forest species that change their behavior according to climatic conditions and the seasons, especially the deciduous species.

References

1. Mickelson, J.G., Civco, D.L., Silander, J.A.: Delineating forest canopy species in the northeastern united states using multi-temporal TM imagery. Photogramm. Eng. Remote Sens. **64**, 891–904 (1998)
2. Schmitt, U., Ruppert, G.S.: Forest classification of multitemporal mosaicked satellite images. Int. Arch. Photogramm. Remote Sens. **31**, 602–605 (1996)
3. Walsh, S.J.: Coniferous tree species mapping using LANDSAT data. Remote Sens. Environ. **9**, 11–26 (1980)
4. Madonsela, S., et al.: Multi-phenologyWorldView-2 imagery improves remote sensing of savannah tree species. Int. J. Appl. Earth Obs. Geoinf. **58**, 65–73 (2017)
5. Xie, Y., Sha, Z., Yu, M.: Remote sensing imagery in vegetation mapping: a review. J. Plant Ecol. **1**, 9–23 (2008)
6. Griths, P., et al.: Forest disturbances, forest recovery, and changes in forest types across the carpathian ecoregion from 1985 to 2010 based on landsat image composites. Remote Sens. Environ. **151**, 72–88 (2014)

7. Ballanti, L., Blesius, L., Hines, E., Kruse, B.: Tree species classification using hyperspectral imagery: a comparison of two classifiers. Remote Sens. **8**, 445 (2016)
8. Ghosh, A., Fassnacht, F.E., Joshi, P.K., Kochb, B.: A framework for mapping tree species combining hyperspectral and LiDAR data: role of selected classifiers and sensor across three spatial scales. Int. J. Appl. Earth Obs. Geoinf. **26**, 49–63 (2014)
9. Dudley, K.L., Dennison, P.E., Roth, K.L., Roberts, D.A., Coates, A.R.: A multi-temporal spectral library approach for mapping vegetation species across spatial and temporal phenological gradients. Remote Sens. Environ. **167**, 121–134 (2015)
10. Fassnacht, F.E., et al.: Review of studies on tree species classification from remotely sensed data. Remote Sens. Environ. **186**, 64–87 (2016)
11. Leckie, D.G., et al.: Production of a large-area individual tree species map for forest inventory in a complex forest setting and lessons learned. Can. J. Remote. Sens. **43**(2), 140–167 (2017)
12. Pimple, U., Sitthi, A., Simonetti, D., Pungkul, S., Leadprathom, K., Chidthaisong, A.: Topographic correction of Landsat TM-5 and Landsat OLI-8 imagery to improve the performance of Forest classification in the mountainous terrain of Northeast Thailand. Sustain. (Switzerland) **9**(2), 1–26 (2017)
13. Viña, A., Liu, W., Zhou, S., Huang, J., Liu, J.: Land surface phenology as an Indicator of biodiversity patterns. Ecol. Indic. **64**, 281–288 (2016)
14. Sheeren, D., et al.: Tree species classification in temperate forests using Formosat-2 satellite image time series. Remote Sens. **8**(9), 734 (2016)
15. Hill, R.A., Wilson, A.K., George, M., Hinsley, S.A.: Mapping tree species in temperate deciduous woodland using time-series multi-spectral data. Appl. Veg. Sci. **13**(1), 86–99 (2010)
16. Immitzer, M., Vuolo, F., Atzberger, C.: First experience with sentinel-2 data for crop and tree species classifications in central Europe. Remote Sens. **8**, 166 (2016)
17. Persson, M., Lindberg, E., Reese, H.: Tree species classification with multi-temporal sentinel-2 data. Remote Sens. **10**, 1794 (2018)
18. Wessel, M., Brandmeier, M., Tiede, D.: Evaluation of different machine learning algorithms for scalable classification of tree types and tree species based on Sentinel-2 data. Remote Sens. **10**, 1419 (2018)
19. Maxwell, A.E., Warner, T.A., Fang, F.: Implementation of machine learning classification in remote sensing: an applied review. Int. J. Remote Sens. **39**(9), 2784–2817 (2018)
20. Paul, T.: Tutoriel d'initiation à la télédétection spatiale sur logiciel libre. Book (2019)
21. Rouse, J.W., Haas, H.R., Deering, D.W., Schell, J.A., Harlan, J.C.: Monitoring the vernal advancement and retrogradation (green wave effect) of natural vegetation. NASA/GSFC Type III Final Report. Greenbelt, MD, vol. 371 (1974)
22. Huete, A.R.: A soil-adjusted vegetation index (SAVI). Remote Sens. Environ. **25**, 295–309 (1988)

Cyber Security, Database and Language Processing for Human Applications

When Microservices Architecture and Blockchain Technology Meet: Challenges and Design Concepts

Idris Oumoussa[1(✉)], Soufiane Faieq[2], and Rajaa Saidi[1]

[1] SI2M Laboratory, National Institute of Statistics and Applied Economics (INSEA), Rabat, Morocco
{ioumoussa,r.saidi}@insea.ac.ma

[2] LRIT Associated Unit to CNRST (URAC 29), Faculty of Sciences, Mohammed V University in Rabat, Rabat, Morocco

Abstract. Microservices Architecture Style (MSA) and Blockchain technology are rapidly gaining ground as popular research topics in computer science, capturing the interest of researchers to explore new horizons for new applications. Integrating MSA and blockchain technology is not straightforward as the MSA faces challenges of confidentiality and functional analysis in order to bring about the emergence of autonomous and strongly cohesive functional units. Due to its embedded encryption, decentralizing feature and digital signature, the latest advances in blockchain possess the potential ability to tackle MSA design challenges. There are two primary advantages of integrating blockchain with MSA: firstly, the blockchain has the ability to handle the major challenges of MSA design; and secondly, MSA assist the blockchain thrive. This paper discusses the design concepts for creating blockchain applications adopting an MSA that may assist in mitigating these challenges, and addresses pertinent topics in a setting such as Blockchain as a Service (BaaS) architecture. Further, it also covers the open issues of using blockchain in conjunction with MSA.

Keywords: Microservices architecture · Blockchain · Smart contracts · Decentralized application · Design pattern

1 Introduction

In recent years, there has been a tremendous change in the approach used to develop and deliver applications or services. The MSA has gained a foothold in the software development industry and has become one of the latest architectural trends in software engineering [1]. Microservice architecture is an approach of application development in which a large application is broken down as suites of independently deployable services, and these services are often specialized in a single Task [2]. On another side, blockchain is essentially a distributed and decentralized database, which contains blocks of connected transactions [3]. A

R. Saidi et al. (Eds.): ICATH 2021, LNDECT 110, pp. 161–172, 2022.
https://doi.org/10.1007/978-3-030-94188-8_16

blockchain contains six layers: data, service, application, smart contract, consensus, and network. The data layer and the network layer are where data is obtained, licensed and controlled [4]. The smart contract, consensus protocol, and incentive structure functionality are part of the smart contract layer and consensus mechanism. The blockchain has five basic principles, which are the distributed database, pseudonym transparency, peer-to-peer transfer, record irreversibility and computational logic [5]. All parties have the option of including new information on the network, but the data is never erased or updated. A blockchain can be used to build a distributed system of records that have been performed and shared among the participating parties [6]. There is no central database and it is a peer-to-peer system where transactions are kept. Nobody has this information. The audit trail is immutably recorded in the blockchain.

The paradigms of microservices and blockchain share many similarities; while scaling computing resources relies heavily on microservices, blockchain acts as an enabler for real-time management and operations. The combination of blockchain and microservices will then inspire companies to maintain a higher volume of partnerships and work as a bridge between partners, without compromising the integrity and security of products and services. Blockchain can solve the complexity by working as a delegate for trust relationships. This article discusses the design concepts for creating blockchain applications adopting a microservices architectural approach. We initially overview the core principles for microservices, blockchain and smart contracts. We propose three major design principles for implementing blockchain softwares based on an architectural style of microservices. Finally, we will identify open challenges in the domain.

The remainder of this paper is organized as follows: Sect. 2 introduces the fundamental concepts of blockchain, smart contracts and microservices architecture. Section 3 delves into the design principles, and discusses relevant topics in a context such as Blockchain as a Service (BaaS). Section 4 examines the challenges of integrating these two technologies. Finally, we conclude this paper in Sect. 5.

2 Background Knowledge

2.1 Microservices Architecture

The microservices architecture style as a new computing paradigm has gained momentum in the software engineering industry. Microservices break down traditional monolithic applications into a set of fine-grained services that can be developed, tested, and deployed independently [1]. It improves the application's dependability and scalability, and establishes simple alliances, streamlines service integration, without obstructing partners or clients. According to [7], microservices are "*lightweight, independent services that perform unique functions by collaborating with other similar services through a well-defined interface*". Microservices components are built as a result of application granularity across functional boundaries and adhere to the principle of single responsibility. Each microservice communicates with each other through a lightweight mechanism, such as representational state transfer (REST) or an asynchronous message bus [8]. Each service is fully autonomous and full stack. The functional decomposition of an application helps build a well-executed microservices architecture with loose coupling and strong cohesion (multiple services can be integrated to form higher-level services or applications). Important characteristics of microservices include fine granularity, which means that each microservice can be developed using a range of frameworks such as programming languages or resources, and loose coupling implies that the functions of the components of the microservices are independent of each other [9].

The architectural style of microservices addresses the technological issues involved in achieving business process modeling goals. The MSA has been investigated in more intelligent solutions to enhance the security and scalability of applications. [10]. It was used to implement an intelligent blockchain-enabled system to support predictive analytics of personalized fitness data in an IoT environment [11]. The microservice architecture has been adopted to design smart microservices for processing IoT fitness data using blockchain approach to solve issues like data security, identity, etc. Microservices, although distributed, face privacy and security risks. The recent advancements of blockchain in microservices architecture present risk-reduction opportunities, increase efficiency through its built-in encryption as well as digital signature schemes, decentralization feature, and intrinsic incentive mechanisms [12].

2.2 Blockchain

Blockchain, introduced by [13] through the Bitcoin, enabled the implementation of new software architecture for decentralized computing that does not rely on a central trust unit. Essentially, the blockchain is a decentralized ledger system, which stores groups of transactions (blocks) and then performs transaction list binding and sequencing using cryptography. These can be public, like Bitcoin, or can be developed privately or by associations. More formally, the blockchain is a database replicated on a set of entities connected to a peer-to-peer network. Since there is no central authority (e.g. administrator), it is these entities that are in charge of maintaining the network and the database, where each entity has a unique identifier and exchanges information through transactions. The authors cryptographically sign their entities in order to guarantee their authenticity (i.e. who creates the transaction) and its integrity (i.e. ensure its inviolability).

Given the nature of the network, the sender transmits his transaction to his neighbors, with whom he is directly connected. When the latter receive it, they ensure the validity of the information before transmitting it, if necessary, in the same way to their close counterparts. This process is repeated, thus only allowing the dissemination of correct information on the network. All transactions are then ordered and grouped into blocks, and the entire blockchain network records them. Through a consensus algorithm, the entities agree on the order and content of the blocks, namely the data, the cryptographic link to the previous block and the proof of validity. Redundancies in blocks and consensus mechanisms ensure that transaction manipulation cannot be performed. Disseminated in the same way as transactions, any block considered invalid is deleted. Iteratively, the entities condition their blocks following this block, thus forming a blockchain. The blockchain has an intriguing property: once information is held in a blockchain, it is exceedingly impossible to alter it.

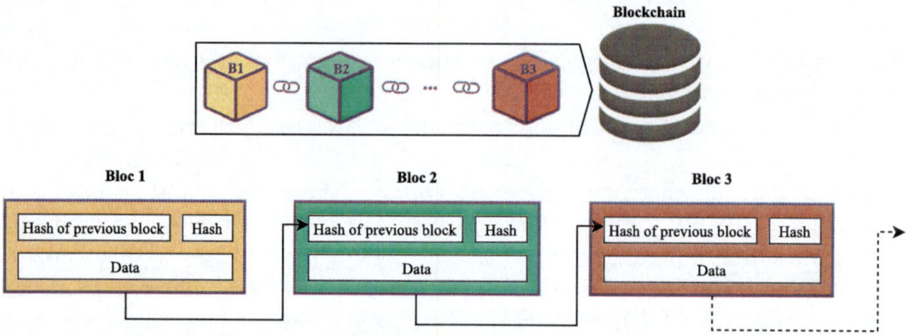

Fig. 1. General structure of a blockchain and its blocks.

In general, the structure of a blockchain and its blocks is summarized in Fig. 1. Each block contains some data, the hash of the block, and the hash of the previous block. The data that is stored inside a block depends on the type of blockchain. The bitcoin blockchain, for illustration, preserves transaction data such as sender, receiver, and coin amounts. A hash, like a fingerprint, identifies a block and all of its contents and is always unique. When a block is produced, its hash is computed; modifying anything within the block will cause the hash to change. Hashes, in other words, are extremely useful for identifying modifications to blocks. The hash of the preceding block is the third element within each block. This essentially forms a chain of blocks, and it is this mechanism that ensures the security of a blockchain.

2.3 Smart Contracts

Blockchains are constantly evolving, one of the most recent developments is the creation of smart contracts. The term *smart contracts* was first used by Nick Szabo [14]. These contracts are simple programs that are stored on the blockchain and can be used to create a system that does not require a trusted third party, as well as to automatically trade currencies based on certain conditions. While smart contracts are stored on a blockchain, they inherit interesting properties, they are immutable, and they are distributed. The fact that a smart contract is immutable implies that it can never be modified again, and that it is distributed means that the contract's output is vetted by everyone in the network. Tampering with smart contracts is becoming almost impossible. Banks, for example, could use it to issue loans or to offer automatic payments. There are a handful of blockchains that support smart contracts, today's most popular of which is Ethereum [15]. It was specifically created and designed to support smart contracts. It should be noted that Bitcoin also supports smart contracts although it is much more limited compared to Ethereum.

3 Blockchain-Based Application Architecture Design

Designing blockchain applications as microservices presents an opportunity to reduce risk, increase efficiency, and automate the execution of business logic with smart contracts. Microservices and blockchain smart contracts share many similarities. They are both supposed to run independently (on-chain) and connect with the outside (off-chain) via a message-based channel. Due to the complexity of implementing blockchain technology in various applications, the development of such applications is typically not an easy task. This section aims to present design principles that can help minimize these challenges for building blockchain applications using a microservice style of architecture, and discusses relevant topics in a context such as BaaS.

3.1 Domain-Driven Decentralized Design

To accomplish particular operations within a smart contract, blockchain applications require user and device identification as well as authorization. This approach is known as "*Decentralized Domain-Driven Design*", or DDDD, introduced by [16]. The DDDD approach proposes the notion of *delimited context.* According to this approach, a business domain can be divided into subdomains within which reside entities internal to the context, entities exchanged between contexts, processes and properties of a blockchain system. Microservices can be designed by adhering to this credo, which allows units to be self-sufficient and independent. Smart contracts are workflows that represent the blockchain application's business logic. Each step in a workflow is characterized by a state (the contract's arguments, represented as a message delivered to a function, and the contract's internal state) and a message sender (who is responsible for a specific task in the contract) [17]. The execution of a smart contract function may have an impact on other parties. When the internal state of a contract changes, an event is generated to notify other sections of the smart contract or external applications of the change.

3.2 Acquisition of Events and CQRS Pattern

Smart contracts with accountability for only one capacity are strongly suggested. While capability-driven design is an important technique for separating smart contracts, it isn't enough to assure that they can be deployed independently. Despite the fact that smart contracts are implemented in isolation, they might share a similar data model in the system domain. However, this creates a reliance between smart contracts, necessitating the use of a data modeling approach to prevent data sharing across smart contracts. We may store events that led to the present state of the application as a sequence of immutable events instead of storing structures that model the state of the domain. This modeling method is known as event sourcing [2], which is a persistent notion utilized in event architectures. The storage of facts is what event provisioning is all about. A fact is a representation of an event's worth. It enables for a comprehensive replay of all events that have transpired since the event began to be recorded. Because data is immutable, a new command or a new event is always issued to compensate for, rather than alter, an entity's status. This method uses CRAB acronym (Create, Retrieve, Append, Burn) [18], which is precisely what a blockchain does: it adds to the chain rather than updating or deleting data. The constancy of the blockchain is harmed when something is removed from it, however, the transfer of assets may be halted by *burning* the recipient's address.

Performance is an initial focus of this strategy. If a state value is a function of events, every access to the value needs the current state from the source events to be recalculated. It would, of course, be quite sluggish. Costly procedures are avoided in event sourcing by employing what is known as a rolling snapshot: a projection of the entity's state at a moment in time. The notion of *immutable single-add log*, which is viewed as one source of truth containing all of the events that have occurred, is shared by blockchain and event provisioning [19]. In this regard, throughout our study, we came across use instances in which blockchain was utilized as a trustless event store [20], as well as use cases in which blockchain transactions were saved in a traditional event store [21]. The Command-Query Responsibility Segregation (CQRS) pattern [22] is frequently referenced in conjunction with event sourcing, because event sourcing nearly always results in some kind of CQRS. This approach promotes a single, effective responsibility as well as the deployment of microservices and, as a result, intelligent contracts [20]. CQRS encourages the separation of concerns as it proposes that the data editing function in different models should be isolated from the data query function. The requirement to access data in various contexts may be removed by utilizing CQRS. Any model state update may be owned and encapsulated by a smart contract, which can then trigger events when that state changes. A separate smart contract can construct a totally isolated and efficient model for scenarios that do not need to be shared with another contract or external service by subscribing to alerts of these occurrences.

3.3 Asynchronous Communication

In a microservices design, asynchronous messaging is critical for preserving loose coupling [23]. For example, it enables the use of a message broker to send event notifications asynchronously, avoiding point-to-point connections that rely on the availability and message format of each endpoint. Similarly, appropriate communication protocol must be established for smart contracts. Asynchronous communication via messaging-based integration promotes smart contracts for incoming (external to the internal blockchain) and outgoing (blockchain to exterior applications) communication. Events can be produced to inform users and transaction systems, as well as to modify a smart contract. Given that smart contracts are a business workflow that connects with external systems and devices, transactions must be able to be launched in a distributed ledger that incorporates data from an external system or device, and external systems must be able to interact with events from smart contracts in the distributed ledger. REST API (Application Programming Interface) and message integration allows sending transactions from external systems to smart contracts integrated in a microservices-based Blockchain application, as well as sending event alerts to external systems based changes to the application [24].

Looking into our literature, the choice of a messaging model is mostly determined by the type of integration. The first type of integration we recommend is unidirectional event delivery from a smart contract to an event consumer. In this scenario, a smart contract experiences an event, such as a state change the performance of a particular sort of transaction. This event will be communicated to downstream customers using an Event Grid and then these consumers take the necessary measures. An event is generated by an external system and sent to a smart contract in the second type of integration, which works in the opposite manner. The transfer of data from financial markets to a smart contract is a popular example. Both messaging models can theoretically be utilized for blockchain integration, but the choice is influenced by the surrounding application components as well as the software stack integration effort. In general, the logic provided at endpoints should be robust regardless of the method used, so that twice the same event does not have any side-effects.

3.4 Blockchain as a Service (BaaS)

Examining recent software development practices, microservices architecture has proved useful in a variety of sectors by promoting the concept of encapsulation of functions in services. This tendency has also reached the blockchain sector, which is now referred to as BaaS. It is about ushering in a new era of cloud-native applications to make complicated systems easier to understand, where cloud service providers are widely being used to host IT infrastructure and applications, ensuring excellent abstraction of functionality, lowering costs, and enhancing operation and oversight. There are several approaches to combining functions as a service, the extensibility and reuse of blockchain can be improved by using it as a service. In order to ease the creation, tests, deployments, and continued administration of blockchain applications, BaaS was developed to enable companies to count on a service supplier to furnish and maintain blockchain infrastructure components.

Although it might be paradoxical to depend on one supplier to maintain a decentralized blockchain based system, because a kind of recentralization is involved (putting faith in those in charge of infrastructure). Provider foreclosure is perilous, too, since it is hard or even very costly (since BaaS does not have standards) to shift to another service provider. Existing BaaS systems, while still in their infancy, were typically implemented in a four-tier design, as depicted in Fig. 2 (from high to low), as follows:

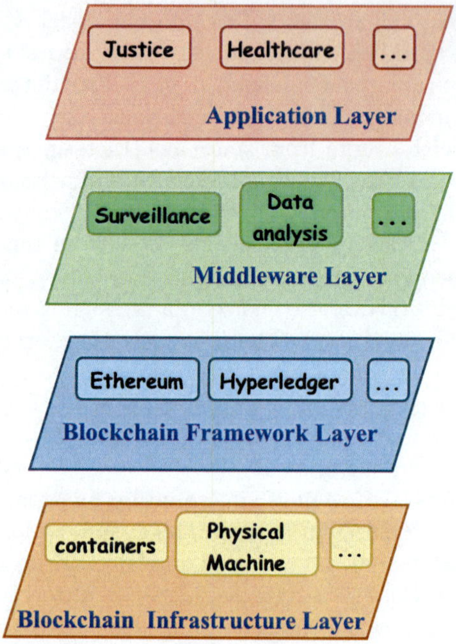

Fig. 2. Baas structure.

- **Layer of blockchain application.** The application layer sits on top of BaaS architecture. Over the core blockchain architecture, applications based on blockchain are created. As a result, developers may concentrate on application core functionality, while BaaS suppliers have to install decentralized applications on top of a blockchain infrastructure.
- **Layer of middleware.** This layer serves as an intermediary between the application and framework layers. Surveillance, and management of resources are among the fundamental system handling functions provided by this layer [25]. Data may be gathered and evaluated for the development of blockchain applications maintained on the blockchain network [26].
- **Layer of blockchain Framework.** The blockchain infrastructure based framework is designed in this level. Functionality may be implemented in particular program codes and operated on top of reliable settings by designers and developers. The BaaS architecture is compatible with major smart contract systems such as Hyperledger Fabric, EtherDelta, and others.
- **Layer of infrastructure blockchain.** To run smart contracts, blockchain is fundamentally a distributed system with matching processing resources. This makes it possible to develop an application with the layer of infrastructure without needing to create the base network from start.

Many renowned IT firms like Amazon and Microsoft, are already providing BaaS options, permitting blockchain nodes to be operated through their own infrastructures [27]. A third-party infrastructure is still being used by a few of them. A comprehensive range of applications for operations, node and smart contract management is also available on many platforms. The level of integration built-in blockchain functionality varies amongst service suppliers. With various inclusion techniques, certain companies enable wide-reaching connection with other efficient services using various inclusion strategies.

4 Future Directions

Despite all the advantages, there is a plethora of factors that not all organizations prefer to opt for the microservices blockchain alliance. The biggest impediment to putting this magnificent combo into action is the disparity between their native surroundings: even being distributed, most centralized platforms continue to be serviced by microservices; blockchain Technology, on the other hand, provides a decentralized network without the need for authority or consent. In terms of security, storage and validity of the information, there are a fair bit of unresolved concerns to be handled, which will be explored in further depth below.

4.1 Validity of the Information

The integrity of data is a serious challenge in microservices architecture. Smart contracts cannot protect the integrity of microservices-generated data; however, they assure blockchain data authenticity. Although microservice systems yield many advantages in terms of dynamic data concordance and efficacious verification via a hashed index authentication process, data consistency in the system is hard to ensure [28]. Data subscribers simply look for a chopped-key-value index from the blockchain in the authentication procedure and verify the index to the computed hash data they received. In order to overcome this challenge, the [29] community created Oracle, a data exchange broker that allows data to be shared throughout the chain. As data carrier, it operates to mediate in the course of inter-organizational business operations activation of smart contracts and external services. Whilst authentication has not yet been properly handled in this agent, numerous suggestions based on Oracle have been proposed.

4.2 Data Storage

Blockchains are used to store all the information that needs to be secured on a network, which may soon overflow storage capacity and lead to the frenetic development of new blocks. The removal of old blocks appears appealing, but it has a significant influence on the sustainability and security of the blockchain, because it relies on previous ones to confirm new blocks. A split of the blockchain can be taken into consideration to alleviate the problem of storage scalability because of the enormous amount of data; the technique consists of repeating the blocks just on a subset of the nodes, rather than on all of them. This reduces storage capacity at the expense of decentralization and security. Another approach makes use of the principles of shards [30] and sidechains [31]. Whereas these strategies are operated for scalability, the local storage capacity is less pressured as only data from shards/sidechains is handled.

4.3 Security

The security risks of blockchain systems designed on the basis of a microservices architecture are increasing due to the virtualization of microservices through containers but also by the availability of on-chain data to everyone, despite the fact that the blockchain provides pseudonyms. Therefore, special attention to privacy and security while ensuring effective data sharing is required during design and implementation [32]. Exploring various techniques such as ring signature or homomorphic encryption is essential for developing solutions to blockchain data privacy issues. In [33] the authors provide a collection of popular security principles that may be used to mitigate typical attack situations as a solution to particular security concerns. A microservice based security method has been suggested to ensure data sharing and payment among participants inside an approved blockchain network for Decentralized Data Marketplaces (BlendSM - DDM) [34]. In order to maintain security while implementing an appropriate learning process for the accessibility of trading data and, consequently, preserving confidentiality of data providers, Zhao et al. [35] introduced a procedure for combining several schemes such as dual authentication signature and ring signature.

5 Conclusion

This article highlights the feasibility and opportunity to develop systems that follow the microservices architecture, using blockchain smart contracts. We investigated state-of-the-art literature for building blockchain applications using a microservice architecture style and summarize three major design principles: i) Decentralized Domain-Driven Design, ii) Event Sourcing and CQRS, and iii) Asynchronous Messaging. The main properties of the emerging blockchain technology have the ability to protect data transmitted across microservices while also allowing the operator to trace the data and eliminate data fraud. Moreover, we briefly introduce BaaS and its architecture. In addition, we also outline open challenges of blockchain-based microservices architecture style. In summary, we believe that blockchain technology and microservices architecture works perfectly towards a future with open and rapid collaboration and limited data risk.

References

1. Alshuqayran, N., Ali, N., Evans, R.: A systematic mapping study in microservice architecture. In: IEEE International Conference on Service-Oriented Computing and Applications (SOCA), pp. 44–51. IEEE (2016)

2. Martin Fowler Homepage. https://martinfowler.com/articles/microservices.html. Accessed 28 Aug 2021
3. Samaniego, M., Deters, R.: Blockchain as a service for IoT. In: 2016 IEEE International Conference on Internet of Things (iThings), pp. 433–436. IEEE (2017)
4. Khan, P.W., Byun, Y.C., Park, N.: A data verification system for CCTV surveillance cameras using blockchain technology in smart cities. Electronics **9**(3), 484 (2020)
5. Iansiti, M., Lakhani, K.R.: The truth about blockchain. Harv. Bus. Rev. **95**(1), 118–127 (2017)
6. Crosby, M., Pattanayak, P., Verma, S., Kalyanaraman, V.: Blockchain technology: Beyond bitcoin, 2nd edn. Applied Innovation, New York (2016)
7. Kecskemeti, G., Marosi, A.C., Kertesz, A.: The ENTICE approach to decompose monolithic services into microservices. In:2016 IEEE International Conference on High Performance Computing and Simulation (HPCS), pp. 591–596. IEEE (2016)
8. Lu, D., Huang, D., Walenstein, A., Medhi, D.: A secure microservice framework for IoT. In: 2017 IEEE Symposium on Service-Oriented System Engineering (SOSE), pp. 9–18. IEEE (2017)
9. Yu, D., Jin, Y., Zhang, Y., Zheng, X.: A survey on security issues in services communication of microservices-enabled fog applications. Concurr. Comput.: Pract. Exp. (2018)
10. Nagothu, D., Xu, R., Nikouei, S.Y., Chen, Y.: A microservice-enabled architecture for smart surveillance using blockchain technology. In: 2018 IEEE International Smart Cities Conference (ISC2). IEEE (2018)
11. Jamil, F., Qayyum, F., Alhelaly, S., Javed, F., Muthanna, A.: Intelligent microservice based on blockchain for healthcare applications. CMC-Comput. Mater. Continua **69**(2), 2513–2530 (2021)
12. Li, X., Zheng, Z., Dai, H.N.: When services computing meets blockchain: challenges and opportunities. J. Parallel Distrib. Comput. **150**, 1–14 (2021)
13. Nakamoto, S.: Bitcoin: a peer-to-peer electronic cash system (2008)
14. Szabo, N.: Smart contracts: building blocks for digital markets. EXTROPY: J. Transhumanist Thought (16) **18** (1996)
15. Go ethereum Homepage. https://github.com/ethereum/go-ethereum/. Accessed 12 Aug 2021
16. Evans, E.: Domain-Driven Design: Tackling Complexity in the Heart of Software. Addison-Wesley Professional, Boston (2004)
17. Zheng, Z., et al.: An overview on smart contracts: challenges, advances and platforms. Futur. Gener. Comput. Syst. **105**, 475–491 (2020)
18. Demchenko, O.: CRAB technology platforms and CRAB technology based smart contracts: benefits, ways of application, legal challenges and future development. Pécs Journal of International and European Law (2018)
19. Wöhrer, M., Zdun, U.: Architecture design of blockchain-based applications. In: 2021 International Conference on Blockchain and Cryptocurrency (2021)
20. Transmute Framework Homepage. https://github.com/transmute-industries/transmute. Accessed 20 Aug 2021
21. Ocean Bounty Homepage. https://explorer.bounties.network/bounty/2146. Accessed 10 July 2021
22. Microservices Patterns Homepage. https://microservices.io/patterns/index.html. Accessed 10 July 2021
23. Joseph, C., Chandrasekaran, K.: Straddling the crevasse: a review of microservice software architecture foundations and recent advancements. Softw. Pract. Exp. **49**(10), 1448–1484 (2019)

24. Tonelli, R., Lunesu, M.I., Pinna, A., Taibi, D.: Implementing a microservices system with blockchain smart contracts. In 2nd International Workshop on Emerging Trends in Software Engineering for Blockchain (2019)
25. Zheng, P., Zheng, Z., Luo, X., Chen, X., Liu, X.: 2018 IEEE/ACM 40th International Conference on Software Engineering: Software Engineering in Practice Track (ICSE-SEIP), pp. 134–143. IEEE (2018)
26. Chen, W., Zheng, Z., Cui, J., Ngai, E., Zheng, P., Zhou, Y.: Detecting Ponzi schemes on Ethereum: towards healthier blockchain technology. In: Proceedings of the 2018 International World Wide Web Conference Committee, pp. 1409–1418. ACM (2018)
27. Kernahan, A., Bernskov, U., Beck, R.: Blockchain out of the Box - where is the blockchain in blockchain-as-a-service? In: 54th Hawaii International Conference on System Sciences (HICSS), pp. 4281–4290 (2021)
28. Safina, L., Mazzara, M., Montesi, F., Rivera, V.: Data-driven workflows for microservices: genericity in Jolie. In: Proceedings of the 30th International Conference on Advanced Information Networking and Applications (AINA), pp. 430–437. IEEE, Crans-Montana (2016)
29. Oraclize Homepage. http://docs.oraclize.it. Accessed 04 Aug 2021
30. Luu, L., Narayanan, V., Zheng, C., Baweja, K., Gilbert, S., Saxena, P.: A secure sharding protocol for open blockchains. In: Proceedings of the 2016 ACM SIGSAC Conference on Computer and Communications Security, Austria, pp. 17–30 (2016)
31. Back, A., et al.: Enabling blockchain innovations with pegged sidechains (2014)
32. Casale, G., et al.: Current and future challenges of software engineering for services and applications. In: Procedia Computer Science, pp. 34–42. Elsevier, Madrid (2016)
33. Wohrer, M., Zdun, U.: Smart contracts: security patterns in the ethereum ecosystem and solidity. In 2018 International Workshop on Blockchain Oriented Software Engineering (IWBOSE), pp. 2–8. IEEE (2018)
34. Xu, R., Ramachandran, G.S, Chen, Y., Krishnamachari, C., Krishnamachari, B.: BlendSM-DDM: blockchain-enabled secure microservices for decentralized data marketplaces. In 2019 IEEE International Smart Cities Conference (ISC2). IEEE (2019)
35. Zhao, Y., Yu, Y., Li, Y., Han, G., Du, X.: Machine learning based privacy-preserving fair data trading in big data market. Inf. Sci. **478**, 449–460 (2019)

Ensuring the Integrity of Cloud Computing Against Account Hijacking Using Blockchain Technology

Assia Akamri(✉) and Chaimae Saadi

Laboratory of System Analysis, Information Processing and Industrial Management, Sale High School of Technology, Mohammed V University, Rabat, Morocco
assia.akamri@uit.ac.ma

Abstract. Cloud computing is a model of the internet that allows us to store our data on a remote server rather than on a local server, It helps us to access data from anywhere with an internet connection. This cloud has several features such as: easy to access, on demand services, high reliability, and elasticity. However, it has some disadvantages, among them are: Not all the data that we access to is secure, also it is easy for the hackers to attack this data. Account hijacking is one of the major attacks in the cloud, and it is considered as one of the most famous technologies that keeps and maintains the integrity of data. Blockchain technology is a digital distributed ledger that collects data in the form of blocks that are linked together with a hash. In this paper, we propose a decentralized cloud that combines two technologies: Blockchain and Private Cloud Computing in order to ensure the data integrity of the user by encrypting the authentication, and the document before send it to the cloud.

Keywords: Cloud computing · Blockchain · Distributed ledger · Security · Session hijacking · Data integrity

1 Introduction

Cloud computing is a model for enabling access to computing resources that evolved in information Technology, and has become a dominant business model for delivering IT infrastructure, components, and applications [1]. In consonance with the National Institute of Standards and Technology defines cloud computing as a model for enabling convenient, on demand network access to a shared pool of customizable computing resources that can be rapidly provisioned and released with minimal management effort or service provider interaction [2] as you can see in the Fig. 1.

R. Saidi et al. (Eds.): ICATH 2021, LNDECT 110, pp. 173–181, 2022.
https://doi.org/10.1007/978-3-030-94188-8_17

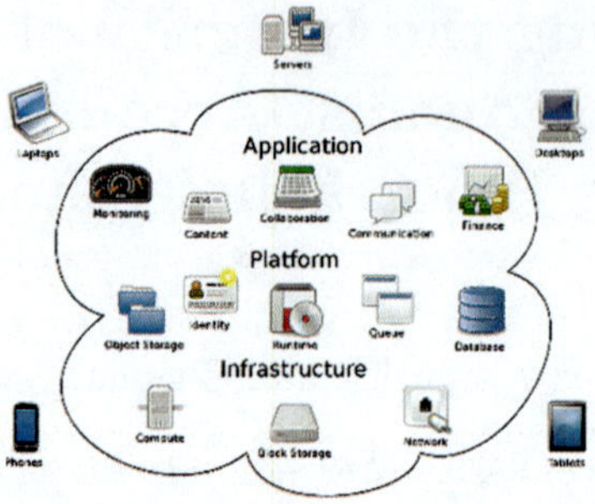

Fig. 1. Cloud computing architecture [3]

Despite all these features, anything connected to the internet is not totally secure, even servers with invincible security are vulnerable to attacks [4]. The SaaS cloud computing architecture is the subject of this study. The remainder of the article is structured as follows. Section 2 discusses some relevant works and the current cloud storage security solution proposed by various researchers. In Sect. 3, we present our planned work, which includes combining Blockchain technology with cloud computing to assure data integrity in the cloud, as well as certain testing and findings. Finally, we present a summary of this paper as well as some future plans in Sect. 4.

2 Related Works

This section describes the existing solution proposed by various researchers for integrating blockchain technology with cloud computing. Then we provide a table that compares this solution, as well as the disadvantages and advantages of each article. Finally, we discuss our proposed architecture in terms of both technologies.

We begin with the enigmatic [Satoshi Nakamoto], the Inventor of Bitcoin who proposed the first Cryptocurrency without relying on a third party replaced by asymmetric cryptography, so as to resolve the problem of "Double spending" using a peer-to-peer Network [5].

The following article The authors [Nikita Sanghi], [Gaganjot Kaur], [Rupali Bhatnagar] and [Vinay Jain] focused on the anonymity of user data in the Bitcoin Cryptocurrency's electronic wallet. The issue arises when users log into their wallet using addresses in order to delete their account. The account has not been permanently deleted. Because hackers can recover all of their information and use it to their advantages, the authors of this article solve this problem by preserving the anonymity of the user by unifying two technologies [4].

The paper's authors [Ashok Gupta],[Shams Tabrez Siddiqui] and [Shadab Alam] discussed the overall architecture of Blockchain technology, with a particular focus on the security of electronic wallet Blockchain, as well as data transmission, saving, and integrity and privacy in cloud computing. The integration of both technologies provides a means to erase the remaining user data effectively and securely when the e-wallet is not utilized [6].

To strengthen cloud data security, the authors [Lawanya Shri] and [Seifedine Kadry] combined both technologies in this paper. In the new proposition, users can connect to web applications via the application layer. When the user makes a transaction in his/her account, the transaction's details are grouped together in a block, the network nodes will then check this block [7] (Table 1).

Table 1. Comparison between the articles [4–7].

Paper Title	Advantage	Disadvantage
Bitcoin: A Peer to Peer Electronic Cash System	• Without third party	• Mining operations require high performance
	• First Cryptocurrency	
	• More Secure	
	• Transaction control	
	• Transparent	
	• Transparent	
BlockCloud Blockchain with Cloud Computing	• Permanent deletion of an account	• 51% attack
Cloud Computing Security using Blockchain	• The security of the electronic wallet in the mobile devices	• There is no architecture proposed
Blockchain Based Cloud Computing: Architecture and Research Challenges	• Integration of cloud Blockchain technology	• The implementation is difficult and takes a long time

In cloud computing, when a user connects to a web application to check his data, a session_id is created at that time, then the hacker may quickly be able to access the user's servers, and determines the id of that user in order to act as the owner of this session, then accesses to sensitive information like an email account then falsifies whatever he wants consequently, there is a loss of the integrity.

The figure [4] depicts the user attempting to connect to the cloud in order to access his data. When he connects, an attack is sent to the same server request with the same id as when the legitimate user connects. As a user, he connects to the cloud and can perform any operation (Fig. 2).

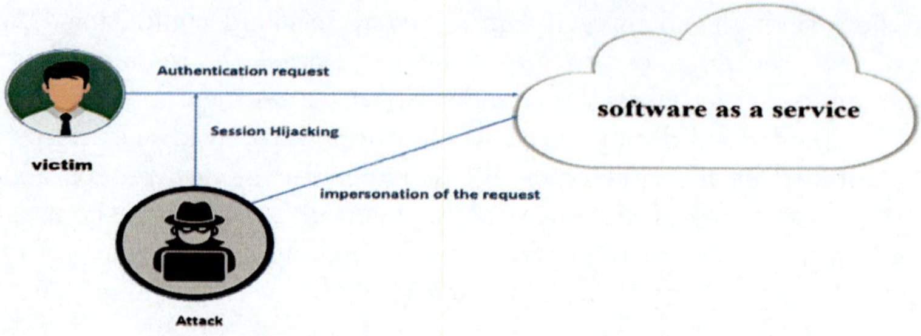

Fig. 2. Session hijacking in cloud computing

3 Proposed Work

3.1 Blockchain Technology

Blockchain is the technology behind Cryptocurrency Bitcoin which appeared in 2008 [8]. It's a digital distributed ledger that records all the transactions in the peer-to-peer network without the need for a third party to validate the transactions [5]. Instead, cryptography is used to validate the transactions, the network's participants and nodes, known as miners, are responsible for validating and creating new blocks using the POW consensus algorithm (Proof of work). When new blocks are added, they are reflected in all network copies of the ledger. The p2p network governs the Blockchain, there is no centralized database [6]. The following characteristics are associated with Blockchain technology:

- **Disintermediation:** The Blockchain eliminates the third party that allows this trusted third party to be subrogated through a mechanism of consensus or validation of transactions that circulates within its network.
- **Enhance Security:** All information in the Blockchain is encrypted. Moreover, every block in Blockchain links with other blocks with a hash. Then any change in the ledger means changing all the block [9].
- **Immutable or irreversible**: when data is validated, and it is added to the Blockchain, it is impossible to change or delete it [9].
- **Distributed:** Using a distributed database means the data is replicated on all nodes in the network [9].
- **Transparency:** Because the public Blockchain is open-source software, any node can view transactions and the source code. They can even use the code to create new applications and suggest code improvements. Suggestions are accepted or rejected with the help of consensus [10].
- **Anonymity:** no other information is required unless The Blockchain address of the miner, that leads to the protection of the user's identity [10].

3.2 Blockchain Architecture

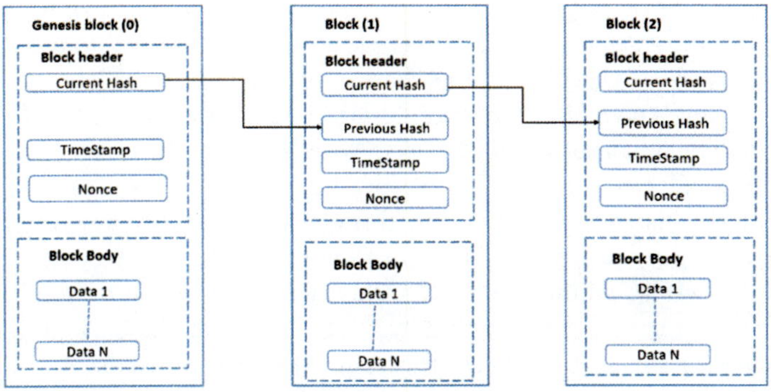

Fig. 3. Blockchain architecture

Each block within the Blockchain is linked to the other blocks by a hash [11]. Blockchain block headers include four attributes:

- **"Current Hash"**: the hash of Block.
- **"Previous Hash"**: this hash contains all the information of the preceding block [11].
- **"timestamp"**: the date of the creation of the block.
- **"Nonce"**: is A random number that used in mining processes [11] (Fig. 3).

The block body is composed of a List of Data (Fig. 4).

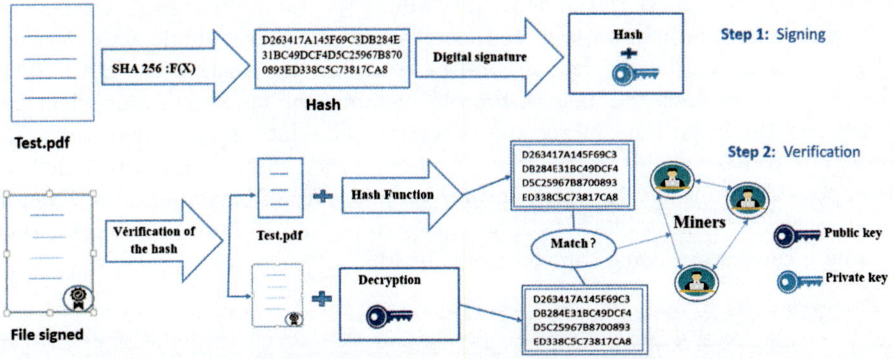

Fig. 4. How does a Blockchain work

Digital signatures are used in Blockchain technology to safeguard the communication channel between the sender and receiver of data in order to demonstrate that the message which is received from the sender is in its original form, and that it had not been tampered with, its role is to prevent hacking, as a result, to ensure the data's integrity, this technique consists of two steps, as you can see on the figure [4] the first one is the use of mathematical hash function, its role is to convert the string file into a shorter file with a fixed-length, [12] this latter represents the original file, The following step explains how the sender can encrypt this Hash by using his/her private key, so the file is signed,

The last step is to verify the data through nodes of network p2p, the node uses the public key to decrypt the data then recalculates the hash of this data that leads to two possibilities: if the hash matches with the original hash then the sender's data is valid, so the node will add it into the blockchain. If the hash does match with the original hash, then the node will refuse it.

3.3 Integration of Cloud and Blockchain

The solution is to create a decentralized cloud, which functions similarly to a centralized cloud but replaces the third party with asymmetric encryption. D'Apps (a decentralized application) data storage is created by developers within an organization using the new Blockchain technology, which is built on a peer-to-peer network. Network peers are taking advantage of their unused computing resources by acting as cloud service providers. A decentralized storage solution based on the Blockchain encourages users with unused hard drive space to host customer data. Customers are the individuals who pay for storage. Peers, also known as miners who control the transactions of the customers in their network, are the people who rent out their storage and host data for the client. Before sending this data to the cloud, a customer first encrypts the files on their computer with the private key and signed it. The files are then broadcast to the network As you can see in the figure [5]. Miners (peers) validate with the help of user public key verification, miners (peers) verify the user's public keys. To save the data block to the Blockchain, you have to find the hash of the block. The first Miner who finds the hash of the block claims the other pairs for which he has found the hash, then he receives a reward. The data is appended to all the pairs of database copies in the form of blocks. Once the document is added, it is impossible to modify it, which ensures the integrity of customer data. Even if an attack will falsify a block, it is necessary to find all the previous blocks, this is where Blockchain comes out on top (Fig. 5).

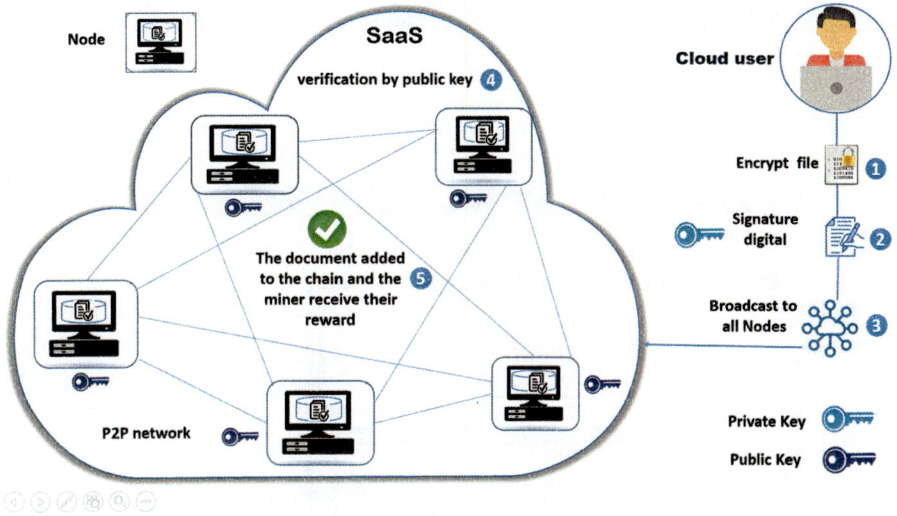

Fig. 5. Integration Blockchain with Cloud

4 Test and Results

To prevent the session hijacking attack, we build a distributed cloud based on Blockchain technology. This distributed cloud requires the client to have an account, which he must create if he does not already have one. This account page is Fig. 6 to encrypt the user password through authentication, this authentication is based on the Secure Hash Algorithm (Sha-256) to control the session id of the user.

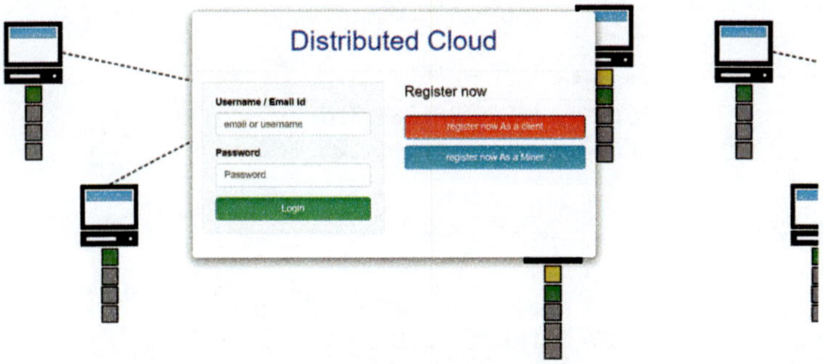

Fig. 6. Login page

Before submitting the file to the networks to be registered in the Blockchain, the client encrypts it first Fig. 7, then the file's hash appears in a column in form of numbers, and finally the client signs the file using digital signature then they send it to the miners.

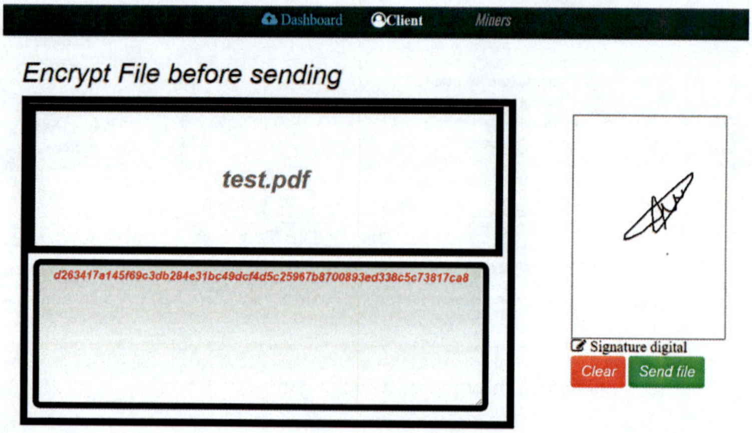

Fig. 7. The client Space

The miners check the file that the client sends with the help of the public key to decrypt the data then recalculates the hash of this data that leads to two possibilities: if the hash matches with the original hash then the sender data is valid so they will add it into the Blockchain. If the hash does match with the original hash, then the miners will refuse it.

Table 2. Test and results

File title	File Hash	Digital Signature
Test.pdf	d263417a145F69c3db284e31bc49dcf4d 5c25967b8700893ed338c5c73817ca8v	
Blockchain.pdf	6b662367fa36e748a9d5af261623b9b10d5c 03f4c7509bf4d298e122188795	
Article.pdf	03f4c7509bf4d298e122188795edb 50bb7ef8f32df5ac58a4829f121b0e82904	

Table 2 shows the Digital Signature results of Blockchain. All of these outcomes are satisfied, ensuring data integrity, because each file has its own signature as well as its own hash that is distinct from the others.

5 Conclusion and Future Work

This work discussed in the first section: the definition of cloud computing including the SaaS service model and some related works in the next section added to this we proposed a solution that combined both technologies Cloud computing with Blockchain technology to enhancing security in the cloud computing specifically the integration of data to prevent account hijacking for future work, implementation of the proposed solution.

References

1. Suwanyukabordin, P.: Cloud Computing, pp. 195–236 (2020). https://medium.com/@ponlawatsuwanyukabordin/cloud-computing-12484d88c93f
2. Count, C., Citations, R., Rank, A.J.: SP 800–145 the NIST definition of cloud computing, pp. 8–9 (2021)
3. Erl, T.: Cloud computing? Cloud Computing?, vol. 17, no. I, p. 55 (2012). https://en.wikipedia.org/wiki/Cloud_computing#/media/File:Cloud_computing.svg
4. Sanghi, N., Bhatnagar, R., Kaur, G., Jain, V.: BlockCloud: blockchain with cloud computing. In: Proceedings - IEEE 2018 International Conference Advances in Computing, Communication Control and Networking, ICACCCN 2018, pp. 430–434 (2018). https://doi.org/10.1109/ICACCCN.2018.8748467
5. Monti, M., Rasmussen, S.: RAIN: a bio-inspired communication and data storage infrastructure. Artif. Life **23**(4), 552–557 (2017). https://doi.org/10.1162/ARTL_a_00247
6. Srilakshmi, K., Bhargavi, P.: Cloud computing security using cryptographic algorithms. Asian J. Comput. Sci. Technol. **8**(S3), 76–80 (2019). https://doi.org/10.51983/ajcst-2019.8.s3.2082
7. Murthy, C.V.B., Shri, M.L., Kadry, S., Lim, S.: Blockchain based cloud computing: architecture and research challenges. IEEE Access **8**, 205190–205205 (2020). https://doi.org/10.1109/ACCESS.2020.3036812
8. Atzori, M.: Blockchain technology and decentralized governance: is the state still necessary? J. Gov. Regul. **6**(1), 45–62 (2017). https://doi.org/10.22495/jgr_v6_i1_p5
9. Iredale, G.: 6 key blockchain features you need to know now. 101 Blockchains, pp. 1–11 (2020). https://101blockchains.com/introduction-to-blockchain-features/
10. Joshi, A.P., Han, M., Wang, Y.: A survey on security and privacy issues of blockchain technology. Math. Found. Comput. **1**(2), 121–147 (2018). https://doi.org/10.3934/mfc.2018007
11. Hmimou, H.: Hicham Hmimou La Blockchain: Applications Dans (2018)
12. Chowbe, V.S.: Digital signature. Digit. Enterp. 257–283 (2021). https://doi.org/10.1201/9781003203131-23

Lightweight-Blockchain for Secured Wireless Sensor Networks: Energy Consumption of MAC Address-Based Proof-of-Authentication

Yves Frédéric Ebobissé Djéné[1,2](✉), Mohammed Sbai EL Idrissi[3], Pierre-Martin Tardif[3], Brahim El Bhiri[2], Youssef Fakhri[1], and Younes Karfa Bekali[4]

[1] LRI, Ibn Tofail University, Kenitra, Morocco
{yesfrederic.ebobissedjene,fakhri}@uit.ac.ma
[2] SMARTiLab, EMSI, Rabat, Morocco
[3] Department of Computer Science, Sherbrooke University, Sherbrooke, QC, Canada
{mohammed.sbai.el.idrissi,pierre-martin.tardif}@usherbrooke.ca
[4] LCS, Faculty of Sciences, Mohammed V University in Rabat, Rabat, Morocco
y.bekali@um5r.ac.ma

Abstract. A wireless sensor network is a self-organized infrastructure network connected to a large number of devices. These devices are cheap power constrained sensors, which collect data from supervised areas and transmits it in real-time to the base station for processing and decision-making. Due to constrained resources of sensor nodes, WSN has become a vulnerable target of many security attacks. Thus, security and confidentiality in WSN systems are an important topic of research. This study presents a robust authentication technique to prevent WSN attacks. Using the security advantages provided by blockchain, the present study proposes Lightweight-Blockchain for Secured Wireless Sensor Networks (LBSWSN) using Contiki Cooja. The impact of cryptographic techniques used in blockchain was also evaluated in order to provide security in centralized communication models. The use of LBSWSN slightly increased the mean cumulative energy in the network, compared with normal unicast communications. The increase remained stable at 4% for a network of 20 nodes and 2% for 50 nodes for each round. The FDN (First Dead Node) and LDN (Last Dead Node) were also close to the value of normal unicast messaging.

Keywords: WSN · MAC address · Blockchain · Authentication · Security · Confidentiality · Integrity · Proof-of-Authentication · Contiki · Cooja

---

Y. F. Ebobissé Djéné—The authors would like to thank SMARTiLab/EMSI and Public Safety Canada for support and infrastructure.

R. Saidi et al. (Eds.): ICATH 2021, LNDECT 110, pp. 182–192, 2022.
https://doi.org/10.1007/978-3-030-94188-8_18

1 Introduction

Wireless Sensor Networks (WSNs) have been rapidly developed and widely deployed due to their low cost and adaptability to a variety of challenges. WSNs are used in diverse fields of civil applications. They are rapidly installed and used to control the environment in smart industries [1,2] energy-efficient systems [4], smart cities, health infrastructures, traffic control, buildings, smart homes, etc. based on autonomous devices [3]. The WSN contains a large number of sensor nodes, which are spatially dispersed in the environment. The objective of this network is to collect the measurements of the physical environment, then process and transmit the collected data to the sink. Many problems related to WSNs are with respect to energy consumption, computation complexity, communication, storage and security requirements.

WSN security is increasingly gaining the attention of many researchers. It is considered to be a sensitive and important subject due to specific requirements, in particular, minimal energy consumption, low computing and storage resources, with limited hardware functionalities. In addition, WSN security is crucial because, after deployment, nodes are hardly maintained manually or visually inspected. Therefore, breaches in the communication network may occur. WSN vulnerability is critical because data needs to remain confidential and when it is not the case, authorized users can access accumulated data using a simple request to a sensor node. On the other hand, the shared wireless medium significantly increases the probability of malicious attacks [5]. These networks are considered vulnerable environments, especially because sensors have limited resources in terms of computing, memory and network capacity. They were not initially meant to be exposed to the internet, which often makes them vulnerable to cyber-attacks, such as the execution of unauthorized commands, MITM "Man In The Middle" attacks, DoS, and replay attacks [6].

Many studies currently present issues related to security in WSN [7,8]. Traditional security mechanisms require large computation and communication and therefore can hardly be applied to wireless sensor networks. Thus, WSNs face new security challenges compared with traditional networks [9]. The traditional blockchain provides a safer architecture based on Proof of Work (PoW) as a consensus process. However, this process is costly in terms of computational resources, thus requiring high speed hardware and a substantial amount of energy and cannot be implemented in a distributed network with limited resources [10]. The use of the new consensus algorithm based on Proof of Authentication (PoAh) which uses Mac address Hashes as a means of authentication, makes blockchain lightweight suitable for resource-constrained devices [11]. The combination of this new method with symmetric cryptography - considered lighter and widely accepted for infrastructure with limited resources - increases security of wireless sensor networks without a major impact on energy consumption or latency.

2 Methodology

The main objective of this paper is to provide authentication as well as data integrity and confidentiality in a WSN using MAC addresses of nodes and cryptographic algorithms. Operations are performed on each node and on the sink. Each node performs 3 basic operations:

- **MAC Address Generation:** each node is required to have a unique address MAC. This is critical as the sink maintains a table of trusted nodes. An attacker must not be able to forge an already assigned MAC.
- **Hash Generation:** Each node computes a hash of its MAC Address and of the data sent to the sink. In case the data is altered during transmission, the sink can check the data and MAC integrity by computing the hash of received data and comparing it with the received hash. If the data has not been altered, the comparison will match. The same principle is used for the MAC hash to ensure authentication. During simulations, a 64 byte packet size was used (Fig. 1).
- **Data Encryption:** to ensure data confidentiality between a node and the sink, the final data (a concatenation of MAC Hash, Message Hash) is encrypted on each node and sent to the sink.

Fig. 1. Data format

On the other side, the sink performs the following actions:

- **Data decryption:** Received data is deciphered using the reverse function of the encryption algorithm
- **MAC Address hash and Data extraction:** MAC hash and other data are extracted for future use
- **Authentication:** the sink checks if the sending node characteristics (node Id, MAC address and Hash value) matches entry in the list of trusted nodes. In that case, the message is accepted. In the contrary scenario, the message is considered as issued by an untrusted node. A node is also considered as malicious if it attempts to hijack a trusted node's MAC hash.

3 Evaluation and Results

3.1 Environment and Simulation

Simulations were ran using Contiki 3.0 on Ubuntu 18.04 LTS virtual machine (VMware). Several configurations, based on unicast messages between nodes and a sink, were implemented with various parameters. The algorithm was simulated

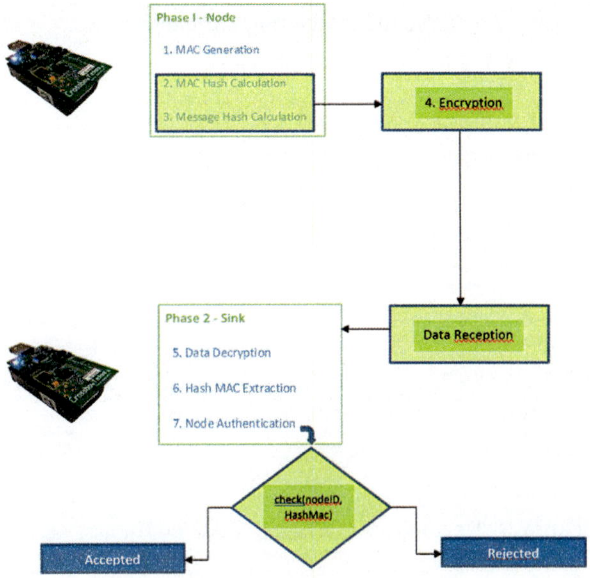

Fig. 2. LBSWSN algorithm

with 20 and 50 nodes randomly positioned (Figs. 3 and 4) in a 100 m × 100 m area. Sinks (id = 21 for 20 nodes and id = 51 for 50 nodes) were set at the center of the area in order to reach each node within the transmission range (50 m radius). In every simulation, the list of trusted nodes (nodeid, MacAddress and Mac Address hash of the node) was hard-coded at sink level. This was used to verify the trustworthiness of the node (Fig. 2).

In order to validate the results, simulations were ran 10 and 4 times for 20 and 50 node models respectively, each simulation consisting of at least 400 rounds. Graphs were plotted using average values of measurements. It is important to note that sink values were excluded from our evaluation because the sink was considered as a limitless resource hardware. Detailed simulation parameters are provided in Table 1. In order to evaluate the impact of the algorithm on the network, 2 communication models were considered:

- Normal or clear communications: Each node sends unicast messages to the sink without encryption, hash or authentication at the sink level.
- Hashed and Encrypted communications: MD5 was used to hash MAC and messages and AES256CBC to encrypt and decipher communications between a node and the sink.

Contiki's generated MAC addresses were used instead of implementing a specific MAC Address generation algorithm. At start-up, the sink loaded a list of predetermined trusted nodes, their MAC addresses and node IDs as well as the corresponding hash of the MAC address as shown in Table 2.

Table 1. Simulation parameters

Contiki version	3.0
Number of nodes	20, 50
Communication model	Unicast messages
Id trusted nodes	3, 5, 15, 20
Type of nodes	SkyMote
MAC addresses	Contiki generated
Area	100 m × 100 m
Radio range	50 m
Hash algorithms	None, MD5 (adapted version of [12])
Encryption algorithm	None, AES256 CBC
Max packets size	64

Table 2. Example of nodes ID, MAC addresses, hashes

ID	MAC address	MAC hash
1.0	0012740100010101	4f86b1c02fefed7e31f026528aabce3c
2.0	0012740200020202	e244cb083266f73112c8ae811b75da3d
3.0	0012740300030303	c2002693c211c45a10276e7d68f9b7a8
4.0	0012740400040404	95fcc776e7172fb0c7d5ae470acdfef9

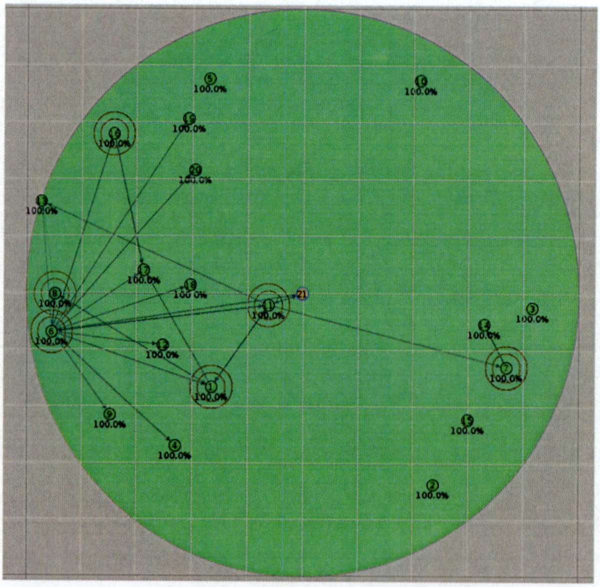

Fig. 3. 20-node model

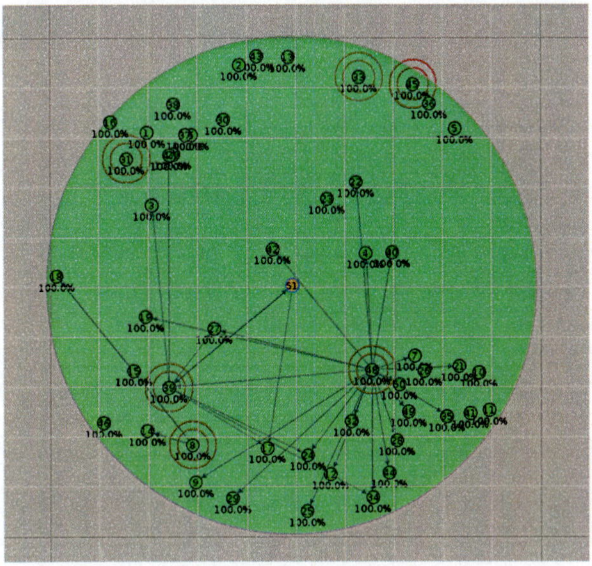

Fig. 4. 50-node model

3.2 Results and Discussion

Adding security features such as hash, encryption and authentication had an impact on energy consumption compared with normal (centralized) unicast messaging. The LBSWSN cumulated energy in the network overtime was higher compared with the normal unicast as shown in Fig. 5 & 6. The difference was significant as the simulations reached the 400th round. The increase was 4% in each round. For 50 nodes, the increase was 2% in each round and the LBSWSN was closer to the normal mode as the number of nodes increased. This should be verified with larger numbers of nodes. The FDN (first dead node) metric was analyzed based on the number of alive nodes over time (Fig. 7 & 8). Upon data analysis, the 20-node model had a greater life span than the 50-node model. Therefore, different initial energy values were affected for nodes in each model. In the 20-node model, using LBSWSN, the first node died at round 261 and at round 267 for the normal communication. With more than twice initial energy in the 50-node model, the first dead node was recorded at round 190 for LBSWSN and round 195 for normal communications. That the gap between the number of alive nodes increased over time in Fig. 7 but remained tight in Fig. 8. On the other hand, the last node died at round 329 in normal communication and 349 for LBSWSN, in the 50-node model.

To evaluate the effectiveness of the algorithm compared to normal communications, the sink printed several results. As shown in Fig. 9, all incoming messages were accepted in normal mode. When authentication was introduced in LBSWSN, only messages from trusted nodes were accepted, messages from nodes not listed as trusted nodes were rejected (Fig. 10).

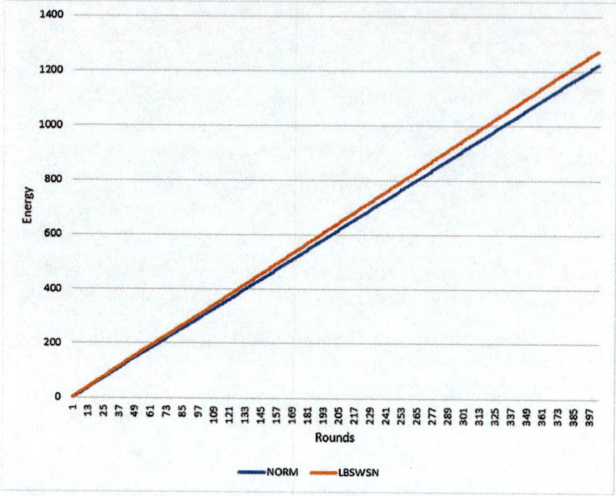

Fig. 5. Network cumulative energy over time 20-node model

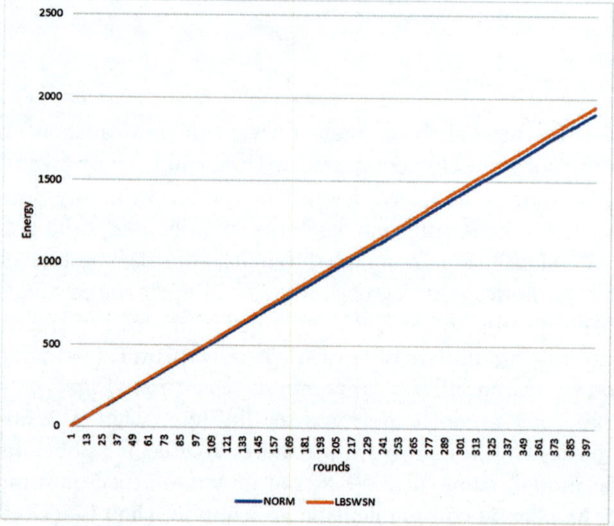

Fig. 6. Network cumulative energy over time 50-node model

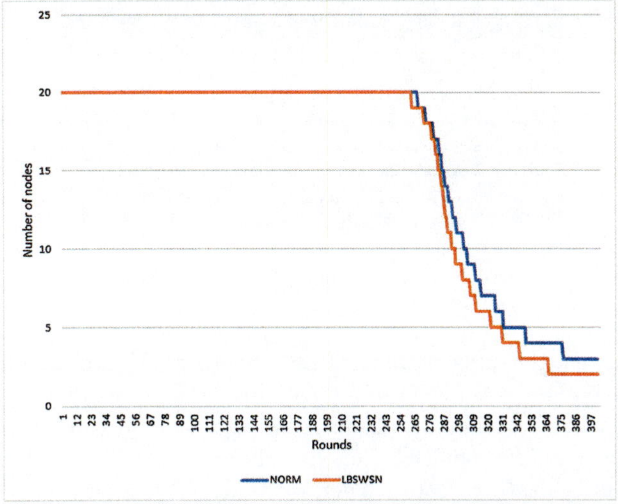

Fig. 7. Alive nodes evolution over time (20-node model)

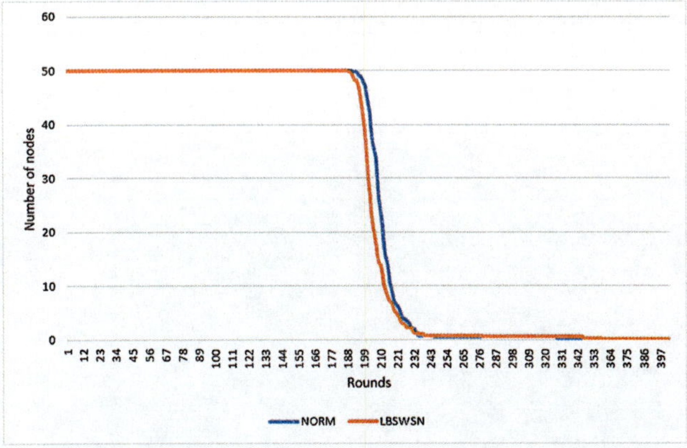

Fig. 8. Evolution of alive nodes over time (50-node model)

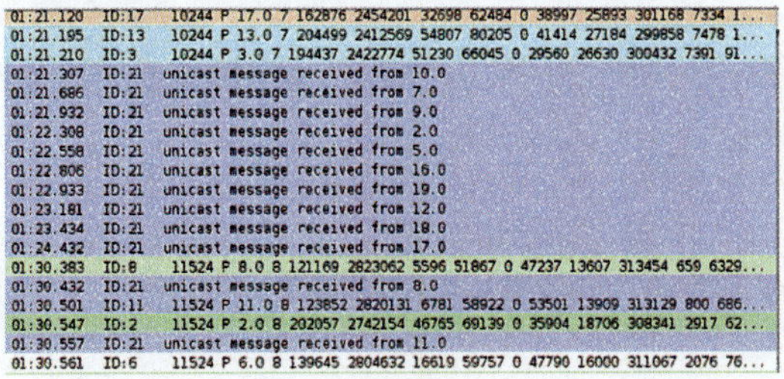

```
01:21.120  ID:17   10244 P 17.0 7 162876 2454201 32698 62484 0 38997 25893 301168 7334 1...
01:21.195  ID:13   10244 P 13.0 7 204499 2412569 54807 80205 0 41414 27184 299858 7478 1...
01:21.210  ID:3    10244 P 3.0 7 194437 2422774 51230 66045 0 29560 26630 300432 7391 91...
01:21.307  ID:21  unicast message received from 10.0
01:21.686  ID:21  unicast message received from 7.0
01:21.932  ID:21  unicast message received from 9.0
01:22.308  ID:21  unicast message received from 2.0
01:22.558  ID:21  unicast message received from 5.0
01:22.806  ID:21  unicast message received from 16.0
01:22.933  ID:21  unicast message received from 19.0
01:23.181  ID:21  unicast message received from 12.0
01:23.434  ID:21  unicast message received from 18.0
01:24.432  ID:21  unicast message received from 17.0
01:30.383  ID:8    11524 P 8.0 8 121169 2823062 5596 51867 0 47237 13607 313454 659 6329...
01:30.432  ID:21  unicast message received from 8.0
01:30.501  ID:11   11524 P 11.0 8 123852 2820131 6781 58922 0 53501 13909 313129 800 686...
01:30.547  ID:2    11524 P 2.0 8 202057 2742154 46765 69139 0 35904 18706 308341 2917 62...
01:30.557  ID:21  unicast message received from 11.0
01:30.561  ID:6    11524 P 6.0 8 139645 2804632 16619 59757 0 47790 16000 311067 2076 76...
```

Fig. 9. Sink messages (normal mode)

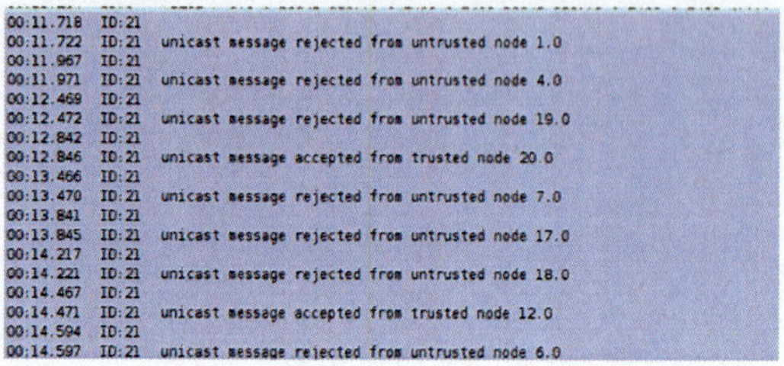

Fig. 10. Sink messages using LBSWSN

In order to test the integrity of the LBSWSN (Fig. 11), a malicious node (node 22) that spoofed the MAC hash of a trusted node was introduced. In reality, to accomplish such, the malicious node must decipher encrypted messages of a trusted node using AES256CBC. This means that node 22 was able to crack the key and the initialization vector. Then, it analyzed the data and collected the MAC hash of a trusted node. Even if node 22 used the collected hash, it was detected by the sink and was considered a malicious node as shown in Fig. 12. In case the encrypted message is tampered and sent to the sink by another node, values of MAC and Message hashes will be altered. The sink can therefore check the integrity of the data received and reject the malicious message.

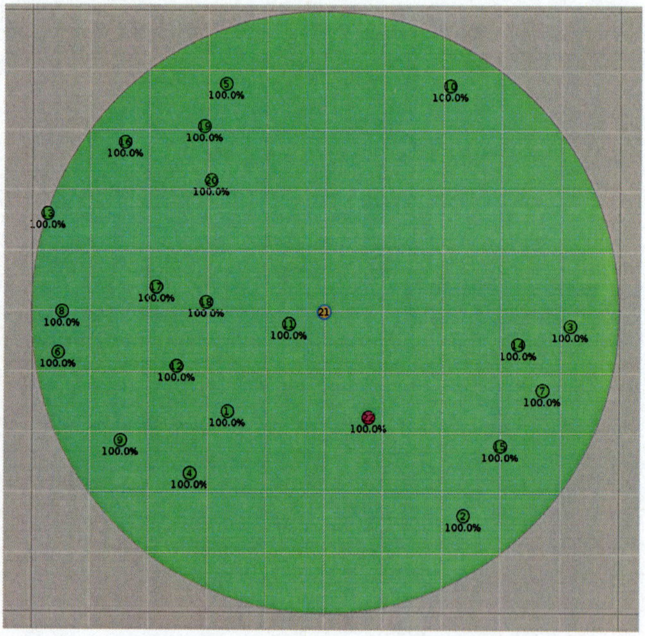

Fig. 11. Malicious node (20-node model)

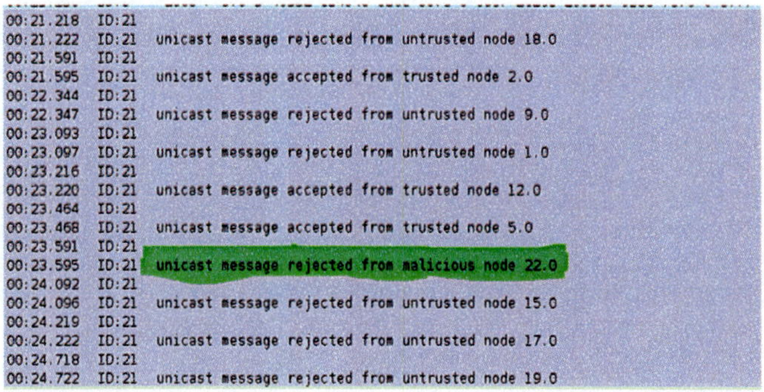

Fig. 12. Malicious node detection using LBSWSN (sink)

Conclusion

The Wireless Sensor Network security is increasingly gaining the attention of many researchers. These networks are considered vulnerable environments, especially because sensors have limited resources in terms of computing, memory and network capacity. The present study highlighted the importance of security in WSN applications and presented LBSWSN, a Lightweight Blockchain for Secured Wireless Network Sensors. LBSWSN is a new concept that combines hashing methods, symmetric cryptography used in blockchain in order to ensure confidentiality, integrity and authentication of nodes in a WSN with low energy consumption. The use of LBSWSN slightly increased the mean cumulative energy over time in the network compared with normal unicast communications. At each round, the increase remained stable at 4% for the network of 20 nodes and 2% for 50 nodes. The analysis of FDN also showed that LBSWSN remained close to the normal communication model as the event (FDN) occurred at rounds 261 and 267 for the normal mode and the LBSWSN mode respectively in the 20-node model. Likewise, FDN was observed at rounds 190 and 195 for the normal mode and the LBSWSN mode respectively in the 50-node model. The LDN (Last Dead Node) was also recorded at rounds 329 and 349 (normal, LBSWSN) for the 50-node model.

Future studies may include implementing and comparing other hashing and cryptographic techniques in other to select energy efficient techniques. In this study, all nodes were within the sink radio range with a single hop. Implementing a multiple-hop solution will provide a deeper view of the impact of these techniques on that network model. This study also used a predetermined list of trusted nodes for the simulations, thus trust scores or levels could be assigned to nodes, and the score could be changed based on the activity or behavior of the node overtime. MAC address generation and assignment could also be studied. To ensure integrity, the used techniques could be more efficient if unique MAC addresses were generated regardless of the sensor manufacturer. Finally, the performance of the present work with regards to routing, clustering techniques and protocols in WSN should also be analyzed.

References

1. Sandra, P.S., Sandeep, C.M., Nair, V., Vindhuja, M.V., Nair, S.S., Raja, M.P.: WSN based industrial parameter monitoring using smartwatch. In: 2017 International Conference on Circuit, Power and Computing Technologies (ICCPCT), pp. 1–6, April 2017
2. Kumar, A., Ovsthus, K., Kristensen, L.M.: An industrial perspective on wireless sensor networks a survey of requirements, protocols, and challenges. IEEE Commun. Surv. Tutor. **16**(3), 1391–1412 (2014)
3. Chi, Q., Yan, H., Zhang, C., Pang, Z., Da Xu, L.: A reconfigurable smart sensor interface for industrial WSN in IoT environment. IEEE Trans. Ind. Inf. **10**(2), 1417–1425 (2014)
4. Das, K., Zand, P., Havinga, P.: Industrial wireless monitoring with energy harvesting devices. IEEE Internet Comput. **21**(1), 12–20 (2017)
5. Kolias, C., Kambourakis, G., Stavrou, A., Gritzalis, S.: Intrusion detection in 802.11 networks: empirical evaluation of threats and a public dataset. IEEE Commun. Surv. Tutor. **99**, 184–208 (2015)
6. Maitra, S., Yanambaka, V.P., Puthal, D., Abdelgawad, A., Yelamarthi, K.: Integration of Internet of Things and blockchain toward portability and low-energy consumption. Trans. Emerg. Telecommun. Technol. **32**(6), e4103 (2021). https://doi.org/10.1002/ett.4103
7. Tomic, I., McCann, J.A.: A survey of potential security issues in existing wireless sensor network protocols. IEEE Internet Things J. **4**(6), 1910–1923 (2017)
8. Merlo, A., Migliardi, M., Caviglione, L.: A survey on energy-aware security mechanisms. Pervasive Mob. Comput. **24**, 77–90 (2015)
9. Sen, J.: A survey on wireless sensor network security. arXiv:1011.1529 [cs] (2010)
10. Caesarendra, W., et al.: An AWS machine learning-based indirect monitoring method for deburring in aerospace industries towards industry 4.0. Appl. Sci. **8**(11), 2165 (2018). https://doi.org/10.3390/app8112165
11. Puthal, D., Mohanty, S.P., Nanda, P., Kougianos, E., Das, G.: Proof-of-authentication for scalable blockchain in resource-constrained distributed systems. In: 2019 IEEE International Conference on Consumer Electronics (ICCE), pp. 1–5 (2019). https://doi.org/10.1109/ICCE.2019.8662009
12. Dahler, B.: MD5 implementation in C. https://github.com/crossbowerbt/md5/blob/master/md5.c

Virtual OBDA Mechanism Ontop for Answering SPARQL Queries Over Couchbase

Hakim El Massari(✉), Sajida Mhammedi, Noreddine Gherabi, and Mohammed Nasri

National School of Applied Sciences, Lasti Laboratory, Sultan Moulay Slimane University, Khouribga, Morocco
h.elmassari@usms.ma

Abstract. In the last decade, the database field has become substantially diversified, as a consequence, a number of non-relational databases (known also as NoSQL) have been developed, e.g., key-value stores and JSON-document databases, XML, and graph databases. Several issues associated with big data were addressed as a result of the rise of this new generation of data services. However, in the rush to address the problems of big data and vast numbers of active users, NoSQL dropped certain of the core features of databases that make them highly performant and usable such as the global view, that permits users to access data without needing to know how they are logically organized or physically stored in their sources. We address, in this article, the challenge of how to fill the gap between NoSQL and the Semantic web, in order to enable access to such databases and integration of non-relational data sources. However, we extend the well-known framework for ontology-based data access (OBDA), intending to allow a mediating ontology to query arbitrary databases. We instantiate this framework to a popular JSON-document database called Couchbase, and implement a prototype extension of the virtual OBDA mechanism Ontop to answer SPARQL queries over Couchbase.

Keywords: NoSQL · OBDA · Ontop · Ontology · Couchbase

1 Introduction

For many decades, relational data management has been the most widely used technique for storing and manipulating structured data. However, the development of increasingly large-scale applications highlighted the limitation of relational data management to handle the storing and querying large volumes of data in an efficient and horizontal way.

This sparked a paradigm shift, requiring a new generation of databases capable of handling huge volumes of data without losing query efficiency by decreasing query expressivity and accuracy. A large variety of so-called non-relational or NoSQL (not only SQL) databases appeared. (e.g., Neo4j, Cassandra, Couchbase, MongoDB… etc.).

Moreover, this extensive choice of DBMS provides the opportunity to deal with the requirements of a diversity of modern applications and to match more closely their

R. Saidi et al. (Eds.): ICATH 2021, LNDECT 110, pp. 193–205, 2022.
https://doi.org/10.1007/978-3-030-94188-8_19

differing needs with respect to data management, enabling more flexible data schemas, for instance, or more efficient (though simple) queries. Although this heterogeneity contributed to one of the core aspects of Big Data challenges: variety, as databases grow in size and heterogeneity, accessing data using native query languages is becoming more challenging and getting a more and more involved task for users.

In order to facilitate this form of access, the OBDA [1, 2] was proposed to allow users to create high-level ontological queries that will be automatically converted into low-level queries for the conventional virtual method utilized by DB engines. In practice, the distinction between the conceptual and the DB levels has been shown success, particularly where data sources have a rather complex structure and end-users have expertise in data management [3]. The OBDA method connects a database to an ontology using a declarative specification in the form of mappings that link ontology concepts to SQL views over the data. The ontology is typically described using the OWL2 QL profile of the OWL2 [4], SPARQL is used to write queries, and the database is regarded as relational [5]. Ontop takes place with the goal of making the OBDA approach practicable in such cases by automating the procedure, it converts the queries that users raise over the ontology into queries that are performed efficiently over legacy databases.

In our work, we concentrate our efforts on Couchbase, a document-based database management system that is one of the most widely used NoSQL databases today. N1QL is used to query Couchbase, and can be interpreted as a type of SQL injection into a NoSQL database, N1QL is an expressive and powerful language and a complete SQL dialect for querying, transforming, and handling JSON data. Consequently, OBDA over Couchbase can leverage this advantage to efficiently answer queries, while at the same time offering a more user-friendly query language. Accordingly, we introduce an approach to using OBDA on NoSQL databases, by instantiating the generalized OBDA framework over Couchbase as an extension of the OBDA system Ontop, the rest of the paper is structured as follows.

We begin in Sect. 2 by presenting related work from previous researches in the associated fields. To extend the well-known ontology-based data access (OBDA) framework with NoSQL Systems, we give in Sect. 3, our proposal system architecture to query large volumes of data. Then, to evaluate its feasibility, in Sect. 4, we apply this approach in an OBDA that employs a document-oriented NoSQL database in Couchbase. In Sect. 5, we perform experiments and discuss evaluation results. Finally, we draw the conclusions and present future works.

2 Related Work

The RDF [6, 7] is becoming more common as a pivot format for integrating heterogeneous data sources. It provides a single data model that allows building upon large number of existing vocabularies and domain ontologies while still taking advantage of the Semantic Web's reasoning capability. Also, it enables the use of the Web of Data, which is a rapidly expanding global knowledge base.

RDF data is increasingly being released on the Web, notably following the Linked Data principles [8, 9]. This data is often sourced from heterogeneous silos that are unavailable to data integration systems and search engines. As a result, converting legacy data from disparate formats into RDF representations is a first step toward allowing RDF-based data integration.

In the past fifteen years, a lot of research has gone into figuring out how to convert popular databases and data formats into RDF. The main emphasis was on relational databases. The primary focus was on relational databases [7, 10], together with a set of data formats including XML [11] and CSV [12]. Besides that, with the introduction of numerous non-relational models, the database landscape has become significantly more diverse. NoSQL databases, which were initially designed as the backbone of Big Data Web applications, have gained traction and are now being used as general-purpose, commonplace databases. Nowadays, companies and organizations are using NoSQL to store large volumes of data. These data, on the other hand, are often unavailable to RDF-based data integration systems, and hence unseen to the Web of Data. Despite the fact that releasing their data could open up new integration possibilities and propel the Web of Data forward.

Over the past several years, there has been a lot of research into exposing legacy data as RDF, with two main approaches: materialization (i.e. all legacy data is converted into an RDF graph at once), or on-the-fly conversion of SPARQL queries into the query language required. When dealing with large datasets, materialization can be challenging and expensive, particularly when data freshness is on the line. Numerous methods for achieving SPARQL access to relational data have been suggested, whether in the context of RDB-backed RDF stores or in the case of RDF stores [13–15] or using arbitrary relational schemas [16–19]. R2RML [20], the W3C RDB-to RDF mapping language recommendation is a well-accepted standard many SPARQL-to-SQL rewriting techniques depend on it [17, 19, 21]. Other alternatives seek to map XML [22–24] or CSV data to RDF. RML [1] tackles the mapping of heterogeneous data formats such as CSV/TSV, XML and JSON. xR2RML [25] is an R2RML and RML extension that addresses the mapping of a wide variety of databases to RDF.

In [26], the authors suggest a method to take on the issue of querying vast quantities of statistical RDF data. To support the analysis of such data, this method relies on pre-aggregation strategies. Particularly, the authors describe a conceptual model representing original RDF data with multidimensional structure aggregates.

In another interesting work, the authors [27] have developed a SPARQL to MongoDB query mapping tool, which converts the legacy databases into an easily accessible source of data. A Virtual RDF database can be shown with all stored documents as RDF triples. The conversion takes two phases: the SPARQL query is converted into an abstract query by using mappings from MongoDB documents to RDF written in an intermediate language called xR2RML, and then the query is rewritten as a concrete MongoDB query. Consequently, they demonstrated that rewriting a query to obtain accurate answers is often feasible.

In line with the use of OBDA in NoSQL the authors in [28], study the problem of ontology-mediated query answering over key value stores. The authors create a rule-based ontology language in which keys are used as unary predicates and rules are applied at the record stage (a record is a set of key-value pairs).

Considering the fact that queries are a mixture of get and check operations that, given a path, return a set of values that can be gotten via that path. The authors examine the challenge of answering these queries using a set of rules. Due to the lack of mappings and, as a result, no difference between user and native database query languages, this work is still outside of the OBDA framework. It's also worth noting that their ontology and query languages don't follow any Semantic Web standards. On the OBDA over NoSQL side there have been a lot of attempts, [29] proposes integrating ontology-based data access into NoSQL stores, emphasizing the importance of using ontology to search for data inconsistencies and, as a result, increase data quality in NoSQL repositories. The integration is accomplished by rewriting SPARQL queries into the native query language of the NoSQL database. The authors give eight examples of queries and how they could be optimized into queries for document or columnar stores.

Additional work in this field [3] introduced a detailed and comprehensive architecture for an OBDA solution in a Big Data scenario is proposed in the Optique project. The key factors in this work are the usability and manageability of an OBDA system. Present OBDA systems, according to the authors, have significant shortcomings such as the use of a formal query language like SPARQL and complicated mapping management. Optique aims to improve the user experience when it comes to querying and handling ontology-based access to vast amounts of data from various sources.

In a somewhat different approach, [21] extend the Ontop Ontology-Based Data Access (OBDA) system to support R2RML mappings. A Datalog program is created by converting a SPARQL query and an R2RML mapping graph. This structured representation is used for integrating and applying logic programming and SQL query optimization techniques. The optimized program is then converted into a SQL query that can be executed.

To the best of our knowledge, little work has investigated how to extend the well-known ontology-based data access (OBDA) framework, in order to allow a mediating ontology to query arbitrary databases especially heterogeneous and non-relational. The works that are based on ontology are related to the mapping and representation of data in OWL format, such as approaches related to measuring the semantic similarity of concepts [30], or approaches based on segmentation or classification [32].

More in line with our work, authors in [31], suggested that the OBDA concepts be applied to MongoDB. They explain a two-step rewriting process of SPARQL queries into the MongoDB aggregate query language, their work is an extension of Ontop [25], which is an OBDA system for relational databases. authors explain a two-step rewriting process of SPARQL queries into the MongoDB aggregate query language. Using a document-oriented MongoDB database, the latest proposed architecture was tested. In a previous work, authors have provided a systematic assessment of a subset of MongoDB data access queries. This assessment revealed that creating a fully generic framework capable of querying any NoSQL DBMS is extremely difficult. NoSQL DMBS share few query patterns, which require the use of a query translator for any NoSQL DBMS, as opposed to relation databases with a SQL (common query language).

In another approach [33] which is comparable in spirit to ours, in that it also seeks to delegate query execution to a NoSQL source engine, and relies on an object-oriented (OO) intermediate representation, which is similar to our "relational view". However, instead of mapping from the source DB to the ontology vocabulary, the mapping is from the ontology vocabulary to the OO layer.

3 ODBA with NoSQL Databases

3.1 OBDA Over Couchbase

Ontology-Based Data Access (OBDA) has been a common technique since the mid-2000s to resolve the issue of accessing current data sources through scalable methods that are both effective and efficient [10].

A conceptual layer in OBDA establishes a common vocabulary, builds the domain, covers the data source structure, and improves the context data of incomprehensive knowledge. Thus, users don't need to know about the data sources, their relationships, or how the data is encoded since queries are queried over this high-level conceptual view. The data sources and ontology are linked via a declarative specification expressed in terms of mappings that bind ontology (properties, classes ...) to data views (SQL). The R2RML [20] W3C standard was developed with the intent of offering a language for specifying mappings in an OBDA environment. The ontology and mappings from a virtual RDF graph that can be queried with SPARQL (the Semantic Web's standard query language).

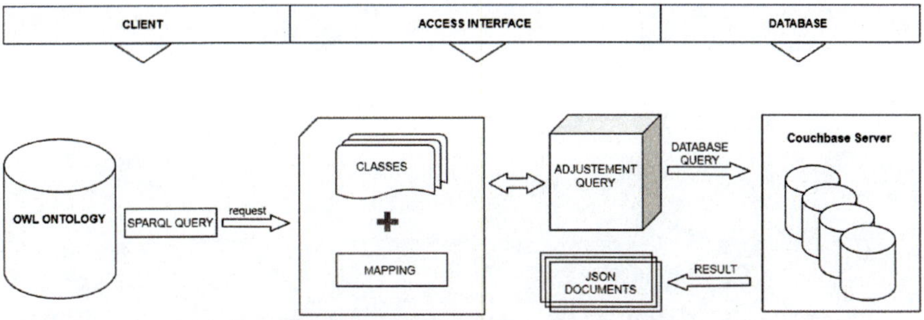

Fig. 1. The Ontop-CB project's architecture

In the context of the Semantic Web, query answering is essential because it offers a mechanism for users and applications to engage with ontologies and data. For this reason, several query languages have been developed, including SeRQL, RDQL, and most recently, SPARQL. The World Wide Web Consortium (W3C) standardized the SPARQL query language in 2008, and most RDF triple stores now support it, the thing that led us to choose it.

3.2 Ontop-CB System

We adopted in this article, the Ontop OBDA system [5], which is an open-source system that is actually being used in a number of projects. Ontop supports all W3C OBDA guidelines, including OWL2 QL, SWRL, R2RML, SPARQL, as well as support for all existing relational databases. Ontop is available as a SPARQL endpoint via Sesame Workbench, a Protégé plugin, and a Java library supporting OWL API and Sesame API. Ontop allows for RDFS and OWL2QL [5] as ontology languages. OWL2QL is built on the DL-Lite family of compact description logics [34, 35], which ensures that ontology queries can be rewritten into database queries equivalently.

To present the different notions and concepts cited in this article, we suggest the use of the OBDA model composed of ontology and mappings as well as an intermediate conceptual layer, in order to access the data of a NoSQL database. We present the Ontop-CB project which implement the query translation method based on the Ontop system, which allows to query NoSQL database, Couchbase in our case, in order to generate a set of JSON Document as a result. As illustrated in Fig. 1, the following are the key components of the onto-CB project: an OWL Ontology, an Access Interface, mappings, a NoSQL database, a SPARQL to NoSQL query adjustment, and a JSON export.

Ontology. An ontology called University Fig. 2 was created with the information systems of two universities describing students, academic staff and courses, based on University database that contains two universities named "uni1" and "uni2".

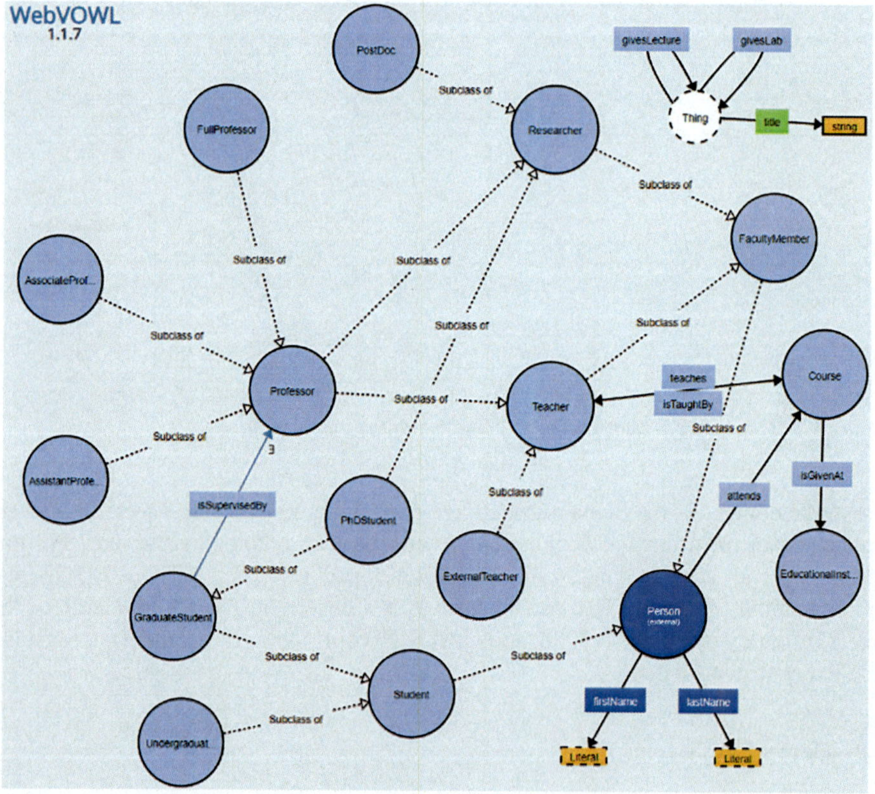

Fig. 2. The graphical representation of "University" ontology

Access Interface. This Interface is a module capable of translating SPARQL query and responding the json document from database, the model was developed using java programming language based on Ontop API and Couchbase API. Our java program takes as input "owlFile" (Classes, Object Properties, Data Properties), obdaFile (Mappings), propertyFile (connection to database). Given an OWL file, OWLReasoner will check for consistency in ontologies, find subsumption relationships between classes, and even more [15]. The mapping assumption is comprised of two parts: A source, which is a SQL query that retrieves values from the database, and a target, that is a collection of RDF triples containing values from the source (Fig. 3). Our Java program's classes, combined with the mappings, reveal a virtual RDF Graph, which will be queried using SPARQL Fig. 3(a) by converting SPARQL queries into SQL queries Fig. 3(c).

The generated SQL queries are not necessarily efficient and cannot be directly executed by our DB engine. Hence, we have to adjust the SQL Fig. 3(d) syntax by adding the adjustment query phase in order to generate a N1QL query, taking into account that N1QL is considered as SQL for JSON since it looks very much like a SQL query.

It is designed to work with both structured and semi-structured data, and it is based on the original SQL with extensions that it can work with JSON document database by relaxing its restrictions on the data model. Thus, the query language retains the advantages of SQL, including its high-level (declarative) nature, while enabling it to deal the more flexible structures typically found in the semi-structured world. Based on that and Since our DB engine does not support slightly generated SQL dialect, we have to adjust the SQL syntax accordingly Fig. 3(c). For instance, the operator for string concatenation is || in Couchbase and the concat function in other relational databases; another example In Couchbase, we used backtick instead of double quotation marks; and owing to the fact that Couchbase does not support CAST function lead us to eliminate it.

Finally, the adjusted SQL query is executed over Couchbase database and retrieve json document as results.

Database. The Ontop-CB project uses Couchbase a document-oriented NoSQL database for storage. The database contains two universities named "uni1" and "uni2".

The University data was generated randomly with java method Based on the relational schema of an excited composed of 8 tables (Student, academic, courses, etc.). We generated a two million of json documents divided between both universities.

Our approach aims to exploit Ontop answers end-user's SPARQL queries by rewriting them into SQL queries and delegating their execution to the database. To do so, we established an intermediate model layer using classes in java programming language as an intermediate layer between owl ontology and Couchbase.

The Ontop system disclose relational databases as virtual RDF graphs (VRG) by connecting the terms of the ontology to the data sources through mappings. This VRG can then be queried using SPARQL by converting the SPARQL queries into SQL queries over the relational databases.

In our system particularly in the access interface, we aim to adopt the methodology of Ontop, by retrieving the generated SQL query that is then used to query our database in Couchbase. N1QL can be used to query Couchbase Server as an expressive, effective and full SQL dialect to query, update, and manipulate json data [34]. Contrast to other NoSQL databases, Couchbase supports SQL-like query language which makes the transition to Couchbase from RDBMS much easier.

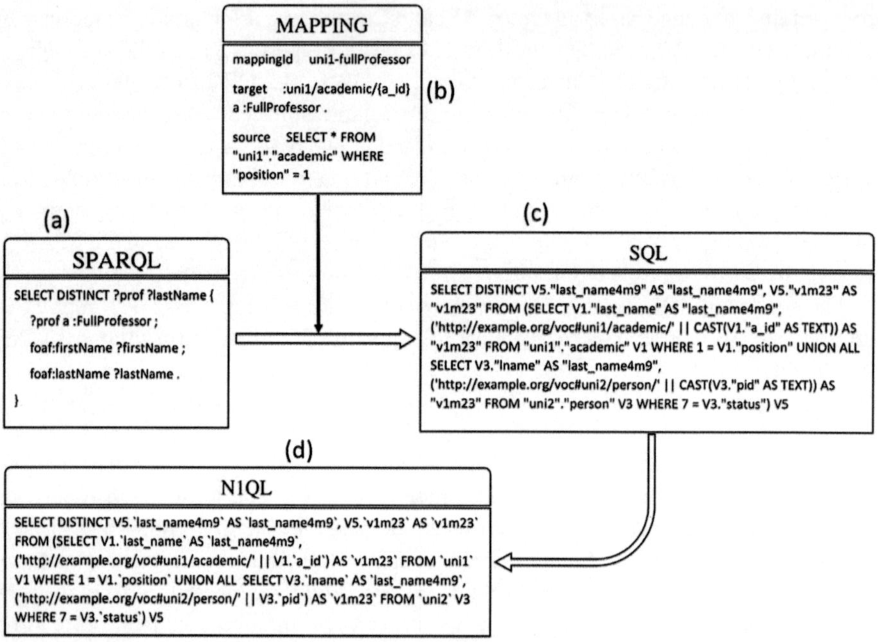

Fig. 3. Adjustment query process (a) Example of SPARQL query (b) Sample of mappings (c) The generated SQL query (d) the adjustment of generated SQL query

4 Evaluation and Environment

An evaluation has been carried out to evaluate whether OBDA over Couchbase is a practical performance solution and, in particular, whether it is capable of leveraging the document structure of Couchbase collections. We implemented an access interface for a query answering system using SPARQL as the query language, Ontop for translating and Couchbase as NOSQL DATABASE.

To realize the proposed system, "University" database is imported to cluster named "University" in Couchbase. The "University" database is available in the CSV format (the format of the CSV is unique to Couchbase).

We developed a java method to dump "University" data based on the relational schema to Couchbase. With the intention to cover all Couchbase constructions, we had to make adjustments in the database design.

Within the tests we used five different SPARQL queries. The listings for the queries can be found online. Here we describe them shortly:

- **FullProfessor:** in this query we searched for a professor with position = 1 for the university uni1 and status = 7 for uni2.
- **FacultyMembers:** this query retrieves all member of faculty.
- **PersonNames:** all person in the university database.
- **Teachers:** all teachers of the university database.
- **Courses:** names of students attended in a course

All experiments were conducted on a PowerEdge R740 server which has Intel(R) Xeon(R) Silver 4110 CPU @ 2.10 GHz with 8 Core/16 threads and 16 GB DDR4-SDRAM, with a 1.63 TB 10K RPM SAS 12 Gbps as RAID-0 hard drive cluster. The RAID controller is Dell PERC H740P Mini (integrated).

This work is done by Java programming language and ONTOP API 4.0.2 version. Couchbase-java-client:3.0.8 is used to access Couchbase. For the graphical representation, we used "WebVOWL" [36, 37] a web application for the interactive visualization of ontologies.

The database and all configuration files are available online, together with the SPARQL queries and mapping are also provided1, in order for the experiment can be reproduced.

5 Result and Discussion

We consider a University database with 8 tables that contains information about two universities 'uni1' and 'uni2' stored into two backets in Couchbase. The results are summarized in Table 1, and we show the impact of number of documents over execution time in Fig. 5 and Fig. 4 respectively for the 5 Ontop-CB queries.

Table 1 reports the execution times for our system Ontop-CB w.r.t number of documents returned. We did not include in this table the query rewriting time (SPARQL to SQL), due to its small time (<1000 ms) resulting from using ONTOP API under Java programming language.

We now focus on query evaluation times. Several considerations can be made when looking at the results of our experimentation, but the most important one is that no relation between the number of documents returned and query execution time, which is observed in queries 3, 4 and 5 we got a very close execution time Fig. 4, and confirmed if we compare q3 and q5, query 3 outperforms query 5 even if it returns a large amount of documents Fig. 5.

Table 1. Query answering times by the number of documents over Couchbase (in milliseconds)

Query	Q1	Q2	Q3	Q4	Q5
Execution time	21572	41498	49105	50480	51030
Number of docu-ments returned	473	1400730	3000000	644230	502675

The execution time has a relationship with the complexity of the SQL, even though Couchbase supports a number of data processing capabilities including filtering, deep traversal of nested documents, querying through relationships using JOINs or subqueries, grouping, combining result sets using operators, sorting, aggregation, and more. The main reason for that is the system Ontop produces rewriting containing complex subqueries, composed of unions of several select-project-join queries and these types of queries are not evaluated efficiently [5]. This is explained by the fact of having a negative effect on the query execution time.

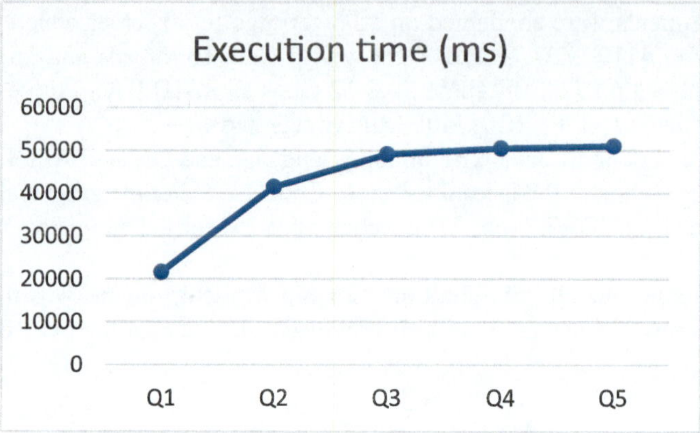

Fig. 4. Execution time in milliseconds

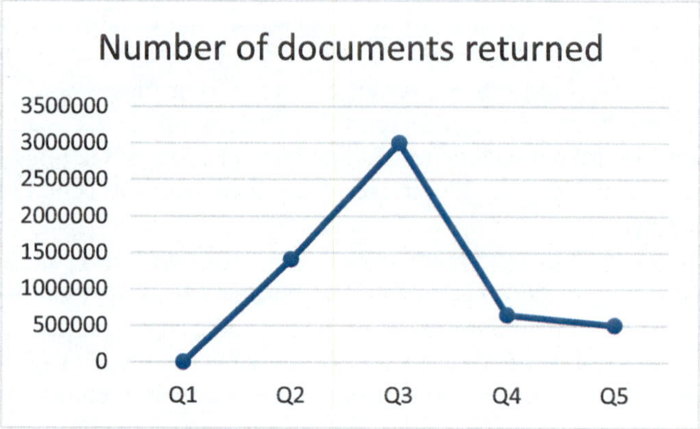

Fig. 5. The number of documents returned for each query

Our system shows that in NoSQL databases, OBDA can include some of the most popular features. The features covered in this paper are the extension of the well-known ontology-based data access system for the appropriate management of data that allows high level conceptual integration.

Nonetheless, we have shown that the use of OBDA offers functionality that go beyond the standards of most developers and users of NoSQL, such as Querying NOSQL Databases over semantic schemes.

The approach taken in this study has advantages for the research community, mainly since it uses Ontop answers SPARQL queries by rewriting into SQL queries and assigning them to the database. By programming the whole project in JAVA language using the APIs (application program interfaces), we made this process easier to OBDA Systems.

6 Conclusion

In this article we aim to encourage the creation of SPARQL interfaces to heterogeneous databases, which we agree is critical to the advancement of the Web of Data. Especially, we believe that this will aid in bridging the gap between the Semantic Web and the NoSQL database family.

A practical architecture for a virtual OBDA approach is proposed that allows SPARQL queries to be answered via arbitrary data sources. As an extension (called Ontop-CB) of the OBDA system Ontop, we've implemented this framework in the particular case of Couchbase, and we conducted an experimentation based on the real-world use case of a document reference stored in a Couchbase database. The evaluation we have carried out shows that Ontop-CB is able to generate queries and retrieve json document as results.

7 Future Work

As we prove from analysis, we found that the complexity of the query affect time of the execution, which conduct us to Improve query rewriting by reducing complex subqueries.

As a continuation of this work, we intend to minimize execution time in the following directions:

(i) Eliminate phase of adjustment by placing an optimizer layer next to the resulting SQL.
(ii) Integrate the phase of adjustment query within Ontop project to optimize the generated queries, more than that produce and delegate N1QL query directly to Couchbase.
(iii) We plane to compare execution time of the two proposed solutions (i) and (ii).

Furthermore, we'll develop the GUI interfaces and add functionality, for ease of use by end users of Ontop-CB.

Footnotes

1 shorturl.at/eBT06.

References

1. Dimou, A., Vander Sande, M., Colpaert, P., Verborgh, R., Mannens, E., Van de Walle, R.: RML: a generic language for integrated RDF mappings of heterogeneous data. In: Bizer, C., Heath, T., Auer, S., and Berners-Lee, T. (eds.) Proceedings of the 7th Workshop on Linked Data on the Web, Korea (2014)
2. Bereta, K., Papadakis, G., Koubaraki, M.: OBDA for the web: creating virtual RDF graphs on top of web data sources. arXiv:2005.11264v1 (2020)
3. Calvanese, D., et al.: Optique: OBDA solution for big data. In: Cimiano, P., Fernández, M., Lopez, V., Schlobach, S., Völker, J. (eds.) ESWC 2013. LNCS, vol. 7955, pp. 293–295. Springer, Heidelberg (2013). https://doi.org/10.1007/978-3-642-41242-4_48

4. Hogan, A.: Web ontology language. In: Hogan, A. (ed.) The Web of Data, pp. 185–322. Springer, Cham (2020). https://doi.org/10.1007/978-3-030-51580-5_5
5. Calvanese, D., et al.: Ontop: answering SPARQL queries over relational databases. Semant. Web **8**, 471–487 (2017). https://doi.org/10.3233/SW-160217
6. Buron, M., Goasdoué, F., Manolescu, I., Mugnier, M.-L.: Obi-Wan: ontology-based RDF integration of heterogeneous data. Proc. VLDB Endow. **13**, 2933–2936 (2020). https://doi.org/10.14778/3415478.3415512
7. Sequeda, J.F., Tirmizi, S.H., Corcho, O., Miranker, D.P.: Survey of directly mapping SQL databases to the semantic web. Knowl. Eng. Rev. **26**, 445–486 (2011). https://doi.org/10.1017/S0269888911000208
8. Po, L., Bikakis, N., Desimoni, F., Papastefanatos, G.: Linked data visualization: techniques, tools, and big data. Synthesis Lect. Semant. Web: Theory Technol. **10**, 1–157 (2020). https://doi.org/10.2200/S00967ED1V01Y201911WBE019
9. Zangeneh, P., McCabe, B.: Ontology-based knowledge representation for industrial megaprojects analytics using linked data and the semantic web. Adv. Eng. Inform. **46**, 101164 (2020)
10. Gherabi, N., Addakiri, K., Bahaj, M.: Mapping relational database into OWL Structure with data semantic preservation. arXiv:1205.5922 [cs] (2012)
11. Calvanese, D., Liuzzo, P., Mosca, A., Remesal, J., Rezk, M., Rull, G.: Ontology-based data integration in EPNet: production and distribution of food during the Roman Empire. Eng. Appl. Artif. Intell. **51**, 212–229 (2016). https://doi.org/10.1016/j.engappai.2016.01.005
12. Spanos, D.-E., Stavrou, P., Mitrou, N.: Bringing relational databases into the semantic web: a survey. Semant. Web **3**, 169–209 (2012). https://doi.org/10.3233/SW-2011-0055
13. Bilidas, D., Koubarakis, M.: Efficient duplicate elimination in SPARQL to SQL translation. In: Description Logics (2018)
14. Chebotko, A., Lu, S., Fotouhi, F.: Semantics preserving SPARQL-to-SQL translation. Data Knowl. Eng. **68**, 973–1000 (2009). https://doi.org/10.1016/j.datak.2009.04.001
15. Elliott, B., Cheng, E., Thomas-Ogbuji, C., Ozsoyoglu, Z.M.: A complete translation from SPARQL into efficient SQL. In: Proceedings of the 2009 International Database Engineering & Applications Symposium, pp. 31–42. Association for Computing Machinery, New York (2009)
16. Matentzoglu, N., Leo, J., Hudhra, V., Sattler, U., Parsia, B.: A survey of current, stand-alone OWL reasoners. In: ORE (2015)
17. Vavliakis, K.N., Grollios, T.K., Mitkas, P.A.: RDOTE – publishing relational databases into the semantic web. J. Syst. Softw. **86**, 89–99 (2013). https://doi.org/10.1016/j.jss.2012.07.018
18. Priyatna, F., Corcho, O., Sequeda, J.: Formalisation and experiences of R2RML-based SPARQL to SQL query translation using morph. In: Proceedings of the 23rd International Conference on World Wide Web, pp. 479–490. Association for Computing Machinery, New York (2014)
19. Rodríguez-Muro, M., Kontchakov, R., Zakharyaschev, M.: Ontology-based data access: ontop of databases. In: Alani, H., et al. (eds.) ISWC 2013. LNCS, vol. 8218, pp. 558–573. Springer, Heidelberg (2013). https://doi.org/10.1007/978-3-642-41335-3_35
20. Das, S., Sundara, S., Cyganiak, R.: R2RML: RDB to RDF mapping language. Working group recommendation, World Wide Web Consortium. W3C (2012)
21. Rodríguez-Muro, M., Rezk, M.: Efficient SPARQL-to-SQL with R2RML mappings. J. Web Semant. **33**, 141–169 (2015). https://doi.org/10.1016/j.websem.2015.03.001
22. Gherabi, N., Bahaj, M.: Outline matching of the 2D shapes using extracting XML data. In: Elmoataz, A., Mammass, D., Lezoray, O., Nouboud, F., Aboutajdine, D. (eds.) ICISP 2012. LNCS (LNAI), vol. 7340, pp. 502–512. Springer, Heidelberg (2012). https://doi.org/10.1007/978-3-642-31254-0_57

23. Bikakis, N., Tsinaraki, C., Gioldasis, N., Stavrakantonakis, I., Christodoulakis, S.: The XML and semantic web worlds: technologies, interoperability and integration: a survey of the state of the art. In: Anagnostopoulos, I.E., Bieliková, M., Mylonas, P., Tsapatsoulis, N. (eds.) Semantic Hyper/Multimedia Adaptation: Schemes and Applications, pp. 319–360. Springer, Heidelberg (2013). https://doi.org/10.1007/978-3-642-28977-4_12
24. Yang, D., Yan, L.: Transforming XML to RDF(S) with Temporal Information. J. Comput. Inf. Technol. **26**, 115–129 (2018). https://doi.org/10.20532/cit.2018.1004005
25. Bagosi, T., et al.: The Ontop framework for ontology based data access. In: Zhao, D., Du, J., Wang, H., Wang, P., Ji, D., Pan, J.Z. (eds.) CSWS 2014. Communications in Computer and Information Science, vol. 480, pp. 67–77. Springer, Heidelberg (2014). https://doi.org/10.1007/978-3-662-45495-4_6
26. Ravat, F., Song, J., Teste, O., Trojahn, C.: Efficient querying of multidimensional RDF data with aggregates: comparing NoSQL, RDF and relational data stores. Int. J. Inf. Manage. **54**, 102089 (2020). https://doi.org/10.1016/j.ijinfomgt.2020.102089
27. Michel, F., Faron-Zucker, C., Montagnat, J.: A mapping-based method to query MongoDB documents with SPARQL. In: Hartmann, S., Ma, H. (eds.) DEXA 2016. LNCS, vol. 9828, pp. 52–67. Springer, Cham (2016). https://doi.org/10.1007/978-3-319-44406-2_6
28. Mugnier, M.-L., Rousset, M.-C., Ulliana, F.: Ontology-mediated queries for NOSQL databases. In: AAAI Conference on Artificial Intelligence, Phoenix, USA, pp. 1051–1057 (2016)
29. Curé, O., Kerdjoudj, F., Faye, D., Duc, C.L., Lamolle, M.: On the potential integration of an ontology-based data access approach in NoSQL stores. In: 2012 Third International Conference on Emerging Intelligent Data and Web Technologies, pp. 166–173 (2012)
30. Daoui, A., Gherabi, N., Marzouk, A.: A new approach for measuring semantic similarity of ontology concepts using dynamic programming. J. Theor. Appl. Inf. Technol. **95**(17), 4132–4139 (2017)
31. Botoeva, E., Calvanese, D., Cogrel, B., Rezk, M., Xiao, G.: OBDA beyond relational DBs: a study for MongoDB. In: Proceedings of the 29th International Workshop on Description Logics (DL 2016), p. 12 (2016)
32. Nejjahi, R., Gherabi, N., Marzouk, A.: Towards classification of web ontologies using the horizontal and vertical segmentation. In: Noreddine, G., Kacprzyk, J. (eds.) ITCS 2017. AISC, vol. 640, pp. 70–81. Springer, Cham (2018). https://doi.org/10.1007/978-3-319-64719-7_7
33. Araujo, T.H.D., Agena, B.T., Braghetto, K.R., Wassermann, R.: OntoMongo - ontology-based data access for NoSQL. In: Proceedings of the X Seminar on Ontology Research, Brazil, pp. 55–66 (2017)
34. Calvanese, D., De Giacomo, G., Lembo, D., Lenzerini, M., Rosati, R.: Tractable reasoning and efficient query answering in description logics: the DL-Lite family. J. Autom. Reason. **39**, 385–429 (2007). https://doi.org/10.1007/s10817-007-9078-x
35. Artale, A., Calvanese, D., Kontchakov, R., Zakharyaschev, M.: The DL-Lite family and relations. J. Artif. Int. Res. **36**, 1–69 (2009). https://doi.org/10.1613/jair.2820
36. Lohmann, S., Negru, S., Haag, F., Ertl, T.: Visualizing ontologies with VOWL. Semant. Web **7**, 399–419 (2016). https://doi.org/10.3233/SW-150200
37. Lohmann, S., Link, V., Marbach, E., Negru, S.: WebVOWL: web-based visualization of ontologies. In: Lambrix, P., et al. (eds.) EKAW 2014. LNCS (LNAI), vol. 8982, pp. 154–158. Springer, Cham (2015). https://doi.org/10.1007/978-3-319-17966-7_21

Toward a Smart Approach of Migration from Relational Database System to NoSQL System: Analyzing and Modeling

Abdelhak Erraji[1](✉), Abderrahim Maizate[2], and Mohamed Ozzif[2]

[1] RITM ESTC Laboratory Hassan II University, ENSEM Casablanca, Casablanca, Morocco
[2] RITM ESTC Laboratory Hassan II University, ESTC Casablanca, Casablanca, Morocco

Abstract. In recent years, data has taken a strategic and decisive position in the existence and evolution of different organizations, and in this perspective several institutions have used to store and manage its data from several sources. At the time, the relational model and systems reigned in the management and exploitation of this data thanks to their strict mathematical principles, efficiency and profitability. The web2.0 revolution has caused an exponential increase in the amounts of data received and coming from many sources and in many forms by building Big Data, and as long as the data is more important, then the related data processing systems, will be more reliable. But relational systems have major limitations in managing and exploiting Big Data, which has given rise to the emergence of NoSQL systems to overcome these limitations. This situation pushes institutions to adopt the new system, without losing their data accumulated for years, and builds their present and their future, expressing the need for the migration of their databases from the relational system to another NoSQL. With this in mind, several studies and approaches have been created to meet this need, but their effectiveness and closely depend on its analytical logic and the methodology applied. In this paper, we will propose a methodology and analysis method to identify the elements of relational databases, that we need to migrate to the NoSQL system, as well as the relational and NoSQL systems models and meta-models, which will be used in a proposed approach for migration, respecting sequence based on layers and parallelism.

Keywords: Migration · NOSQL · Database · Analyzing · Modeling · Big Data

1 Introduction

Currently, data is doubling its volume every two years [1], knowing that an amount of data will be created in 2025 in the order of 163 trillion gigabytes [2]. This gives rise to what is called BigData, which can play a very important role in analytical systems. However, this enormous mass of data poses a major problem when it is managed by relational systems. Relational systems implement the ACID principles (Atomicity, Consistency, Isolation, and Durability) and ensure that data is consistent at all times, by applying a set of rules and barriers at the level of data manipulation and data structure. Among these barriers at the level of data manipulation we cite: the referential integrity constraints

R. Saidi et al. (Eds.): ICATH 2021, LNDECT 110, pp. 206–217, 2022.
https://doi.org/10.1007/978-3-030-94188-8_20

which prohibit the addition of a value as a foreign key without it being existing as the value of primary key in the database, also prohibited removing a value from the primary key as long as it is used elsewhere as a foreign key. Also among the constraints of relational systems at the structure level, we cite the dependence between fields, the obligation of the existence of the primary key, the obligations of the relational system such as access rights, authentication, and the internal link between physical files; which are frozen by the system and no longer by the database designers, etc.…

These constraints with the mathematical model which organizes the tables, their structures, their data, and their relations, form a real obstacle to storing and managing Big Data. We must know that Big Data can receive heterogeneous data in several forms or structures, with a considerable speed of reception, and sometimes it goes beyond the main structure and forms structured, semi-structured, or even unstructured databases. This situation requires a simple and flexible data management system in order to manage this type of data and at the same time, it favors the horizontal scalability of the system by adding as many database servers as we want, building linked servers between them in a specific order to build a single distributed data container, which will contain distributed and replicated data according to a simple and flexible principle. Also, this new system must guarantee the availability of data at all times, thus it must increase the performance of the system in terms of response time and in terms of advanced data analysis. This situation goes far beyond relational systems, which offer a rigid, fixed, and constrained data structure. Thus they do not offer horizontal scalability by adding other servers, on the contrary, they offer vertical scalability, which impacts data availability and increased response time. Relational databases are designed to operate on a single server. Relational databases cannot meet the increasing demands of data processing [3]. To overcome these circumstances, NoSQL systems have been designed to handle big data and to overcome the limits of relational systems, by adopting horizontal scalability, flexible and simple schemes to accommodate and analyze heterogeneous data, and also adopting a set of tools to increase the efficiency of these systems. These foundations form the power of NoSQL systems, which are distinguished by four categories: key-value store, document store, column store, and graph store. Each one adopts a specific model of data representation, and several tools that use this representation. In the world, there are several organizations that which require a means for the management of their databases; these databases can be small or giant, but in continuous evolution, which shows the challenge to adopt NoSQL systems. At this level, it is legitimate that these organizations want to migrate to the NoSQL system, and at the same time keep and conserve the old data stored in the old relational systems, accumulated by many ten years of exercises in the market and advanced studies. This data plays a very important and strategic role for these companies, especially in the management of business processes, in the control of the various operations carried out, in its management, its governance, and in decision support. Therefore several studies and approaches are developed to succeed and implement effective tools to migrate this important data from the relational system to another NoSQL.

Even though NoSQL systems offer a better solution for Big Data than relational systems but, for small DataBases, the relational system remains very effective especially in data consistency, optimization in terms of storage, the possibility of making simple or complex calculations operations, extraction, and derivation of new data. For a complete migration to NoSQL systems, the approach must migrate most or all of the data stored in the relational system and its functionality, which helps to measure the success and effectiveness of the migration approach.

2 Relational Database and NoSQL Database Overview

2.1 Relational Database Management Systems

Relational databases are designed to store, manage and exploit data following a strict mathematical model, which Aim to the validity of the data entered, their consistency and dependence between them, the optimization in terms of storage capacity, the simplicity of operations, and the existence of a set of formalities to calculate and extract new information's. All these advantages are designed on the basis of a strict model, called MCD, comprising entities, associations, and the multiplicity of their objects which relate to each other according to the associations designed. Constraints and programs play an important role in perfecting the validity and consistency of the data stored inside the tables. The Relational Database Management System (RDBMS), implements all these concepts of relational databases and offers a powerful tool playing the most important role in the success of many business solutions. With the evolution of data to very large sizes, the RDBMS presented several limitations and problems, especially in the evolution of their system by horizontal scalability, the availability of data at all times, the response real-time, and the processing of incomplete information or overflowing the entity structures. This situation has pushed organizations, using relational databases, to overcome these difficulties, by changing the data management system without losing their data, which they have accumulated for all years ago, and without losing the advantages of DBMSR. Knowing that the data stored by these organizations can play an important role in the management and control of the business processes and services of these organizations, also it can play decision support management assistance and the governance of these organizations and their Information Systems. An RDBMS is responsible for managing the data and making it available. Client applications that use RDBMS do not need to implement this management. The physical structure of the data has no influence on the code of the client application. Data processing requests do not systematically change the code of an application, especially thanks to the views. An RDBMS verifies the integrity constraints established by the user and thereby guarantees the consistency of the data. An RDBMS manages concurrent accesses to avoid inconsistencies during modification. This type of relational system implements the ACID principles, then presented by [4]:

- Atomicity
- Consistency
- Isolation
- Durability

2.2 NoSQL Databases Systems

NoSQL database management systems are distinguished from relational DBMSs (RDBMS), particularly by their performance. Records are no longer tabular but usually in the form of collections managed by a key/value system. They make it possible not to restrict the base to its schema. The system is simplified and responds to the problems of Big Data thanks to its scalability principle (it is possible to maintain performance by increasing the number of servers). It also allows greater portability on different platforms (Android, Windows, and Linux) while using fewer machine resources (memory and processor). NoSQL eliminates a certain structure in advance. We are no longer constrained by tables and data whose type is specified. It is also possible to distribute the data more easily by adding new servers. These databases also meet the needs for rapid change, especially for developers who do not have the exact structure of their data during the various phases of development [5].

The NoSQL movement represents a set of technologies offering new architectures for databases. These new types of databases have been stipulated to align with the current technological evolution related especially to the concepts of Big Data and Cloud Computing. The NoSQL databases allow one side to make the system much more powerful especially with the sharp increase in data volumes, and on the other hand, they also make the system more resistant to failures by taking advantage of the possibilities of Cloud Computing. This type of relational system implements the BASE principles, then presented by [4]:

- Basically Available
- Soft State
- Eventually Consistent

3 Related Work

Currently, scientific research rarely discusses how to effectively validate the best way to migrate data from RDBMS to NoSQL databases to ensure the quality of data migrated. Therefore, those needs become a major concern and an interesting research topic. According to [6] data has become an essential hub for information processing, authors present a framework, named NoSQLayer, based on two parts; a migration module and a mapping module, the first one is a set of methods enabling seamless migration between RDBMS and the second one provides a persistence layer to process database requests to translate and execute requests in any RDBMS. in 2016, Rania Yangui et al. [7] are presenting an automatic transformation of a multidimensional conceptual model into two NoSQL databases; column-oriented and document-oriented models, and to validate the transformation, they implemented four data warehouses using Cassandra (as a column-oriented NoSQL system) and MongoDB (as a document-oriented NoSQL system), results show that MongoDB with hierarchical transformation is more suitable when dealing with OLAP queries. In the same year 2016, Ying-Ti Liao et al. [8] have proposed a flexible and highly modularized data adapter for hybrid database system by using a general SQL layer accepting queries from application services, it also controls query flow during database transformation, therefore, they can provide a seamless mechanism to use RDBMS and NoSQL database at the same time. NoSQL databases are designed to solve data processing problems in volume [6]. Cloud computing also has enabled the database as a service model to manage large volumes of user-generated data using NoSQL data repositories, authors in [9], have presented a NoSQL data migration framework to foster data portability across cloud-based heterogeneous NoSQL data repositories; results show that the document-based database supports efficient migration of data, whereas the graph database ensures effective management of large volumes of data [10]. Data integration of heterogeneous database systems raises many interesting questions, in 2017, Agnes et al. [11] have proposed a meta-model based data merging approach, which allows querying of data from heterogeneous database systems simultaneously, results of database queries are translated to JSON objects and data merging is done by concatenating separate JSON files, and the result can be represented in any kind of data visualization form (tables, CSV files, etc.). Big data also has become a crucial issue and one of the most important technologies in the modern world, in [12], authors have presented a study by proposing a document-oriented data model for big data and then applies this model to migrate relational databases applications to NoSQL.

4 Analyzing and Modeling Systems

4.1 Analysis and Modeling Methodology

In this section, we will present the most interesting part of our article; this is the basis for correct modeling, which will subsequently be the fundamental and main element in our migration approach. In this article, we intend to identify all of the elements constituting the relational database, classify them according to their nature: structure, semantics, and data. Then we propose to follow the steps of our methodology for a smart approach to complete migration from relational Databases to NoSQL systems. To do this, we will proclaim our analysis and modeling methodology, which consists of carrying out the following eleven steps:

- The first one consists in specifying the System or the NoSQL tool which will be the destination System of the migration.
- The second consists in identifying all the elements constituting the destination system of the migration.
- The third consists of proposing a diagram summarizing these elements and producing a representative model.
- The fourth consists in presenting the meta-model of the destination system.
- The fifth consists in identifying all the elements constituting the relational databases.
- The sixth consists in classifying these elements according to the natures: Data, structure and semantics, to have a vision on the role of each element and its importance.
- The seventh consists of reclassifying these elements according to the categories: element already exists in the destination system (supported by system), element already guaranteed by the destination system (to be redeployed), element that cannot migrate to the destination system because it is contradictory with its foundations (not supported by system), elements which has no added value in the destination system and the elements to be effectively migrated with an explanation of its role (to be redefined

in the system), using the elements of the second step of this methodology, and in the form of a table.

- The eighth is to determine the components of the relational database to migrate.
- The ninth consists in modeling the elements to be migrated to the NoSQL system, by proposing a summary diagram.
- The tenth is to produce a Meta model of the relational databases for migration to the destination system based on the elements of step 9, chained by its meta-model.
- The eleventh step consists in carrying out a correspondence between the source and destination models, starting from step 9 and using step 3. This correspondence must also respect the meta-models carried out in steps 4 and 10. The result of this step produces the model which represents the structure of the database after its transformation. This last model will be used in the algorithms of the approach to go from a relational database to a NoSQL database as a result.

To better understand our methodology and at the same time to show the dependence and the order of the different stages of our methodology, we propose the following explanatory diagram, which will formulate the following sections of our analysis and modeling:

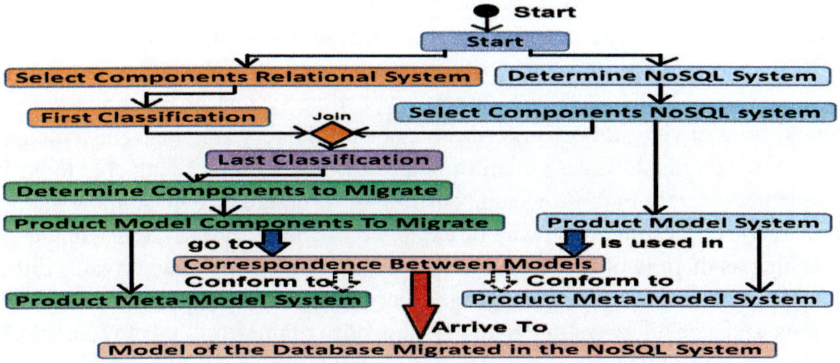

Fig. 1. Steps of our methodology to product models in our approach

4.2 Category and Tool of the NoSQL Destination System

To complete our migration approach study, it is necessary to specify the destination category, which the old databases will be transformed. The authors of this paper have shown that Document-oriented NoSQL databases are the most suitable for transforming relational databases [7] because it's similar in their structure to the Relational system. And their structure can be favorite exploitation and extraction of the new information from their data. Also, she's the only one who presented data in row information that is the same as a relational system. Also, she's the only one who offered similar functions, like triggers and joins operations. MongoDB is the favorite system as a destination system because it implements the principles of the Document oriented NoSQL category and it provides high performance in its category NoSQL.

MongoDB stores data in JSON or Binary JSON (BSON) files, as collections, which are structured in their own files as a list of objects. Each object stored can be named 'Document', and respects JSON's form. The object in JSON has a reference and an object, which contains a several attributes with its values. This structure is very similar to the projection of a table on one row in the relational system. So one row in a table can be transformed into an object JSON, and one table stores a list of rows that can be transformed by a collection of objects. So we can transform all tables to JSON files. In another hand, the stored functions can be transformed to programs JavaScript, to might manipulate JSON files. Other data, which control structure in a relational system, can be transformed and stored in Mongoose, which controls the structure of data in MongoDB. MongoDB can use a framework named "Mongoose", which defines the structure of documents in their schemas, which will be stored in JSON files. In the following section, we will see the components of the relational system that we must migrate to MongoDB.

4.3 MongoDB System and Components

MongoDB is one of the celebrated tools in all NoSQL systems; it stores their Data in the collection, which contains several documents, which respect the JSON's form. It uses in their storage Data a binary JSON (BSON) for their simplicity, usefulness, and format as a JSON structure. It is endowed with several features which make it very efficient, very famous, powerful, scalable, secure, and other positive system characteristics. From a functional point of view, it is endowed with a powerful query language and a mongoose framework which makes it very interesting as a flexible system, being able to go from non-structure to a very interesting and constraining structure. We must know that when the structure is interesting then it can be exploited by the analysis systems, on the other hand as long as there is no structure then the exploitation of the data remains difficult. MongoDB implements a query language, that proves the speed of the most CRUD operations, and also allows to integrate the mongoose Framework, this Framework uses the NodeJS technology to develop certain JavaScript functions at the server level, which want to define schemas of the recorded documents and to enrich the system with a set of functionalities to be developed by the developers. These schemas follow JSON notation and define necessary and optional components in documents through their schemas and a set of attributes such as required and validator. These attributes will have the values true or false or a function that returns true or false according to a specific treatment. This schema also allows, thanks to their validators, to validate or invalidate CRUD operations [13].

4.4 Destination System Modeling

In the following figure, we present a simple schema of MongoDB and Mongoose, which present their interesting component and their functioning:

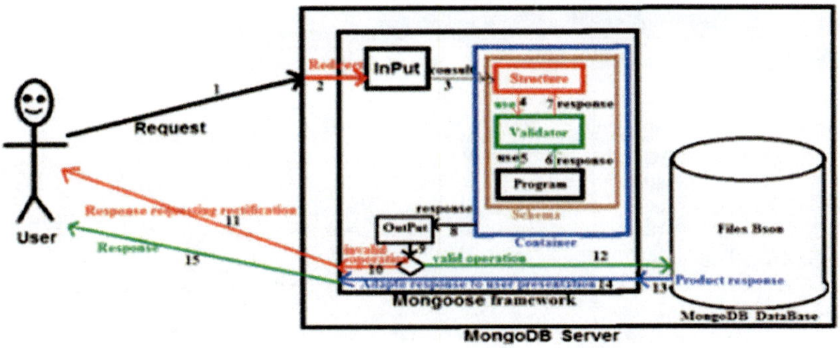

Fig. 2. Simple diagram of MongoDB and Mongoose: structure and mechanism

In the following figure, we present a model of MongoDB as Document Store Database.

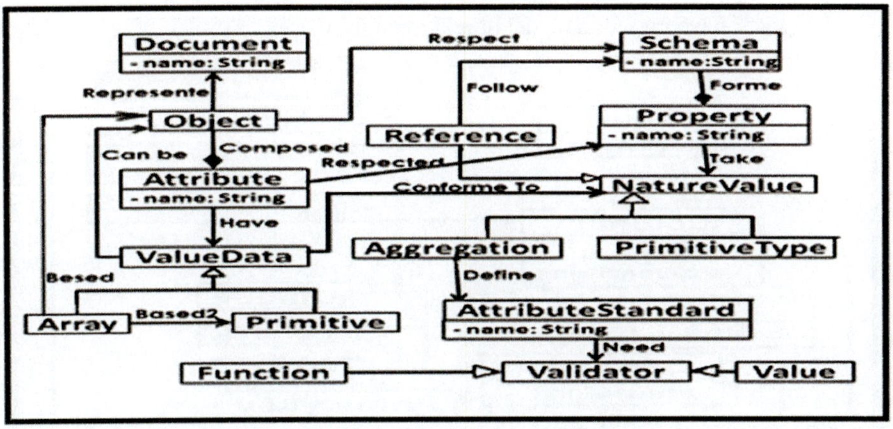

Fig. 3. Model for Document Store Database

4.5 Analyzing Relational System: Components and Modeling

To define a framework of a complete and smart migration from the relational Database system to the NoSQL Database system, it's necessary to specify all the constituents of the relational system, so that the approach can define their transformation rules. In this study, we present, in the Fig. 4, an exhaustive list of the constituents of the relational system to migrate to the NoSQL system, classified in their three natures that presented by: structure, Data, and semantic:

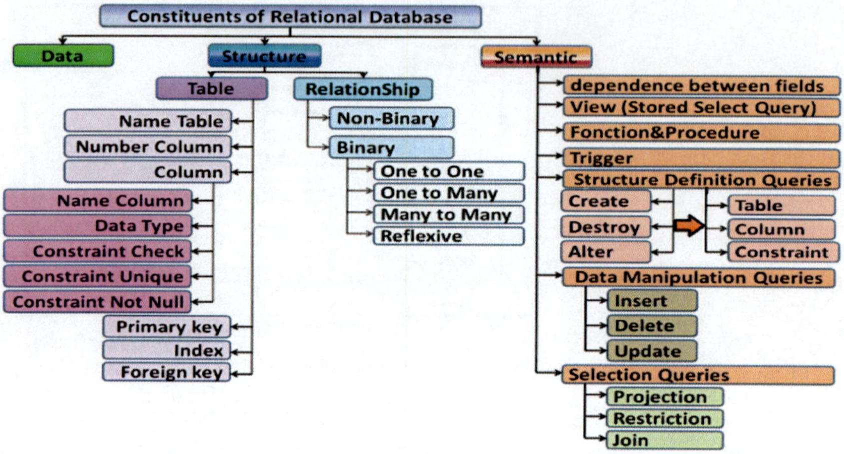

Fig. 4. List of all components of relational database

In this section we present a model of relational databases.

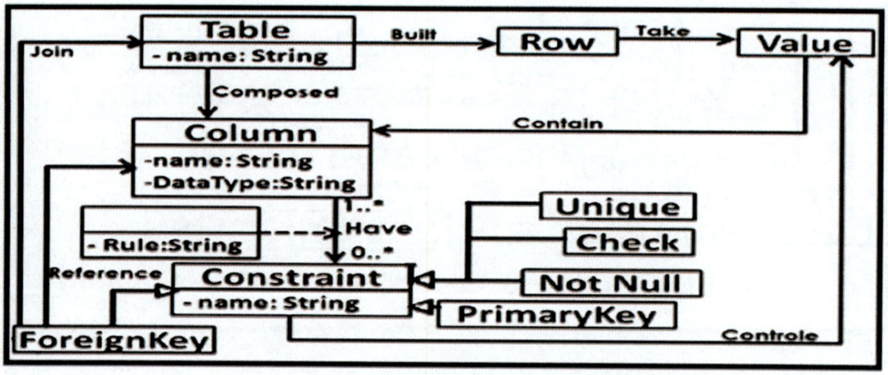

Fig. 5. Model of relational databases

4.6 Correspondence Between Relational Databases and MongoDB

This section is the result of strict analysis work, the objective of which is to determine the scope of the migration, which determines the elements to migrate and at the same time the correspondence between the different elements of the relational databases to those who build MongoDB Databases. Thus the models are to be respected during operations of transformation of structures and data migration. This introduces the stage of transformation and implementation of the entire approach. In the following table, we present this correspondence:

Table 1. Correspondence between the components of the RDBMS and mongoDB

Natures of data	Elements in RDBMS	Elements in MongoDB
Data	Records (values)	Object (values)
Structure data	Table	Collection
	Row	Document
	Name of table	Name of collection
	Column	Field
	Column Name	Field Name
	Data type	Data type
	Check constraint	Mongoose + schema + attribute Validator + function JavaScript
	Unique constraint	Mongoose + schema + attribute unique
	Not null constraint	Mongoose + schema + attribute required
	Primary key	Mongoose + schema + attribute required + attribute unique
	Index	index
	Foreign key	To migrate
	relational Ship	To migrate
Semantic data	Dependence between fields	Schema in Mongoose
	View (stored select query)	To migrate
	Function & procedure	To migrate
	Trigger	To migrate
	Structure definition queries	not supported and without added value
	CRUD operation	Insert, delete, update and find operations
	join tables	To migrate
	usual functions	aggregation, time management and string functions

The queries that define and manipulate structure in relational data base system are not supported by NoSQL system, because NoSQL system implements a simple and flexible structure of their information in their files. Indeed we can store the document {"FirstName": "James", "LastName": "gizze", "age": 18} in the same file with documents {"userName": "James33", "Email": "j.gizze@gamil.com", "age": 18} and {"FirstName": "James", "LastName": "gizze"} without any changing.

After watchful reading this table, we can resume that our approach of migration, can transform directly some components, and must transform and migrate someone to their according roles in the destination system. In the following figure, we present the components that need to transform to another form, to be adequate with the new system and with their roles:

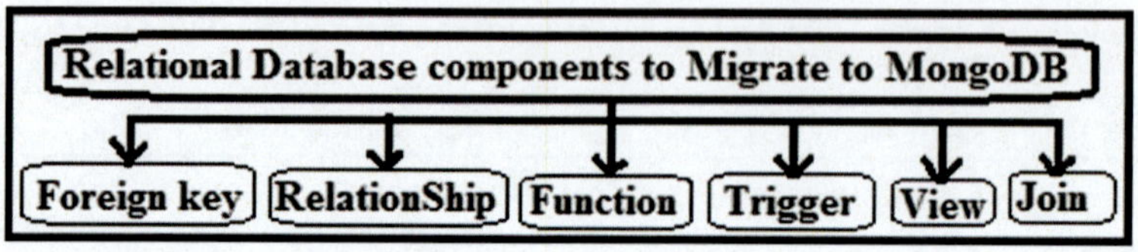

Fig. 6. The elements of relational database to Migrate to MongoDB system

4.7 Architecture of the Transformation and Migration Mechanism

In this section, we propose an architecture of our transformation and migration mechanism, in the way to implement our proposal approach:

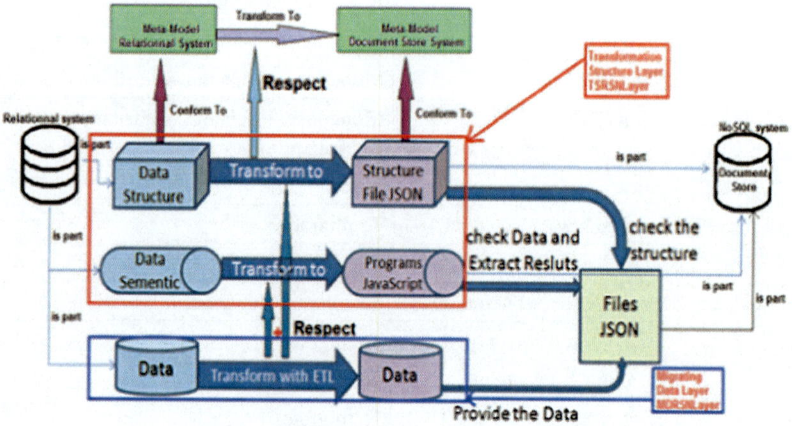

Fig. 7. Architecture of our proposed approach

5 Conclusion

The purpose of this paper is to propose a new methodology of analyzing and modeling the principal's steps and tools to define our approach, called "TMSDRDND Approach", for complete migration from a relational database to a document-oriented NoSQL database in a distributed environment. This analysis was successful in product a set of list step of our methodology; to define a several models and meta-models in profit of our approach for resolves many problems in the last same approaches. Our approach has two phases: the first will transform the structure and the semantic Data from a relational system to two files in a NoSQL system, in a layer called "TSRSNLayer" and according to many transformation rules, which control their structure and propose some functions. The second uses an ETL tool, which we develop according to the first layer, to extract, transform and load data. This step represents the second layer called "MDRSNLayer". In the future work, we will present an implementation for "TSRSNLayer" and "MDRSNLayer", by the definition of a heuristic, toward evaluating the "TMSDRDND Approach".

References

1. The digital universe of opportunities. https://www.emc.com/leadership/digitaluniverse/2014iview/executive-summary.htm
2. Data is eating the world. https://whatsthebigdata.com/2017/04/18/idc-163-trillion-gigabytesof-data-will-be-created-in-2025/
3. Solanke, G.B., Rajeswari, K.: Migration of relational database to MongoDB and Data Analytics using Naïve Bayes classifier based on Mapreduce approach. In: 2017 International Conference on Computing, Communication, Control and Automation (ICCUBEA), pp. 1–6. IEEE, August 2017
4. Copeland, R.: MongoDB Applied Design Patterns: Practical Use Cases with the Leading NoSQL Database. O'Reilly Media, Inc. (2013)
5. Db-engines.com, DB-Engines - Knowledge Base of Relational and NoSQL Database Management Systems (2015). https://db-engines.com/en/
6. Rocha, L., Vale, F., Cirilo, E., Barbosa, D., Mourão, F.: A framework for migrating relational datasets to NoSQL. Procedia Comput. Sci. **51**, 2593–2602 (2015)
7. Yangui, R., Nabli, A., Gargouri, F.: Automatic transformation of data warehouse schema to NoSQL data base: comparative study. Procedia Comput. Sci. **96**, 255–264 (2016)
8. Liao, Y.T., et al.: Data adapter for querying and transformation between SQL and NoSQL database. Futur. Gener. Comput. Syst. **65**, 111–121 (2016)
9. Bansel, A., González-Vélez, H., Chis, A.E.: Cloud-based NoSQL data migration. In: 2016 24th Euromicro International Conference on Parallel, Distributed, and Network-Based Processing (PDP), pp. 224–231. IEEE, February 2016
10. Atzeni, P., Bugiotti, F., Cabibbo, L., Torlone, R.: Data modeling in the NoSQL world. Comput. Standards Interfaces **67**, 103149 (2020)
11. Vathy-Fogarassy, Á., Hugyák, T.: Uniform data access platform for SQL and NoSQL database systems. Inf. Syst. **69**, 93–105 (2017)
12. Hamouda, S., Zainol, Z.: Document-oriented data schema for relational database migration to NoSQL. In: 2017 International Conference on Big Data Innovations and Applications (Innovate-Data), pp. 43–50. IEEE, August 2017
13. Mongoose v6.0.12: Schemas. https://mongoosejs.com/docs/guide.html

Sentence Generation from Conceptual Graph Using Deep Learning

Mohammed Bennani(✉) and Adil Kabbaj

INSEA, Rabat, Morocco
{m.bennani,akabbaj}@insea.ac.ma

Abstract. Deep learning (DL) is increasingly used in different fields of research in several domains, among which is Natural Language Generation (NLG). The objective of this work is to study the feasibility of using DL for sentence generation from a Knowledge Representation Formalism, namely Conceptual Graphs (CG). As a first step, we realized an extensive bibliographic work to identify the different aspects and uses of NLG that could be related to our work. Then we trained a sequence-to-sequence model, using a corpus made up of sentences paired with conceptual graphs, that aims to generate sentences from new conceptual graph inputs. The results were evaluated using Bilingual Evaluation Understudy (BLEU) scoring a high value of 0.96.

Keywords: Natural language generation · Conceptual graph · Seq2seq

1 Introduction

Natural Language Generation (NLG) is a relevant topic in Communication Systems, Natural Language Processing (NLP) and Artificial Intelligence in general. It could be defined as the "process by which thought is rendered into language" (McDonald 2010). Concretely, NLG aims to produce understandable coherent texts in human language from different sorts of input (image, text, graph). It can be used for several purposes such as automatic synthesis, translation, image capturing, systems for answering questions and other communication systems.

Many researches focused on generating text from a semantic representation (Flanigan et al. 2016; Konstas et al. 2017; Ferreira et al. 2017; Puzikov and Gurevych 2018, in Cabezudo and Pardo 2019). For a long time, grammar-based generation was the most widely adopted approach (Cheyer and Guzzoni 2007; Mirkovic and Cavedon 2011). This type of model is known for its robustness and suitability (Wen et al. 2015) which - according to Stent and Molina (2009) - characterizes a good generator. However, a key limitation to this approach is that it can induce errors since it automatically generates the alignments between surface tokens and the graph nodes (Wang et al. 2020). More recent researches used a neural sequence-to-sequence model representation and obtained promising results (Konstas et al. 2017; Beck et al. 2018). However, the application of this approach remains very limited or even non-existent for the generation of text from Conceptual Graph (CG).

R. Saidi et al. (Eds.): ICATH 2021, LNDECT 110, pp. 218–227, 2022.
https://doi.org/10.1007/978-3-030-94188-8_21

Conceptual Graph is a graphic language for knowledge representation, which is based on the formalism of semantic networks in artificial intelligence, semantic of Natural Language Processing, semantic for Databases and the existential graphs of Charles Sanders Peirce (Sowa 1992). According to Sowa (1992), the goal of the CG is to express meaning with precision while being easily read by humans and calculable (Sowa 1992). The CG formalism is used in many areas (data base semantics, natural language processing, knowledge-based systems, information systems, communication systems, multi-agent systems, etc.). A CG can be the result of a reasoning process or a complex natural language process, and it is useful to communicate this result to a human user in natural language instead of a formal language. Hence the importance to use Deep Learning in sentence generation from CG, and the reason of choosing it for this work.

CG is generally made up of two types of nodes: concept nodes (also called concepts) and relationship nodes (also called conceptual relations). An example of a CG corresponding to the sentence "the brave knight fights the evil dragon" in linear and graphic form is shown in Fig. 1 and 2 respectively.

```
[knight: *p1 ] -
        -propertyOf->[state_brave],
        <-agentOf-[ action_fight : *p2 ]-objOf->[tomato : *p3 ]-propertyOf->[ evil]
```

Fig. 1. Linear form of conceptual graph

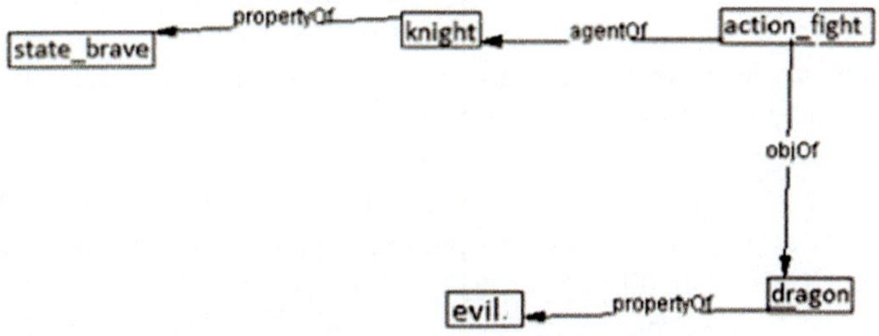

Fig. 2. Graphic form of conceptual graph

In this work, we used a sequence-to-sequence model (seq2seq), generally used in translating text from one language to another, to generate text from CG. The approach we adopted is to consider the CG as a language (with its own grammar and vocabulary) and to train the model to "translate" graphs into sentences in a natural language. Before training the model, we represented the CG in its linear form and we applied transformations to it so that it is written in one line that would be interpreted as a sentence in the CG language.

The paper is organized as follows: after the introduction in the first section, the second section covers related work. The third section presents the methods, mainly the Sequence-to-Sequence model used. The fourth section is related to the results obtained, the performance evaluation and samples of sentences generated, followed by the future work planned and a conclusion.

2 Related Work

Many researchers have worked on sequence-to-sequence learning and their models showed great performance across a broad range of applications. Nevertheless, a very few numbers of these applications have been applied to parsing and generating semantic representations. Konstas et al. (2017), proposed a Seq2Seq model for parsing and generation text from Abstract Meaning Representation (AMR), the main problem was the non-sequential nature of AMR graphs and the lack of labeled data (Konstas et al. 2017). In addition to that, the high cost of annotating training data in AMR limits neural network models use (Misra and Artzi 2016; Peng et al. 2017; Barzdins and Gosko 2016; Konstas et al. 2017).

To resolve this particular problem, Konstas et al. (2017) proposed a training procedure that uses millions of unlabeled sentences and accurate preprocessing of the AMR graphs after a phase of linearization and anonymization. Through their work, Konstas et al. (2017) trained seq2seq models using graph-isomorphic linearization and reduced sparsity using unlabeled text.

Besides Konstas et al. (2017), Pourdamghani et al. (2016) also used linearization approaches, which showed that graph simplification and anonymization are key to good performance. However, compared to the work done by Beck et al. (2018), linearization incurs in loss of information (Beck et al. 2018).

These more recent approaches transform the graph into a linearized form and use off-the-shelf methods such as phrase-based machine translation (Pourdamghani et al. 2016) or neural sequence-to-sequence models (Konstas et al. 2017) to bypass limitations related to the fact that alignments between graph nodes and surface tokens are required. These alignments are usually automatically generated so they can propagate errors when building grammar. Nevertheless, such approaches do not provide a global knowledge of the graph structure, key information (Beck et al. 2018).

To solve these problems, Beck et al. (2018) proposed an encoder-decoder architecture for graph-to-sequence learning using Gated Graphs Neural Network (GGNN). In particular, the approach solved issues related to graph-based networks using graph transformations without modifying the underlying architecture.

However, this architecture has two major limitations: (i) The fixed number of layers in GGNNs, despite graphs having a variable size in the number of nodes and edges; (ii) Since edge labels are represented as nodes, they end up sharing the same vocabulary and so the same semantic space, even though they are different entities (Beck et al. 2018).

Besides these limitations, Beck's approach is not used often because of its complexity, we will thus base our research more on the work of Konstas et al. (2017) as it's the one that achieves the best performance in our case.

3 Methods: Sequence-to-Sequence Model

We used in our study a sequence-to-sequence neural architecture (Sutskever et al. 2014), published by Google in 2014, typically employed in Neural Machine Translation (NMT). Seq2Seq consists of training models to convert (translate) sequences from one domain (source language) to sequences in another domain (target language), by modeling the conditional probability P(S|G) of translating a source sequence $G = (g_1, \ldots, g_n)$ to a target sequence $S = (s_1, \ldots, s_m)$. The strength of this model is that the input and the output can be of different sizes and categories.

During the training phase, the model is fed a corpus training constituted of input-output pairs, and it settles its parameters to maximize the probability of generating a correct output given a new input (Albarino 2019).

The basic form of Seq2Seq model, called Encoder-Decoder Sequence to Sequence Model, includes two Long Short-Term Memory (LSTM) models: (i) an Encoder that reads the input sequence, computes (encodes) algebraic representations (called the hidden state and cell state vectors). These representations summarize the information of each element of the source sequence and his relation with the elements that precede it; (ii) And a Decoder, that has the final states of the Encoder as input, and generates (decodes) one target word at a time. It, hence, decomposes the conditional probability as shown below.

$$P(S|G) = P(s_1|G)\,P(s_2|s_1, G)\,P(s_3|s_1, s_2, G) \ldots \underbrace{P(s_T|s1,\ldots,sT\text{-}1,G)}_{\substack{\text{Probability of next element in}\\ \text{the target sequence, given target el-}\\ \text{ements so far and source sequence}}} \tag{1}$$

Bidirectional Encoder Representations from Transformers (BERT) (Devlin et al. 2018) is a more sophisticated and recent model, descripted on a paper published by Google in 2018. It revolutionizes Machine Learning for Natural Language Processing. BERT's main strength is that it applies a bidirectional training of Transformer to language modelling. As opposed to Encoder-Decoder model, which read the text input sequentially (from left to right or vice versa step by step) BERT reads the entire sequence at once. In fact, BERT is more non-directional than directional model. The main goal of BERT is to generate a context of the input and make sure that the output is near enough of this context (Horev 2018). Hence, BERT's area of predilection is Classification Tasks, Question Answering tasks and a Named Entity Recognition. It can also be used with NMT to refine the translation: for example, in general we will tend to translate the English word "Mouth" by the French word "Bouche" but if we are talking about an animal the adequate translation would be "Gueule". Despite its great potential, BERT will not be useful in our case, as we operate a transformation of representation in the same linguistic domain. However, it could be used for parsing a CG expressed in one language into a sentence in another language.

4 Results

4.1 Corpus Building

To carry out our study, we had to have a learning corpus made up of two sets: a set of sentences *S* and a set of conceptual graphs *G*, where each sentence of *S* is associated with a conceptual graph of *G*. Corpora that associate a large number of conceptual graphs with natural language sentences are very rare or even non-existent.

To remedy this problem, we built a referential made up of 100,000 sentences formulated in a simplified English called "Attempto Controlled English" (ACE), a controlled natural language developed as part of the research project "Attempto" by the University of Zurich (Fuchs et al. 2005). We made sure that the majority of ACE grammar was covered by these sentences. Then, by adding modifications on a semantic analyzer integrated into Amine platform (Kabbaj 2009; Nasri et al. 2013), we were able to generate for each sentence of the referential a corresponding CG represented in its linear form (Amine platform offers the possibility to generate CG in different forms). The next step was to apply an automatic processing that we developed in order to write each graph in a single line. Furthermore, the writing of the graphs was optimized to minimize the vocabulary defined for the set of graphs, and facilitate the learning phase.

We thus obtained a corpus made up of two sets: a set of sentences on ACE language and a set of CGs. This allowed us, after the training phase, to transform a conceptual graph into a sentence in ACE and vice versa.

The grammar structure of the training set can be summarized as follows:

- Declarative sentence:

$$S = NP + VP + \text{"."} \tag{2}$$

- Interrogative sentence:

$$Q = (\text{"What"} \mid \text{"Why"} \mid \text{"How"}) + \text{"does"} + VP + \text{"?"} \tag{3}$$

With:

$$NP = (\text{Article} + [\text{adjective}] + \text{Noun}) \mid \text{proper noun} + [(\text{"who"} \mid \text{"that"} \mid \text{"which"}) + [VP] \tag{4}$$

$$VP = \text{Verb} + NP + \text{"."} \tag{5}$$

- NP: Noun Phrase.
- VP: Verb Phrase.
- [X] means X is optional.
- X | Y means X or Y.

Examples of generated sentences:

- Sami eats.
- Sami eats an apple.
- Sami eats the delicious apple.
- The knight eats the delicious apple.
- The young knight eats the delicious apple.
- Sami meets the young knight who eats the delicious apple.
- Sami who eats the delicious apple meets the young knight who fights the villain dragon.

To ensure the good quality of these sentences, we used the BLEU scoring.

4.2 Performance Evaluation

As mentioned above, we formed 100,000 pairs of sentences - of different structures - and their corresponding conceptual graphs. We randomly selected 70,000 of these 100,000 formed pairs, 80% of which were used for the learning phase and 20% for the automatic test. The remaining 30,000 sentences were used as a reference corpus to calculate the BLEU score of the model. BLEU is the most used reference score to evaluate the quality of the sentences generated that varies from 0 to 1 (1 being the highest score). We obtained a score of 0.96.

4.3 Samples

Below, Tables 1, 2 and 3 show examples of the sentence generation process of different structures. For each example, we provide the CG used as an input, the linearized form of the CG and the generated sentence.

Table 1. Example of a declarative sentence generation

CG input	[action_smell : *p1] - -agentOf->[person_man :Saad], -objOf->[lemon]
Linearized CG	[action_smell:*p1] - -agentOf-> [person_man:Saad] , -objOf-> [lemon]
Generated sentence (declarative sentence)	saad smells a lemon.

Table 2. Example of an interrogative sentence generation

CG input	[pp_cg : [write : *p1] - -agentOf->[person_man :Hamza], -locOf->[location]]-propertyOf->[cg_question :where]
Linearized CG	[pp_cg: [write:*p] - -agentOf-> [person_man:Hamza] , -locOf-> [location]] -propertyOf-> [cg_question:where]
Generated sentence (interrogative sentence)	where does Hamza write?

Table 3. Example of a sentence generation with relative pronouns

CG input	[pp_cg : [object_knight : *c1] - -propertyOf->[state_brave], <-agentOf-[devour : *p1]-objOf->[apple : *p2]-propertyOf->[state_green]]-conjOf->[pp_cg : [action_meet : *p3] - -agentOf->[person_man :Layla], -objOf->[object_knight : ?c1]] [pp_cg : [object_knight : *c1] - -propertyOf->[state_brave], <-agentOf-[devour : *p1]-objOf->[apple : *p2]-propertyOf->[state_green]]-conjOf->[pp_cg : [action_meet : *p3] - -agentOf->[person_man :Layla], -objOf->[object_knight : ?c1]]
Linearized CG	[pp_cg: [object_knight:*c1] - -propertyOf-> [state_brave] , <-agentOf- [devour:*p2] -objOf-> [apple:*p1] -propertyOf-> [state_green]] -conjOf-> [pp_cg: [action_meet:*p3] - -agentOf-> [person_man:Layla] , -objOf-> [object_knight:?c1]]
Generated sentence: sentence with relative Pronouns	Layla meets the brave knight who devours the green apple.

5 Future Work

Future work will consist of the realization of a semantic analyzer of a simplified Arabic language - a controlled language mainly intended for primary school students. Since Arabic language is very rarely the subject of NLG with neural approach, this semantic analyzer will allow us to form a corpus that will be used in training a seq2seq model to generate Arabic sentences from CGs.

6 Conclusion

In this work, we trained a sequence-to-sequence model to the task of Sentence generation. The main objective was to study the feasibility of generating comprehensible sentences from conceptual graphs, which we verified thanks to the BLEU score obtained (0.96). This score is related to a referential we built, which was made up of 100,000 sentences and formulated in a simplified English ACE. Due to the lack of corpora pairing conceptual graphs to sentences, we were left with a limited number of sentence structures. Nevertheless, we obtained very encouraging results. Future work will be dedicated to training a seq2seq model to the generation of sentences in simplified Arabic starting from a conceptual graph.

References

Albarino, S.: Does Google's BERT Matter in Machine Translation? (2019). https://slator.com/does-googles-bert-matter-in-machine-translation/

Artzi, Y., Lee, K., Zettlemoyer, L.: Broadcoverage CCG semantic parsing with AMR. In: Proceedings of the 2015 Conference on Empirical Methods in Natural Language Processing, Lisbon, Portugal, pp. 1699–1710. Association for Computational Linguistics (2015). http://aclweb.org/anthology/D15-1198

Barzdins, G., Gosko, D.: RIGA at SemEval-2016 task 8: impact of Smatch extensions and character-level neural translation on AMR parsing accuracy. In: Proceedings of the 10th International Workshop on Semantic Evaluation, San Diego, California, pp. 1143–1147. Association for Computational Linguistics (2016). http://www.aclweb.org/anthology/S16-1176

Bahdanau, D., Cho, K., Bengio, Y.: Neural machine translation by jointly learning to align and translate. In: Proceedings of the 2015 International Conference on Learning Representations, San Diego, California. CBLS (2015). http://arxiv.org/abs/1409.0473

Beck, D., Haffari, G., Cohn, T.: Graph-to-sequence learning using gated graph neural networks. In: Proceedings of the 56th Annual Meeting of the Association for Computational Linguistics (Volume 1: Long Papers), pp. 273–283. Association for Computational Linguistics (2018)

Brandt, L., Grimm, D., Zhou, M., Versley, Y.: ICL-HD at SemEval-2016 task 8: meaning representation parsing - augmenting AMR parsing with a preposition semantic role labeling neural network. In: Proceedings of the 10th International Workshop on Semantic Evaluation, San Diego, California, pp. 1160–1166. Association for Computational Linguistics (2016). http://www.aclweb.org/anthology/S16-1179

Bjerva, J., Bos, J., Haagsma, H.: The meaning factory at SemEval-2016 task 8: producing AMRs with boxer. In: Proceedings of the 10th International Workshop on Semantic Evaluation, San Diego, California, pp. 1179–1184. Association for Computational Linguistics (2016). http://www.aclweb.org/anthology/S16-1182

Cabezudo, M.A.S., Pardo, T.: Natural language generation: recently learned lessons, directions for semantic representation-based approaches, and the case of Brazilian Portuguese language. In: Proceedings of the 57th Annual Meeting of the Association for Computational Linguistics: Student Research Workshop, pp. 81–88, July 2019

Cheyer, A., Guzzoni, D.: Method and apparatus for building an intelligent automated assistant. US Patent App. 11/518,292 (2007)

Damonte, M., Cohen, S., Satta, G.: An incremental parser for abstract meaning representation. In: Proceedings of the 15th Conference of the European Chapter of the Association for Computational Linguistics, Valencia, Spain, pp. 536–546. Association for Computational Linguistics (2017). http://www.aclweb.org/anthology/E17-1051

Devlin, J., Chang, M.W., Lee, K., Toutanova, K.: Bert: pre-training of deep bidirectional transformers for language understanding. arXiv preprint arXiv:1810.04805 (2018)

Ferreira, T.C., Calixto, I., Wubben, S., Krahmer, E.: Linguistic realisation as machine translation: comparing different MT models for AMR-to-text generation. In: Proceedings of the 10th International Conference on Natural Language Generation, Santiago de Compostela, Spain, pp. 1–10. Association for Computational Linguistics, September 2017

Flanigan, J., Thomson, S., Carbonell, J., Dyer, C., Smith, N.: A discriminative graph-based parser for the abstract meaning representation. In: Proceedings of ACL (2014)

Flanigan, J., Dyer, C., Noah, S., Carbonell, J.: Generation from abstract meaning representation using tree transducers. In: Proceedings of the Conference of the North American Chapter of the Association for Computational Linguistics: Human Language Technologies, San Diego California, USA, pp. 731–739. Association for Computational Linguistics (2016)

Fuchs, N.E., Höfler, S., Kaljurand, K., Rinaldi, F., Schneider, G.: Attempto controlled English: a knowledge representation language readable by humans and machines. In: Eisinger, N., Małuszyński, J. (eds.) Reasoning Web. LNCS, vol. 3564, pp. 213–250. Springer, Heidelberg (2005). https://doi.org/10.1007/11526988_6

Goodman, J., Vlachos, A., Naradowsky, J.: UCL+Sheffield at SemEval-2016 task 8: imitation learning for AMR parsing with an alphabound. In: Proceedings of the 10th International Workshop on Semantic Evaluation, San Diego, California, pp. 1167–1172. Association for Computational Linguistics (2016). http://www.aclweb.org/anthology/S16-1180

Horev, R.: BERT Explained: State of the art language model for NLP (2018). https://towardsdatascience.com/bert-explained-state-of-the-art-language-model-for-nlp-f8b21a9b6270

Kabbaj, A.: An overview of amine. In: Hitzler, P., Schärfe, H. (eds.) Conceptual Structures in Practice, pp. 321–348. Chapman & Hall/CRC (2009)

Konstas, I., Iyer, S., Yatskar, M., Choi, Y., Zettlemoyer, L.: Neural AMR: sequence-to-sequence models for parsing and generation. arXiv preprint arXiv:1704.08381 (2017)

Luong, T., Pham, H., Manning, C.: Effective approaches to attention-based neural machine translation. In: Proceedings of the 2015 Conference on Empirical Methods in Natural Language Processing (2015)

McDonald, D.D.: Natural language generation. Handb. Nat. Lang. Process. **2**, 121–144 (2010)

Misra, D., Artzi, Y.: Neural shift-reduce CCG semantic parsing. In: Proceedings of the 2016 Conference on Empirical Methods in Natural Language Processing, Austin, Texas, pp. 1775–1786. Association for Computational Linguistics (2016). https://aclweb.org/anthology/D16-1183

Mirkovic, D., Cavedon, L.: Dialogue management using scripts. EP Patent 1,891,625 (2011)

Nasri, M., Kabbaj, A., Bouzoubaa, K.: Integration of a controlled natural language in an intelligent systems platform. J. Theor. Appl. Inf. Technol. **56**(2) (2013)

Pourdamghani, N., Gao, Y., Hermjakob, U., Knight, K.: Aligning English strings with abstract meaning representation graphs. In: Proceedings of the 2014 Conference on Empirical Methods in Natural Language Processing, Doha, Qatar, pp. 425–429. Association for Computational Linguistics (2014). http://www.aclweb.org/anthology/D14-1048

Pourdamghani, N., Knight, K., Hermjakob, U.: Generating English from abstract meaning representations. In: Proceedings of the 9th International Natural Language Generation Conference, Edinburgh, UK, pp. 21–25. Association for Computational Linguistics (2016). http://anthology.aclweb.org/W16-6603

Sowa, J.F.: Conceptual graphs summary. Conceptual Struct.: Curr. Res. Pract. **3**, 66 (1992)

Stent, A., Molina, M.: Evaluating automatic extraction of rules for sentence plan construction. In: Proceedings of SIGDial. Association for Computational Linguistics (2009)

Sutskever, I., Vinyals, O., Le, Q.V.: Sequence to sequence learning with neural networks. In: Advances in Neural Information Processing Systems, pp. 3104–3112 (2014)

Peng, X., Wang, C., Gildea, D., Xue, N.: Addressing the data sparsity issue in neural AMR parsing. In: Proceedings of the 15th Conference of the European Chapter of the Association for Computational Linguistics, Valencia, Spain, pp. 366–375. Association for Computational Linguistics (2017). http://www.aclweb.org/anthology/E17-1035

Vinyals, O., Kaiser, L., Koo, T., Petrov, S., Sutskever, I., Hinton, G.: Grammar as a foreign language. In: Proceedings of the 28th International Conference on Neural Information Processing Systems, pp. 2773–2781. MIT Press (2015). http://papers.nips.cc/paper/5635-grammar-as-a-foreign-language.pdf

Wang, C., Pradhan, S., Pan, X., Ji, H., Xue, N.: CAMR at SemEval-2016 task 8: an extended transition-based AMR parser. In: Proceedings of the 10th International Workshop on Semantic Evaluation, San Diego, California, pp. 1173–1178. Association for Computational Linguistics (2016). http://www.aclweb.org/anthology/S16-1181

Wang, T., Wan, X., Jin, H.: Amr-to-text generation with graph transformer. Trans. Assoc. Comput. Linguist. **8**, 19–33 (2020)

Wen, T.H., Gasic, M., Mrksic, N., Su, P.H., Vandyke, D., Young, S.: Semantically conditioned LSTM-based natural language generation for spoken dialogue systems. arXiv preprint arXiv:1508.01745 (2015)

Puzikov, Y., Kawahara, D., Kurohashi, S.: M2L at SemEval-2016 task 8: AMR parsing with neural networks. In: Proceedings of the 10th International Workshop on Semantic Evaluation, San Diego, California, pp. 1154–1159. Association for Computational Linguistics (2016). http://www.aclweb.org/anthology/S16-1178

Puzikov, Y., Gurevych, I.: E2E NLG challenge: neural models vs. templates. In: Proceedings of the 11th International Conference on Natural Language Generation, pp. 463–471. Association for Computational Linguistics (2018)

Wu, Y., et al.: Google's neural machine translation system: bridging the gap between human and machine translation. CoRR abs/1609.08144. http://arxiv.org/abs/1609.08144 (2016)

Zhou, J., Xu, F., Uszkoreit, H., Qu, W., Li, R., Gu, Y.: AMR parsing with an incremental joint model. In: Proceedings of the 2016 Conference on Empirical Methods in Natural Language Processing, Austin, Texas, pp. 680–689. Association for Computational Linguistics (2016). https://aclweb.org/anthology/D16-1065

LSTM-CNN Deep Learning Model for French Online Product Reviews Classification

Nassera Habbat(✉), Houda Anoun, and Larbi Hassouni

RITM Laboratory, CED ENSEM Ecole Superieure de Technologie Hassan II University, Casablanca, Morocco
nassera.habbat@ensem.ac.ma, houda.anoun@univh2c.ma, larbi.hassouni@univh2c.ac.ma

Abstract. Sentiment analysis (SA) is one of the most popular areas for analyzing and discovering insights from text data from various sources, including Facebook, Twitter, and Amazon. It plays a pivotal role in helping enterprises actively improve their business strategies and better understand customers' feedback on products. In this work, the dataset has been taken from Amazon, which contains French reviews. After preprocessing and extracting the features using contextualized word embedding for French-language including ELMO, ULMFiT, and CamemBERT, we applied deep learning algorithms including CNN, LSTM, and combined CNN and LSTM to classify reviews as positive or negative. The results show that the combined model (LSTM+CNN) using CamemBERT achieved the best performance to classify the French reviews with an accuracy of 93.7%.

Keywords: Sentiment analysis · Deep learning · CNN · LSTM · Product reviews · ELMO · ULMFiT · CamemBERT

1 Introduction

Because there are many brands in the market; Choosing one is an arduous task for consumers. The development of e-commerce affects consumers' buying habits. Consumers make the decisions they want according to the reviews present in E-commerce.

Through SA, customers and manufacturers will have positive and negative emotions for each product, so that the company can make necessary modifications to the product to better respond to the needs of customers. Then improve their products as needed.

SA is one of the main tasks of Natural Language Processing (NLP), is used to well-identified Critics' emotions or attitudes as negative or positive. Therefore, all product reviews are summarized and sentiments are classified.

Among languages most investigated in the majority of SA implementations, we find English, Meanwhile, a little number of researchers extend focusing to other languages such as French.

French is one of the seven most used languages in the world, spoken by more than 260 million speakers [1]. To predict the sentiment of a French piece of text, SA performs a series of steps and determines the polarity of reviews. Firstly, it obtains reviews from the

R. Saidi et al. (Eds.): ICATH 2021, LNDECT 110, pp. 228–240, 2022.
https://doi.org/10.1007/978-3-030-94188-8_22

web, then preprocesses them, and finally classifies reviews using Deep neural networks that have shown the right performance in this area over the traditional ML approaches.

In this paper, we will focus on performing Sentiment Analysis with Deep Learning on French Amazon reviews and next apply the best algorithm to classify reviews in Jumia as Moroccan e-commerce.

We used different Neural Network models; Long-Short Term Memory Neural Networks (LSTMs), Convolutional Neural Networks (CNNs), Long-Short Term Memory Neural Networks (LSTMs), and a combined model that aims to combine the LSTM with CNN. Furthermore, we also compare our models using different contextualized word embedding; Universal Language Model Fine-tuning for Text Classification (ULMFiT), Embeddings from Language Models (ELMO), and French Pretrained language model (CamemBERT) in terms of accuracy, precision, recall, and F1-score. The principal contributions of this work are as follows:

1. We propose a promising ASA method using different contextual word embeddings to tackle the French language.
2. We inquire about the pre-trained CamemBERT by comparing it with static word embedding models, Secondly; we exploit CamemBERT as a feature extraction model merged with various classifiers.
3. We conduct diverse comparative experiments to show that the fine-tuned CamemBERT model combined with a hybrid network (LSTM+CNN) obtains high performance.

This paper first presents a brief literature review in Sect. 2. Section 3 illustrates the architecture and describes different used models. The results of the experiments are presented in Sect. 4. Section 5 concludes the study and outlines the future work.

2 Related Work

Sentiment Analysis (SA) has attracted extensive attention in recent years by proposing different techniques and solutions and analyzing different languages. This section exhibits different SA approaches.

To predict the sentiment of a piece of text, recent works rely on Deep learning algorithms, such as CNN and LSTM, which are compared in [2] using the Sentiment dataset for the Twitter US Airline for training English models (around 14500 tweets concerning six principal companies of US airline). As result, they found that LSTM is slightly better than CNN for detecting the sentiment of tweets in terms of accuracy and F1-score. Also integrating CNNs and LSTMs achieves better results, In [2, 3], combining the two models achieves an accuracy between 81% and 93% for Arabic SA on different datasets.

In [4], the authors proposed an analysis of Toxic Comments in Wikipedia using different text representation techniques; such as Term Frequency-Inverse Document Frequency (TF-IDF), BERT, XLNet tokenizer, and DistilBERT, and also they used Glove, word2vec, and Fasttext in the representation of text. Concerning the classification phase, they implemented four deep learning models: FeedForward Neural network (FFNN),

CNN, Gated Recurrent Unit (GRU), and LSTM. Also, they used a combination of bidirectional GRU, LSTM, and convolutional layer and two composed architectures. The 1st one was composed of an embedding layer, bidirectional LSTM layer, and a 1D convolution layer, whereas in the second one; they used the GRU layer instead of the LSTM layer. As a result of their experiments, they found that the most suitable representation is the Glove pre-trained embedding without standard pre-processing.

Adding more contextual information to the models using semantic embeddings enriches the representations and advances the state of the art for various NLP tasks including SA, and Bidirectional Encoder Representations from Transformers (BERT) is the most outstanding of these models. In [5], the authors used the BERT model for sentiment analysis and they got 94% in terms of accuracy evaluated on Twitter data, and using the pre-trained word embeddings with the CNN model [6] obtained the best results after only 3 epochs with an accuracy of 84.1%.

English has been investigated in the majority of SA implementations and research. Only some recent studies extend focusing on other low-resource languages such as French. In this context, the authors in [7] compare different French Pre-Trained Models like multilingual BERT (mBERT), FlauBERT, and CamemBERT using different classifiers like CNN, LSTM, and Conditional Random Field (CRF) applied on SemEval2016 French datasets about museums and restaurants. The results show the higher performance of FlauBERT and CamemBERT as monolingual French models compared to the multilingual model (mBERT) with accuracy exceeding 80%.

3 Methodology

We will describe in this section, the word embedding representations and deep learning models. But first, we will present the overall architecture.

The global architecture presented in Fig. 1 is composed of four principal steps:

1. Data Acquisition: This component consists of collecting and preprocessing datasets,
2. Text representation: Extraction of features using different contextual word embedding,
3. Deep learning models: Sentiment analysis using different classifiers,
4. Evaluation: Assessment of different models using the standard measurements.

3.1 Word Embedding

Feature selection is a significant task in NLP; this task does have a tremendous impact on the success of text analysis. In this part, we present the used contextualized word embedding:

ELMO. Embeddings from Language Models (ELMO) [5] developed by Allen NLP, ELMO embedding is the most advanced pre-training model on the TensorFlow hub. It learns from the bidirectional LSTM internal state and represents the context of features of input text, which means that ELMO embeddings are context-sensitive, and generates different representations for words with the same spelling but different meanings. It is better than embedding pre-training words such as glove and word2vec in various NLP tasks.

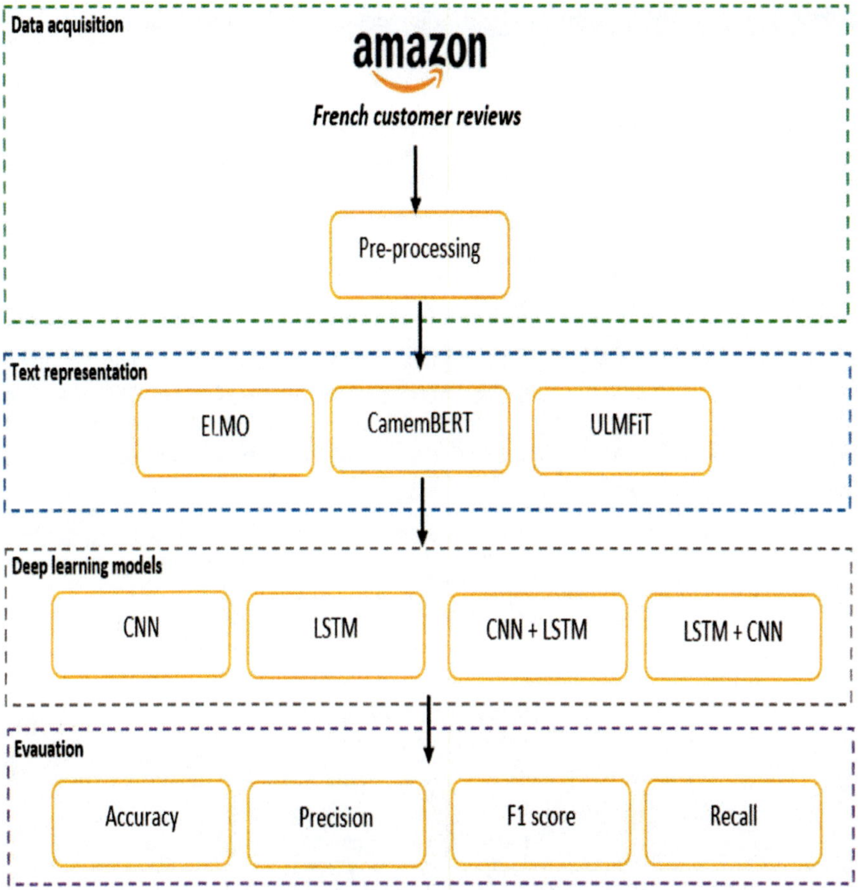

Fig. 1. The overall architecture.

In our study, we employed the pre-trained ELMo Representations for Many Languages [8].

ULMFiT. It is an effective sample-based transfer learning method [9], which can be applied to all NLP tasks, including emotion analysis, it was evaluated on the IMDb binary movie review dataset and the binary and five-class version of a dataset of the Yelp review. ULMFiT presents essential techniques for fine-tuning a language model.

In our study, we used an implementation of the ULMFiT model in the French language, this model uses the fastai v1 library based on Pytorch [10].

CamemBERT. It is a French Pretrained language model [11] trained on 138GB of the French text of the newly available multilingual corpus Open Super-large Crawled Aggregated coRpus (OSCAR). Similar to BERT [12], CamemBERT is a bidirectional multi-layer Transformer and uses the initial BERT configuration: 768 hidden dimensions, 12 layers, 12 attention heads, which equals 110M parameters.

In our study, we employed the implementation of French CamemBERT for classification NLP tasks based on Pytorch.

3.2 Deep Learning Models

This part defines the famous classifiers CNN, LSTM, and outlines the combination of the two neural networks: CNN and LSTM.

CNN. A Convolutional neural network (CNN) is a feature extraction network that can detect local predictors in large structures.

The basic principle of CNN is to introduce multidimensional data (such as images, word embeddings, etc.) into the convolution layer composed of several filters, which will learn different features. These filters can be considered as a sliding kernel on the vector representation, and the same operation is performed on each element until all vectors are overcast.

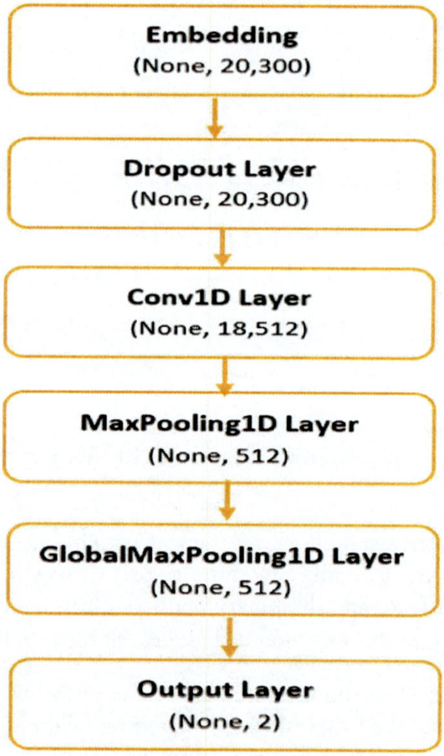

Fig. 2. CNN implementation.

As shown in Fig. 2, we constructed a one-dimensional convolution network (1D-CNN), in which the first layer is a convolution network containing different parameters,

These feature detectors use a text representation matrix to indicate specific features using the ReLU function, this layer comprising word embeddings with a fixed length of 20 words and of size 300D. The convolution output is pooled using the Max Pooling layer and Global Max Pooling layer with a filter size of 512. Then, we use dropout to stop overfitting and increase its score to 0.8. Finally, at the output level, we use the sigmoid activation function with 2 units to divide the text into predefined classes.

LSTM. It is a recurrent neural network (RNN) designed for sequence processing and to remember previously read values at any given time.

LSTM is usually composed of three gates to control the flow in and out of memory:

1. The input gate: controls the input of new information in memory.
2. The forgetting gate: controls the time when some values remain in memory.
3. The output gate: controls how much the value stored in memory affects the block output activation.

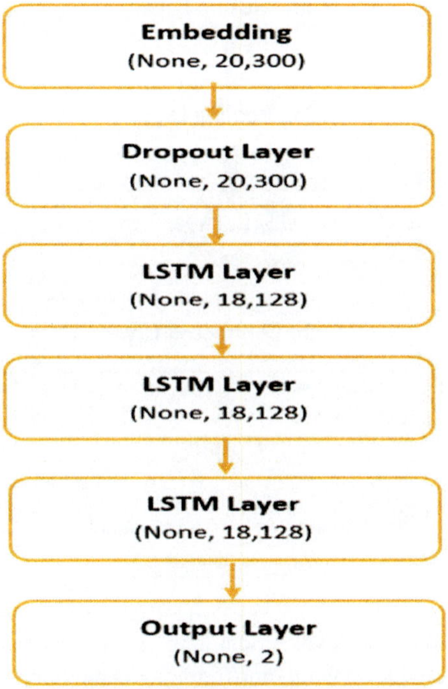

Fig. 3. LSTM implementation.

As shown in Fig. 3, we constructed an LSTM with three consecutive layers, one of which generated the result with 128 units. We provide a representation matrix that contains the embedding vectors generated by contextual word embeddings of 20 words and with a size of 300D. The dropout score is set to 0.6. Then the result vector of the upper layer is transmitted to the fully connected network. The sigmoid activation function is used in the output layer to obtain the appropriate class with 2 units.

CNN+LSTM. The combination of the CNN → LSTM model (Fig. 4) is composed of an initial convolution layer, which takes the received word embedding as the input. The output vector of the maximum pool layer becomes the input of the LSTM network to compute the long-term correlation of feature sequences. LSTM output vectors are connected in series and the activation function is used to produce the final output: negative or positive The working of the model is that the convolution layer extracts local features, after that, the LSTM layer can use the order of features to understand the order of input text.

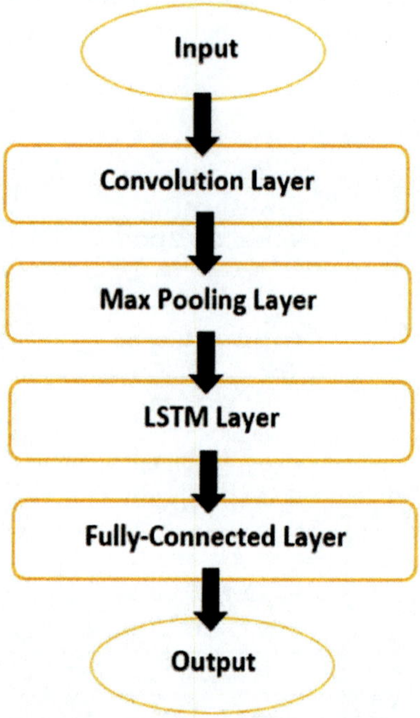

Fig. 4. CNN + LSTM model.

LSTM+CNN. The LSTM → CNN model (Fig. 5) is composed of an initial LSTM layer, that takes as the input, the word embedding of each token in the received dataset, and its output token intuitively saves the information about any previous token, plus the information of the initial token; In other words, the LSTM layer produces a new code for the initial input, the output of the LSTM layer is sent to the convolution layer, and the convolution layer needs to extract local features. The output of the convolution layer is combined into a smaller dimension, and finally output as positive and negative labels.

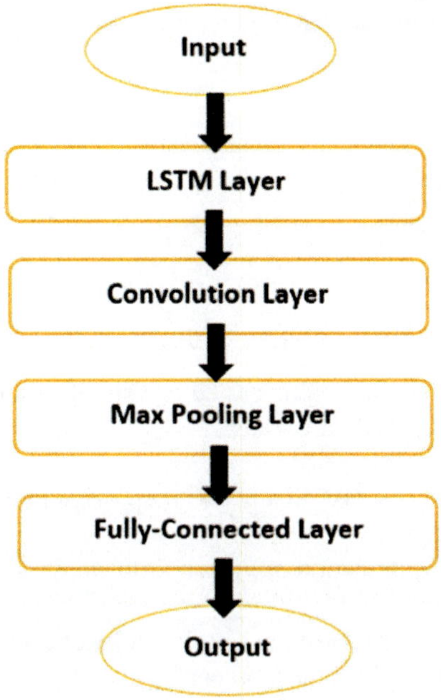

Fig. 5. LSTM + CNN model.

4 Experiments and Results

We describe in this section the datasets and different used preprocessing steps, then we define the parameters of our implementation, the used evaluation metrics, and finally, we expose the results of our experiments.

4.1 Datasets

Amazon Customer Reviews Dataset. This dataset [13] contains more than 130 million client reviews and is available in English and French; this dataset is collected between November 1, 2015, and November 1, 2019.

In our study, we used a French dataset; there are 200,000, 5,000, and 5,000 reviews in the training, test, and development sets respectively, other details are in the following table (Table 1):

Jumia Dataset. After getting the best model, we applied it to Jumia [14] which is one of the biggest E-Commerce stores in Africa, and to pull French reviews from Jumia, we used web scraping techniques with Python language (Version 3.8) by using external libraries designed for automatic browsing specifically BeautifulSoup4 (Version 4.9.3) and Requests (Version 2.25.0). In our study, we collected 23,210 reviews published from 2017 and 2021.

Table 1. Training French corpus statistics.

Number of products	183,345
Number of reviewers	157,922
Average characters/review	159,4
Average characters/review title	19,1

4.2 Data Preprocessing

As the word-embedding phase, preprocessing is a significant task in NLP. In our case, we used the following preprocessing steps:

1. Tokenization: it is the first task in any text preprocessing cycle that breaks running texts into short text entities.
2. Stop word Removal: Elimination of French Stop words, such as: "le", "un", "pour" which means in English: "the", "a": "for" respectively.
3. Stemming: It is used to decrease the dimension by converting words to their roots, for example: "spreading" becomes "spread".
4. Lemmatization: it is the process of converting every word in a text to its originating structure. For example: "worst" becomes "bad".
5. Data transformation: this step consists of changing data to a format that corresponds to the fine-tuned models using CamemBERT.

4.3 Experimental Setup

We used in our experiments, the following parameters:

- Programming language: Python 3.8
- Implementation library: Pytorch Library (Version 1.6.0)
- Optimizer Function: Adam
- Learning rate: 2e-3
- Batch size: 64.
- Epochs: 20.

4.4 Evaluation Metrics

NLP systems are usually evaluated based on their performance on task-specific test data sets. and the accuracy (1) is the common evaluation measure for binary classification, which is defined as follows:

$$Accuracy = \frac{True\_Pos + True\_Neg}{True\_Pos + True\_Neg + False\_Pos + False\_Neg} \quad (1)$$

Where True_Pos, True_Ned, False_Pos, and False_Neg are respectively the number of true positives, true negatives, false negatives.

To observe the performance per class we need more detailed metrics. The F1-score (4) is used, which is the harmonic average of precision (2) and recall (3):

$$Precision = \frac{True\_Pos}{True\_Pos + False\_Pos} \tag{2}$$

$$Recall = \frac{True\_Pos}{True\_Pos + False\_Neg} \tag{3}$$

$$F1 - score = \frac{2 \times Recall \times Precision}{Recall + Precision} \tag{4}$$

4.5 Results Analysis

We show in this part, the results of our experiments comparing the effect of different contextualized word embedding techniques on the classification of French reviews using CNN, LSTM, and combining CNN and LSTM, we compared those models using the evaluation metrics described in the previous part.

Concerning word embedding techniques, as you can see in Fig. 6, the pre-training word embedding CamemBERT outperforms other contextualized word embeddings using different classifiers.

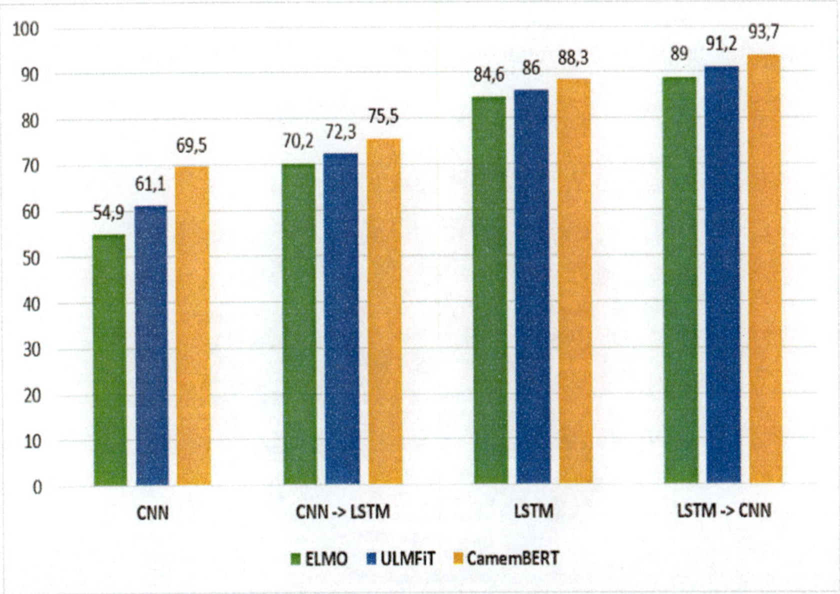

Fig. 6. Average accuracy of different neural network models using different word embeddings.

Concerning deep learning models, the CNN-LSTM model attained an accuracy of 6% better than the CNN model but 12.8% lower than the LSTM model. However, The

Table 2. Performance results of different models

Classifers	Word embedding	Accuracy	Precision	Recall	F1-score
CNN	ELMO	0,549	0,652	0,461	0,101
	ULMFiT	0,611	0,669	0,407	0,267
	CamemBERT	0,695	0,673	0,392	0,340
LSTM	ELMO	0,846	0,867	0,721	0,423
	ULMFiT	0,860	0,879	0,963	0,367
	CamemBERT	0,883	0,890	0,529	0,390
CNN+LSTM	ELMO	0,702	0,710	0,567	0.510
	ULMFiT	0,723	0,753	0,548	0,427
	CamemBERT	0,755	0,766	0,519	0,301
LSTM+CNN	ELMO	0,898	0,903	0,731	0,540
	ULMFiT	0,912	0,921	0,691	0,672
	CamemBERT	0,937	0,958	0,751	0,527

LSTM-CNN model obtained 24.2% higher than a CNN model and 5.4% higher than an LSTM model. The details of performance results are summarized in Table 2.

Compared with the LSTM-CNN model, the performance of the CNN-LSTM model is low. The LSTM-CNN model lookalike to be the best because its initial LSTM layer is as an encoder, so each token in the input has an output token, which contains not only the information of the original token but also the information of all other previous tokens, so it can better understand the input because the network will remember that it has been

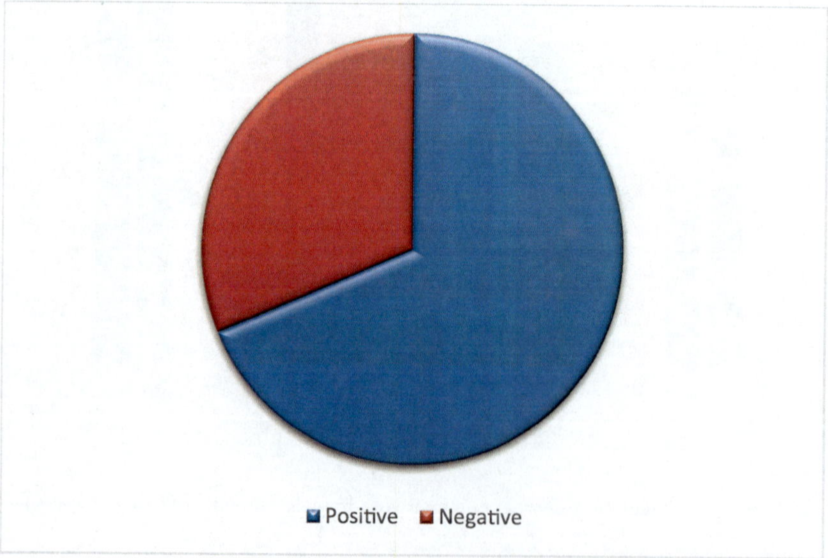

Fig. 7. Sentiment analysis on the Jumia dataset.

read before. Then, the CNN layer will use richer original input representations to find local patterns for higher accuracy (0.937).

After getting the best model (LSTM+CNN+CamemBERT), we applied it in the Jumia dataset, and as shown in Fig. 7 the users post more Positive reviews with 68,6%.

5 Conclusion and Future Work

We have investigated in this study, the use of Deep Learning architectures for French SA on Amazon and Jumia datasets. For that, we have presented deep learning models; CNN, LSTM, and combined CNN and LSTM neural networks to achieve better performance on sentiment analysis tasks by a substantial margin. The LSTM-CNN model does 5.4% better than a regular LSTM model and 24,2% better than a CNN model with an accuracy of 93.7%. Furthermore, we have done an in-depth exploration of the effects of contextualized word embedding techniques in used models, and CamemBERT was better than ELMO and ULMFiT.

As future work, we aim to try different types of RNNs apart from the LSTM model. For example, using bidirectional LSTMs (Bi-LSTM) on the LSTM-CNN model might give an even better result.

References

1. Most spoken languages in the world. Statista. https://www.statista.com/statistics/266808/the-most-spoken-languages-worldwide/. consulté le août 23, 2021
2. Barakat, H., Yeniterzi, R., Martín-Domingo, L.: Applying deep learning models to Twitter data to detect airport service quality. J. Air Transp. Manag. **91**, 102003 (2021). https://doi.org/10.1016/j.jairtraman.2020.102003
3. Alayba, A.M., Palade, V., England, M., Iqbal, R.: A combined CNN and LSTM model for Arabic sentiment analysis. In: Holzinger, A., Kieseberg, P., Tjoa, A.M., Weippl, E. (eds.) CD-MAKE 2018. LNCS, vol. 11015, pp. 179–191. Springer, Cham (2018). https://doi.org/10.1007/978-3-319-99740-7_12
4. Maslej-Krešňáková, V., Sarnovský, M., Butka, P., Machová, K.: Comparison of deep learning models and various text pre-processing techniques for the toxic comments classification. Appl. Sci. **10**(23), 8631 (2020). https://doi.org/10.3390/app10238631
5. Singh, M., Jakhar, A.K., Pandey, S.: Sentiment analysis on the impact of coronavirus in social life using the BERT model. Soc. Netw. Anal. Min. **11**(1), 1–11 (2021). https://doi.org/10.1007/s13278-021-00737-z
6. Lora, S.K.: A comparative study to detect emotions from tweets analyzing machine learning and deep learning techniques. Int. J. Appl. Inf. Syst. **12**, 8 (2020)
7. Essebbar, A., Kane, B., Guinaudeau, O., Chiesa, V., Quénel, I., Chau, S.: Aspect based sentiment analysis using French pre-trained models. In: Proceedings of the 13th International Conference on Agents and Artificial Intelligence, Vienna, Austria, pp. 519–525 (2021). https://doi.org/10.5220/0010382705190525
8. Pre-trained ELMo Representations for Many Languages. 哈工大社会计算与信息检索研究中心 (2021). Consulté le: août 21, 2021. [En ligne]. Disponible sur: https://github.com/HIT-SCIR/ELMoForManyLangs
9. Howard, J., Ruder, S.: Universal Language Model Fine-tuning for Text Classification. ArXiv180106146 Cs Stat, mai 2018, Consulté le: août 09, 2021. [En ligne]. Disponible sur: http://arxiv.org/abs/1801.06146

10. Welcome to fastai. fast.ai, 2021. Consulté le: août 20, 2021. [En ligne]. Disponible sur: https://github.com/fastai/fastai
11. Martin, L., et al.: CamemBERT: a tasty French language model. In: Proceedings of the 58th Annual Meeting of the Association for Computational Linguistics, pp. 7203–7219 (2020). https://doi.org/10.18653/v1/2020.acl-main.645
12. Devlin, J., Chang, M.-W., Lee, K., Toutanova, K.: BERT: pre-training of deep bidirectional transformers for language understanding. ArXiv181004805 Cs, mai 2019, Consulté le: janv. 14, 2021. [En ligne]. Disponible sur: http://arxiv.org/abs/1810.04805
13. Keung, P., Lu, Y., Szarvas, G., Smith, N.A.: The Multilingual Amazon Reviews Corpus. ArXiv201002573 Cs, October 2020, Consulté le: août 22, 2021. [En ligne]. Disponible sur: http://arxiv.org/abs/2010.02573
14. Jumia Maroc | Téléphones, TV, PC, Vêtements, Maison, Beauté et plus encore. Jumia Maroc. https://www.jumia.ma/. consulté le août 22, 2021

Amazigh Handwriting Recognition System—Multiple DCNN Strategies Applied to New Wide and Challenging Database

Abdellah Elzaar[1(✉)], Rachida Assawab[1], Ayoub Aoulalay[2], Lahcen Oukhoya Ali[1], Nabil Benaya[1], Abderrahim El Mhouti[3], Mohammed Massar[2], and Abderrahim El Allati[1]

[1] Laboratory of R&D in Engineering Sciences, Faculty of Sciences and Techniques Al-Hoceima, Abdelmalek Essaadi University, Tetouan, Morocco
abdellah.elzaar@etu.uae.ac.ma

[2] PMIC Laboratory, FST Al-Hoceima, Abdelmalek Essaadi University, Tetouan, Morocco

[3] Faculty of Science Tetouan, Abdelmalek Essaadi University, Tetouan, Morocco

Abstract. In this research work we proposed multiple strategy based on the Convolutional Neural Network (CNN) algorithm for Amazigh handwritten character recognition. The Amazigh language consists of many confusing characters that provide a similar feature map. Thus, classifying these characters is a challenging task. The most existing works do not lead to satisfactory results for practical applications, and have much lower recognition performance. To improve the performance of Amazigh handwritten recognition process, we investigate in Deep Learning (DL) techniques by implementing three learning methods based on Deep Convolutional Neural Network (CNN): training the CNN from scratch, fine tuning CNN and combined CNN-SVM. CNN algorithm revolutionized the field of handwritten recognition and shown excellent performance, but has not been attempted for Amazigh handwritten characters. We build our Model based on CNN and Support Vector Machine (SVM) classifiers and we took into consideration the effectiveness and the complexity of our architecture also the speed of the training process. A new dataset consisting of about 50000 characters and 33 classes is collected and used for training and testing. We were able to achieve an excellent recognition rate of 98.23%.

Keywords: Handwritten Amazigh recognition · Character classification · Deep Learning · Deep Convolutional Neural Network · Amazigh-33 database

1 Introduction

Despite the successful works and applications of Handwritten Character Recognition, handwritten Amazigh character recognition (HACR) is still an active

R. Saidi et al. (Eds.): ICATH 2021, LNDECT 110, pp. 241–252, 2022.
https://doi.org/10.1007/978-3-030-94188-8_23

area of research and present a major challenge. The recognition of Handwritten Amazigh or Tifinagh character is becoming very necessary, especially in countries where Tamazight is an official language, and where it is studied at schools. The Amazigh language is a very old language that has a history and numerous historical manuscripts. Accordingly, using computer vision to recognize Amazigh language, will help to study and understand Tamazight, also to protect and analyse old documents. In the literature, several works have been conducted in the field of handwritten character recognition. For handwritten Arabic recognition based machine learning techniques, we present the work done by **Althobaiti et al.** [1]. They developed an optical character recognition system based on a support vector machine for classification and a combination of normalized central moment and local binary model for feature extraction, the proposed method achieves a recognition rate of 96.79%. In the other side, **Altwaijry et al.** [2] proposed an Arabic handwriting recognition system using convolutional neural network, the CNN was trained using children handwritten Arabic characters dataset and achieves a score of 97%. We observe that approaches based Deep Learning provides better results than Machine Learning based approaches, especially in large datasets. For Chinese handwritten characters, the authors of [3], showed that increasing the depth of a CNN while having fewer parameters increases the performance of GoogleNet based models. They also show that handcrafted features can still be useful to improve performance. Their proposed system achieves an accuracy of 96.74%. For Devanagari character recognition, in their work [4] the authors introduced a new dataset which contains 92 thousand images, this dataset is used to train a CNN with dropout and data augmentation and achieved 98.47% accuracy recognition system achieves high accuracy and satisfactory results. In case of Amazigh language, a few works are done to deal with the recognition problem. **Sadouk et al.** [5] proposed deep learning architecture using CNN to recognize printed Tifinagh characters, the proposed system achieves an accuracy of 98%. Printed characters are less challenging than Handwritten characters, also the dataset of printed characters did not provide the diversity of patterns and shapes of characters. **YE Saady et al.** [6], presented a system of Amazigh handwriting recognition based on horizontal and vertical centerline of the character. The system requires a preprocessing step. The text is then segmented into lines and then into characters and finally fed to Artificial Neural Network (ANN) the proposed system achieves a recognition rate of 96%. The ANN algorithm is considered as machine learning based approach, Therefore, it requires a large dataset and a preprocessing step to provide high performance results. **Rachidi** [7], used Hidden Markov Models (HMM) to recognize Amazigh text. The approach was trained using a dataset of 2220 characters and provides an accuracy of 90%. **EW Dadi** [8], used a dataset of 3366 characters to train the CNN. The input requires images of size 28×28. After the preprocessing phase, the system provides an accuracy of 94%. The complexity of Amazigh handwritten and the multiple confusing characters made the recognition problem difficult and challenging. Figure 4 presents some examples of similar Amazigh characters that are confused by the recognition system. The tifinagh letters are separated and not connected like Latin and Arabic letters. In addition to that, the special shape of Amazigh language makes it harder to differentiate them. The presented datasets in literature are not diverse and wide, and not suitable for practical applications. The state-of-the-art performances in term of accuracy are 93.68% and 96.32%, corresponding to [5] and [6]. Therefore, we investigated in deep learning to perform a powerful and effective recognition system. In this work we proposed three learning strategies based on the Convolutional Neural Network (CNN) for Amazigh handwritten character recognition. In the first strategy, we used CNN as feature extractor then we applied Support Vector Machine (SVM) as a classifier, we apply also the CNN for transfer learning and finally we used CNN trained from scratch as a baseline to compare the performances of three strategies. We used the Amazigh handwritten character dataset (AHCD) collected by us. The dataset contains about 50000 images of handwritten characters. Our results provided high accuracy of 99.23% for CNN-SVM based strategy, 98.13% for CNN learned from scratch and 96% for transfer learning. The rest of the paper is organized as follows: Sect. 2 describes the proposed approach and experiments. Achieved results are discussed in Sect. 3. Conclusion and coming work in Sect. 4.

2 Proposed Approach

In our case, we investigate in deep learning by proposing three learning methods for handwritten Amazigh character recognition. In the first method we train CNN from scratch, we kept this strategy as a reference to other strategies and also for performance comparison. The second strategy aims to combine the key characteristics of both CNN and SVM classifiers, CNN is used to extract features from input images and SVM is implemented as multiclass classifier. Finally, we used a pretrained CNN to perform the transfer learning on our collected dataset. An overview of our proposed method is illustrated in Fig. 1.

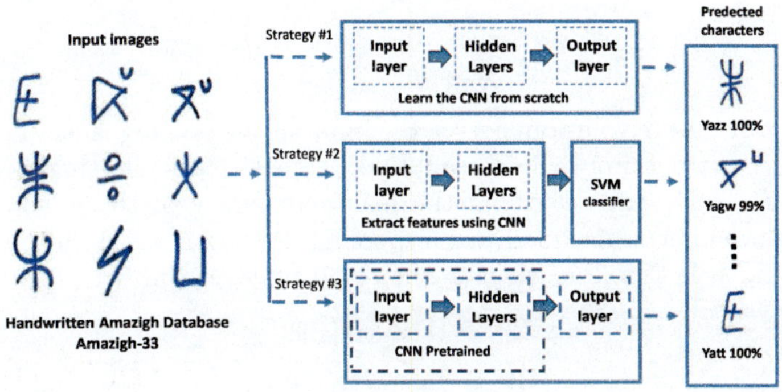

Fig. 1. Proposed approaches for Amazigh handwritten characters.

We take advantage to train and test our dataset on the three learning strategies using the proposed CNN architecture. Before feeding our dataset images to our architecture, we firstly perform size normalization as preprocessing step. The size of our dataset images is 93×96, and the input of our CNN architecture expects an input of 32×32 pixels.

2.1 CNN Architecture

The main difference between the traditional Fully Connected Neural Network (FCNN) and CNN, is that the connection between the layers of FCNN is fully connected (the data is fed successively from a layer to another, from the input to the output), while the connection between the layers of CNN are organized with sparse (carefully designed) [9]. Each layer in the convolutional network is a 3-dimensional grid structure, which has a height, width, and depth. The Convolutional Neural Network consists of three (03) main layers: Convolutional layer, pooling layer and fully connected layer. Our CNN architecture is inspired on VGG [10] with four (04) convolution layers, two (02) max-pooling layers and three (03) fully connected layers. The Fig. 2 illustrates Our proposed CNN architecture for AHCR dataset. The convolution operation is a dot product between an input layer and the filter. The filters (Kernels) are three dimensional parameters $K_n \times K_n \times d_n$ which represents the network parameters. The dimensions (length and width) of the output layer after performing the convolution operation is:

$$L(n+1) = L_n - K_n + 1 \tag{1}$$

$$W(n+1) = W_n - K_n + 1 \tag{2}$$

The depth of the output layer is defined by the number of the used filters. The convolutional operation from the nth layer to the $nth + 1$ is defined as follows:

$$M_{ijz}^{n+1} = \sum_{r=1}^{K_n} \sum_{s=1}^{K_n} \sum_{k=1}^{d_n} w_{rsk}^{(z,n)} h_{i+r-1,j+s-1,k}^{(n)} \tag{3}$$

$$\forall i \in \{1 \ldots L_n - K_n + 1\}$$
$$\forall j \in \{1 \ldots W_n - K_n + 1\}$$
$$\forall z \in \{1 \ldots d_{n+1}\}$$

The output of the convolution layer is the input of the pooling layer. The pooling operation is efferent from the convolution, it consists of selecting a small size region of $P_n \times P_n$ and returning the maximum value of the selected region. Another parameter is included in the pooling operation which is the stride. If the stride $S_n > 1$, the dimensions of the resulting layer will be:

$$L_n - P_n + 1/S_n + 1 \tag{4}$$

$$W_n - P_n + 1/S_n + 1 \tag{5}$$

The final layer in the Convolutional Neural Network is the fully connected layer. This layer functions exactly like a traditional feed-forward neural network, and the data is forwarded from the input to the output.

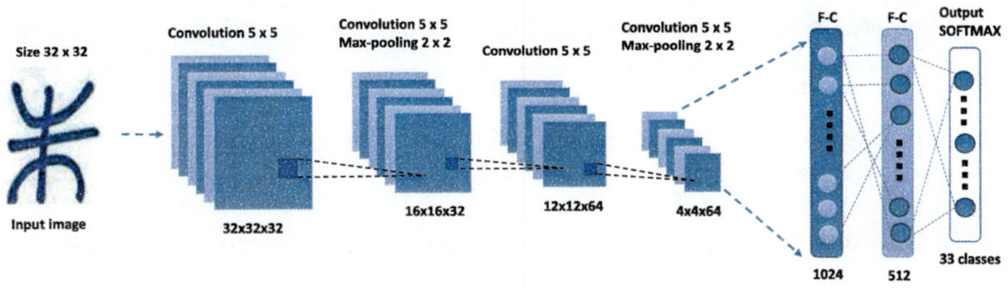

Fig. 2. CNN architecture for AHCR.

2.2 Method.I-CNN from Scratch

In the first method of training, we choose to train our CNN architecture from scratch. The images are fed to the input of the model passing by convolution and max-pooling layers, and then to the fully connected layers for classification and prediction. The data are split into Train and Test sets and the training process was initialized with 100 epochs. This strategy of training provides less performances in case of small data sets.

2.3 Method.II-CNN-SVM

The strategy CNN-SVM aims to combine the key characteristics of Convolutional Neural Network (CNN) and Support Vector Machine (SVM) algorithm. CNN is used to extract features from Amazigh Handwritten characters by using only convolution and pooling layers. The final layer (Convolutional layer) is replaced by the SVM classifier. The output of conv-pool layers is then feeded to SVM inputs for classification. SVM classifier represents the classes of the dataset in a space, theses classes are then separated by a line which called hyperplane. SVM classifier can find with high precision the optimal hyperplane that separates between classes.

2.4 Method.III-Fine Tune CNN

Most of presented Amazigh datasets in the literature are small. For that reason we choose also to work with the transfer learning method. Transfer learning [11, 12] is the process of transferring knowledge from a source model trained on a source dataset, to another target model with a target dataset and task. Consider a source domain and a target domain Ds and Dt, with a feature space X, a marginal probability distribution $P(X)$ defined by $D = X, p(X)$, a source task and a target task Ts and Tt with a label space Y, and a prediction function f. Transfer learning aims to improve the prediction performance of a target function Ft on a target task Tt and a target domain Dt by using the knowledge of the source function Fs trained on a source task and a domain Ts and Ds.

Using this method, as shown in Fig. 1, the convolutional layers of the pre-trained model are transferred to our model to classify Amazigh handwritten characters. We use fine-tuning of a pre-trained VGG16 [10] model on ImageNet. The first convolutional layers of a pre-trained VGG16 model are used to extract general discriminative features [13]. The final convolutional layers are initialized with the weights of the pre-trained network and are updated during the training of the network to be able to extract specific features about our dataset. To initialize the fully connected layers, we first randomly initialized them and then trained them for a small number of epochs while freezing the convolutional layers of the pre-trained VGG16 source model.

3 Experiments and Results

In this section, we will discuss and present in details the results achieved by our proposed approach using the three learning strategies. We will also compare our work and achievements with previous works done in the field of Handwritten Amazigh Characters recognition.

3.1 Amazigh-33 Dataset

The most challenging part of any handwritten character recognition system is the dataset. Deep learning systems require wide and diverse datasets to perform perfectly and to provide accurate recognition. The existing Amazigh datasets in the literature are not wide and diverse enough to perform an effective deep learning recognition system. For this reason, and to improve the credibility of our proposed method, we collected the Amazigh dataset. The dataset contains 48840 characters written by 148 participants, 85% of them are right-handed and the age range is between 14 and 60. The Amazigh characters are 33 (from 'ya' to 'yazz'), Each participant wrote each character 10 times. The collected characters are then segmented automatically using Python 3 code to determine classes. The final dataset is splitted into two sets: a training set with (41250 characters '1250 samples per class'), and a test set with (7590 characters '230 samples per class'). The dataset Amazigh-33 is available for free for researchers, and can be accessed through this link: https://drive.google.com/file/d/1hNJFxVXw3oHfHnYB4WgJtQ29uK2kxEp4/view?usp=sharing.

3.2 Experiments and Results

The CNN algorithm was implemented using Python language. We used several libraries that contains visualisation and image processing functions such as $Tensorflow^{TM}$, $Keras^{TM}$ Models and $Sklearn^{TM}$. We run our CNN model on Microsoft Azure virtual machine with six (06) core processor and 56 GB of RAM. Working with Convolutional Neural Networks with a high number of epochs, also visualising a large amount of data can be time-consuming process, For this reason we used TESLA K80 NVIDIA GPU. GPU boost gives superior performance to our Model and accelerates Libraries during the training process. Our models were trained using our collected Amazigh dataset for 100 epochs.

Experiments on Amazigh-33 Dataset. First, we randomly divide our dataset into two parts: 80% for training and the remaining 20% for testing to avoid the well-known phenomenon of over-fitting. The hyper-parameter technique is used for a dropout optimization in both the convolutional layers and fully connected layers. A Mini Batch size of 32 images is selected, the number of epochs for each method was: 100 epochs. The metric used to evaluate the performance of our model is the accuracy. This metric consists of dividing the number of correct predictions by the total number of predictions.

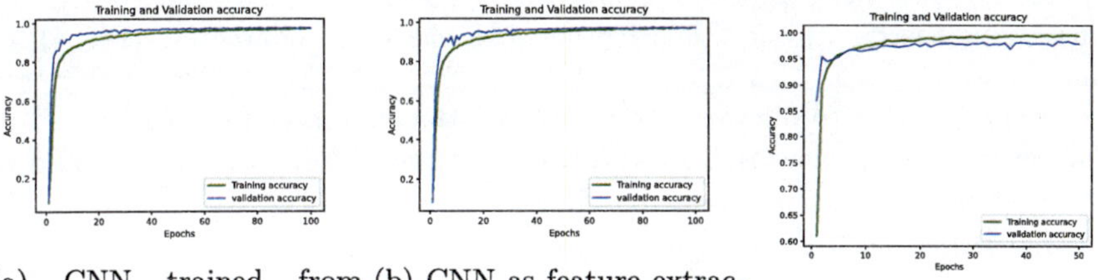

(a) CNN trained from scratch (b) CNN as feature extractor (c) Fine tune CNN

Fig. 3. Accuracy of our proposed approach

The Fig. 3 illustrates the accuracies obtained using the proposed three strategies during the training process. It can be seen that our proposed method performs perfectly on Amazigh dataset, the accuracy metric divides the number of the correct predictions by the total number of predictions. Table 1 presents the accuracies achieved by our proposed method in different tests. It can be seen that using CNN as feature extractor (Combined CNN-SVM) with pre-processing step achieves better accuracy with 98% compared to the same strategy without pre-processing step which provides 97.66%. This can be explained by the ability of machine learning algorithms like SVM to work with pre-processed data better than non pre-processed data. The pre-processing step refers to the technique of cleaning and organizing data to make it suitable for training machine learning models. With the fine tuning strategy, when we feed unprocessed data to the pre-trained model we can hold an accuracy of 97.18% and 97.80% with preprocessing. This can be justified by the capacity of Deep learning models to work with wide and unprocessed data. For the learning from scratch strategy, our CNN architecture achieves a satisfactory recognition rate with 98.12% and outperforms all other strategies using Amazigh-33 dataset. The perfect score

Table 1. Comparison of test accuracy of the three learning strategies using CNN

Learning strategy	Without preprocessing	With preprocessing	Score (%)
CNN trained from scratch	**98%**	**98.23%**	**98.12%**
CNN as feature extractor (CNN-SVM)	97.66%	98%	97.83%
Fine-tune CNN	97.18%	97.80%	97.49%

obtained by this strategy shows its capacity to work with such images. The confusion matrices presented in Figs. 6, 5, 7 evaluate our method using the three learning strategies without preprocessing step. By observing the diagonal of the confusion matrices, we observe that all its maximum values are equal to 100 which is the number of images in each class in the test set. In addition, in the confusion matrix of the CNN-SVM strategy Fig. 6, we observe that some characters like (yak), (yaj) and (yaz) are less recognized than other characters. The character (yak) was confused six times with (yakw) and five times with (yaq). This is due to the strong similarity between the three characters (Fig. 4), this confused situation is also repeated in two other strategies, and it is logical as it can also happen to humans. The same for the character (Ya) with (yas) and (yar) were strongly confused by the two strategies CNN from scratch and fine tuning. This is normal because the difference between the three letters is just a dot that is placed in the cavity of the character (yas).

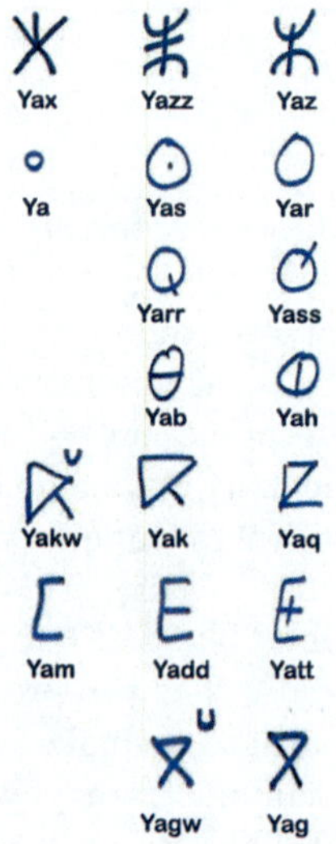

Fig. 4. Amazigh characters shape similarity.

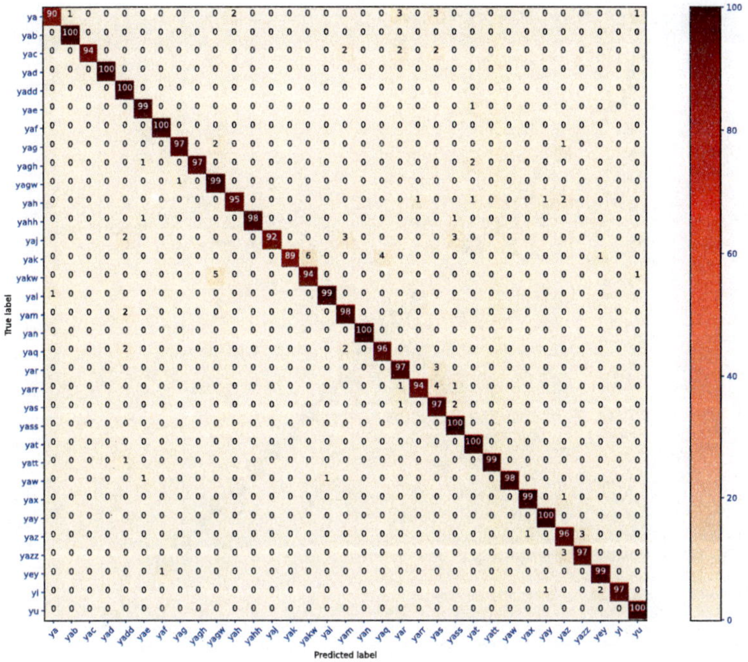

Fig. 5. Confusion matrix using CNN trained from scratch without preprocessing

3.3 Comparison and Analysis

In this section, we briefly present some comparisons between our proposed approach and recent state-of-the-art handwritten Amazigh recognition systems. Table 2 shows the results achieved by previous works and our work. The table presents the used method, the year of publication, the employed database and the recognition rate achieved by each system.

Table 2. Comparison with previous works on Amazigh handwritten character recognition

Authors	Method	Database	Classes	Train and test data size	Accuracy on test data
Present approach	CNN from scratch CNN as feature extractor (CNN-SVM) Fine tune CNN	Amazigh-33	33	50000 images 40000 train images 10000 test images	**98.23%** 98% 97.80%
Sadouk et al. [5]	CNN based LeNet	AMHCD	31 without (yakw) and (yagw)	24180 images	95.47%
Es-saady et al. [6]	ML horizontal and vertical centerline of character and ANN	AMHCD	31 without (yakw) and (yagw)	20150 images	96.32%
Dadi [8]	CNN from scratch based LeNet	IRCAM-Tifinagh	33	3366 images	94%

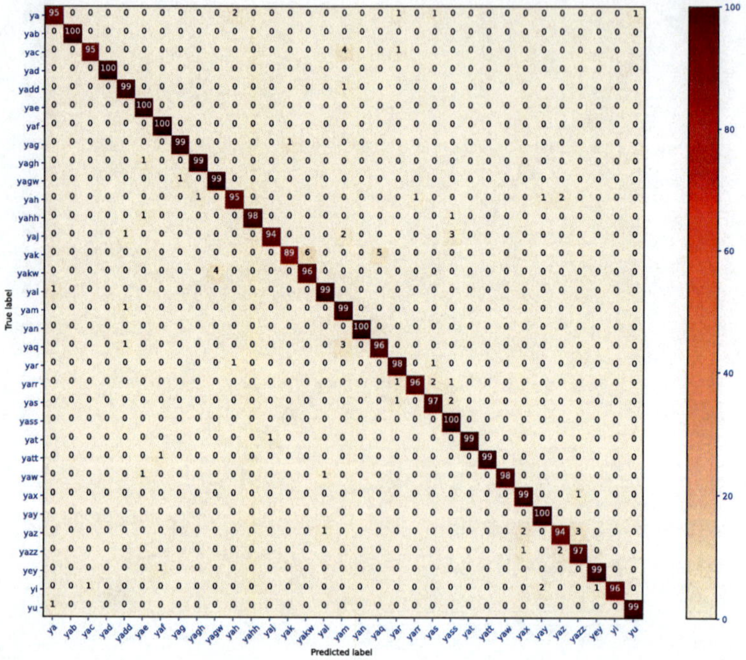

Fig. 6. Confusion matrix using CNN as feature extractor without preprocessing

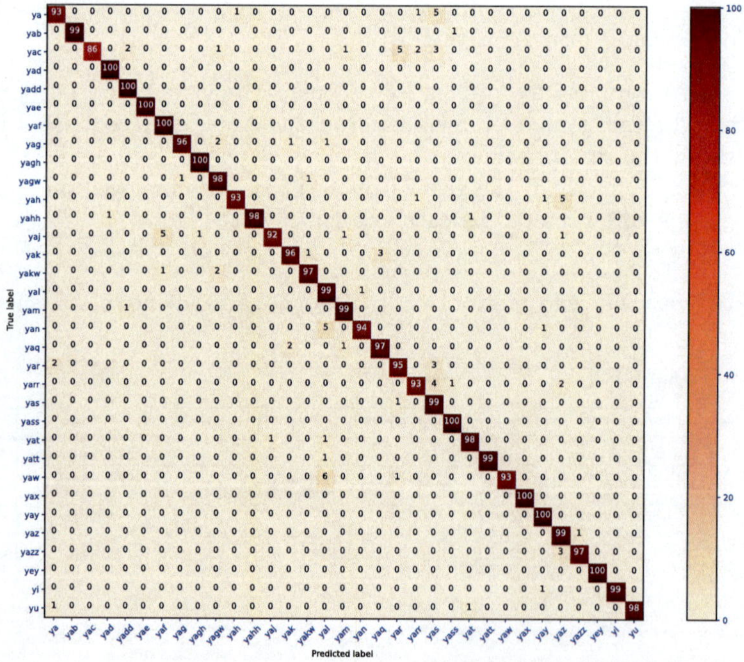

Fig. 7. Confusion matrix using fine tuning CNN without preprocessing

Our results on Amazigh-33 using the three strategies (Accuracy on test data: 98.23%, 98%, 97.80%) shows higher results compared with the previous works. The results on Amazigh-33 dataset using CNN learned from scratch (Test accuracy: 98.23%) are better than those found by (EW Dadi [8]) using the same learning strategy. This can be explained by the strength Deep Learning system when it is dealing with wide databases. It can be seen also that our model gives better recognition rate using 33 classes data than (YE Saady et al. [6]) and (Sadouk et al. [5]) works that use only 31 classes without (yakw) and (yagw). This demonstrates the effectiveness of our method to work with challenging databases with high similarity of characters.

4 Conclusion

Handwritten character recognition field still an important research issue. It is widely used in large application such as education, medicine, financial companies and industry. Deep learning is now the key solution to develop the field of Handwritten character recognition. In this paper, we discussed the use of different CNN strategies with a new wide and challenging dataset. We trained and tested our proposed approach using Amazigh-33 dataset 33 classes of characters, contrary to the existing state-of-the-art databases which contains 31 classes of characters. Obtained results were very satisfactory with an accuracy of 98.23%. In future studies, we plan to perform Amazigh dynamic word recognition. In addition, we intend to combine several methods for Amazigh text processing and recognition.

References

1. Althobaiti, H., Lu, C.: Arabic handwritten characters recognition using support vector machine, normalized central moments, and local binary patterns. In: Proceedings of the International Conference on Image Processing, Computer Vision, and Pattern Recognition (IPCV), The Steering Committee of The World Congress in Computer Science, Computer Engineering and Applied Computing, pp. 121–127 (2018)
2. Altwaijry, N., Al-Turaiki, I.: Arabic handwriting recognition system using convolutional neural network. Neural Comput. Appl. **33**(7), 2249–2261 (2020). https://doi.org/10.1007/s00521-020-05070-8
3. Zhong, Z., Jin, L., Xie, Z.: High performance offline handwritten Chinese character recognition using GoogleNet and directional feature maps. In: 2015 13th International Conference on Document Analysis and Recognition (ICDAR), pp. 846–850. IEEE (2015)
4. Acharya, S., Pant, A.K., Gyawali, P.K.: Deep learning based large scale handwritten Devanagari character recognition. In: 2015 9th International Conference on Software, Knowledge, Information Management and Applications (SKIMA), pp. 1–6. IEEE (2015)
5. Sadouk, L., Gadi, T., Essoufi, E.H.: Handwritten Tifinagh character recognition using deep learning architectures. In: Proceedings of the 1st International Conference on Internet of Things and Machine Learning, pp. 1–11 (2017)

6. Saady, Y.E., Rachidi, A., El Yassa, M., Mammass, D.: Amazigh handwritten character recognition based on horizontal and vertical centerline of character. Int. J. Adv. Sci. Technol. **33**(17), 33–50 (2011)
7. Rachidi, A.: Reconnaissance automatique de caractères et de textes amazighes: état des lieux et perspectives (2014)
8. Dadi, E.W.: Tifinagh-IRCAM handwritten character recognition using deep learning. arXiv preprint arXiv:1912.10338 (2019)
9. Aggarwal, C.C.: Neural Networks and Deep Learning. Springer, Cham (2018). https://doi.org/10.1007/978-3-319-94463-0
10. Simonyan, K., Zisserman, A.: Very deep convolutional networks for large-scale image recognition. arXiv preprint arXiv:1409.1556 (2014)
11. Pan, S.J., Yang, Q.: A survey on transfer learning. IEEE Trans. Knowl. Data Eng. **22**(10), 1345–1359 (2010)
12. Torrey, L., Shavlik, J.: Transfer learning. In: Handbook of Research on Machine Learning Applications and Trends: Algorithms, Methods, and Techniques, pp. 242–264. IGI Global (2010)
13. Yosinski, J., Clune, J., Bengio, Y., Lipson, H.: How transferable are features in deep neural networks? arXiv preprint arXiv:1411.1792 (2014)

Toward an End-to-End Voice to Sign Recognition for Dialect Moroccan Language

Anass Allak[1(✉)], Imade Benelallam[1,2], Hamdi Habbouza[1], and Mohamed Amallah[1]

[1] SI2M Laboratory, Institut National de Statistiques et d'Économie Appliquée, B.P. 6217, Rabat, Morocco
{aallak,i.benelallam,hhabbouza,mamallah}@insea.ac.ma
[2] AIOX-LABS, 5 Rue Bouiblane, Agdal, Rabat, Morocco
ibenelallam@aiox-labs.com

Abstract. Automatic Voice to sign recognition language is a challenging task. Building a system able to generate sign language from voice has the potential to help deaf people. However, most of the research on this topic neglects to include less used sign languages like Moroccan sign language (MSL). In this work, we present a holistic approach for building an automated voice to sign system that can be applied to Moroccan Dialect (Darija) based on pose estimation. The system was created using a consolidated dataset, namely Dvoice and newly created Moroccan sign language dataset.

Keywords: Moroccan sign language · Pose estimation · Mediapipe · Visual generation

1 Introduction

For the deaf community, The primary means of communication is the sign language. According to the World Health Organization, 5% of the global population is deaf [21]. Sign languages are as diverse as spoken languages. In Morocco and in the rest of the world, deaf people still encounter difficulties in professional and administrative context. The work carried out on this subject mainly focuses on the most widely used languages in particular American (ASL) [17], Chinese (CSL) [8] and Indian Sign Language (ISL) [14,16]. However research, on under-represented language, like the Moroccan sign language is still lacking. In this context, our work aims to reduce the gape by providing a system carted to the need of the Moroccan deaf community.

This paper is presented as follow, an overview of the latest progress insign language, exploring the methods used for data collection and the task of recognition, and the pipeline used to create the avatar. This work makes the following contributions:

R. Saidi et al. (Eds.): ICATH 2021, LNDECT 110, pp. 253–262, 2022.
https://doi.org/10.1007/978-3-030-94188-8_24

- Creation of a new dataset for MSL.
- The use of Speech to text for Darija in the context of MSL.
- The creation of a corpus of MSL signs using avatars.

2 Related Works

2.1 Moroccan Sign Language

A sign language can be categorized following the type of movement [1]:

- **Static sign language** is the usage of sign without movement. It is used mainly for numbers and letter spellings, but its scope is limited in use since it takes a lot sign to spell a single word.
- **Isolated sign language** represents a word in sign language. The word is a sequence of basic signs.
- **Continuous sign language** represents multiple isolated words, in continuous sign language changing from a word to another might require intermediately movement that doesn't add to the meaning.

Like other sign languages, Moroccan sign language has a visual and spacial component: Sign languages are not solely composed of hand movements, but a combination of hand position, orientation, body position, facial expressions, and body movement as well. Moroccan sign language derived from French sign language (LSF) and ASL with local influences [9].

2.2 Sign Language Translation

The task of machine translation has seen a huge improvement in recent years, but unlike conventional translation tasks, sign language translation (SLT) is inherently visual, requiring the creation of visual content. In sign recognition, the team working on WLASL [10] have proposed a multitude of methods for single word recognition. They found 3D convolution networks I3D (An inflated or generalized Inception-v1 into 3D space) had performed better than models used in video classification. Other work like (Maraqa et al.) [12] and (Bantupalli et al.) [3] teams have proposed a recurrent neural (RNN) network model for recognizing sign language.

In sign generation eSign [23], Tessa [6] are systems that use large avatar corpus. These avatars where animated using labeled data. Text2sign is a system that generated sign without the need of a graphical avatar. They have based their work on a visual generation using Generative adversarial Network (GAN) [18].

Some of these methods have shown a great deal of promises. However, according to our research, some of these methods suffer from the need of acceleration hardware (high end GPU). While others show the need of specialist in animation that can reproduce the animation. Furthermore, these methods are focused on the most used languages.

3 Proposed Method

3.1 Data Collection

Sign Language Dataset: Datasets concerning Moroccan sign language are scarce. To solve this issue, we build our own dataset in collaboration with El Shorouk Association for the Deaf and Hard of Hearing, (Ouarzazate) which is a charity specialized in teaching young deaf children.

The recording was done using a high-resolution smartphone and took place in the local of the Association in a non controlled environment. The main focus was to record of common word in Darija using MSL. 150 words were selected with input from the teachers. 6 adults volunteer teachers (5 women and one man) agreed to recording. The recordings were shot at a resolution of 1280×720 using the frontal body to facilitate the keypoint detection, zooming, cropping and downscaling for processing.

Darija Dataset. Dvoice [7] is a new and open dataset that contains 2992 audio files of Darija speech and their transcriptions. The distribution is:

- 2392 training files
- 600 testing files

3.2 Speech to Text

Darija is considered one of the most spoken Arabic dialects and its vocabulary is derived especially from modern standard Arabic. Darija has many differences that make it unique: the influence of Amazigh, Spanish and French add foreigner sounds and words in day to day Darija [19].

Darija data is limited, so we focused on fine tuning a pre-trained model for transcription, we choose the model Wav2vec2 [2] especially the XLSR [5] variant. The model strength comes from two characteristics:

- The usage of a two stages network where the first encode the audio into a latent space and the next one uses attention to create context.
- The initial training on multiple languages which can outperform certain monolingual models after the fine tuning.

3.3 Openpose

With the Openpose library, users can detect human pose from a single image. The library has the capability to detect up to 135 key points. Built on a two-branch multi-stage CNN, Openpose is mainly used for Real-time 3D single-person keypoints detection and Real-time 2D multi-person keypoints detection (Fig. 1).

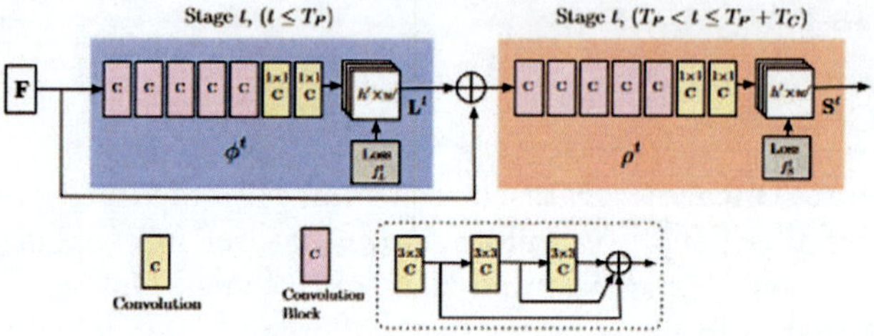

Fig. 1. Architecture of the Openpose network [4]

3.4 Mediapipe

Mediapipe [11] is a multi-platform framework that implements a graph-based approach for building multi modal audio, video or time series data pipelines applied to machine learning. Mediapipe offers plenty of customizable machine learning solutions like hand tracking, human pose detection, and tracking, hair segmentation, object detection, face detection, 3D object detection, and many other technologies and on a wide variety of different hardware platforms like android and iOS. Among its most famous solutions is Mediapipe Holistic, which provides human pose topology with up to 540 keypoints. It enable fast and near instantaneous performance on mobile devices (Fig. 2).

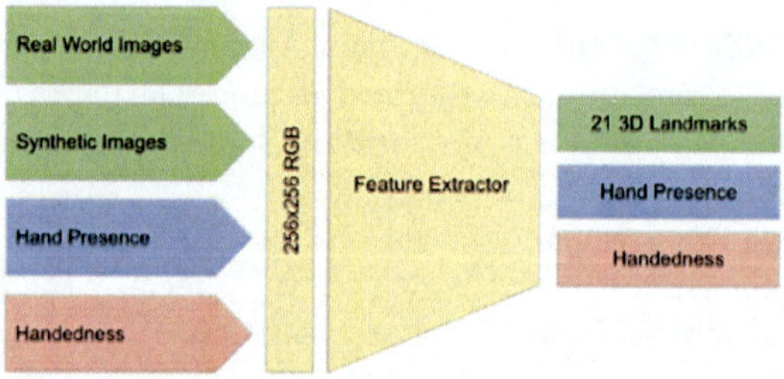

Fig. 2. Mediapipe hand landmark architecture [22]

4 Experiment

Our goal from the implementation is to create a Machine learning pipeline that can map the spoken Darija language to a generated sign language avatar in order to reduce the communication barrier among the deaf or hard of hearing communities. The process of translation contain two parts: First a module capable of transcribing the Dariija speech and the second is a visual module for the avatar generation. The system works as it follows:

1. Choosing the videos that contain the targeted word in this case "Honney" in Darija.
2. Extract the keypoint using Mediapipe.
3. Mapping the movement to an avatar.
4. Saving the result.
5. Detecting the utterance of the word "Honey" in Darija.
6. Returning the avatar corresponding to the word (Fig. 3).

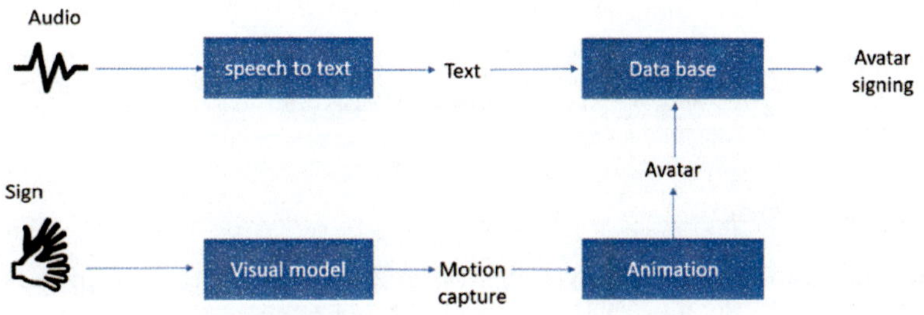

Fig. 3. Process of our system

4.1 Speech to Text

We fine-tuned the model Wav2vec on Dvoice dataset. Since the Dvoice data was limited, we used the data augmentation to increase the efficiency of our model. In the context of audio and speech, data augmentation consists of a slight modification to the waveform. These transformations consist of: time dropout, frequency dropout, speed perturbation, time dropout, frequency dropout, augmentation lobe and clipping which the library Speechbrain [13] provided. We trained our model using the Huggingface's transformer [20] using the GPU provided by Google Colaboratory.

In speech recognition, the word error rate or as it is known WER is the metric used to assess the performance of the model. The lower the WER the more accurate is the model. The WER formula:

$$WER = \frac{S + R + I}{N} \tag{1}$$

S constitute the number of swapped words, R the number of removals, I the new added words and N the number of reference words.

The results were encouraging for the amount of data we feed the model with. We got a Word error rate of 0.38% after training for 10 epochs (Fig. 4).

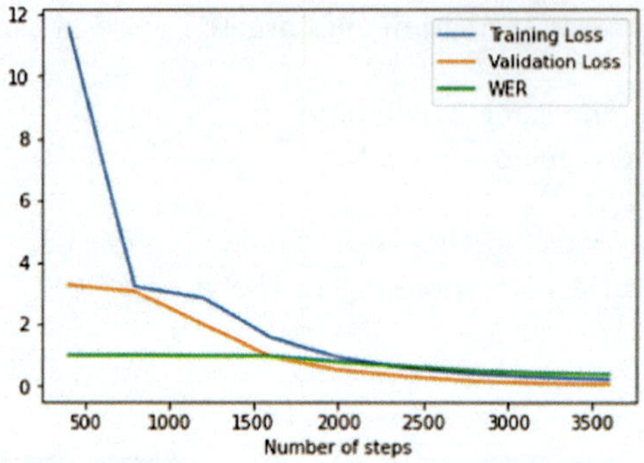

Fig. 4. Evolution of the WER and loss

4.2 Visual Model

The avatar approach has many benefits, one of them is that it preserves the anonymity of the speakers that are in the dataset, since the dataset will most likely be composed of multiple signers, the avatar approach is also necessary in order to normalize the motion capture data to account for the speakers different body shapes and different sizes, to get multiple signers to sign one sentence their skeletons must be normalized first. Translating spoken language into avatar animation, require converting videos to avatar. For this problem to be solved, we use the following general pipeline: we first extract keypoints from a video then convert them into motion capture data and then use Blender (A free and open-source 3D graphics program) to apply the motion to an avatar. We used two different sets of tools and approaches for key points extraction.

The Openpose Approach. For this approach, we first ran the Openpose motion capture on one of our dataset videos in Google Colab. It outputs a folder that contains a set of JSON files representing the landmarks for all the video frames (each JSON file represents a single frame). The next step was to use MocapNET2 [15] to create a motion avatar.

The Mediapipe Approach. For this approach, we use Google's Mediapipe Holistic model for the landmark detection. The model outputs data that can be used directly in Blender. In Fig. 5 we extract the landmarks from a sign video that represent the word honey in MSL.

Fig. 5. Landmark extraction for the word honey in Darija

4.3 Animation

After the creation of motion capture, we use Blender to animate an avatar. The animation has two stages:

1. Creating a medium animation using basic structure called bones
2. Aligning the basic structure "bones" a chosen avatar.

The bone creation Fig. 6 was done using the distance between the landmarks. Since most of the videos were of the upper body so there was no need to create bones for the rest of the body.

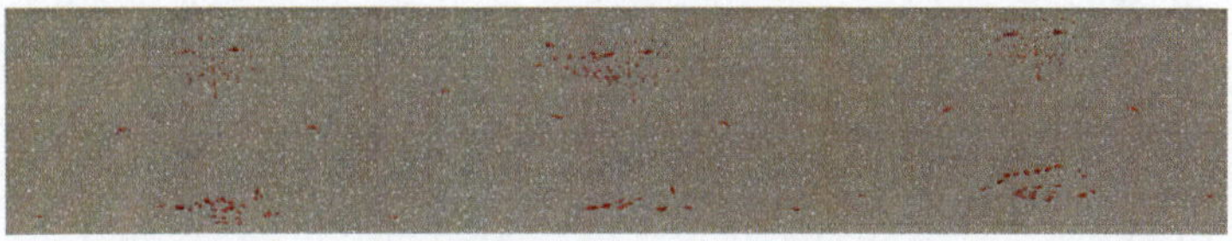

Fig. 6. Bone animation

The avatar used was a free asset Fig. 7 created using the DAZ (3D modeling software) and imported to Blender.

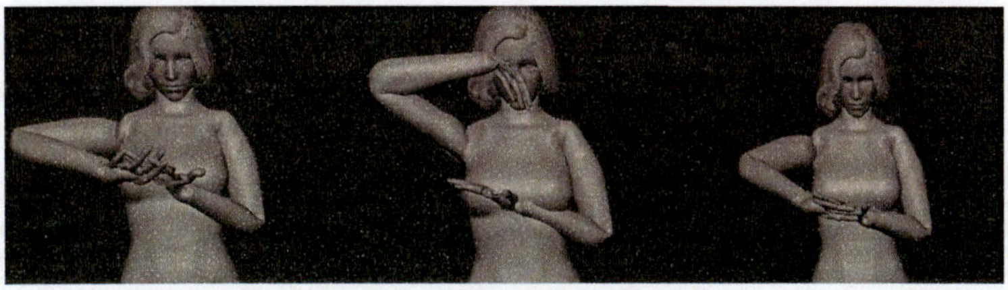

Fig. 7. Avatar animation

5 Discussion

During the recording we noticed that sometimes the sign used in the dataset is diffident from one signer to another. This suggests the existence of synonyms in the Moroccan sign language. For the purpose of this work, variation was set aside for future endeavors.

The motion capture using Openpose results were overall good and promising, but not satisfying, due to the many challenges that this approach faced: First working with Openpose, it required us to use videos that capture the full body from head to feet and not just the upper body, otherwise the resulting motion capture would have random animations associated with it. In our case, the dataset has mostly upper body videos which means this method was not optimal. Second, working with Openpose was very slow: motion extraction takes a relatively long time even on decent machines. Using it on a dataset that has hundreds of videos will be impractical. So, we decided to rely on Mediapipe for landmark extraction.

Table 1. The difference between Openpose and Mediapipe

	Num of body point	Num of face point	Num of hand point	Inference time
Mediapipe	33	468	2×21	21.59 s
Openpose	25	70	2×21	6862.30 s

The Table 1 shows the amount of data and the time it took for the two libraries to detect the pose, hand and face landmark on a single 5 s videosign using a Ryzen six cores CPU.

While our system has shown great result, it manifest shows some defects: The resulting avatar might twitch while preforming the sign. Which occurred due to: One the failure to capture the necessary keypoints. The second is the results of a wrong depth detection. This is can be remedied by the size of the dataset, the more sample the dataset has, the more probable to find videos that don't contain this problem.

6 Conclusion

Research centered around MSL is still in its infancy. This work hopes to further drive the research in this filed. We proposed a system for Moroccan sign language translation. This architecture has shown promising results, however, there is plenty aspect for improvement. Our immediate objective for now is the enlarging of our dataset and taking into consideration regional variation, lowering the WER of our speech to text and finally create a system able to recognizes and translate dialectical sign with acceptable accuracy.

References

1. Agrawal, S., Jalal, A., Tripathi, R.: A survey on manual and non-manual sign language recognition for isolated and continuous sign. Int. J. Appl. Pattern Recogn. **3**, 99 (2016). https://doi.org/10.1504/IJAPR.2016.079048
2. Baevski, A., Zhou, H., Mohamed, A., Auli, M.: wav2vec 2.0: a framework for self-supervised learning of speech representations (2020)
3. Bantupalli, K., Xie, Y.: American sign language recognition using deep learning and computer vision. In: 2018 IEEE International Conference on Big Data (Big Data), pp. 4896–4899 (2018). https://doi.org/10.1109/BigData.2018.8622141
4. Cao, Z., Hidalgo Martinez, G., Simon, T., Wei, S., Sheikh, Y.A.: OpenPose: real-time multi-person 2D pose estimation using part affinity fields. IEEE Trans. Pattern Anal. Mach. Intell. **43**(1), 172–186 (2019)
5. Conneau, A., Baevski, A., Collobert, R., Mohamed, A., Auli, M.: Unsupervised cross-lingual representation learning for speech recognition (2020)
6. Cox, S.: TESSA, a system to aid communication with deaf people, p. 205, January 2002. https://doi.org/10.1145/638286.638287
7. Benelallam, I., Naira, A.M., Allak, A.: Dvoice: an open source dataset for automatic speech recognition on Moroccan dialectal Arabic (2021). https://doi.org/10.5281/zenodo.5482551
8. Jiang, X., Satapathy, S.C., Yang, L., Wang, S.-H., Zhang, Y.-D.: A survey on artificial intelligence in Chinese sign language recognition. Arab. J. Sci. Eng. **45**(12), 9859–9894 (2020). https://doi.org/10.1007/s13369-020-04758-2
9. LeMaster, B.: Moroccan sign language: a language of Morocco (2018)
10. Li, D., Rodriguez, C., Yu, X., Li, H.: Word-level deep sign language recognition from video: a new large-scale dataset and methods comparison. In: The IEEE Winter Conference on Applications of Computer Vision, pp. 1459–1469 (2020)
11. Lugaresi, C., et al.: Mediapipe: a framework for building perception pipelines (2019)
12. Maraqa, M., Abu-Zaiter, R.: Recognition of Arabic sign language (ArSL) using recurrent neural networks. In: 2008 First International Conference on the Applications of Digital Information and Web Technologies (ICADIWT), pp. 478–481 (2008). https://doi.org/10.1109/ICADIWT.2008.4664396
13. Ravanelli, M., et al.: Speechbrain: a general-purpose speech toolkit (2021)
14. Nair, A.V., Bindu, V.: A review on Indian sign language recognition. Int. J. Comput. Appl. **73**, 33–38 (2013)
15. Qammaz, A., Argyros, A.A.: Occlusion-tolerant and personalized 3D human pose estimation in RGB images. In: IEEE International Conference on Pattern Recognition (ICPR 2020), January 2021, to appear. http://users.ics.forth.gr/argyros/res_mocapnet_II.html
16. Sahoo, A.K.: Indian sign language recognition using machine learning techniques. Macromol. Symp. **397**(1), 2000241 (2021). https://doi.org/10.1002/masy.202000241
17. Shivashankara, S., Srinath, S.: A review on vision based American sign language recognition, its techniques, and outcomes. In: 2017 7th International Conference on Communication Systems and Network Technologies (CSNT), pp. 293–299 (2017). https://doi.org/10.1109/CSNT.2017.8418554
18. Stoll, S., Camgoz, N.C., Hadfield, S., Bowden, R.: Text2sign: towards sign language production using neural machine translation and generative adversarial networks. Int. J. Comput. Vision **128**(4), 891–908 (2020). https://doi.org/10.1007/s11263-019-01281-2

19. Tachicart, R., Bouzoubaa, K., Jaafar, H.: Lexical differences and similarities between Moroccan dialect and Arabic. In: 2016 4th IEEE International Colloquium on Information Science and Technology (CiSt), pp. 331–337 (2016). https://doi.org/10.1109/CIST.2016.7805066
20. Wolf, T., et al.: Transformers: state-of-the-art natural language processing. In: Proceedings of the 2020 Conference on Empirical Methods in Natural Language Processing: System Demonstrations, pp. 38–45. Association for Computational Linguistics, October 2020. https://www.aclweb.org/anthology/2020.emnlp-demos.6
21. WHO/NMH/PBD: Millions of people in the world have hearing loss that can be treated or prevented (2013)
22. Zhang, F., et al.: Mediapipe hands: on-device real-time hand tracking. arXiv:2006.10214 (2020)
23. Zwitserlood, I., Verlinden, M., Ros, J., Schoot, S.: Synthetic signing for the deaf: Esign, January 2005

Renewable and Sustainable Energies

A Comparative Study of LSTM and RNN for Photovoltaic Power Forecasting

Mohammed Sabri[1](✉) and Mohammed El Hassouni[2]

[1] LRIT, Mohammed V University in Rabat, 1014, Rabat, Morocco
mohammed_sabri3@um5.ac.ma
[2] FLSH, Mohammed V University in Rabat, 1014, Rabat, Morocco
mohamed.elhassouni@flsh.um5.ac.ma

Abstract. Photovoltaic power generation is one of the most efficient renewable energy generation techniques. To ensure photovoltaic (PV) safe operation and cost-effective integration into smart grids. As a result, a realistic PV power forecast method can aid in mitigating PV power generation's drawbacks, which is critical for power plant maintenance and repair. In this research, we compared two Deep Learning models which are both highly advantageous, namely a long short-term memory (LSTM) and a recurrent neural network (RNN). A case study employing an actual dataset obtained from 1B DKASC, Alice Springs, Australia, is used to demonstrate the performance of LSTM and RNN. The forecasts generated by the LSTM are compared with the results of RNN using MAE, MSE, MBE, R^2 and RMSE for all the forecast horizons. The results of the study demonstrate that the LSTM based forecast outperforms RNN and has the potential to improve the accuracy and stability of the prediction.

Keywords: Deep Learning · Long Short-Term Memory (LSTM) · Recurrent neural network (RNN)

1 Introduction

Globally, energy and environmental challenges have become a widespread source of worry. With the eventual exhaustion of fossil fuels and the severity of environmental pollution growing, making full use of renewable energy will become a necessity to address energy and environmental issues [1]. PV energy has risen to prominence as the most environmentally sustainable, pollution-free, and limitless energy source, PV power has emerged as the greatest option for industrial and domestic power generation [2]. Solar power generation is generally unstable and intermittent makes integrating it into current energy networks extremely difficult. PV prediction accuracy is a good solution to overcome these issues [3].

At present, Statistical approaches and artificial neural network (ANN)-based methods are two main types of load forecasting methods. To forecast, statistical approaches use time series prediction [4], mainly including auto-regression

R. Saidi et al. (Eds.): ICATH 2021, LNDECT 110, pp. 265–274, 2022.
https://doi.org/10.1007/978-3-030-94188-8_25

(AR), auto-regressive moving average (ARMA), and multiple linear regression (MLR). Traditional statistical approaches are unable to understand these nonlinear data since there are many inherent nonlinear patterns in the massive data [5]. Artificial intelligence techniques have been extensively and efficiently used to classification and prediction problems, including support vector machine [6] and artificial neural network (ANN) [7]. Many researchers have tried and succeeded in applying this concept to load forecasting and new energy power forecasting [8–10]. The authors in [11] utilized a deep learning architecture based on long-short term memory (LSTM) recurrent neural network (RNN) to predict the PV power. For load forecasting, Salah et al. [12] used an LSTM model that outperformed traditional machine learning approaches. Deep LSTM networks were utilized to develop a new model for predicting PV power one hour in advance [13], which can capture abstract notions in PV power sequences. The recurrent neural network (RNN) algorithm can consider the dependencies along consecutive time steps [14,15]. RNN architecture structures were utilized to model time series in which neurons are fully connected with cycles feeding the activation functions from past time steps as inputs, and the relationship between historical data and current status could be defined using this structure [16,17]. The RNN-based LUBE approach is suggested in [18] for directly constructing optimal PIs for wind power forecasting. However, experts continue to stress the need for more precise and dependable load forecasting techniques.

Although PV power output is seasonal and periodic, large changes are prevalent on overcast and wet days, which are difficult to represent using solely historical PV data. The performance of RNN and LSTM models in reducing error rates is compared in this research. The RNN was chosen because it can capture the dynamic of time-series data by storing data from previous computations in its internal memory. Because of its ability to preserve and train the characteristics of given data over a longer period of time, the LSTM approach is used. The key contributions of this paper are:

- Examine the performance of deep learning-based algorithms for prediction procedures through empirical research and analysis.
- Evaluate the performance of LSTM and RNN in terms of the error rate minimizing obtained in prediction. According to the findings, LSTM outperforms RNN.

This paper is organized as follows: Sect. 2 gives an overview of our comparative methodology. In Sect. 3, we present the results of PV power forecasting by both LSTM and RNN models and will have a discussion on that. Finally, the conclusions and future work are summarized in Sect. 4.

2 Theoretical Background

2.1 Recurrent Neural Networks and Long Short-Term Memory

The recurrent neural network (RNN) is a type of neural network that is used to forecast sequential data and whose output depends on the input [19]. By preserving knowledge from earlier computations in the internal memory, the RNN is able to capture the dynamism of time-series data [20]. RNN has been used in a situation where the output's previous values have a substantial impact on the future. Because of its capacity to analyze sequential data of various lengths, it is mostly utilized in forecasting applications. RNN operates by taking into account the input of hidden layer neurons that accepts input from neurons from the preceding time step. [21]. To this purpose, they use cells represented by gates to control the output, which is generated using past data observations. RNN excels at understanding the dynamic temporal properties that emerge in time series data [22]. Figure 1 illustrates a comprehensive overview of recurrent neural networks.

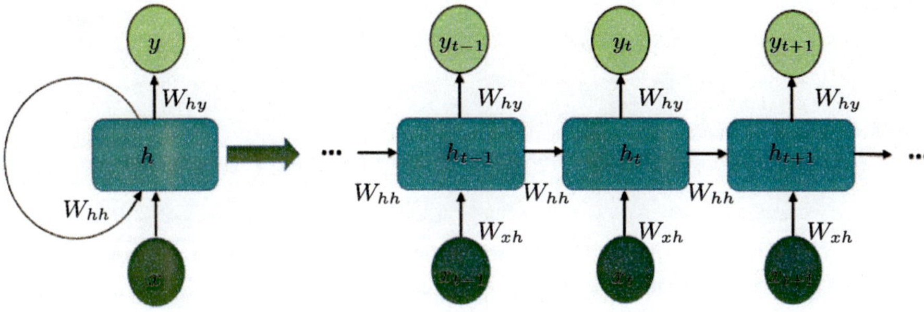

Fig. 1. General overview of recurrent neural networks

For a particular input sequence x_t, the hidden neuron h_t gets information feedback from other neurons in the preceding time step multiplied by a W_{hh}, which is the weight of the preceding hidden state h_{t-1}, which can be determined sequence by Eq. 1. The output state y_t is calculated according to the Eq. 2.

$$h_t = f(W_{hh}h_{t-1} + W_{xh}x_t + b_h) \tag{1}$$

$$y_t = g(W_{hy}h_t + b_y) \tag{2}$$

Where W_{xh} is the weight of the actual input state, W_{hy} is the weight at the output state, x_t is the input at instant time t, f and g represent hidden layer activation function and output layer activation function.

RNNs have a gradient vanishing problem. It indicates that the weights moved forward and backward.

The LSTM is a special kind of RNN, due to gradient disappearance, the long-term reliance problem, which makes the model is inappropriate for learning sequence data maybe solved by inserting a memory cell into each neuron in the hidden layer and using the input gate, output gate and forget gate to regulate the memory cell's state. [23]. The architecture of LSTM neuron is shown in Fig. 2.

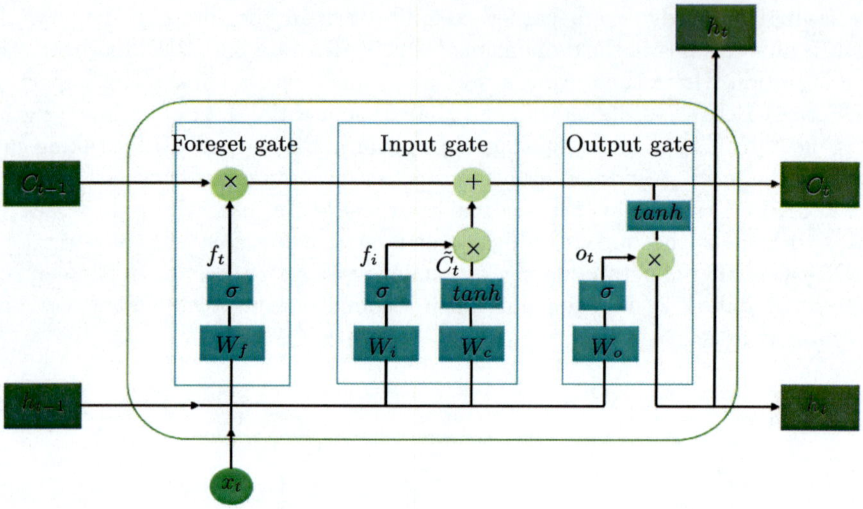

Fig. 2. The architecture of LSTM

Together, the memory cell and the hidden state can memorize the sequence data historical information. The information in the memory cell is controlled by three gate units. Per the previous moment's hidden state h_{t-1} and the current moment's input x_t, forget gate deletes the information in the memory cell. Forget gate is calculated as follows:

$$f_t = \sigma(W_f.[h_{t-1}, x_t] + b_f) \tag{3}$$

Per the previous moment's hidden state h_{t-1} and the current moment's input x_t, the input gate adds information to the memory cell. i_t represents the input gate, C_t is the cell state and $\tilde{C_t}$ is the value for calculating the current cell state C_t.

$$i_t = \sigma(W_i.[h_{t-1}, x_t] + b_i) \tag{4}$$

$$\tilde{C_t} = tanh(W_c.[h_{t-1}, x_t] + b_c) \tag{5}$$

The following equation is used to update the memory cell once forget gate and output gate has been calculated.

$$C_t = f_t.c_{t-1} + i_t.\tilde{C_t} \tag{6}$$

The current hidden h_t state is determined by the output gate based on the hidden state of the previous moment h_{t-1}, the input of the current moment x_t, and the updated memory cell C_t.

$$o_t = \sigma(W_o.[h_{t-1}, x_t] + b_o) \tag{7}$$

$$h_t = o_t.\tanh(C_t) \tag{8}$$

Where W_f represents forget gate weight; W_i, W_c represents input gate weights; and W_o represents output weight; b_f, b_i, b_c and b_o represent the biases. σ is the sigmoid function.

2.2 Performance Metric

To examine the performance of the LSTM and RNN models for PV power prediction, we applied five performance metrics, including mean absolute error (MAE), mean square error (MSE), root mean square error (RMSE), mean bias error (MBE), and goodness of (R-Square). The expression of these evaluation indexes are presented as follows

$$\mathrm{MAE} = \frac{1}{\mathrm{n}} \sum_{\mathrm{i}=1}^{\mathrm{n}} |\mathrm{y_i} - \tilde{\mathrm{y_i}}| \tag{9}$$

$$\mathrm{MSE} = \frac{1}{\mathrm{n}} \sum_{\mathrm{i}=1}^{\mathrm{n}} (\mathrm{y_i} - \tilde{\mathrm{y_i}})^2 \tag{10}$$

$$\mathrm{RMSE} = \sqrt{\frac{1}{\mathrm{n}} \sum_{i=1}^{\mathrm{n}} (\mathrm{y_i} - \tilde{\mathrm{y_i}})^2} \tag{11}$$

$$\mathrm{MBE} = \frac{1}{\mathrm{n}} \sum_{\mathrm{i}=1}^{\mathrm{n}} (\tilde{\mathrm{y_i}} - \mathrm{y_i}) \tag{12}$$

$$\mathrm{R}^2 = 1 - \frac{\sum_{\mathrm{i}=1}^{\mathrm{n}} (\mathrm{y_i} - \tilde{\mathrm{y_i}})^2}{\sum_{\mathrm{i}=1}^{\mathrm{n}} (\mathrm{y_i} - \bar{\mathrm{y_i}})^2} \tag{13}$$

where n is the total number of y_i series, $\bar{y_i}$ is the average of the measured power in the test set, y_i is the measured output power and $\tilde{y_i}$ is the predicted output power. The predictive model is more accurate when R^2 is near 1. The shorter are the values of MAE, MSE, MBE and RMSE, the higher is the performance evaluation.

3 Experimental Setup

With the objective of analyzing the performance of two very promising artificial neural network models namely LSTM and RNN, the authors conducted various studies on some selected time series data. In this work, the main research questions are investigated as follows: Which algorithm, RNN or LSTM, could produce better forecasting of PV power data? Do the different types of season and weather conditions, in deep learning-based techniques affect the accuracy of the trained model?

3.1 Datasets Description

In this research, the PV power dataset is supplied by the "Desert Knowledge Australia Solar Center", the historical data selected from 1B DKASC, Alice Springs [24]. The dataset has a 5-min resolution and lasted from March 1, 2020 to February 28, 2021. The data includes diffuse horizontal radiation ($W/m^2 * sr$), active power (KW), wind direction ($\hat{A}°$), weather relative humidity (%), current phase average (A), weather temperature Celsius ($°C$), global horizontal radiation ($W/m^2 * sr$), etc. The data set is separated seasonally into four situations, namely autumn (March–May), winter (June–August), spring (September–November) and summer (December–February), each season contains three months of data, The training dataset is made up of the first two months of data, while the testing dataset is made up of the remaining data.

In order to determine, which deep-learning model, RNN or LSTM would be a better choice for PV generation forecasting, both LSTM and RNN have the same number of layers and inputs to make the comparison consistent. The first LSTM/RNN model we utilized had one input layer, followed by dropout layer, one output layer that provides a single value forecasting. The hidden layer was initially given 100 memory units, and the model was compiled with the ADAM optimizer. Table 1 lists all of the parameters that were used in this study.

Table 1. Specification of parameters for training.

Parameters	Values
Hidden layer	1 LSTM/RNN layer with 100 units
Dropout layer	1 with (0.1 dropout rate)
Output layer	1
Number of epochs	100
Batch size	80
Loss function	MSE

In each of all seasons winter, spring, summer and autumn, the MAE, MSE, MBE, RMSE and R^2 metrics are shown in Table 2. Under the same conditions, our results show that the performances of LSTM and RNN are statistically different from each other, the MAE of the LSTM runs from 0.1039 to 0.2531, with an average of 0.1887. While the average MAE of RNN is 0.2096. The average MBE of the LSTM model is −0.1207, better than −0.1576 of RNN. The RMSE index obtained from the LSTM model varies from 0.1632 to 0.2651, with an average of 0.2233 and the RMSE index obtained from the RNN model varies from 0.1093 to 0.2860, with an average of 0.2344. In all seasons, the MSE of the LSTM ranges from 0.0266 to 0.0703, with an average of 0.0514, When compared to the RNN, it has a minimum of 0.0119 and a maximum of 0.0817, with an average of 0.0602. The information is previously shown in Table 2 is intuitively shown in Fig. 3.

Table 2. The results of the forecasting model.

Season	Models	MAE	MSE	R^2	RMSE	MBE
Winter	RNN	0.0914	0.0119	0.9986	0.1093	0.0740
	LSTM	0.1039	0.0266	0.9968	0.1632	0.0930
Spring	RNN	0.2410	0.0665	0.9946	0.2580	−0.2408
	LSTM	0.2531	0.0703	0.9943	0.2651	−0.2530
Summer	RNN	0.2536	0.0817	0.9931	0.2860	−0.2469
	LSTM	0.1830	0.0461	0.9961	0.2148	−0.1651
Autumn	RNN	0.2527	0.0808	0.9883	0.2843	−0.2170
	LSTM	0.2149	0.0626	0.9909	0.2503	−0.1586
Average	RNN	0.2096	0.0602	0.9936	0.2344	−0.1576
	LSTM	0.1887	0.0514	0.9945	0.2233	−0.1207

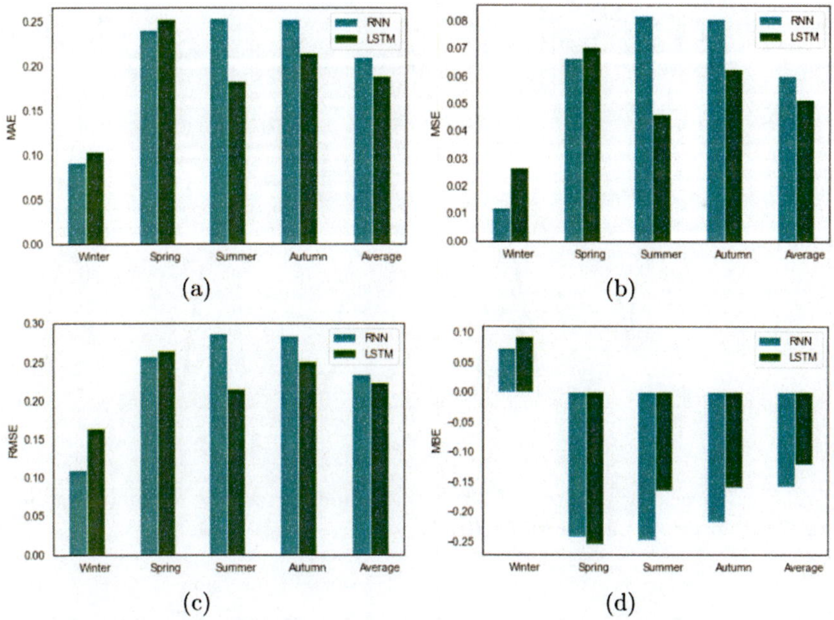

Fig. 3. The comparison of MAE (a), MSE (b), RMSE (c) and MBE (d) of two models for each season.

To verify the accuracy and stability of LSTM and RNN, R^2 is used to measure which model indicates better performance in the four seasons. The R^2 values of the LSTM are 0.9968 in winter, 0.9943 in spring, 0.9961 in summer and 0.9909 in autumn, The improvement grows from 0.9883 to 0.9986 when compared to typical R^2 metrics from RNN algorithms. In comparison to the RNN, the experimental findings reveal that the LSTM prediction approach has the

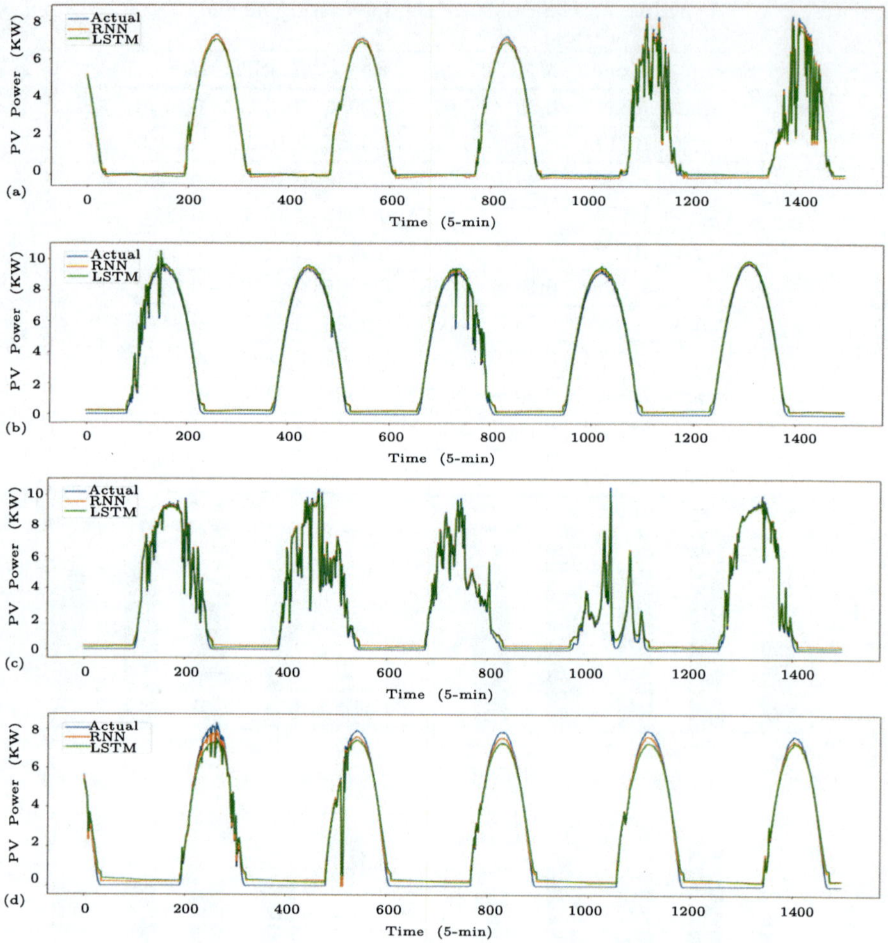

Fig. 4. The forecasting results of LSTM, RNN and actual PV power in winter (a), spring (b), summer (c), and autumn (d).

best prediction accuracy because the unique structure of LSTM with additional features to memorize the sequence of data. This indicates that the LSTM model has a higher prediction accuracy in different weather types and performs better on PV power forecasting.

Five days from different seasons were chosen at random for further examination. The prediction results of these twenty days via the LSTM model and RNN model are shown in Fig. 4. It can be seen from the model's exhibit satisfactory forecasting performance, it was observed that the results of the LSTM and RNN models follow the same pattern as the actual values, but the LSTM model is the closest to the actual forecasting result.

Deep learning methods, such as LSTM and RNN, are popularly used in many applications. In this paper, the LSTM model can achieve a more accurate prediction result than RNN. Each LSTM is constituted of cells, or system modules, that capture and store data streams. The cells resemble a transport line (the top line in each cell) that links one module to the next, transferring data from the past and collecting it for the present. In this study, the dataset is divided into four cases by season, to test the accuracy and stability of LSTM and RNN. The results prove that the LSTM produced a lower average MAE, MSE, MBE and RMSE than the RNN. However, both models appear to have produced robust results.

4 Conclusion

With recent advances in the development of sophisticated machine learning-based approaches, especially deep learning algorithms, these approaches, are increasing favor among academics from a variety of disciplines. When compared to old procedures, the main question is how accurate and strong these newly offered methodologies are. This paper compares the accuracy of LSTM and RNN, these two algorithms were implemented and applied on a sample of PV power data and the result reveals that LSTM outperformed RNN. Particularly, as compared to RNN, the LSTM-based approach increased prediction by 99.45% on average. However, one disadvantage of this study is that we did not include other variables such as wind speed, weather daily rainfall, radiation global tilted, radiation diffuse tilted, etc. As a result, the conclusion about the superior performance of the LSTM model over the RNN is not definitive. So, as future research, we would like to evaluate the enhancement in Short and Medium-term PV power forecasting by adding those parameters.

References

1. Lee, C.T., Hashim, H., Ho, C.S., Van Fan, Y., Klemeš, J.J.: Sustaining the low-carbon emission development in Asia and beyond: sustainable energy, water, transportation and low-carbon emission technology. J. Clean. Prod. **146**, 1–13 (2017)
2. Koo, C., Hong, T., Jeong, K., Ban, C., Oh, J.: Development of the smart photovoltaic system blind and its impact on net-zero energy solar buildings using technical-economic-political analyses. Energy **124**, 382–396 (2017)
3. Blaga, R., Sabadus, A., Stefu, N., Dughir, C., Paulescu, M., Badescu, V.: A current perspective on the accuracy of incoming solar energy forecasting. Prog. Energy Combust. Sci. **70**, 119–144 (2019)
4. Pang, C., Bao, T., He, L.: Power system load forecasting method based on recurrent neural network. In: E3S Web of Conferences, vol. 182, p. 02007. EDP Sciences (2020)
5. Chen, G.-Y., Gan, M., Chen, G.-L.: Generalized exponential autoregressive models for nonlinear time series: stationarity, estimation and applications. Inf. Sci. **438**, 46–57 (2018)

6. Barman, M., Choudhury, N.B.D.: Season specific approach for short-term load forecasting based on hybrid FA-SVM and similarity concept. Energy **174**, 886–896 (2019)
7. Shrivastava, S., Chaturvedi, K.T.: A review of artificial intelligence techniques for short term electric load forecasting. Int. J. Adv. Res. Electr. Electron. Instrum. Eng. **7**(5), 2241–2247 (2018)
8. Xu, L., Wang, W., Zhang, T., Yang, L., Wang, S., Li, Y.: Ultra-short-term wind power prediction based on neural network and mean impact value. Autom. Electr. Power Syst. **41**(21), 40–45 (2017)
9. Liu, R., Huang, L.: Wind power forecasting based on dynamic neural networks. Autom. Electr. Power Syst. **36**(11), 19–22 (2012)
10. Hernández, L., et al.: Artificial neural network for short-term load forecasting in distribution systems. Energies **7**(3), 1576–1598 (2014)
11. Abdel-Nasser, M., Mahmoud, K.: Accurate photovoltaic power forecasting models using deep LSTM-RNN. Neural Comput. Appl. **31**(7), 2727–2740 (2019)
12. Bouktif, S., Fiaz, A., Ouni, A., Serhani, M.A.: Optimal deep learning LSTM model for electric load forecasting using feature selection and genetic algorithm: comparison with machine learning approaches. Energies **11**(7), 1636 (2018)
13. Gensler, A., Henze, J., Sick, B., Raabe, N.: Deep learning for solar power forecasting-an approach using autoencoder and LSTM neural networks. In: 2016 IEEE International Conference on Systems, Man, and Cybernetics (SMC), pp. 002 858–002 865. IEEE (2016)
14. Gers, F.A., Schraudolph, N.N., Schmidhuber, J.: Learning precise timing with LSTM recurrent networks. J. Mach. Learn. Res. **3**(Aug), 115–143 (2002)
15. Qureshi, A.S., Khan, A., Zameer, A., Usman, A.: Wind power prediction using deep neural network based meta regression and transfer learning. Appl. Soft Comput. **58**, 742–755 (2017)
16. Koschwitz, D., Frisch, J., Van Treeck, C.: Data-driven heating and cooling load predictions for non-residential buildings based on support vector machine regression and NARX recurrent neural network: a comparative study on district scale. Energy **165**, 134–142 (2018)
17. Taniar, D., Safar, M., Tran, Q.T., Rahayu, W., Park, J.H.: Spatial network RNN queries in GIS. Comput. J. **54**(4), 617–627 (2011)
18. Shi, Z., Liang, H., Dinavahi, V.: Direct interval forecast of uncertain wind power based on recurrent neural networks. IEEE Trans. Sustain. Energy **9**(3), 1177–1187 (2017)
19. Alzahrani, A., Shamsi, P., Dagli, C., Ferdowsi, M.: Solar irradiance forecasting using deep neural networks. Procedia Comput. Sci. **114**, 304–313 (2017)
20. Alzahrani, A., Shamsi, P., Ferdowsi, M., Dagli, C.: Solar irradiance forecasting using deep recurrent neural networks. In: 2017 IEEE 6th International Conference on Renewable Energy Research and Applications (ICRERA), pp. 988–994. IEEE (2017)
21. LeCun, Y., Bengio, Y., Hinton, G.: Deep learning. Nature **521**(7553), 436–444 (2015)
22. Wang, H., Lei, Z., Zhang, X., Zhou, B., Peng, J.: A review of deep learning for renewable energy forecasting. Energy Convers. Manage. **198**, 111799 (2019)
23. Bengio, Y., Boulanger-Lewandowski, N., Pascanu, R.: Advances in optimizing recurrent networks. In: 2013 IEEE International Conference on Acoustics, Speech and Signal Processing, pp. 8624–8628. IEEE (2013)
24. http://dkasolarcentre.com.au/locations/alice-springs?source=1B

Solar Energy Resource Assessment Using GHI and DNI Satellite Data for Moroccan Climate

Omaima El Alani[1,2](✉), Hicham Ghennioui[1], Mounir Abraim[1,2], Abdellatif Ghennioui[2], Philippe Blanc[3], Yves-Marie Saint-Drenan[3], and Zakaria Naimi[2]

[1] Faculty of Sciences and Technologies, Sidi Mohamed Ben Abdellah University, Route d'Immouzer, Fez, Morocco
omaima.elalani@usmba.ac.ma

[2] Green Energy Park, Km 2 Route Régionale R206, Benguerir, Morocco

[3] O.I.E. Centre Observation, Impacts, Energy, MINES ParisTech, PSL - Research University, Sophia Antipolis, France

Abstract. Solar energy is in rapid development in Morocco. Thanks to its geographical and climatic location, the Kingdom of Morocco is privileged to develop large-scale solar energy exploitation. The knowledge of the solar potential in a given site is crucial for most solar energy applications. The present study aims to evaluate the energy potential in several cities in Morocco to provide guidance to solar energy users about the available amount of energy. The assessment was done using 10 years of data from the Helioclim3 satellite database. In addition to the solar potential analysis, we performed a classification of the days to determine the dominant climatology based on the clear sky index. The results showed the dominance of clear days toward all study sites, and selected sites may be a promising location for solar projects with annual global horizontal irradiation reaching 2075 kWh/m^2 and direct normal irradiation reaching 2463 kWh/m^2.

Keywords: Solar resource assessment · Global horizontal irradiation · Direct normal irradiation · Morocco · HelcioClim3

1 Introduction

Energy is a crucial factor for the economic and social development of countries. Energy is at the heart of all sectors, including industry, transport, commerce, agriculture, health, residential, etc. A reliable and sufficient energy supply is necessary to meet the various energy needs and increase productivity. However, fossil energy sources, in addition to being soon exhausted and unsustainable for future development, they present serious threats to the environment. About two-thirds of the world's carbon dioxide ejections come from these fuel sources, whose current share of energy production.

Morocco is the largest net importer of energy in the North African region. Morocco's energy profile has been dominated for a long time by importing its energy resources [1]. The Moroccan energy mix is composed of more than 90% of fossil fuel of the total

R. Saidi et al. (Eds.): ICATH 2021, LNDECT 110, pp. 275–285, 2022.
https://doi.org/10.1007/978-3-030-94188-8_26

primary energy supply and 80% of the electricity supply. To overcome this dependence, the country is making great efforts in developing renewable energy and integrating it into existing and possibly emerging networks. Morocco has launched a plan to increase the share of renewable energy in the energy mix. By 2030 the country has committed to reducing greenhouse gas emissions by 17% from baseline levels and achieving 52% of installed electricity capacity from renewable sources. Currently, the Moroccan energy mix is quite diversified, with a remarkable contribution from renewable sources. In 2018, Morocco installed 3700 MW (1170 MW from hydro, 1220 MW from wind, and 710.8 MW from solar), and is anticipated rising 12.900 MW by 2030, which will be distributed as follows: 3100 MW from hydro, 5000 MW from wind, and 4800 MW from solar [2, 3].

Solar energy is the most plentiful source of energy on the planet [4]. Solar energy can be exploited in several forms: direct conversion into electrical energy by photovoltaic cells; direct conversion into thermal energy by solar thermal collectors; thermodynamic conversion into electrical energy by combining solar thermal collectors, turbines or thermal engines and electrical generators; and conversion into chemical energy by photochemical means [5].

For all these applications, solar radiation is a very important variable. The information on solar radiation is also essential for feasibility study and selection of favorable sites, optimization of the energy produced through accurate forecasting of solar radiation, estimation of the financial cost of any new solar project. The solar resource can be evaluated through meteorological stations installed on the ground and equipped with various sensors. The best accuracy and data quality are obtained through these ground sensors. However, the density of measuring stations is still insufficient due to their high costs [6]. To overcome this lack of weather stations, satellite imagery is used as an alternative approach for solar resource estimation [7].

This study evaluated the solar potential by analyzing two solar components, the global horizontal irradiation (GHI) and the direct normal irradiation (DNI) at nine sites in Morocco. In addition, we performed an analysis of the climatology regarding the clear sky based on the clear sky index calculation K_t. For these purposes, irradiation data and typical meteorological years (TMY) from the HelioClim3V5 database were used.

2 Study Sites and Data

2.1 Study Sites

The Kingdom of Morocco is a country located in northwest Africa geographically and climatically favored to develop large-scale exploitation of renewable energy. Solar energy is considered a promising renewable alternative for Morocco. The country has a strong potential of sunshine with an average of global radiation of about 5.3 kWh/m^2/year with an annual insolation duration between 2700 h in the north and up to 3500 h in the south [8]. In this study, we selected several sites in Morocco to evaluate their solar potential. The selected sites are: Benguerir, Erfoud, Missour, Zagora, Tantan, Oujda, Benguerir, Ouarzazate, Oujda, Ain Bni Mathar, and Taza. We selected these sites to cover several locations, including some EnerMENA stations (Erfoud, Missour, Zagora, Tantan, and

Oujda) that were installed in addition to six other stations by the German Aerospace Center (DLR) in the Mena region to assess their solar potential [9].

The map of Fig. 1 represents the geographical location of the study sites, the size of bubbles represents GHI intensity estimated from SolarGIS [10]. Table 1 represents the geographical coordinate and the yearly estimated GHI and DNI. It can be seen that the annual GHI varies between 1876 kWh/m^2 and 2157 kWh/m^2, and the DNI varies between 1786 kWh/m^2 and 2473 kWh/m^2. High irradiation intensities are observed at Ouarzazate, Erfoud, Missour, Zagora, and Benguerir.

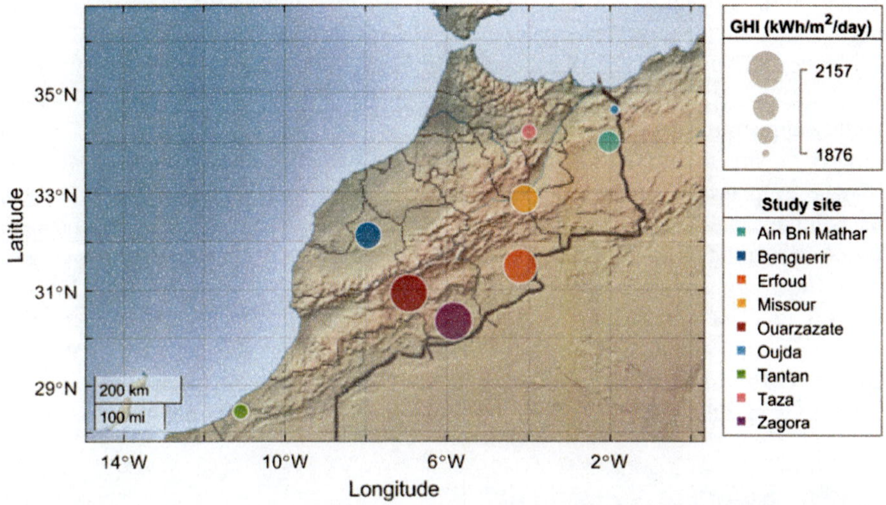

Fig. 1. Geographical location of the study sites. The size of the bubbles corresponds to the intensity of the annual average of GHI estimated from SolarGis.

Table 1. Geographic information and yearly of GHI and DNI from SolarGis [10].

Site	Latitude (°) N	Longitude (°) W	Yearly GHI kWh/m^2/day	Yearly DNI kWh/m^2/day
Benguerir	32.12	7.94	2004	2084
Erfoud	31.49	4.21	2098	2182
Missour	32.86	4.10	2020	2189
Zagora	30.34	5.84	2154	2290
Tantan	28.43	1.09	1909	1786
Oujda	34.65	1.9	1876	1930
Ouarzazate	30.93	6.93	2157	2473
Ain Beni Mathar	34.00	2.03	1957	2060
Taza	34.21	3.998	1904	1967

To assess the solar resource at different sites, solar radiation data were collected from HelioClim3 (HC3v5) version 5 database via soda portal [11]. For the clear-sky climatology study, we calculated the clear sky index using the clear sky radiation derived from the McClear model. The choice of HelioClim3 and McClear was based on previous validations of these two databases at several stations in Morocco where good performances were obtained [12–14].

HC3 [15] is a database providing records of solar irradiation in the field of view of the Meteosat satellite covering Europe, Africa, the Atlantic Ocean, and the Middle East [16] for different time steps ranging from 1 min to 1 month, from 2004 up to the day before the current day. These data are derived from processing Meteosat satellite images using the Heliosat-2 method [17, 18], that combines a clear sky model with a cloud index.HC3v4 and HC3v5 are two versions of HC3. The HC3v4 version uses the European Solar Radiation Atlas model (ESRA) [19, 20] as a clear sky model with a monthly climatology of the linke turbidity [21] as input, while the HC3v5 [22, 23] is a proposed version to improve HC3v4 by combining it with the clear sky model McClear [24] whose inputs are information on the clear atmosphere content (aerosol properties, the total column content in water vapor and ozone) produced by the Monitoring Atmosphere Composition and Climate Services (MACC) [25].

McClear [24] is a physical model to estimate solar irradiation under clear sky conditions. It exploits atmospheric composition datasets provided by MACC projects such as aerosol partial and total optical depths -AOD- at different wavelengths, total water vapor, and ozone contents. The model provides time series at any location since 2004, with a delay of 2 days, with a temporal resolution up to 1 min (interpolation).

3 Solar Resource Assessment

Solar radiation or irradiation is the incident energy received per unit area during a given period (hour or day), measured in kWh/m^2 or J/m^2. As it passes through the atmosphere, solar radiation interacts with the various components of the atmosphere. The interactions of solar radiation with the Earth's atmosphere result in three fundamental broadband components of interest to solar energy conversion technologies. DNI is the solar irradiation coming directly from the solar disk, not absorbed and not scattered. Its value weakens with the presence of clouds. It is of interest to concentrated solar power (CSP) and concentrated photovoltaic (CPV) technologies. DHI is diffuse horizontal irradiation resulting from the scattering of the irradiation by the atmosphere components, and GHI is the total irradiation incident on a horizontal surface. It is the sum of direct and diffuse irradiation. GHI is interesting for photovoltaic technologies (PV). In this study, we will focus on the analysis of these two components, the GHI and the DNI.

3.1 Monthly and Yearly Values

To quantify and evaluate the solar potential for a given location, the calculation and analysis of monthly and yearly values is an approach commonly applied [26, 27]. It allows the evaluation of available solar potential at a given location, and to make comparisons between several sites. To evaluate the solar potential, we used the TMY (Typical

Meteorological Year) generated using 10 years of data from 2011 to 2020. The TMY is a synthetic year of solar and meteorological parameters at an hourly time step which is representative of a meteorological scenario at a given site. Comparison results of monthly averages for GHI and DNI for the nine study sites are shown in Fig. 2 and Fig. 3, respectively. The annual values and monthly averages of daily GHI and DNI for each site are summarized in Table 2 and Table 3, respectively.

The results shown in the tables show that the annual accumulated solar irradiation is significant for all sites. For GHI, annual values of 2075 kWh/m^2, 2025 kWh/m^2, and 2031 kWh/m^2 are obtained for Ouarzazate, Zagora, and Erfoud, respectively. The lowest values are obtained for Oujda with an annual GHI of 1840 kWh/m^2 and Tantan with an annual GHI of 1849 kWh/m2. The annual DNI is also important; the most significant values of DNI are also obtained for Ouarzazate, Erfoud, Missour, and Taza, with values of 2436 kWh/m^2, 2339 kWh/m^2, 2312 kWh/m^2 and 2252 kWh/m^2. While the lowest annual average in terms of DNI is obtained for Tantan.

According to the analysis of the bar plots of Fig. 2 and Fig. 3, the GHI is more important during the summer compared to other seasons. The high value of GHI is detected in Erfoud during June with a value of 7.6 kWh/m^2, Ouarzazate, Taza, and Zagora sites also have high monthly GH values of 7.5 kWh/m^2, 7.49 kWh/m^2, and 7.43 kWh/m^2 respectively. The lowest value of GHI for June was recorded at Tantan with a value of 6.05 kWh/m^2. For the whole month the sunniest sites are Ouarzazate, Erfoud, Missour, Zagora, and Benguerir. The sites of Taza, Oujda, and Ain Bni Mathar represent the lowest radiation values of radiation during January, February, October, November, and December.

The monthly averages of DNI are high during all seasons. The maximum values of DNI are recorded at Ouarzazate with values of 7.84 kWh/m^2, 7.9 kWh/m^2, 7.6 kWh/m^2, and 8.12 kWh/m^2 in March, April, May, and June, respectively. Benguerir, Missour, Erfoud, and Zagora represent a strong DNI varying between 7.8 kWh/m^2 and 4.02 kWh/m^2. During summer the lowest DNI was obtained in the Tantan site with a value of 4.39 kWh/m^2. Taza, Ain Bni Mathar and Oujda share a similar behavior during the majority of months.

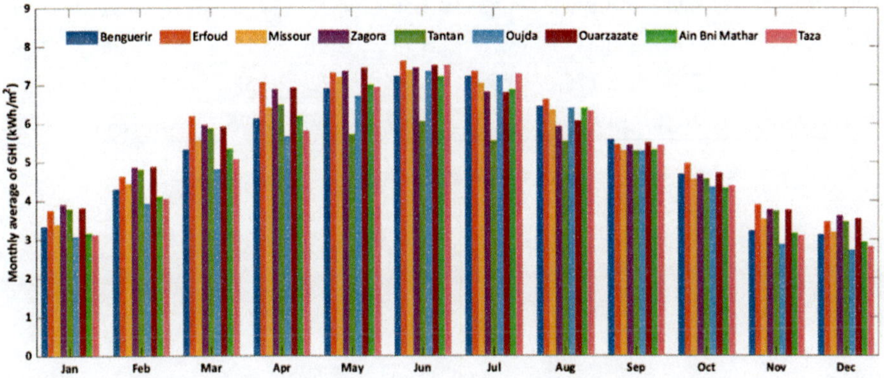

Fig. 2. Comparison between the monthly averages of daily GHI (kWh/m2) for all study sites.

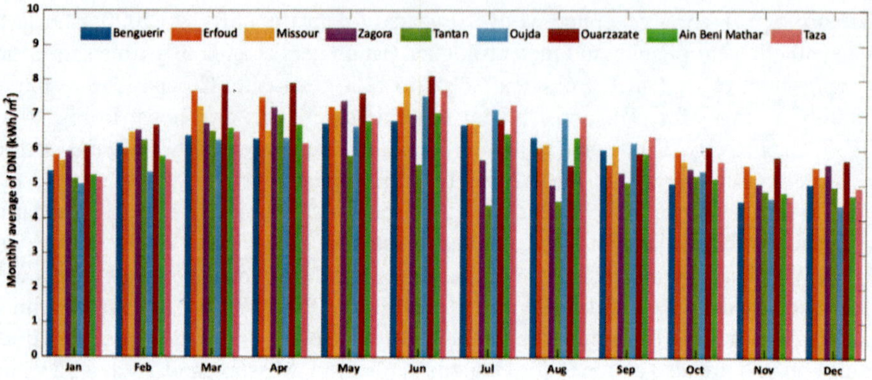

Fig. 3. Comparison between the monthly averages of daily DNI (kWh/m^2) for all study sites.

Table 2. Monthly and annual averages of daily GHI (kWh/m^2) for study sites.

	Bengu-erir	Erfoud	Missour	Zagora	Tantan	Oujda	Ouarz-azate	Ain Bni Mathar	Taza
Jan	3.34	3.75	3.38	3.91	3.79	3.08	3.81	3.16	3.11
Feb	4.30	4.62	4.43	4.86	4.81	3.94	4.87	4.11	4.04
Mar	5.32	6.18	5.54	5.96	5.88	4.81	5.91	5.34	5.05
Apr	6.13	7.06	6.40	6.88	6.48	5.67	6.92	6.19	5.80
May	6.91	7.31	7.19	7.35	5.72	6.70	7.44	7.00	6.92
Jun	7.22	7.61	7.37	7.44	6.05	7.35	7.50	7.21	7.50
Jul	7.22	7.34	7.03	6.81	5.54	7.24	6.79	6.88	7.27
Aug	6.44	6.60	6.33	5.91	5.53	6.40	6.06	6.40	6.30
Sep	5.57	5.44	5.28	5.43	5.28	5.28	5.49	5.30	5.41
Oct	4.67	4.95	4.54	4.66	4.56	4.35	4.70	4.32	4.37
Nov	3.22	3.89	3.51	3.76	3.73	2.87	3.75	3.16	3.09
Dec	3.12	3.44	3.17	3.60	3.46	2.72	3.52	2.93	2.80
Year	1932	2031	1953	2025	1849	1840	2075	1888	1877

Table 3. Monthly and annual averages of daily DNI (kWh/m^2) for study sites.

	Bengu-erir	Erfoud	Missour	Zagora	Tantan	Oujda	Ouarz-azate	Ain Bni Mathar	Taza
Jan	5.37	5.85	5.68	5.93	5.14	4.99	6.08	5.25	5.20
Feb	6.16	6.03	6.48	6.55	6.26	5.34	6.69	5.79	5.70
Mar	6.39	7.67	7.23	6.75	6.51	6.26	7.85	6.60	6.50
Apr	6.29	7.49	6.53	7.20	6.98	6.32	7.90	6.70	6.18
May	6.73	7.22	7.10	7.39	5.82	6.65	7.61	6.80	6.89
Jun	6.80	7.23	7.81	7.01	5.57	7.52	8.11	7.04	7.70
Jul	6.69	6.75	6.73	5.70	4.39	7.14	6.86	6.44	7.28
Aug	6.34	6.06	6.15	4.97	4.51	6.89	5.55	6.34	6.94
Sep	5.99	5.58	6.11	5.34	5.06	6.19	5.90	5.88	6.39
Oct	5.03	5.95	5.67	5.46	5.25	5.38	6.07	5.16	5.66
Nov	4.50	5.56	5.30	5.02	4.80	4.59	5.79	4.76	4.66
Dec	5.00	5.50	5.26	5.58	4.93	4.40	5.70	4.68	4.91
Year	2167	2339	2312	2215	1980	2181	2436	2173	2252

3.2 Daily Average

Another analysis could be performed by analyzing the daily maximum and average components of GHI (Fig. 4) and DNI (Fig. 5) for the Benguerir site as an example. For this site, significant values of irradiance are obtained during the summer season. The maximum and average values of GHI are 1067 w/m^2 and 362.16 w/m^2 detected during May. DNI values are significant for almost all seasons; the maximum value is obtained during May with a value of 965 w/m^2 for an average of 423.41 w/m^2.

3.3 Clear Sky Characterization

To investigate the distribution of clear days at the study sites, we used the clear sky index (Eq. (1)) defined by the ratio between GHI and the clear sky GHI_{cls} derived from the clear sky model McClear. Figure 5 illustrates the daily distribution of the clear sky index over the period from 2011 to 2020. According to [28, 29] clear days correspond to values of $K_t \geq 0.7$.

$$K_t = \frac{GHI}{GHI_{cls}} \tag{1}$$

It can be seen that the majority of K_t variations are between 0.6 and 0.8. Zagora and Ouarzazate are characterized by the highest percentage of clear days that correspond to 95%. Erfoud, Missour, Ain Bni Mathar, Benguerir, Oujda, Taza, and Tantan also have significant percentages of clear days of 93%, 92%, 90%, 89%, 87%, 86%, and

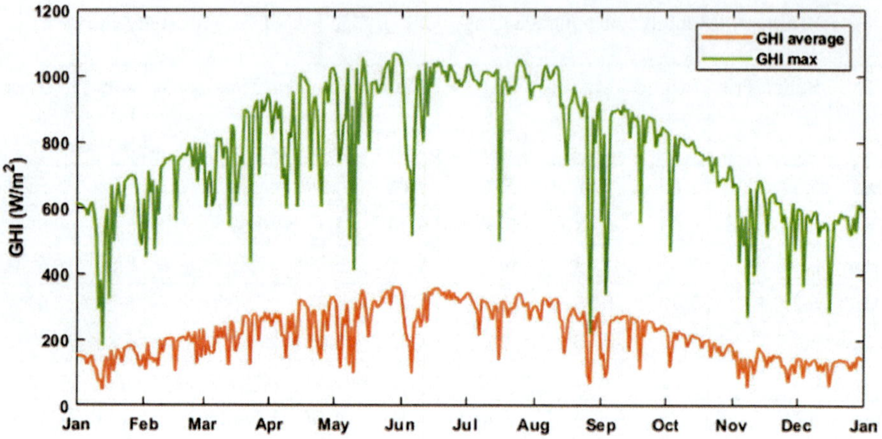

Fig. 4. Daily average and maximum of GHI for Benguerir site as an example.

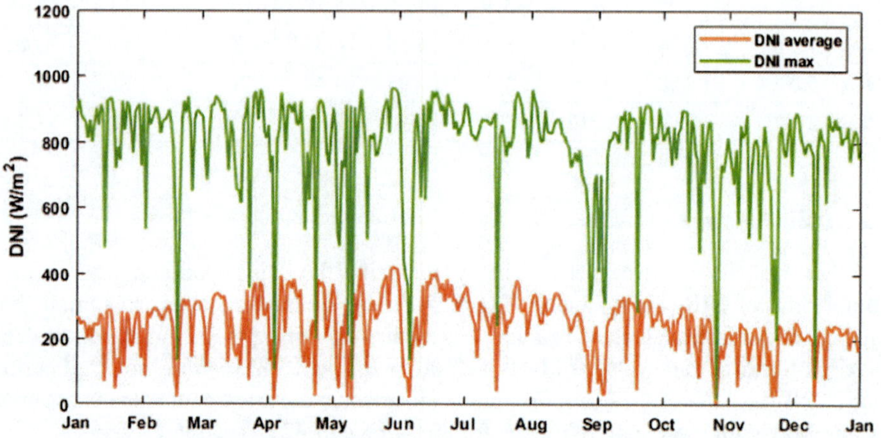

Fig. 5. Daily average and maximum of DNI for Benguerir site as an example.

85%, respectively. This justifies the intensity of radiation observed in these sites;indeed, clouds tend to reduce solar irradiation via the phenomenon of absorption and under clear sky conditions the radiation is less attenuated [30] (Fig. 6).

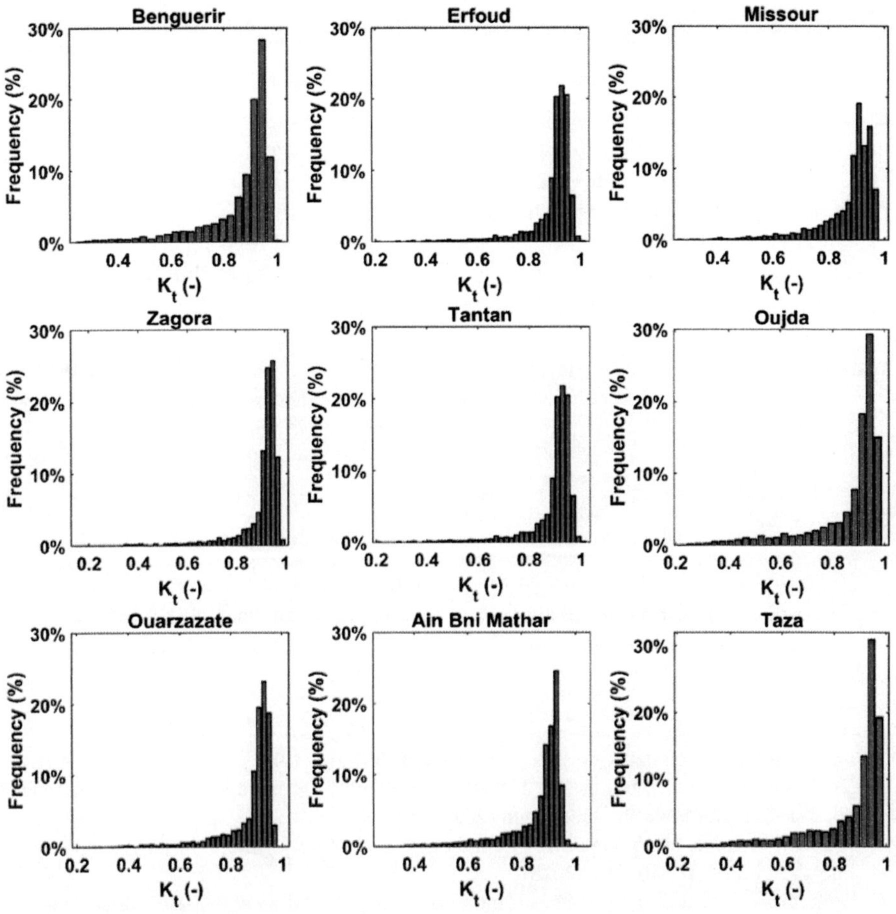

Fig. 6. Frequency distribution of clear sky index Kt using HC3 data from 2011 to 2020.

4 Conclusion

This study aimed to assess the solar potential of nine sites in Morocco (Benguerir, Erfoud, Missour, Zagora, Tantan, Oujda, Ain Bni Mathar, and Taza) using TMY for GHI and DNI generated using 10 years of data from the HelioClim3 satellite database. The purpose of the analysis is to provide a preliminary insight to engineers and project developers about the solar potential available in various cities in Morocco for solar applications, energy efficiency and feasibility analysis of solar projects either for PV through GHI or CSP through DNI analysis, and also includes an analysis of the prevailing climatology regarding clear sky conditions.

The analysis showed that almost all locations are characterized by the dominance of clear days with percentages that exceed 90% over the analysis period between 2011 and 2020. The sites are characterized by high annual irradiation ranging between 1840 kWh/m^2 and 2075 kWh/m^2 for GHI and between 1980 kWh/m^2 and 2436 kWh/m^2 for

DNI. The monthly irradiation is also significant, the monthly evolution of GHI showed that most of the sites are characterized by a high GHI, especially during May, June and July that exceeds 7 kWh/m^2, while during these months, the low values of GHI are obtained for Tantan site with values of 6 kWh/m^2. For DNI, the highest value of radiation is obtained for the site of Ouarzazate with an average of 8.11 kWh/m^2 during June, and the minimum values recorded are 4.3 kWh/m^2 for Oujda in December and Tantan during August and July. The results presented in this paper indicate that solar energy at the sites studied is a promising alternative to fossil fuels and these sites could be candidate sites for strategic solar energy projects. Upcoming studies will focus on the assessment of wind resources for wind energy.

Acknowledgement. The authors thank the Green Energy Park, research platform by IRESEN (Research Institute in Solar Energy and New Energies). The authors would like to express their gratitude to MINES ParisTech for providing access to the solar irradiation data.

References

1. Kousksou, T., Allouhi, A., Belattar, M., Jamil, A., El Rhafiki, T., Zeraouli, Y.: Morocco's strategy for energy security and low-carbon growth. Energy **84**, 98–105 (2015)
2. Choukri, K., Naddami, A., Hayani, S.: Renewable energy in emergent countries: lessons from energy transition in Morocco. Energy Sustain. Soc. **7**(1), 1–11 (2017). https://doi.org/10.1186/s13705-017-0131-2
3. Ameur, A., Berrada, A., Loudiyi, K., Aggour, M.: Analysis of renewable energy integration into the transmission network. Electricity J. **32**, 106676 (2019)
4. Deolalkar, S.P.: Solar power. In: Deolalkar, S.P. (ed.) Designing Green Cement Plants, pp. 251–258. Butterworth-Heinemann (2016)
5. Ahmadi, M.H., et al.: Solar power technology for electricity generation: a critical review. Energy Sci. Eng. **6**, 340–361 (2018)
6. Sengupta, M., et al.: Best Practices Handbook for the Collection and Use of Solar Resource Data for Solar Energy Applications. IEA Solar Heating and Cooling Programme (2015)
7. Das, U.K., et al.: Forecasting of photovoltaic power generation and model optimization: a review. Renew. Sustain. Energy Rev. **81**, 912–928 (2018)
8. Kousksou, T., et al.: Renewable energy potential and national policy directions for sustainable development in Morocco. Renew. Sustain. Energy Rev. **47**, 46–57 (2015)
9. Schüler, D., et al.: The enerMENA meteorological network – solar radiation measurements in the MENA region. In: AIP Conference Proceedings, vol. 1734, p. 150008 (2016). https://doi.org/10.1063/1.4949240
10. s.r.o, © Solargis: Solargis :: iMaps. https://solargis.info/imaps/
11. Home. http://www.soda-pro.com/
12. Marchand, M., Ghennioui, A., Wey, E., Wald, L.: Comparison of several satellite-derived databases of surface solar radiation against ground measurement in Morocco (2018)
13. El Alani, O., Ghennioui, A., Ghennioui, H., Saint-Drenan, Y.-M., Blanc, P.: Evaluation of 24-hours forecasts of global solar irradiation from IFS, GFS and McClear models. Presented at the AIP Conference Proceedings (2020)
14. Alani, O.E., Ghennioui, A., Merrouni, A.A., Ghennioui, H., Saint-Drenan, Y.-M., Blanc, P.: Validation of surface solar irradiances estimates and forecast under clear-sky conditions from the CAMS McClear model in Benguerir, Morocco. Presented at the AIP Conference Proceedings (2019)

15. Espinar, B., et al.: HelioClim-3: a near-real time and long-term surface solar irradiance database (2012)
16. Schmetz, J., et al.: An introduction to Meteosat second generation (MSG). Bull. Am. Meteor. Soc. **83**, 977–992 (2002)
17. Rigollier, C., Lefèvre, M., Wald, L.: The method Heliosat-2 for deriving shortwave solar radiation from satellite images. Sol. Energy **77**, 159–169 (2004). https://doi.org/10.1016/j.solener.2004.04.017
18. Albarelo, T., Marie-Joseph, I., Primerose, A., Seyler, F., Wald, L., Linguet, L.: Optimizing the Heliosat-II method for surface solar irradiation estimation with GOES images. Can. J. Remote Sens. **41**, 86–100 (2015). https://doi.org/10.1080/07038992.2015.1040876
19. Diabaté, L., Blanc, P., Wald, L.: Solar radiation climate in Africa. Sol. Energy **76**, 733–744 (2004). https://doi.org/10.1016/j.solener.2004.01.002
20. Rigollier, C., Bauer, O., Wald, L.: On the clear sky model of the ESRA—European Solar Radiation Atlas—with respect to the Heliosat method. Sol. Energy **68**, 33–48 (2000)
21. Diabaté, L., Remund, J., Wald, L.: Linke turbidity factors for several sites in Africa. Sol. Energy **75**, 111–119 (2003). https://doi.org/10.1016/j.solener.2003.07.002
22. Qu, Z., Gschwind, B., Lefèvre, M., Wald, L.: Improving HelioClim-3 estimates of surface solar irradiance using the McClear clear-sky model and recent advances in atmosphere composition (2014)
23. Thomas, C., Wey, E., Blanc, P., Wald, L., Lefèvre, M.: Validation of HelioClim-3 Version 4, HelioClim-3 Version 5 and MACC-RAD using 14 BSRN stations. Energy Procedia **91**, 1059–1069 (2016). https://doi.org/10.1016/j.egypro.2016.06.275
24. Lefèvre, M., et al.: McClear: a new model estimating downwelling solar radiation at ground level in clear-sky conditions. Atmos. Measur. Tech. **6**, 2403–2418 (2013)
25. Espinar, B., Hoyer-Klick, C., Lefèvre, M., Homscheidt, M.S., Wald, L.: User's Guide to the MACC-RAD Services on solar energy radiation resources (2013)
26. Abreu, E.F., Canhoto, P., Prior, V., Melicio, R.: Solar resource assessment through long-term statistical analysis and typical data generation with different time resolutions using GHI measurements. Renew. Energy **127**, 398–411 (2018)
27. Tahir, Z., Asim, M.: Surface measured solar radiation data and solar energy resource assessment of Pakistan: a review. Renew. Sustain. Energy Rev. **81**, 2839–2861 (2018)
28. Molteni, F., Buizza, R., Palmer, T.N., Petroliagis, T.: The ECMWF ensemble prediction system: methodology and validation. Q. J. R. Meteorol. Soc. **122**, 73–119 (1996)
29. Li, D.H., Lau, C.C., Lam, J.C.: Overcast sky conditions and luminance distribution in Hong Kong. Build. Environ. **39**, 101–108 (2004)
30. El ALani, O., Ghennioui, H., Ghennioui, A.: Intra-day variability quantification from ground-based measurements of global solar irradiance. Int. J. Renew. Energy Res. (IJRER) **10**, 1576–1587 (2020)

Experimental Validation of Different PV Power Prediction Models Under Beni Mellal Climate

Mustapha Adar(✉), Mohamed-Amin Babay, Souad Taouiri, Abdelmounaim Alioui, Yousef Najih, Zakaria Khaouch, and Mustapha Mabrouki

Industrial Engineering Laboratory, Faculty of Science and Technologies, Sultan Moulay Slimane University, Beni Mellal, Morocco

Abstract. Analyzing the efficiency of Photovoltaic installations in real situations is a difficult task, as there are many models and software packages available. Generally, software and models cannot predict PV output under all weather and geographic conditions. It is necessary to develop a protocol to assess the viability and profitability of potential Photovoltaic installations based on a real-world system. In this paper, the production of 3 PV prediction models was evaluated, and their accuracy was compared to one year of actual measurements of an monocristalline silicon Photovoltaics grid-connected plant. The results show that models M1, M2, and M3 are all adequate to simulate Photovoltaic performance in the climatic conditions of the Beni Mellal region with an advantage for the M1 model.

Keywords: PV modeling · Prediction · PV plant · Performance comparison

1 Introduction

Day after day, solar energy is becoming one of the key candidates to replace traditional non-renewable energy sources [1]. This renewable energy source provides solutions to some of the major problems facing fossil energy sources, such as the exhaustion of existing reserves and environmental harm. Solar photovoltaic (PV) systems are a feasible and affordable option for the transition to sustainable energy systems. Its system can be used in any place and on any scale, giving us a range of ways to meet our energy needs. Trend studies have found an exponential increase in total installed capacity, including a further increase in efficiency and a further decrease in device costs, which draws interest from broader populations around the world and new players in a number of markets [2]. PV installed capacity reached 480 GW in 2018, with an annual growth rate of about 30%, and is expected to exceed 1 TW by 2021–2022 [3]. Performance analysis of these photovoltaic installations is very important and is a daunting challenge as it ensures the monitoring of installations by detecting potential anomalies [4]. That is why several researchers have investigated and analyzed the performance and efficiency of PV systems with a systematic approach that focuses on normalized parameters [5–9]. In general, the research that appears in the literature deals with the effects of weather conditions, the impact of dust, humidity, and degradation due to the duration of exposure to actual operating conditions. In addition, the work is aimed at finding the most cost-effective

R. Saidi et al. (Eds.): ICATH 2021, LNDECT 110, pp. 286–299, 2022.
https://doi.org/10.1007/978-3-030-94188-8_27

photovoltaic technology for a given location [10]. The techniques used are varied but may be limited to the assessment of the performance ratio [11], the PVUSA [12] and SANDIA models [13], the use of neural networks and artificial intelligence [14, 15], and the use of simulation models [16]. There is software that simulates the production and performance of all the current photovoltaic technologies on the market, and there are others that render dimensional studies. The problem with this program, however, is that the models adopted by them are not always suitable for these photovoltaic technologies or for these geographical and meteorological conditions.

There are models that make it possible to predict the output of photovoltaic installations [17–21], the value of which is the assessment of performance, productivity, and compliance with the conditions of the manufacturer's warranty of photovoltaic technologies. Such models involve the power temperature coefficient, given by the manufacturer, and meteorological parameters acquisition, in particular solar irradiation, ambient temperature and wind speed, and humidity. However, PV models typically fall into two major groups, those referred to as physical models based on the representation of equivalent cell circuit (I-V characteristic) [22, 23] and those based on empirical data [24]. Most models are either linear and empirical or complex and non-linear using a multivariate regression equation [25]. In paper [26], for example, the authors proposed a model for predicting PV power using recurrent neural networks in the event of future meteorological data. In paper [27], the suitability of different simplistic PV module yield prediction models is investigated for several PV module technologies. While a deep learning process for efficient prediction of short-term photovoltaic power generation is presented in paper [28].

The main objective of this work is to choose the closest model of photovoltaic production to the geographic and meteorological conditions of the Beni Mellal region. The latter, which is an agricultural pole in Morocco, is experiencing an increasing rate of integration of solar photovoltaic pumping irrigation. Therefore, the prediction of photovoltaic yield for this region is an unavoidable necessity. In this paper, we will only evaluate the empirical models as they are more realistic because they use real data as input. Once the model is chosen, the assessment of the capacity of a given region to host photovoltaic installations, the choice of technology to be implemented, and the cost-effectiveness would be easy to care for.

2 Methodology

2.1 Photovoltaic Systems and Instrumentation

To assess the dependability of several PV prediction methods, the producibility of the mc-si modules is obtained by Bluetooth communication with the SMA 2000 inverter. The total number of panels is 8 mounted in series, providing a nominal output power of 2kWp. The PV modules are fixed at a tilted angle of 30°, South oriented on the rooftop of the Faculty of Science and Technology, Beni Mellal, Morocco. This mini station is grid connected (Fig. 2). Geared towards monitoring meteorological parameters, there is a meteorological station that measures horizontal solar radiation and 30° southward inclination solar irradiation, ambient temperature, the temperature of the photovoltaic modules, speed and direction of the wind (Fig. 1). We used two 20 Wp calibrated polycrystalline silicon (pc-si) modules [29] to measure horizontal and 30° titled angle (in the same plane as the PV panel) solar irradiations. Also, we used two PT100 temperature sensors to measure the temperature of the photovoltaic panels and the ambient temperature. The data from the different measurement sensors is processed by PCDUINO boards at five-minute intervals. Such cards, save the data as CSV files and send it by mail [29].

Fig. 1. Meteorological station

Fig. 2. Mc-si PV plant

2.2 Prediction Models

The majority of models contained in the literature use panel temperature and incident radiation as inputs, as well as a variety of empirical coefficients. The determination of the actual operational power of any PV module under varied climatic circumstances is the reliability of a prediction model.

We'll check the reliability of the three models below:

$$\text{M1: } p = p_{ref}(1 - \beta_{ref}(T - T_{ref}))\ [30] \quad (1)$$

$$\text{M2: } p = p_{25}(1 - (0.0026) * (T - T_{ref}))\ [31] \tag{2}$$

$$\text{M3: } p = p_{25}(1 - \gamma * (T - 25))\ [32] \tag{3}$$

Where β is the module temperature coefficients, γ is the temperature correction coefficients, P_{25} is power normalized at 25 °C, P_{ref} is the reference power of photovoltaic module and, T_{ref} is reference PV module temperature.

2.3 Models' Accuracy Evaluation Indicators

In order to determine the accuracy of the simulated model, statistical performance metrics were used in this analysis to calculate the degree to which the simulation model replicates the data obtained in the field. The evaluation of forecasts depends heavily on how we pick experimental validation data and how we equate these measures to the forecast results. These metrics are commonly used in publications as statistical methods for evaluating simulation power efficiency or current forecasts [33, 34].

In a prediction, the model error can be expressed as the deviation of the measured value from the predicted value, as follows [35]:

$$e = v_{mes} - v_{pred} \tag{4}$$

with v_{mes} and v_{pred} are respectively the ith value measured and the value predicted. With this relation we can define the Performance indicators of the simulation models as follows [36]:

$$MBE = \frac{1}{N}\sum_{i=1}^{N} e \tag{5}$$

The MBE is the arithmetic mean of the error. It provides a measurement of the prediction model's bias. The N is the total number of the measurement in the dataset [36].

$$RMSE = \sqrt{\frac{1}{N}\sum_{i=1}^{N} e^2} \tag{6}$$

The RMSE performs a term-by-term comparison of the actual difference between the measured data and those predicted by the prediction model. This indicator is always positive, and the ideal value should be zero [35].

$$nRMSE = (\frac{1}{N}\sum_{i=1}^{N} \frac{|e|}{x_{mes}}) \times 100 \tag{7}$$

The normalized root mean square error nRMSE gives better information on the quality of the prediction model than the EMB or RMSE [35].

$$TS = \sqrt{\frac{(N-1)MBE^2}{(RMSE^2 - MBE^2)}} \tag{8}$$

TS is a commonly used method to determine whether or not the forecast model varies substantially from the measured results. The lower the TS value, the more accurate the data expected would be [35].

$$SD = \sqrt{\frac{N(RMSE^2 - MBE^2)}{(N-1)}} \tag{9}$$

The standard deviation is the difference between the forecast and the measured data. The measure is always positive and the optimal value is zero.

$$R = \frac{\sum_{i=1}^{N}(x_i - \overline{x})(y_i - \overline{y})}{\sqrt{\sum_{i=1}^{N}(x_i - \overline{x})^2 \sum_{i=1}^{N}(y_i - \overline{y})^2}} \tag{10}$$

Pearson correlation coefficient (R) is also calculated as a correlation coefficient – shows the intensity and direction of the linear relationship between the predicted and calculated data [35].

3 Results and Discussion

3.1 PV Power and Meteorological Measurements

Monthly in panels plane irradiation and average ambient temperature measurements carried out over the year 2017 are depicted in Fig. 3. As it can be observed, the highest value of total on PV array plane irradiance was in July with 199 kWh/m^2 and the lowest values were measured in December with 141 kWh/m^2. This indicates an average insolation per month equal to 169 kWh/m^2. The annual insolation was found to be around 2035 kWh/m^2. The mean ambient temperature throughout the year was 22.8 °C. The

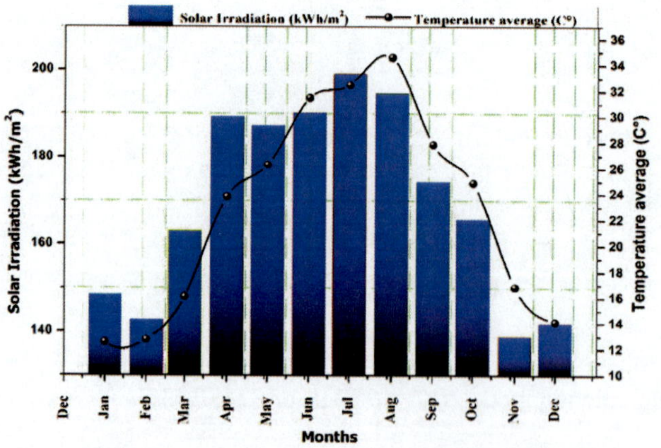

Fig. 3. Total in PV array plan solar irradiation and ambient temperature

hottest month was August, with a high of 34.5 °C, and the coldest was January, with a low of 12 °C.

The production of PV panels according to the conditions presented above is plotted in Figure 4. The total energy values produced throughout the year 2017 are approximately 3612 kWh. This means that the average energy output per month is equivalent to 301 kWh. The seasonal distribution of production is nearly uniform. 27.5% of energy output is generated between April and June, 26.9% between July and September, 22% in autumn, and 23.5% in winter.

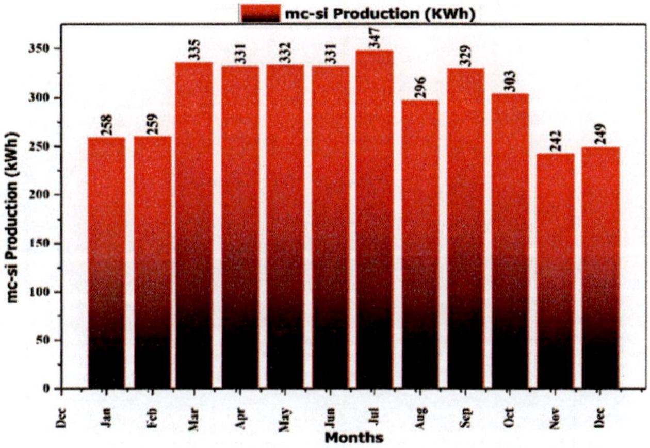

Fig. 4. Distribution of monthly energy production

3.2 Prediction Models Comparison

The accuracy of the prediction models is assessed by examining the correlation between the performance output data and the predicted performance. The results are depicted in Figure 5. In general, all models have a strong linear correlation with actual data with very reasonable correlation coefficients. The M1 ideally fits the experimental results and has a correlation coefficient of 0.975.

To order to properly evaluate the efficiency of the models, a particular method of assessment can be used. We can plot the relative error of each of the models against the real measurements.

Figure 6 shows that, for all models, values below 4 kW are dispersed, which is related to a high forecast error. However, the values would be smoother if they are close to the nominal output of the module or, in other words, if the model error is close to zero. Therefore, and understanding that a positive error means that the expected values are underestimated relative to what they would actually be, we can conclude that all three models underestimate the output of electrical energy. But this underestimation can be considered to be small. However, the M1 model is the best one with an error close to zero, particularly in the optimum production range.

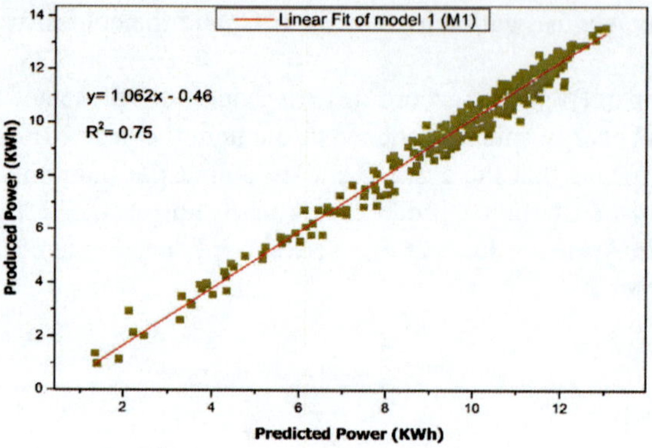

a.

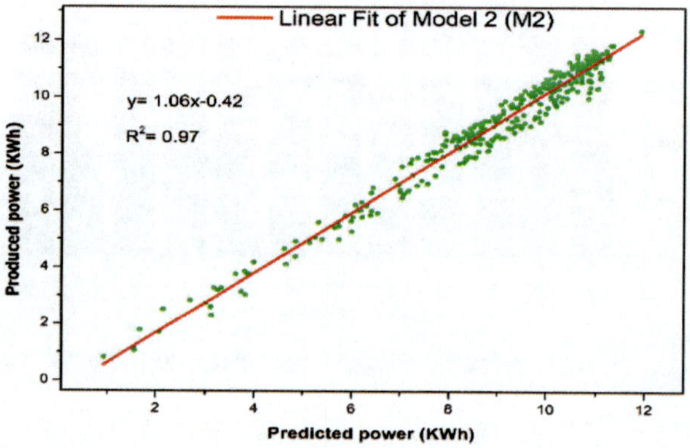

b.

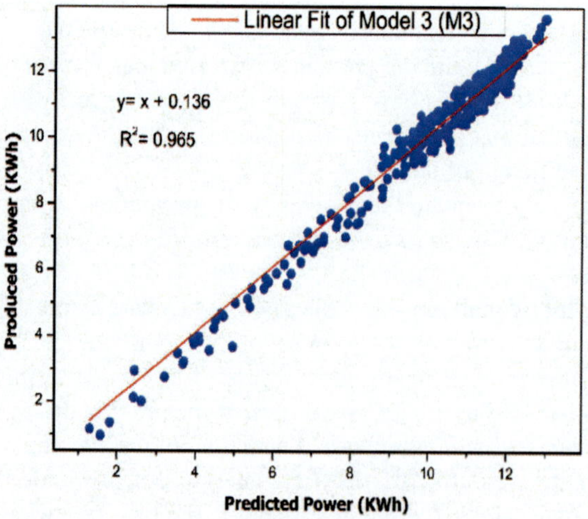

c.

Fig. 5. a. M1's linear fitting of produced and predicted powers, b. M2's linear fitting of produced and predicted powers, c. M3's linear fitting of produced and predicted powers

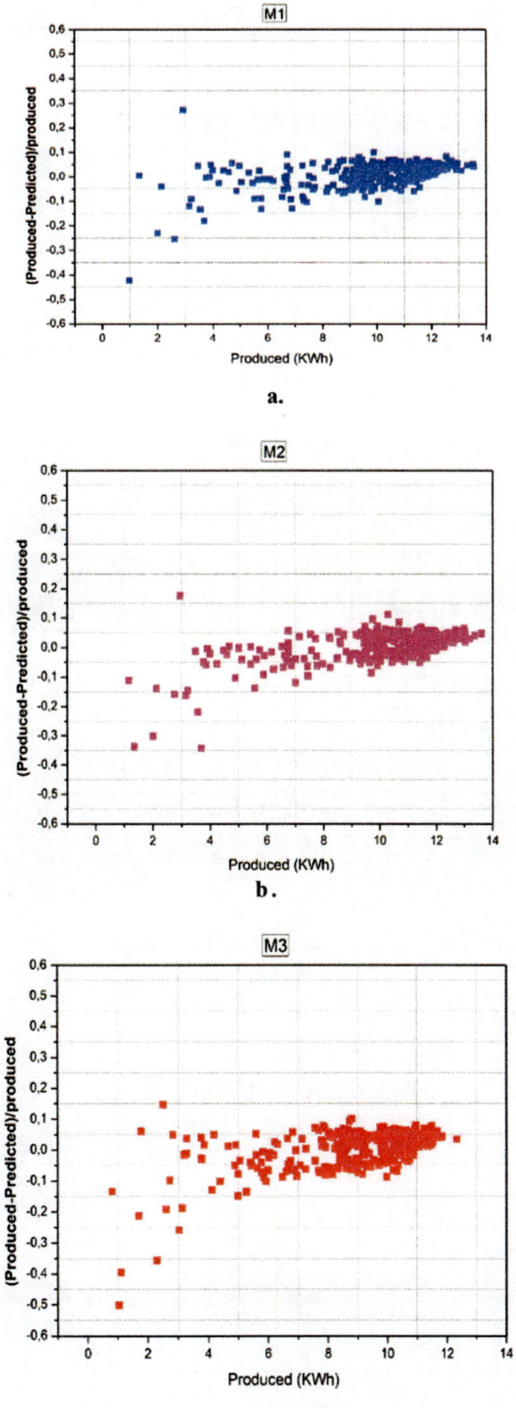

Fig. 6. a. M1 prediction relative error vs. produced power. b. M2 prediction relative error vs. produced power. c. M3 prediction relative error vs. produced power

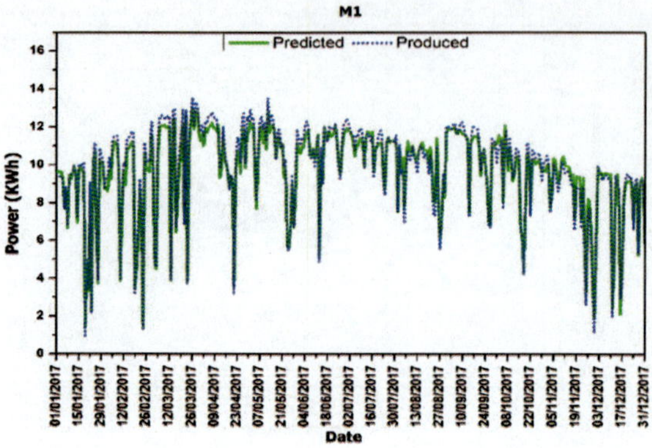

a.

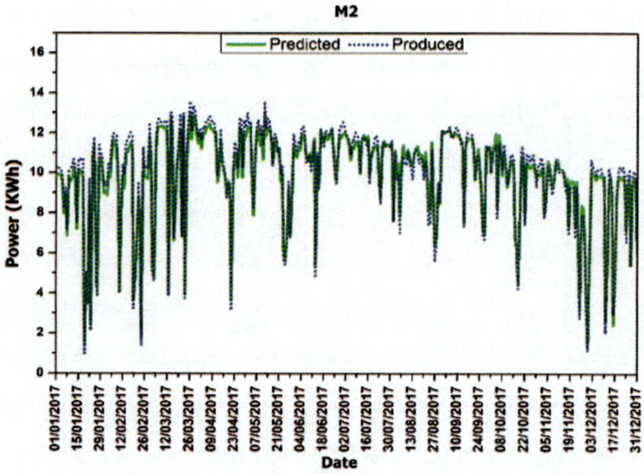

b.

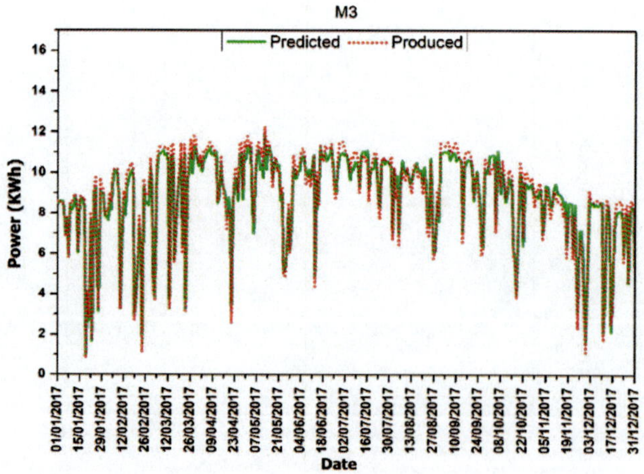

c.

Fig. 7. a. Average daily measured and M1 predicted power for year 2017. b. Average daily measured and M2 predicted power for year 2017. c. Average daily measured and M3 predicted power for year 2017

Figure 7 displays the real and estimated output data from the three predictive models. Such figures display the behavior of the predicted power compared to the measured power for all the days of 2017. For the purpose of quantifying the forecast discrepancies between models, Table 1 presents statistical errors. Table 1 shows the deviations between the computed and measured values of actual power. The main difference appears in the MBE value. The two negative MBE values found for models M1 and M3 imply that the power output predicted by these models is underestimated by −4% for M3 and −1.34% for M1. The positive MBE value found for model M2 shows that it overestimates the real power by 2%. The first two models have a negative value of b, which indicates that the predicted power is negatively affected by the increase in temperature of the PV module. This is true because the temperature coefficient of the PV module power is negative. The positive value of b given by model M3 rounds it off far from reproducing the real PV module behavior. The three models accurately estimate power output, with nRMSEs of 8.3% for the M1, 4.1% for the M2, and 8% for the M3.

From these tests, it can be concluded that the three models are the best models to be used for the simulation of photovoltaic energy generation in the weather conditions of the Beni Mellal region with a slight superiority to Model M1.

Table 1. Statistical parameters for predictive models

Statistical performance indicators							Regression coefficients	
Model	MBE	RMSE	nRMSE	TS	SD	R2	a	b
M1	−4.24	8.73	0.083	9.54	7.6	0.94	1.06	−0.46
M2	3.01	8.75	0.041	6.22	8.3	0.92	1.06	−0.42
M3	−5.98	8.80	0.080	13.0	7.9	0.88	1	0.13

3.3 Comparison with Other Literature Models

Table 2 depicts the accuracy comparison based on MBE, RMSE, and nRMSE of the different forecasting models in the literature. These models were selected with the intention of incorporating actual data in order to prevent meteorological forecasting errors and sensor failures. But the performance of statistical models is always strongly affected by the quality of the actual measurements.

Based on the MBE parameter, we can say that only the two models, M2 and M10, overestimate the photovoltaic power output. The other models underestimate the PV system's power output. According to the nRMSE values, the three models, M4, M5 and M9, estimate the power output of the PV system with good accuracy. The accuracy of model M9 reaches 78.74%. A model's validity and accuracy can be influenced by a variety of factors, including geographical and climatic conditions, as well as the solar module technology under consideration. The choice and reliability of the sensors, used to measure solar irradiation, ambient temperature, and photovoltaic module temperature, can influence the accuracy of a predictive model.

Table 2. Summary of some PV module production predictive models

Symbol	Model	Module reference	MBE	RMSE	nRMSE	Results reference
M1	$P = P_{ref}[1 - \beta ref (T - T_{ref})]$	[30]	−4.24	8.73	8.3	This work
M2	$P = P_{25}[1 - (0.0026)*(T - T_{ref})]$	[31]	3.01	8.75	4.1	This work
M3	$P = P_{25}[1 - \gamma(T - 25)]$	[32]	−5.98	8.80	8	This work
M4	$P = \eta Ht[1 - 0.0045(T - 25)]$	[19]	−30.93	35.72	25	[36]
M5	$P = \eta Ht\ \tau_g P_f[1 - \beta ref\ (T - 25)]$	[18]	−25.20	29.85	21	[36]
M6	$P = S[0.128H - 0.239 \times 10 - 3Ta]$	[37]	−20.11	24.49	17	[36]
M7	$P = Pref\ HCwCeCc[1 + \alpha\ (T - 25)]$	[38]	−12.94	18.41	12	[36]
M8	$P = AC\eta_{NOCT}\ [1 - MPT\ (T_{NOCT} - T)]$	[39]	−4.62	5.60	9.57	[35]
M9	$P = I_0 V_0[1 + (\alpha - \beta)(T - Tref)]$	[40]	−39.30	47.50	78.74	[35]
M10	$P = \eta_{STC} AC[1 - \beta_{STC}(T - T_{STC}) + \gamma \log_{10} H]$	[41]	1.47	2.54	4.75	[35]

4 Conclusion

With the lack of information on the most accurate predictive photovoltaic yield model for a given PV technology and location. This study, by analyzing the monitoring data of the grid-connected photovoltaic plant with the aid of three existing models, comes to make it possible to choose between them. The data used included only the overall in-plane irradiation data of the PV panels and their temperature, measured with the standard irradiation and temperature sensors, respectively, in addition to the inverter output data. A comparison of the estimated and assessed values of the three models reveals that there are no major differences for both models. From the point of view of energy production, it was found that the M1 and M3 models underestimated demand, unlike the M2. The difference between the measured production and the model forecast was almost constant for 2017. The M1 was found to be the best model to use to simulate the PV production in Beni Mellal climatic conditions.

References

1. Islam, M.T., Huda, N., Abdullah, A.B., Saidur, R.: A comprehensive review of state-of-the-art concentrating solar power (CSP) technologies: current status and research trends. Renew. Sustain. Energy Rev. **91**, 987–1018 (2018). https://doi.org/10.1016/j.rser.2018.04.097
2. Fraunhofer: Fraunhofer ISE: Photovoltaics Report (2019)
3. International Renewable Energy Agency (IRENA): Renewable Energy Market Analysis: Southeast Asia (2018)
4. Adar, M., Bazine, H., Najih, Y., et al.: Simulation study of three PV systems. In: 6th International Renewable and Sustainable Energy Conference, IRSEC 201, pp 1–5 (2018)
5. Adar, M., Najih, Y., Gouskir, M., et al.: Three PV plants performance analysis using the principal component analysis method. Energy **207** (2020). https://doi.org/10.1016/j.energy.2020.118315
6. Ascencio-Vásquez, J., Brecl, K., Topič, M.: Methodology of Köppen-Geiger-Photovoltaic climate classification and implications to worldwide mapping of PV system performance. Sol Energy **191**, 672–685 (2019). https://doi.org/10.1016/j.solener.2019.08.072
7. Hachicha, A.A., Al-Sawafta, I., Said, Z.: Impact of dust on the performance of solar photovoltaic (PV) systems under United Arab Emirates weather conditions. Renew. Energy **141**, 287–297 (2019). https://doi.org/10.1016/j.renene.2019.04.004
8. Adar, M., Khaouch, Z., Mabrouki, M., et al.: Performance analysis of PV grid-connected in fours special months of the year. In: Proceedings of 2017 International Renewable and Sustainable Energy Conference, IRSEC 2017 (2018)
9. Lotfi, H., Adar, M., Bennouna, A., et al.: Silicon photovoltaic systems performance assessment using the principal component analysis technique. Mater Today Proc. (2021). https://doi.org/10.1016/j.matpr.2021.04.374
10. Adar, M., Mabrouki, M., Bennouna, A., Chebak, A.: Production study of a grid connected PV plant. In: International Renewable and Sustainable Energy Conference, IRSEC 2016, pp. 116–120 (2017)
11. Anang, N., Syd Nur Azman, S.N.A., Muda, W.M.W., et al.: Performance analysis of a grid-connected rooftop solar PV system in Kuala Terengganu Malaysia. Energy Build **248**, 111182 (2021). https://doi.org/10.1016/j.enbuild.2021.111182
12. Bianchini, G., Pepe, D., Vicino, A.: Estimation of photovoltaic generation forecasting models using limited information. Automatica **113**, 108688 (2020). https://doi.org/10.1016/j.automatica.2019.108688
13. Peng, J., Lu, L., Yang, H., Ma, T.: Validation of the Sandia model with indoor and outdoor measurements for semi-transparent amorphous silicon PV modules. Renew. Energy **80**, 316–323 (2015). https://doi.org/10.1016/j.renene.2015.02.017
14. Yousif, J.H., Kazem, H.A.: Prediction and evaluation of photovoltaic-thermal energy systems production using artificial neural network and experimental dataset. Case Stud. Therm. Eng. **27**, 101297 (2021). https://doi.org/10.1016/j.csite.2021.101297
15. Wang, F., Xuan, Z., Zhen, Z., et al.: A day-ahead PV power forecasting method based on LSTM-RNN model and time correlation modification under partial daily pattern prediction framework. Energy Convers. Manag. **212**, 112766 (2020). https://doi.org/10.1016/j.enconman.2020.112766
16. Georgitsioti, T., Pearsall, N., Forbes, I., Pillai, G.: A combined model for PV system lifetime energy prediction and annual energy assessment. Sol. Energy **183**, 738–744 (2019). https://doi.org/10.1016/j.solener.2019.03.055
17. El Mentaly, L., Amghar, A., Sahsah, H.: The prediction of the maximum power of PV modules associated with a static converter under different environmental conditions. Mater. Today Proc. **24**, 125–129 (2020). https://doi.org/10.1016/j.matpr.2019.07.704

18. Chow, T.T., He, W., Ji, J.: Hybrid photovoltaic-thermosyphon water heating system for residential application. Sol. Energy **80**, 298–306 (2006). https://doi.org/10.1016/j.solener.2005.02.003
19. Jie, J., Hua, Y., Gang, P., et al.: Study of PV-Trombe wall assisted with DC fan. Build. Environ. **42**, 3529–3539 (2007). https://doi.org/10.1016/j.buildenv.2006.10.038
20. Liu, L., Zhao, Y., Chang, D., et al.: Prediction of short-term PV power output and uncertainty analysis. Appl. Energy **228**, 700–711 (2018). https://doi.org/10.1016/j.apenergy.2018.06.112
21. Ni, Q., Zhuang, S., Sheng, H., et al.: An ensemble prediction intervals approach for short-term PV power forecasting. Sol. Energy **155**, 1072–1083 (2017). https://doi.org/10.1016/j.solener.2017.07.052
22. Wolff, B., Kühnert, J., Lorenz, E., et al.: Comparing support vector regression for PV power forecasting to a physical modeling approach using measurement, numerical weather prediction, and cloud motion data. Sol. Energy **135**, 197–208 (2016). https://doi.org/10.1016/j.solener.2016.05.051
23. Ding, K., Zhang, J., Bian, X., Xu, J.: A simplified model for photovoltaic modules based on improved translation equations. Sol. Energy **101**, 40–52 (2014). https://doi.org/10.1016/j.solener.2013.12.016
24. Skoplaki, E., Palyvos, J.A.: On the temperature dependence of photovoltaic module electrical performance: a review of efficiency/power correlations. Sol. Energy **83**, 614–624 (2009). https://doi.org/10.1016/j.solener.2008.10.008
25. Rosell, J.I., Ibáñez, M.: Modelling power output in photovoltaic modules for outdoor operating conditions. Energy Convers. Manag. **47**, 2424–2430 (2006). https://doi.org/10.1016/j.enconman.2005.11.004
26. Lee, D., Kim, K.: PV power prediction in a peak zone using recurrent neural networks in the absence of future meteorological information. Renew. Energy **173**, 1098–1110 (2021). https://doi.org/10.1016/j.renene.2020.12.021
27. Wang, M., Peng, J., Luo, Y., et al.: Comparison of different simplistic prediction models for forecasting PV power output: assessment with experimental measurements. Energy **224**, 120162 (2021). https://doi.org/10.1016/j.energy.2021.120162
28. Abdel-Basset, M., Hawash, H., Chakrabortty, R.K., Ryan, M.: PV-Net: an innovative deep learning approach for efficient forecasting of short-term photovoltaic energy production. J. Clean. Prod. **303**, 127037 (2021). https://doi.org/10.1016/j.jclepro.2021.127037
29. Erraissi, N., Raoufi, M., Aarich, N., et al.: Implementation of a low-cost data acquisition system for "PROPRE.MA" project. Meas. J. Int. Meas. Confed. **117**, 21–40 (2018). https://doi.org/10.1016/j.measurement.2017.11.058
30. Twidell, J., Tony, W.: Renewable Energy Resources, 3rd edn. Routledge (2015)
31. Yamawaki, T., Mizukami, S., Masui, T., Takahashi, H.: Experimental investigation on generated power of amorphous PV module for roof azimuth. Sol. Energy Mater. Sol. Cells **67**, 369–377 (2001). https://doi.org/10.1016/S0927-0248(00)00305-6
32. Parretta, A., Sarno, A., Vicari, L.R.M.: Effects of solar irradiation conditions on the outdoor performance of photovoltaic modules. Opt. Commun. **153**, 153–163 (1998). https://doi.org/10.1016/S0030-4018(98)00192-8
33. Celik, A.N., Acikgoz, N.: Modelling and experimental verification of the operating current of mono-crystalline photovoltaic modules using four- and five-parameter models. Appl. Energy **84**, 1–15 (2007). https://doi.org/10.1016/j.apenergy.2006.04.007
34. Zhou, W., Yang, H., Fang, Z.: A novel model for photovoltaic array performance prediction. Appl. Energy **84**, 1187–1198 (2007). https://doi.org/10.1016/j.apenergy.2007.04.006
35. Nour-eddine, I.O., Lahcen, B., Hassani, O., Amin, B.: Power forecasting of three silicon-based PV technologies using actual field measurements. Sustain. Energy Technol. Assess. **43**, 100915 (2021). https://doi.org/10.1016/j.seta.2020.100915

36. Hajjaj, C., Alami Merrouni, A., Bouaichi, A., et al.: Evaluation, comparison and experimental validation of different PV power prediction models under semi-arid climate. Energy Convers. Manag. **173**, 476–488 (2018). https://doi.org/10.1016/j.enconman.2018.07.094
37. Zervas, P.L., Sarimveis, H., Palyvos, J.A., Markatos, N.C.G.: Model-based optimal control of a hybrid power generation system consisting of photovoltaic arrays and fuel cells. J. Power Sources **181**, 327–338 (2008). https://doi.org/10.1016/j.jpowsour.2007.11.067
38. Wah, W.P., Shimoda, Y., Nonaka, M., et al.: Field study and modeling of semi-transparent PV in power, thermal and optical aspects. J. Asian Archit. Build. Eng. **4**, 549–556 (2005). https://doi.org/10.3130/jaabe.4.549
39. Perlman, J., McNamara, A., Strobino, D.: Analysis of PV system performance versus modeled expectations across a set of identical PV systems. In: Proceedings of the Solar World Congress 2005: Bringing Water to the World, Including Proceedings of 34th ASES Annual Conference and Proceedings of 30th National Passive Solar Conference, pp. 1313–1317 (2005)
40. Patel, M.R.: Wind and Solar Power Systems: Design, Analysis, and Operation, 2nd edn. Taylor & Francis (2005)
41. Notton, G., Cristofari, C., Mattei, M., Poggi, P.: Modelling of a double-glass photovoltaic module using finite differences. Appl. Therm. Eng. **25**, 2854–2877 (2005). https://doi.org/10.1016/j.applthermaleng.2005.02.008

Comparative Study of MPPT Controllers for a Wind Energy Conversion System

Hamid Chojaa[1](✉), Aziz Derouich[1], Yassine Bourkhime[2], Elmostafa Chetouani[3], Billel Meghni[4], Seif Eddine Chehaidia[5], and Mourad Yessef[6]

[1] Laboratory of Technologies and Industrial Services, Higher School of Technology, Sidi Mohamed Ben Abdellah University, 30000 Fez, Morocco
hamid.chojaa@usmba.ac.ma

[2] National School of Applied Sciences, Abdelmalek Essaâdi University, Tetouan, Morocco

[3] Laboratory: Electronics, Instrumentation and Energy, Faculty of Science, University Chouaib Doukkali, Eljadida, Morocco

[4] Engineering Department, LSEM Laboratory, University of Badji Mokhtar, Annaba, Algeria

[5] Research Laboratory of Industrial Risks, Non Destructive Control and Operating Safety, University of Badji Mokhtar, Annaba, Algeria

[6] Laboratory of Engineering, Modeling and Systems Analysis, SMBA University, Fez, Morocco

Abstract. This paper aims to discuss the modeling and control of a wind turbine using the maximum power point tracking technique (MPPT) based on the Tip Speed Ratio (TSR) method to extract the maximum power. A comparative study has been carried out within a four-types control laws. Namely, conventional PI, nonlinear sliding mode, backstepping and finally, artificial neural network controller. To identify which is which to provide the best performances, the proposed control laws are tested under Matlab/Simulink under different operating conditions to check the controller's performances. The performed simulations show that MPPT artificial neural network ensure the best performance compared to other controller because of its ability to map between inputs and outputs and efficiently cope with wind energy conversion system (WECS) nonlinearities.

Keywords: MPPT · Artificial neural network control · Sliding mode control · Backstepping control

1 Introduction

The traditional energy sources have become more dangerous and threatening for both the planet and humanity because of its harmful environmental impact. Hence, the sustainable green energy has gained attention as a clean energies vector avoiding the global warming, through the reduction of polluting gas emissions. There are several kinds of renewable energies, which are promising alternatives to the fossil energy, among these the wind energy is considered as one of the best alternatives to solve this issue. WECS can transform the kinetic energy to mechanical energy and then to electrical energy through the choice of the appropriate to feed the grid [1–3].

R. Saidi et al. (Eds.): ICATH 2021, LNDECT 110, pp. 300–310, 2022.
https://doi.org/10.1007/978-3-030-94188-8_28

The variable speed wind power system based on the Doubly Feed Induction Generator (DFIG) is widely used in onshore wind farms [4–6]. The latest progress in renewable energies proves that the configuration in WECS shown in Fig. 1 is the most demanding for electrical energy production [7]. It allows to extract the power in a wide speed range with a flexible control which reduces the power/cost ratio and the advantages of which are more convincing [8]. The DFIG allows operation over a speed range of ±33% around the synchronous speed, thus guarantee a simple converters configuration and gives more flexibility to control system and task. Consequently, reduce the cost the produced energy [9].

A WECS is a highly nonlinear system, characterized by sudden variations in wind speed. Then the use of an MPPT strategy is essential to improve the extraction of kinetic power from the wind in the wide range of wind speed. The MPPT controllers used to reach and track the MPPT available in the wind by regulating the rotor speed [1, 2].

The aim of this paper is to elaborate and compare serval MPPT control laws; four types of control are proposed to ensure a good rotor speed regulation and mechanical load mitigation. The presented control techniques are respectively:

- PI linear control,
- nonlinear sliding mode control (SMC),
- nonlinear Backstepping control (BAC),
- Artificial neural network (ANN).

This work will be organized as follows: the next part describes the WENCS modeling, then the proposed control strategies are presented. Main results are presented and discussed in Sect. 4 and finally, the conclusion.

2 WECS Modeling

The aerodynamic behavior of wind turbine is the first interesting element in the conversion chain. It used to convert the kinetic power into mechanical one. Therefore, aerodynamic modeling is a key element in the WECS. Its known by its nonlinearity, which requires advanced identification, modeling or estimation technics [3]. Many researchers use different kinds of models according to the wind turbine in use. In the present manuscript, a popular useful model, derived from the reference herein is used [2, 4] WECS based DFIG supplies a grid load through the following chain: aerodynamic wind turbine rotor, gearbox, DFIG, rectifier, and an inverter. Wind turbine converts the kinetic energy to mechanical energy, the gearbox multiplies the rotor speed in order to reach the generation condition and produce electrical energy.

The generated power is given by [2]:

$$P_t = \frac{1}{2}\rho\pi R^2 V^3 C_P(\lambda, \beta) \tag{1}$$

In this work, the variations of $Cp(\lambda, \beta)$ are modeled by [2, 4]:

$$\begin{cases} Cp(\lambda, \beta) = 0.5(\frac{116}{\lambda i} - 0.4\beta - 5)exp(\frac{-21}{\lambda i}) + 0.0068\lambda \\ \frac{1}{\lambda i} = \frac{1}{\lambda + 0.08\beta} - \frac{0.035}{\beta^3+1}; \quad \lambda = \frac{\Omega_t R}{V} \end{cases} \tag{2}$$

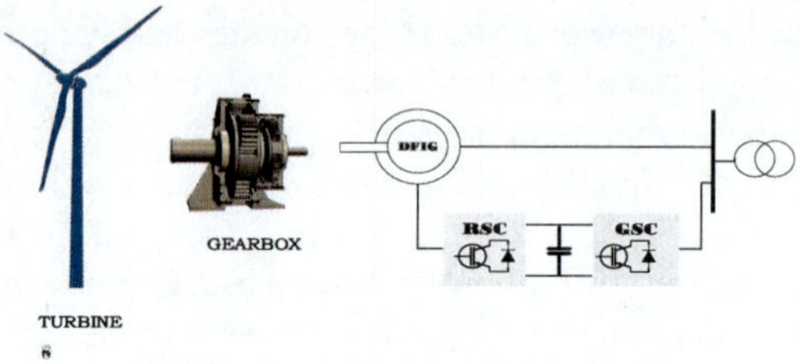

Fig. 1. The WECS based on DFIG.

By applying the fundamental equation of dynamic, one gets:

$$J\frac{d\Omega_g}{dt} = C_{méc} = C_g - C_{em} - C_f \tag{3}$$

Where the C_{em} is the electromechanical torque of generator. We notice that the electromechanical torque is presented as a resistive torque for all system and C_f is proportional to rotational speed of shaft and that is given by (4), such us:

$$C_f = f_v\Omega_g \tag{4}$$

Applying again the Laplace transform (3), we can find the transfer function representing wind turbine mechanical system as shown below:

$$\Omega g(s) = (Cg - Cem)\left(\frac{1}{J.s + fv}\right) \tag{5}$$

Finally, the global model of a wind turbine will be achieved by adding all parts of the transmission chain models and that lead to representing bloc diagram as shown in Fig. 2.

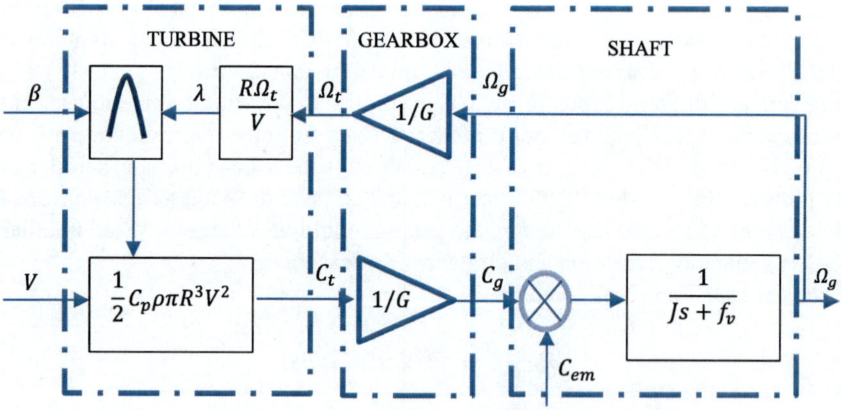

Fig. 2. Global model for mechanical transmission chain

3 MPPT Based TSR Method

In this paper, The MPPT technique has been realized with mechanical speed control as shown in Fig. 3. This control strategy consists of adjusting the electromagnetic torque that is developed by the electrical generator in order to fix it at its reference value. In the next, we present four controllers, being mentioned in Sect. 1, designed to track the generator reference speed $\Omega g*$ according to maximum power value.

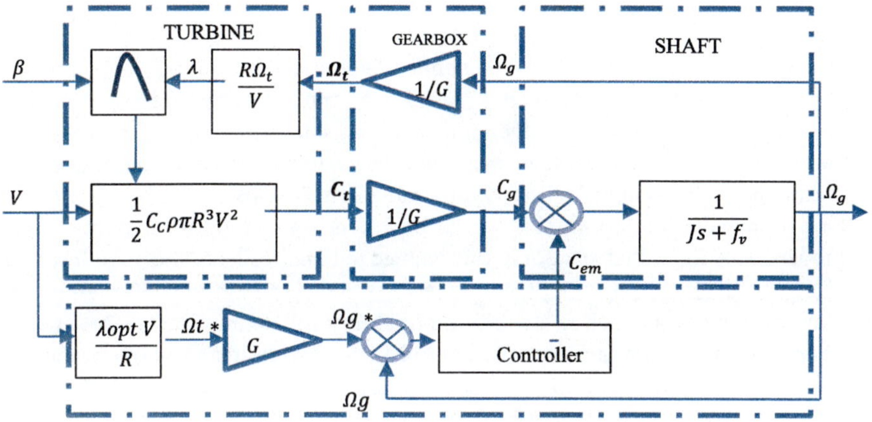

Fig. 3. MPPT technique with speed controller.

3.1 PI Controller

The PI controller is the most popular controller in the industry because of its efficiency and practical implementation; unfortunately, it's less robust and instable in some use circumstances. The PI controller is used in close loop to correct the mechanical transfer function performances parameters such as stability, response time and precision. In most case the choosing factors Kp and Ki depend on the use conditions and preferable performance to respect. To apply this controller the knowledge of the system parameters is required to reach the best results and the best system performances. Figure 4 shows the block scheme for the implementation of the PI controller applied to mechanical shaft equation in close loop.

The closed-loop transfer function is written:

$$\frac{\Omega_{g(s)}}{\Omega_{g^*(s)}} = \frac{2\xi.\omega_n.s + \omega_n^2}{s^2 + 2\xi.\omega_n.s + \omega_n^2} = \frac{\frac{K_i + K_p.S}{J}}{s^2 + \frac{K_p.f_v}{J}.s + \frac{K_i}{J}} \tag{6}$$

The parameters K_p and K_i of the PI controller are given by the expressions:

$$\begin{cases} K_p = 2\xi.\omega_n.J - fv \\ K_i = J.\omega_n^2 \end{cases} \tag{7}$$

The parameters K_p and K_i can be identified by solving the equality of equation above, where the ω_n and ξ is choosing by user according to preferable operating mode.

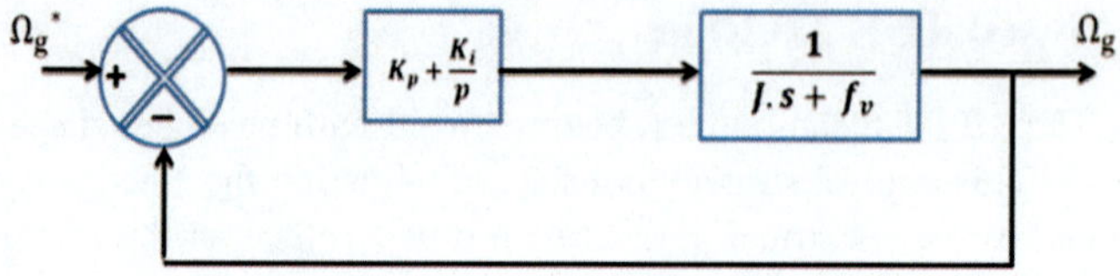

Fig. 4. Typical regulator PI structure.

3.2 Backstepping Controller

The Backstepping methodology can be defined as a way of division of a given system into several cascading subsystems. Based on sliding mode theory, all subsystem has to reach an intrinsic sliding surface, then, maintain the sliding state in order to achieve a global convergence. A control law stabilizing the system is derived from a Lyapunov function to prove the stability of the synthetized control law [6].

One of the advantages provided by the Backstepping method is its ability to keep the properties of the initial system in the synthetized control law. This is in a way the peculiarity of Backstepping compared to other methods [7]. Passivity is linked to other very important concepts such as stability, detectability and optimality. All these notions are necessary for the synthesis of the control laws. For the beign cited reasons, the analysis begins with the study of the Backstepping method to lead to its application to strict-feedback systems. In our study, to design a backstepping speed control, we have to start from the dynamic's equation and define the tracking error of the set point.

$$\mathrm{e}(\Omega_g) = \Omega_g^* - \Omega_g \tag{8}$$

We consider the following Lyapunov function:

$$v(\mathrm{e}) = \frac{1}{2}\mathrm{e}(\Omega_g)^2 \tag{9}$$

The Lyapunov function derivative:

$$\dot{v}(\mathrm{e}) = \mathrm{e}(\Omega_g).\dot{\mathrm{e}}(\Omega_g) = \mathrm{e}(\Omega_g).\left(\dot{\Omega}_g^* + \frac{1}{J}(C_{em} + f_v.\Omega_g - C_g)\right) \tag{10}$$

The stabilizing control of backstepping is defined as follows:

$$C_{em}^* = -J.\dot{\Omega}_g^* - f_v.\Omega_g + C_g - K_1.e(\Omega_g) \tag{11}$$

With: K_1 positive constant.
We replace the expression (11) in (10) we get:

$$\dot{v}(e) = -K_1.\mathrm{e}(\Omega_g)^2 < 0 \tag{12}$$

To ensure system's stability, the condition above must be verified.

3.3 Sliding Mode Controller

The sliding mode control is a variable structure nonlinear control method. The control structure is designed, keeping as primary objective, all system trajectories converge to an desired hypersurface. In our case the application of SMC it done with only one variable which means the application is reduced to a scalar variable C_{em} considered as tracking variable. To make appear the command C^*_{em}, the relative degree of the surface is equal to one. The sliding surface is defined by:

$$S(\Omega_g) = \Omega^*_g - \Omega_g \tag{13}$$

By Applying the following Lyapunov function to slide variable:

$$V(S(\Omega_g)) = \frac{1}{2}S(\Omega_g)^2 \tag{14}$$

The Lyapunov function derivative:

$$\dot{V}(S(\Omega_g)) = S(\Omega_g).\dot{S}(\Omega_g) \tag{15}$$

With: $\dot{S}(\Omega_g) = \dot{\Omega}^*_g - \dot{\Omega}_g$
By combining the above expressions in the last equations, we can write:

$$\dot{S}(\Omega_g) = \dot{\Omega}^*_g + \frac{1}{J}(C_{em} + f_v.\Omega_g - C_g) \tag{16}$$

Replacing the expression of C_{em} by the equivalent commands $(C_{emeq} + C_{emn})$ in Eq. (16) we find:

$$\dot{S}(\Omega_g) = \dot{\Omega}^*_g + \frac{1}{J}((C_{emeq} + C_{emn}) + f_v.\Omega_g - C_g) \tag{17}$$

In steady state we have: $S(\Omega_g) = 0$; $\dot{S}(\Omega_g) = 0$ and $C_{emn} = 0$, from which we extract the expression of the equivalent command C_{emeq}:

$$C_{emeq} = -J.\dot{\Omega}^*_g - f_v.\Omega_g + C_g \tag{18}$$

Replacing in the expressions above gives:

$$\dot{S}(\Omega_g) = \frac{1}{J}C_{emn} \tag{19}$$

To ensure the convergence of Lyapunov's function, we set:

$$C_{emn} = -K_2.\text{sign}(S(\Omega_g)) \tag{20}$$

With: K_2 positive constant.

3.4 Artificial neural networks Controller

In our paper, the MPPT neural networks controller was selected as a static Multi-Layers Perceptron (MLP). The Fig. 5 illustrate the architectural scheme of the used MLP [2, 10]. The proposed controller was composed of an input layer with two neurons, which were mechanical speed and its reference, two hidden layers, with 20 and 10 neurons respectively, an output layer with one neuron, which represents the generated reference torque. The curve of training and test is shown in Fig. 6.

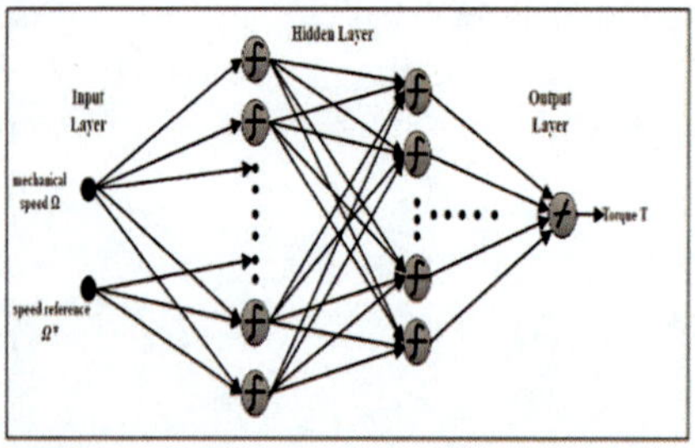

Fig. 5. ANN controller concept MLP.

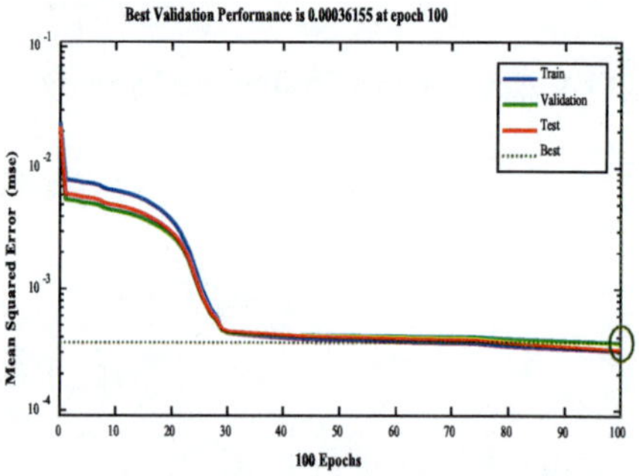

Fig. 6. Training and test performance curves.

4 Results and Discussions

In this section, the performance of the proposed controllers is evaluated and compared in terms of the reference tracking, static error, the dynamic response, system stability and robustness. The overall system considering a wind turbine which has been simulated using Matlab/Simulink, main parameters of the used wind turbine are summirized in Table 1. In addition, for a valuable validation of the proposed control methodologies, two different wind speed scenarios has been applied.

Table 1. Parameters of the used wind turbine

Parameters	Value
Volume density of the air ρ	1.225 $\frac{Kg}{m^3}$
Number of blades	3
Blade radius R	2 m
Pitch angle β	0°

4.1 Robustness Tests

For the best performance evaluation, a robustness test should be carried out on the different control techniques in order to evaluate their respective merit for a radical change in the wind profile as shown in Figs. 7. Figures 8, 9 and 10 shows the evolution of the four MPPT methods with mechanical speed control. Considering variable step wind speed scenario, one can clearly observe that the dynamic performance of the system based on ANNC is very efficient compared to other controllers SMC, PI and BAC. Under these variants conditions, the power coefficient C_p takes a maximum value of $C_{p\text{-}max} = 0.48$ for a pitch angle fixed at its minimum value, $\beta = 0°$.

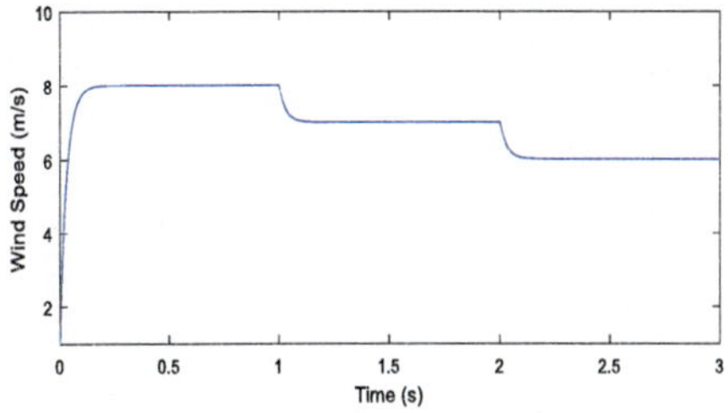

Fig. 7. Wind speed profile

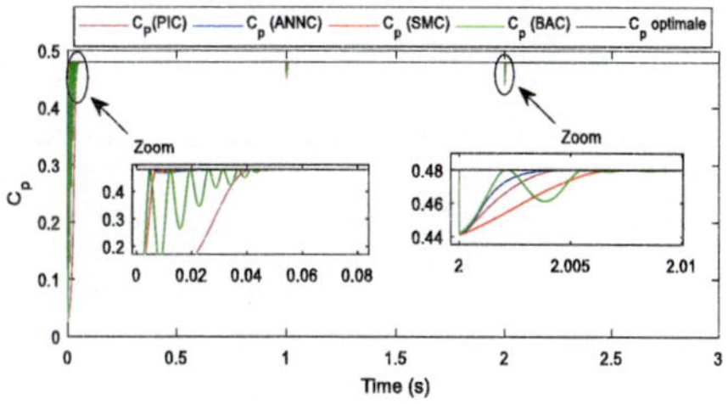

Fig. 8. Power coefficient

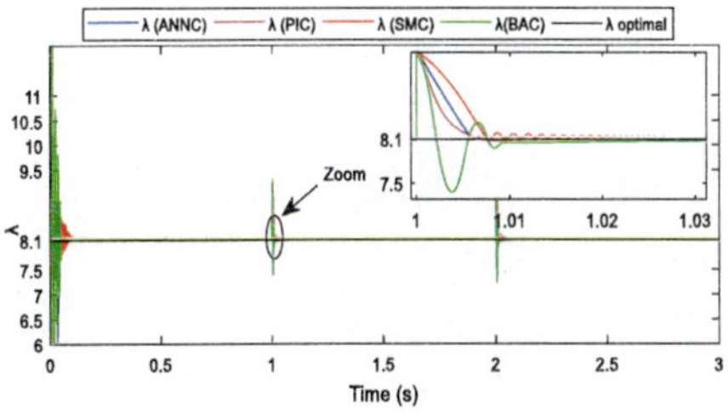

Fig. 9. Tip speed ratio.

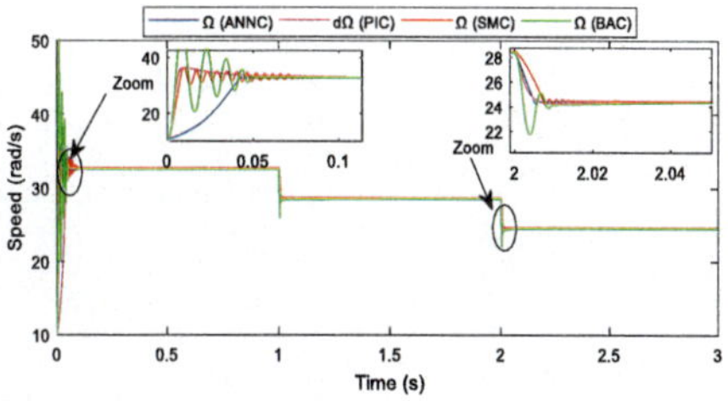

Fig. 10. Mechanical speed.

4.2 Tracking Tests

In this test the wind profile varies between 6 (m/s) and 10.5 (m/s) as shown in Fig. 11. To extract the maximum of the generated power, the speed ratio was set to the value $\lambda_{opt} = 8.1$, which corresponds to the maximum power coefficient $C_{p\text{-}max} = 0.48$ for any variation in wind speed. The results of the MPPT simulation with mechanical speed control by the four proposed controllers (PIC, SMC, BAC, ANNC) show clearly that for each value of wind speed, the mechanical speeds are perfectly following their references for the four MPPT methods with a significant static error for the BAC controller as we can see in Fig. 12. Similarly, it is observed that the ANN controller and SMC methods quickly achieve the static regime with a response time $T_s = 40$ (ms) and a negligible static error. On the other hand, the PI controller and Backstepping controller present a slightly slow response of about 60 (ms) in the dynamics regime with small fluctuations for the BAC controller and a negligible error for the PI controller (Fig. 13).

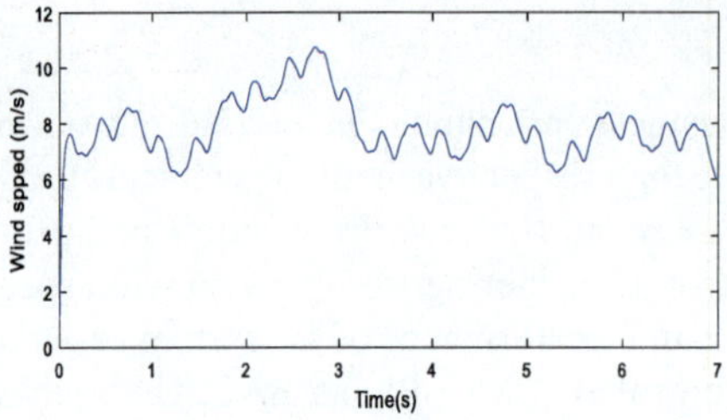

Fig. 11. Wind speed profile

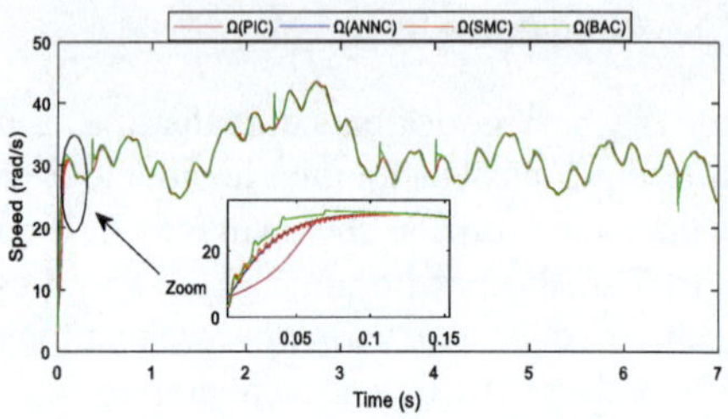

Fig. 12. Mechanical speed.

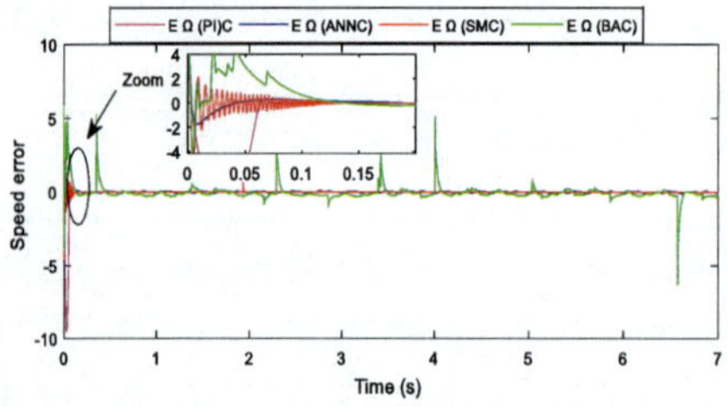

Fig. 13. Speed error.

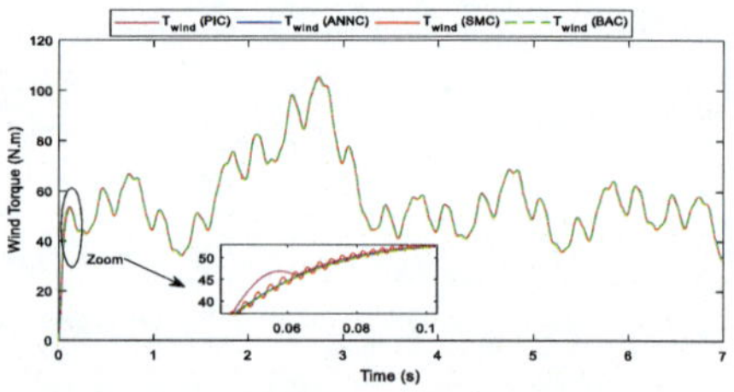

Fig. 14. Aerodynamic torque

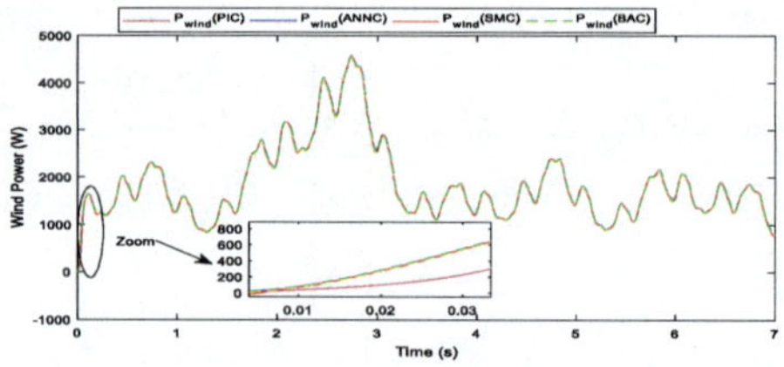

Fig. 15. Mechanical power.

Figures 14 and 15 show that in the four methods, the aerodynamic torque and the mechanical power follow their desired trajectories with a different efficiency. From the analysis of the results, it appears that the ANNC, and BAC present better performance in terms of response time and set point tracking.

The Table 2 below represents a synthesis of the comparison between the PIC, BAC, SMC and ANNC in term of the response time, set point tracking and static error. This table shows remarkable improvements obtained by ANNC. These improvements include an optimization of the static error, response time, set point tracking and robustness.

Table 2. Comparative result between the PIC, BAC, SMC and the ANNC.

Performance	PI	BAC	SMC	ANNC
Response time (s)	0.041	0.020	0.015	**0.0035**
Static errors (%)	0.241	0.175	0.198	**0.103**
Set-point tracking	Medium	Good	Very Good	**Very Good**
Robustness	Not Robust	Robust	Robust	**Robust**

5 Conclusion

This paper presents a comparative analysis of four MPPT control strategy using TSR method for controlling a variable speed wind turbine driven DFIG under low wind speed conditions. The compared controllers are: a conventional PI, BAC, SMC and ANNC. The study shows that the application of the ANNC MPPT control provides better speed regulation performance compared to other controllers, its allows a several advantages such as good tracking of references, robustness, a significant reduction in torque ripples, and faster dynamic response. In conclusion, the use of the ANNC results in the best and efficient control to tracking power.

References

1. Xiong, L., Li, P., Ma, M., Wang, Z., Wang, J.: Output power quality enhancement of PMSG with fractional order sliding mode control. Electr. Power Energy Systems **115** (2020)
2. Chojaa, H., Derouich, A., Chehaidia, S.E., Zamzoum, O., Taoussi, M., Elouatouat, H.: Integral sliding mode control for DFIG based WECS with MPPT based on artificial neural network under a real wind profile. Energy Rep. **7**, 4809–4824 (2021). https://doi.org/10.1016/j.egyr.2021.07.066
3. Zamzoum, O., El, Y., Errouha, M., Derouich, A., El, A.: Active and reactive power control of wind turbine based on doubly fed induction generator using adaptive sliding mode approach. Int. J. Adv. Comput. Sci. Appl. **10** (2019). https://doi.org/10.14569/IJACSA.2019.0100252
4. Hamid, C., Derouich, A., Taoussi, M., Zamzoum, O., Hanafi, A.: An improved performance variable speed wind turbine driving a doubly fed induction generator using sliding mode strategy. In: 2020 IEEE 2nd International Conference on Electronics, Control, Optimization and Computer Science (ICECOCS), pp. 1–8 (2020). https://doi.org/10.1109/ICECOCS50124.2020.9314629
5. Chojaa, H., Derouich, A., Taoussi, M., Zamzoum, O., Yessef, M.: Optimization of DFIG wind turbine power quality through adaptive fuzzy control. In: Motahhir, S., Bossoufi, B. (eds.) ICDTA 2021. LNNS, vol. 211, pp. 1235–1244. Springer, Cham (2021). https://doi.org/10.1007/978-3-030-73882-2_113
6. Meghni, B., Chojaa, H., Boulmaiz, A.: An optimal torque control based on intelligent tracking range (MPPT-OTC-ANN) for permanent magnet direct drive WECS. In: 2020 IEEE 2nd International Conference on Electronics, Control, Optimization and Computer Science (ICECOCS), pp. 1–6 (2020). https://doi.org/10.1109/ICECOCS50124.2020.9314304
7. Fdaili, M., Essadki, A., Nasser, T.: Comparative analysis between robust SMC & conventional PI controllers used in WECS based on DFIG. Int. J. Renew. Energy Res. **7**, 2151–2161 (2017)

8. Chehaidia, S.E., Abderezzak, A., Kherfane, H., Boukhezzar, B., Cherif, H.: An improved machine learning techniques fusion algorithm for controls advanced research turbine (Cart) power coefficient estimation. UPB Sci. Bull. Ser. C: Electr. Eng. Comput. Sci. **82**, 279–292 (2020)
9. Morshed, M.J., Fekih, A.: Integral terminal sliding mode control to provide fault ride-through capability to a grid connected wind turbine driven DFIG. In: 2015 IEEE International Conference on Industrial Technology (ICIT), pp. 1059–1064 (2015). https://doi.org/10.1109/ICIT.2015.7125237
10. Chehaidia, S.E., Abderezzak, A., Kherfane, H., Guersi, N., Cherif, H., Boukhezzar, B.: Fuzzy gain scheduling of pi torque controller to capture the maximum power from variable speed wind turbines. In: 2020 IEEE 2nd International Conference on Electronics, Control, Optimization and Computer Science, ICECOCS 2020 (2020)

Optimization of Energy Consumption of a Thermal Installation Based on the Energy Management System EnMS

Ali Elkihel[1], Amar Bakdid[2](✉), Yousra Elkihel[3], and Hassan Gziri[1]

[1] Laboratory of Engineering, Industrial Management and Innovation, Faculty of Science and Technology of Settat, University Hassan 1er, Casablanca, Morocco
[2] Laboratory of Industrial Engineering and Seismic Engineering, National School of Applied Sciences ENSA-Oujda, Mohammed Premier University, Oujda, Morocco
a.bakdid@ump.ac.ma
[3] Laboratory of Industrial Technologies and Services, EST, University Sidi Mohamed Ben Abdellah, Fez, Morocco

Abstract. Since the industrial revolution, energy consumption is still increasing, this increase is losing companies Thousands of DHs and negatively influence the environment, for this reason companies thought to manage their energy consumption by applying a system called an energy management system. This allows companies to limit energy waste and improve the processes in place in order to reduce the energy bill. The agri-food company is among the companies looking to establish an Energy Management System (EMS) to achieve the objectives mentioned above and to improve its competitiveness and brand image. Now this setting up requires the follow-up of a work methodology imposed by the iso 50001 standard. The objective of our study is to explain what is an EMS, what are the steps to follow for the implementation of an EMS. The expected result of this study is to reduce the energy consumption of an industrial energy company.

Keywords: Energy management system · ISO 50001 · Energy policy

1 Introduction

Industrial energy efficiency is therefore one of the key levers for controlling costs, preserving margins and the competitiveness of companies, both for meeting economic and environmental objectives. The energy audit and the implementation within the company of an energy management system are the main tools necessary to achieve cost reduction objectives [1]. Financial incentive systems with technical support and a financial contribution to investments are available today at the level of several national banks. Manufacturers therefore have every interest in investing in these actions [2–4].

The energy management system is among the most effective solutions, the implementation of an EMS in an organization allows the latter to manage its energy use and in fact improve its productivity [5].

R. Saidi et al. (Eds.): ICATH 2021, LNDECT 110, pp. 311–319, 2022.
https://doi.org/10.1007/978-3-030-94188-8_29

The EnMS involves developing and implementing an energy policy, setting realistic targets for energy use, and designing action plans to achieve them and evaluating progress, including through implementation new energy-efficient technologies, limiting energy waste or improving the processes in place so as to reduce the energy bill. The ISO 50001 standard offers organizations a recognized framework for the implementation of an efficient energy management system [6–9].

Our experimental work is part of an ISO 50001 energy management system. The purpose of an Energy Management System (EMS) is to encourage organizations to subscribe to a voluntary approach. Improvement of the energy performance of their activities. It provides a common thread for developing methodical and lasting energy management [10]. It will help guide strategic decisions, implement cross-functional actions (purchasing, communication, training) and resources, verify their effectiveness, and involve all of the company's stakeholders. In addition, companies involved in a SMEn significantly improve their energy efficiency. The industry is therefore clearly concerned by the implementation of such an approach, being the third most energy-intensive sector of activity [11, 12]. The interest of this work based on the requirements of energy management in industry is:

- help this company to better manage its energy use, and improve its productivity.
- implement an energy policy, set realistic objectives in terms of energy use;
- design action plans to achieve them and assess progress, including through the application of new energy-efficient technologies;
- limiting energy waste or improving the processes in place so as to reduce the energy bill while increasing the efficiency of the thermal installation.

2 Steam and Hot Water Production and Distribution System

For the agrifood industry, the thermal installation that produces the steam plays a very important role for the various production processes.

The boilers that make up this installation are devices under pressure used to warm water or produce steam [13], usually through the energy released by the combustion of a fuel, to supply the installations with steam Fig. 1.

A significant part of energy consumption in industry is used in boilers, which makes it one of the main objectives of optimizing and improving energy efficiency. The main heat losses in steam boilers are losses at the combustion level, losses in pipe and at the processes level. In outraged, certain inevitable losses occur for different reasons (by purging, by radiation, by incomplete combustion, by non-insulated surfaces, etc.).

Different parameters affect the efficiency of the boiler (such as: excess air, the quality of the feed water [14], uninsulated surfaces of the boiler, the quality of the fuel, the smoke temperature….). However, there are many techniques for optimizing heat loss in order to improve boiler yield such as:

- The Setting up heat recovery systems,
- The optimal adjustment of excess air for stoichiometric combustion,
- The insulation of the circuits and surfaces of the boiler,

- The regular maintenance operations to identify failures… etc.

These and other actions help businesses to strengthen their productivity and to improve their competitiveness by targeting optimal maintenance actions that aim to extend component life and optimize expenses.

2.1 Modeling of the Thermal Installation

We modeled the entire thermal system by three units of steam production, distribution, and use in processes. To simplify this modeling, we have chosen a linear presentation which takes into account each area of the company. To do this, we have established a simplified parameterized diagram of the steam production and distribution circuit illustrating an information application developed for the management of this installation in particular:

i - The equipment involved in the production of steam:

For this part we distinguish at the inlet of the boiler the water treatment system and the burner using fuel oil, other elements such as the purge means.

ii - The distribution circuit:

This steam transport circuit is composed of different organ, traps, valves, piping,

iii - The consumers:

Consumption concerns production processes such as heat exchangers, production machines as well as the means of controlling and regulating temperature, pressure, ….

The condensate recirculing system plays the role for the return of the steam.

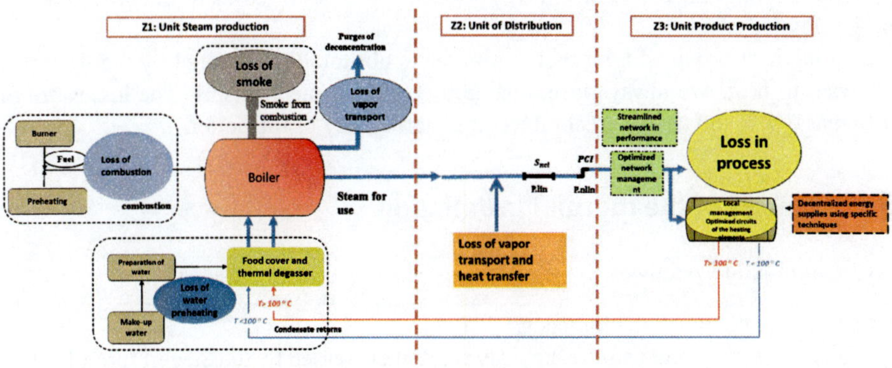

Fig. 1. The three zones of the thermal installation.

The technical specifications of the equipment shown on the diagram.

2.2 Diagnosis and Action Plan

Heat balance of the thermal installation.

The objective is to reduce heat losses at the level of the steam boiler. It is necessary to find the causes responsible for the lowering of the efficiency of the boiler, to seek solutions of improvements and proposals to reduce the losses [15].

The different losses evaluated: Losses by fumes, losses by unburnt materials, losses by purges, losses by walls and loss of management (Fig. 2).

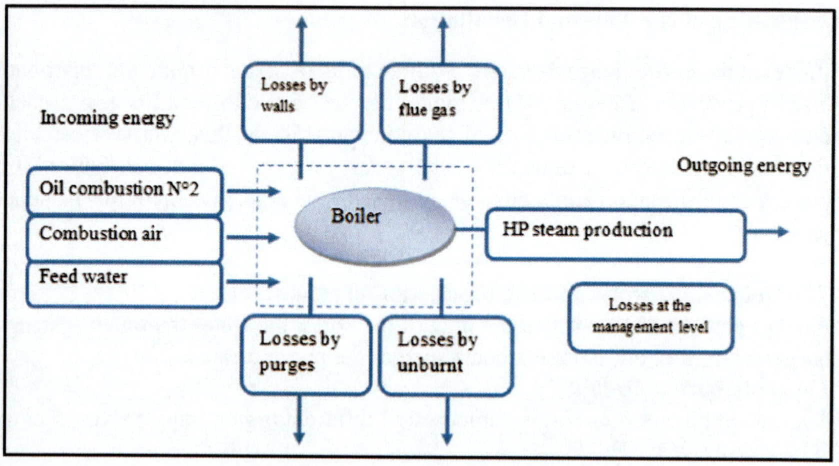

Fig. 2. The various losses at the boiler level.

With:

Input energy = fuel + preheated air + preheated water.

The outgoing energy = steam + smoke losses + wall losses + unburnt losses + purges losses.

So the heat provided by a fuel that burns is not totally recovered by the fluid that we want to heat. We always lose a game by different mechanisms. The losses are of different types, and have a great effect on the efficiency of the boiler.

3 Diagnosis of the thermal installation

3.1 Combustion Analysis

Expression of combustion efficiency.

In practice, the combustion efficiency is often expressed by the Siegert formula [16]:

$$\eta\ comb = 100 - f * (T_{fume} - T_{amb}) / \%CO_2 \qquad (1)$$

Where:

Tfume = the temperature of the flue gas at the boiler outlet [°C].

Tamb = room temperature in the boiler room [°C].

$\%CO_2$ = the CO_2 of the flue gas [%].

f = factor depending mainly on the type of fuel (fuel oil: f = 0.57; natural gas: **fuel oil** = 0.47).

After analysis we found that losses by smoke are more important. This value shows that the smoke losses have a large influence on boiler efficiency. We have found that they can come from excessive excess air which can be due to:

- Incorrect adjustment of the burner.
- Maintenance problems such as poor air distribution or poor fuel oil spraying.
- A clogged boiler: internal (scale) and external (soot) deposits that limit heat transfer between the boiler water and the flue gases.

The flue gas analyzes show that the flue gas temperature is very high. It is of the order of 182 °C, which causes a reduction in the yield. This temperature increase is due to the accumulation of soot and fly ash on the exchange surfaces. As part of preventive maintenance, it is therefore necessary to clean these surfaces, by the sweeping operation which must be carried out regularly.

3.2 Analysis of Thermal Losses Through the Walls

In this case we will take into consideration only the energy losses by radiation. We have made several measurements spaced in time and on the different speeds of the boiler by the thermography instrument, the results are presented in Fig. 3:

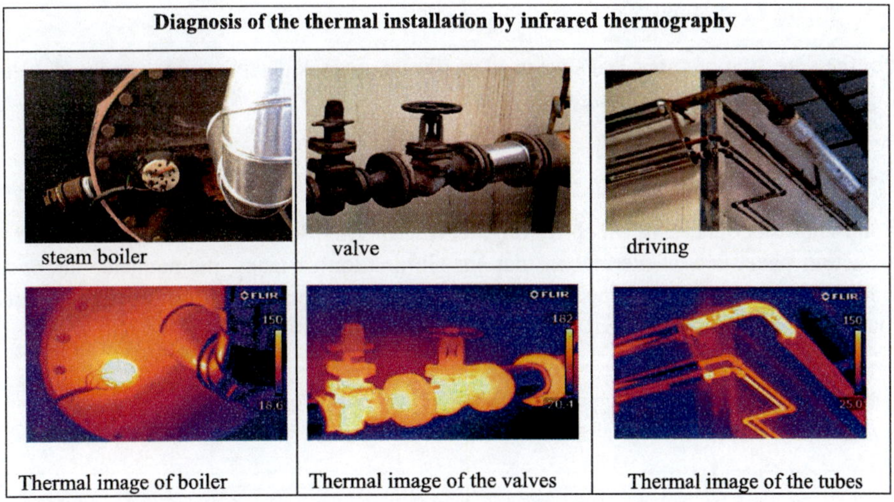

Fig. 3. Diagnosis of the thermal installation by infrared thermography.

Radiation losses are based on the principle that anybody in nature, having internal energy diffuses heat by radiation according to the following law [17, 18]:

$$Q_{ray} = \sigma \cdot S \cdot (T_{body}^4 - T_a^4) \tag{2}$$

With: ε: Thermal emissivity depending on the temperature.
σ: Stephan-Boltzmann constant equal to 5.67. 10^{-8} W/ ($m^2 \cdot K^4$).
S: Radiation surface.
T_{body}: temperature of the wall.
T_a: ambient temperature.

For more precision, we took several temperatures in reference to several points of the wall then we will calculate the average of these. We calculated the area of all the components, approximately 11.5 m^2, which are not insulated and by applying the formula (Eq. 2) Q we calculated exactly the heat losses.

For the purge rate, when operating a boiler, the transformation of water into steam results in a high concentration of minerals in the water, which are the sources of corrosion and scale formation.

Corrosion causes boiler tube failure, scaling of heat transfer surfaces in turn reduces performance and leads to under-deposit corrosion, resulting in downtime and costly maintenance.

The purge makes it possible to control the concentration of minerals within an acceptable limit, by replacing a part of the boiler water containing concentrated minerals with demineralized water, thus considerably increasing the service life and the efficiency of a boiler. The conductivity is the best measure to control the concentration of minerals, it is expressed in ppm TDS (parts per million of total dissolved solids), we have checked the traps and we have analyzed the failures that occur and which are sources of loss heat in the form of vapor.

3.3 Steam Leak Analysis

For the detection of vapor leaks we used an ultrasonic leak detector which enabled us to highlight all these losses along the installation.

3.4 Action Plan

Following this diagnosis which highlighted the thermal losses; we have implemented an action plan such as thermal insulation, elimination of leaks, maintenance actions. Energy indicators are developed for the entire thermal system and improvements are made according to the ISO 51000 standard.

4 Results and Discussion

4.1 Yield Calculation

The evaluation of boiler performance and efficiency is generally carried out by two methods: direct and indirect, as indicated in three boiler performance evaluation codes that will be the subject of our study (ASME PTC4-2008 codes, IS13979: 1994 and BS845-1: 1987).We present below the calculations carried out as part of a measurement campaign, which took place within an agri-food company operating in the beverage industry. The boiler studied is multitubular with a smoke tube with a capacity of 1 T/h working with fuel oil No. 2 as fuel (PCS: 9238.55 kcal/kg). The steam produced is used in manufacturing processes as well as for heating, atomizing and conditioning fuel.

The efficiency of the boiler was evaluated by two methods: direct and indirect. The indirect method is recommended for the accuracy of its results since it is based on detailed and complete information.

The efficiency of the boiler will be calculated by the following indirect formula [16]:

$$\eta = 100 - \sum_i P_i \tag{3}$$

With:

Pi: Heat losses in the boiler.

The direct method defines the efficiency as the ratio between the heat produced and the heat supplied according to the formula below.

$$\eta = \frac{Q_{util}}{Q_{tot}} * 100 \tag{4}$$

4.2 Results and Discussion

According to the direct method, the yield is **85%**. On the other hand, the yield initially calculated by the indirect method is **90.7%** and the final yield following the improvement actions is estimated at **94.8%**.

The majority losses are the losses by dry combustion gas (2.1%) and by the water produced by the combustion of hydrogen contained in the fuel (3.5%) which strongly depend on the excess air and the type of fuel burned. Indeed, the combustion gases leaving the chimney carry away a significant amount of energy, and affect the efficiency of the boiler. These losses are limited to 1.5% and 1.3% by providing an optimum quantity of excess air making it possible to ensure stoichiometric combustion.

In addition, other inevitable losses occur for various reasons, such as radiation and convection losses estimated at 1.8% and which are optimized to 1.1% by deploying insulation of the outer surfaces of distribution circuits and various bare boiler parts that transfer heat to the environment. The thickness of the insulation was calculated based on the losses occurring at the system level, and the costs of installing the insulation.

The results obtained by our method are in perfect agreement with the techniques developed in the works [1, 6, 15].

From an economic point of view, it is better to recover the waste heat and transfer it to a suitable heat sink. The most efficient solution for recovering the flue gases from the boiler is to preheat the combustion air and the feed water. Other solutions exist, but they require larger investments, such as:

- Heat pumps,
- The organic Rankin cycle or the Kalina cycle for generating electricity [19],
- The absorption refrigeration cycle to obtain an air conditioning or cooling effect.

However, the selection criteria for heat recovery technology are based on the temperature and flow rate of the exhaust gases from the boiler.

5 Conclusion

With the objective of reducing energy consumption in a food company, we have applied the SME energy management system, in particular to measure the waste of energy and to propose industrial maintenance techniques.

We carried out a diagnosis on the whole thermal installation by carrying out measurements by various very precise devices in particular thermal camera, leak detector, combustion analyzer,… to identify the thermal losses of the installation at the combustion level, non-insulated pipes and leaks in various components such as valves, purges, etc.

An action plan was developed to remedy all these thermal losses detected and which made it possible to improve the efficiency of the boiler. This improvement results in an efficiency of the thermal system (boiler and distribution means) and an optimization of energy consumption by minimizing the various losses. Improvements are made by low cost industrial maintenance and a proposal for other maintenance techniques which require high investment.

References

1. Meksoub, A., Elkihel, B., Boulerhcha, M.: Etude des Performances Énergétiques des Générateurs de Vapeur. In: 3rd International Conference on Mechanical Materials Structures, vol. 2019, p. 13 (2019)
2. Service public de Wallonie. Economies d'énergie dans l'industrie. Gecinox (2010)
3. Sternlicht, B.: Waste energy recovery: an excellent investment opportunity. Energy Convers. Manag. **22**(4), 361–373 (1982)
4. ECN: Heat Powered Cycles. Alkmaar-Netherlands (2012)
5. Hightower, D.A., Nasal, J.R.: A relative accuracy evaluation of various methods to determine long term coal-burned values for coal pile inventory reconciliation. In: EPRI Heat Rate Improved Conference, 25–27 January, p. 18 (2005)
6. Srinivas, G.T., Kumar, D.R., Venkata, P., Murali, V., Rao, B.N.: Efficiency of a coal fired boiler in a typical thermal power plant. Am. J. Mech. Ind. Eng. **2**(1), 32–36 (2017)
7. Kapre, A.S.: Energy auditing and scope for its conservation in textile industry: a case study. Maharana Pratap University of Agriculture and Technology, Rajasthan (2010)
8. Lang, F.D.: Errors in boiler efficiency standards. In: ASME 2009 Power Conference, no. April, pp. 487–501 (2009)
9. Patro, B.: Efficiency studies of combination tube boilers. Alex Eng. J. **55**(1), 193–202 (2016)
10. IS 13979:1994: Method of calculation of efficiency of packaged boiler, New Delhi (1994)
11. ASME PTC4-2008 Fired steam generators performance test codes, New York (2008)
12. BS 845-1:1987 British standard assessing thermal performance of boilers for steam, hot water and high temperature heat transfer fluids, London (1987)
13. Krishnanunni, S., Paul, J.C., Potti, M., Mathew, E.M.: Evaluation of heat losses in fire tube boiler. Int. J. Emerg. Technol. Adv. Eng. **2**(12), 301–305 (2012)
14. https://energieplus-lesite.be/mesures/chauffage7/mesurer-le-rendement-de-combustion/
15. Cortes-Rodríguez, E.F., Nebra, S.A., Sosa-Arnao, J.H.: Experimental efficiency analysis of sugarcane bagasse boilers based on the first law of thermodynamics. J. Braz. Soc. Mech. Sci. Eng. **39**(3), 1033–1044 (2017)
16. Harimi, M., Sapuan, M., Ahmad, M., Abas, F.: Numerical study of heat loss from boiler using different ratios of fiber to shell from palm oil wastes. J. SciInd. Res. **67**, 440–444 (2008)

17. Barma, M.C., Saidur, R., Rahman, S.M.A., Allouhi, A., Akash, B.A., Sait, S.M.: A review on boilers energy use, energy savings, and emissions reductions. Renew. Sustain. Energy Rev. **79**, 970–983 (2017)
18. Ibrahim, H., Qassimi, M.: Matlab program computes thermal efficiency of fired heater. Period. Polytech. Chem. Eng. **52**(2), 61–69 (2008)
19. Lecompte, S., et al.: Case study of an organic Rankine cycle (ORC) for waste heat recovery from an electric arc furnace (EAF). Energies **10**(5), 1–16 (2017)

Legendre Polynomial Modeling of a Piezoelectric Transformer

D. J. Falimiaramanana[1](✉), H. Khalfi[2], J. Randrianarivelo[1], F. E. Ratolojanahary[1], L. Elmaimouni[2], I. Naciri[2], and M. Rguiti[3]

[1] LAPAUF-Laboratoire de Physique Appliquée de l'Université de Fianarantsoa, Université de Fianarantsoa, 301 Fianarantsoa, Madagascar

[2] ERMAM, Polydisciplinary Faculty of Ouarzazate, Univ. Ibn Zohr, BP 638, 45000 Ouarzazate, Morocco

[3] EA 2443 - LMCPA - Laboratoire des Matériaux Céramiques et Procédés Associés, Université Polytechnique Hauts-de-France, 59313 Valenciennes, France

mohamed.rguiti@uphf.fr

Abstract. A polynomial approach is used for solving wave propagation to forecast the performance of piezoelectric transformers (PTs). The method consists in integrating straightforwardly into the equation of motion via a delta function the continuity and boundary conditions. Resonance and anti-resonance frequencies, electrical and acoustical fields of the PTs are obtained. The energy harvest in the piezoelectric transformer is presented. The validation of the model is done through a comparison of our results against the 2D Finite Element Method ones.

Keywords: Piezoelectric transformers · Polynomial approach · Resonance and anti-resonance frequencies · Modal analysis · Plane stress

1 Introduction

The piezoelectric materials discovered by Pierre and Jacques Curie have already been generally used in the field of miniaturization technology in electronics. The advantage of these materials is the existence of the reversible electromechanical coupling with, notably, a generation of displacement charges proportional to the material deformation [1]. In 1956 C.A. Rosen introduced Piezoelectric Transformer (PT) as a power transferring device [2, 3].

The fabrication of the PT is based by the association of two blocks of piezoelectric materials which are connected together to form a transformer [4–6]. It is an energy conversion component that uses an acoustic vibration to transfer or to convert the electric-to-electric energy [7]. An electrical AC voltage imposed at the primary part element generates a deformation of the whole structure which results in a coupled AC output voltage on the secondary part element [8–10].

To design PT with minimum hardware effort, shorter design cycle and reduced costs, and to better understand the basic phenomena in this component, efficient modeling tools are required.

R. Saidi et al. (Eds.): ICATH 2021, LNDECT 110, pp. 320–330, 2022.
https://doi.org/10.1007/978-3-030-94188-8_30

The development of these efficient modeling tools for PTs is done in the same manner as the transducers and other piezoelectric resonators. Classic models are generally based on the use of an equivalent electrical circuit [11, 12]. Another theoretical approach for the study of piezoelectric transformers is based on the constitutive equations and the dynamic equations from the linear theory of piezoelectricity [13, 14]. However, due both to a strong electromechanical coupling and to the anisotropy necessary for the existence of the piezoelectric phenomenon, there are no exact solutions arising from the 3D equations of piezoelectricity. A numerical analysis, a 3D finite element method, has been proposed in the literature. Moreover, a general energetic approach to establish an accurate electromechanical model of a piezoelectric transformer (PT) using Hamilton's principle has also been dealt with [6].

The previous work [15] gives the results of the two-dimensional semi-analytic model by using the polynomial approach permitting to study the PT, and only the harmonic analysis has been illustrated and the validation has been done by comparing the polynomial results with 3D FEM ones. In the present paper, the extension of the polynomial method for the modal analysis of PT is presented. With our approach, the acoustic and electrical field distributions are easily obtained without meshing.

After the presentation of the studied structure and the description of the method in the second section of this manuscript, the results obtained by the numerical simulation calculated via polynomial approach and compared with 2D finite element method are given in the third section.

2 PT Modeling

In this section the studied structure is described and the theoretical model is presented. Several types of PT are presented in the literature. In this work the studied structure is a Rosen PT in Fig. 1 with dimensions being $L \times l \times h$. $L = L_1 + L_2$, l and h are respectively the length, width and thickness of the PT. L_1 is the length of the primary side that is poled in the thickness direction. The length of the secondary part, poled in length direction and connected with a load resistance R_L at the end, is L_2. The coordinate system used during the resolution is (O, x_1, x_2, x_3); its origin is at the centre of the junction between the two parts.

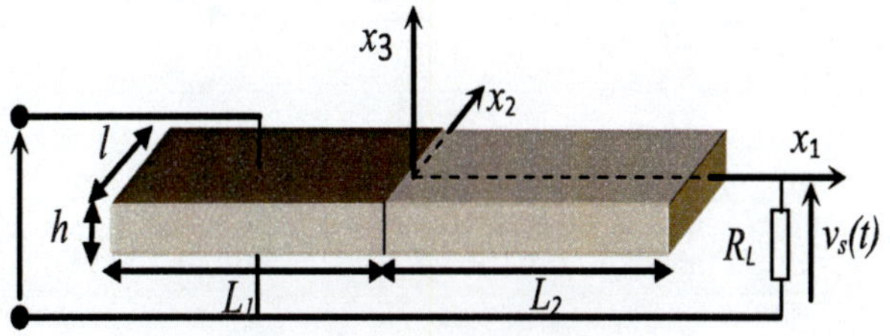

Fig. 1. Studded PT's geometry.

In this paper we assume that the stress over all the outer surfaces of the two parts not including the surface at the junction are mechanically free; with $h \ll l$ and L, a 2D plane-stress condition in the (Ox_1, Ox_2) plane is considered; the electrodes' thickness is neglected. In the polynomial approach the rectangular window functions and the Heaviside function are introduced into the propagation equations [15–17] in order to automatically integrate the limit and the continuity conditions and to describe the studded transformer. The description of the polynomial method is detailed in [15]. It consists in developing the mechanical displacements u and the electric potential ϕ in a double series of orthonormal polynomials with a suitable analytic expression as shown in the following equations [15–17]:

in region 1:

$$u_1^{(1)}(q_1, q_2) = \sum_{n=0}^{+\infty} \sum_{m=0}^{+\infty} Q_m^{(1)}(q_1) Q_n(q_2) p_{1mn}^{(1)}; \tag{1a}$$

$$u_2^{(1)}(q_1, q_2) = \sum_{n=0}^{+\infty} \sum_{m=0}^{+\infty} Q_m^{(1)}(q_1) Q_n(q_2) p_{2mn}^{(1)}; \tag{1b}$$

and in region 2:

$$u_1^{(2)}(q_1, q_2) = u_1^{(1)}(0, q_2) + q_1 \sum_{n=0}^{+\infty} \sum_{m=0}^{+\infty} Q_m^{(2)}(q_1) Q_n(q_2) p_{1mn}^{(2)}; \tag{2a}$$

$$u_2^{(2)}(q_1, q_2) = u_2^{(1)}(0, q_2) + q_1 \sum_{n=0}^{+\infty} \sum_{m=0}^{+\infty} Q_m^{(2)}(q_1) Q_n(q_2) p_{2mn}^{(2)}; \tag{2b}$$

$$\Phi^{(2)}(q_1, q_2) = \left(1 - \frac{q_1}{q_s}\right)\frac{v_p}{2} + \frac{q_1}{q_s} v_s + q_1(q_1 - q_s) \sum_{n=0}^{+\infty} \sum_{m=0}^{+\infty} Q_m^{(2)}(q_1) Q_n(q_2) r_{mn}; \tag{2c}$$

With $q_1 = x_1/L$, $q_2 = x_2/L$, $q_s = L_2/L$, $Q_n(q_2) = \sqrt{(2n+1)/2} P_m(q_2)$ and $Q_m^{(R)} = \sqrt{(2m+1)L/L_R} P_m\big(2q_1 L/L_R - (-1)^R\big)$ where P_m and P_n are the Legendre polynomials of degree m and n. R is the number of each part which is equal 1 at the driver and 2 at the receiver. $P_{kmn}^{(R)}$ and r_{mn} are the expansion coefficients.

Substituting Eqs. 1a–1b and 2a–2c in the constitutive equations injected into the propagation equation, and then exploiting the polynomials' orthogonality, the following system of linear equations with an infinite unknowns $P_{kmn}^{(R)}$ and Ω where Ω is a normalized angular frequency, is obtained:

$$A.P + C.v_s = -\left(\frac{L\pi}{L_1}\right)^2 \Omega^2 E.P + D.v_p \tag{3}$$

The elements of all matrixes and the vector P are given in [15].

From this system of equations, when the magnitude of voltage supplied at the primary part is null, $v_p = 0$, the parallel resonance frequencies are obtained for an important value of the load resistance, and the series resonance frequencies are obtained by letting the receiving electrodes short-circuited. The system (3) can be written as an eigen value equation:

$$E^{-1}A.P = -\left(\frac{L\pi}{L_1}\right)^2 \Omega^2 I_d.P \quad (4)$$

3 Numerical Results

The numerical results obtained in the simulations carried out with the Matlab software are presented in this section. First, the method's convergence is studied, second a comparison of the polynomial results against 2D finite element method results is done and third some results to illustrate the pertinence of the model are given.

To study the convergence of the model we use in simulations a ceramic PT. Its parameters are given in [13] and the dimensions are presented in Table 1. The fields are expanded in an infinite series of orthonormal polynomials. In practice, the infinite summations are truncated to finite values M and N. The problem results in a system of $(M + 1)(N + 1)$ linear equations with $(M + 1)(N + 1)$ eigen modes for the modal analysis. The resonance frequencies are obtained with the method for different values of truncation orders. When M and N are increased, the convergence of the solutions is obtained [15]. For the first four modes of the studded PT, the convergence is appeared from $M = 7$ and $N = 3$.

Table 1. Rosen piezoelectric transformer's geometric parameters

Primary length L_1	12×10^{-3} m
Secondary length L_2	13×10^{-3} m
Width l	5×10^{-3} m
Thickness h	1.7×10^{-3} m

To validate our approach, the results calculated by means of a polynomial method are confronted with a 2D FEM modeling ones. To carry out the finite element analysis (FEA), the multi-physics software Comsol was used. The associated relative accuracy is calculated as follows [16]:

$$\varepsilon_R(\%) = 100 \times \left|\left(R_{fem} - R_{pol}\right)/R_{fem}\right| \quad (5)$$

where R_{fem} and R_{pol} denote FEA and polynomial results, respectively.

Table 2. The anti-resonance frequencies of the ceramic PT.

Mode	2D poly (kHz)	2D FEA (kHz)	Relative error (10^{-2}%)
1	70.46	70.45	0.42
2	141.39	141.39	0
3	206.77	206.76	0.48
4	265.30	265.29	0.37

Table 3. The resonance frequencies of the ceramic PT.

Mode	2D poly (kHz)	2D FEA (kHz)	Relative error (10–2%)
1	61.63	61.63	0
2	124.67	124.66	0.80
3	203.79	203.77	0.98
4	265.31	265.29	0.75

The Table 2 and Table 3 give, respectively, the resonance and anti-resonance frequencies of the PT. It can note that the agreement is good between the results calculated by using 2D polynomial approach and the 2D FEM ones. This validates our method for the ceramic PT. In addition, it is worth to mention that our method can calculate the resonance frequency straightforwardly with a unique code without approximation.

Close to each resonance frequency found above we obtain an approximate expression for each field, mechanical displacement, stress, electrical potential, electrical field, electrical displacement…thanks to the expansion coefficients $P_{kmn}^{(R)}$ calculated through the solving of the eigen value problem. We first calculated the magnitudes of the mechanical displacements to identify the vibration modes and nodal points so important for a non sensitive mounting of the PT on its support. Figure 2 gives the mechanical displacement profiles for the first four longitudinal resonance modes respectively λ/2, λ, 3λ/2 and 2λ.

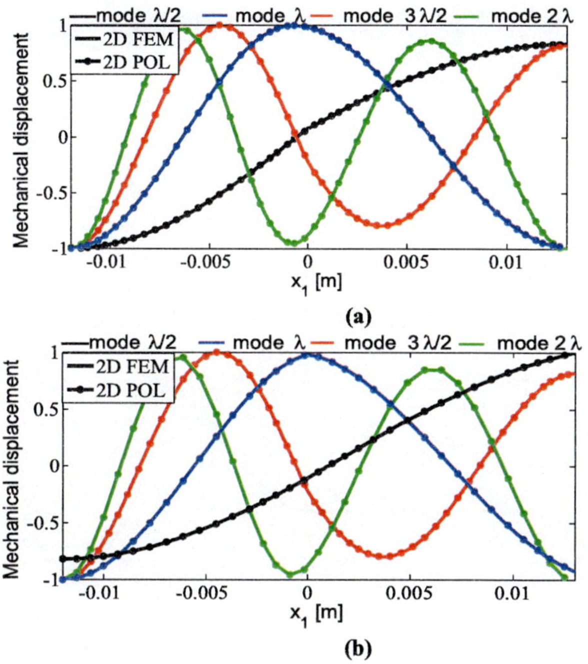

Fig. 2. Mechanical displacement of the Rosen transformer a) secondary open-circuited b) secondary short circuited

Figure 3 allows to view the electrical potential profiles along the x_1 axis for the first four longitudinal vibration modes of the studied PT in the cases (a) secondary open-circuited and (b) secondary short circuited. The polynomial results (dots) and 2D FEM ones (solid lines) overlap and a very good agreement is observed with these two different methods. Moreover, the electrical boundary conditions imposed in the expressions of the electrical potential, constant potential at $x_1 = L_2$ and continuous potential at $x_1 = 0$, are respected.

Also the stress profiles of the PT as shown in Fig. 4 verify the nullity of the mechanical stress at the extremity of primary and secondary sides. Besides, the normal stress continuity at the interface $x_1 = 0$ between the primary and secondary parts is respected. This means that using specific functions to describe the structure in the polynomial method and to automatically incorporate the boundary and continuity conditions into the propagation equations is entirely adequate for PT in the plane stress condition.

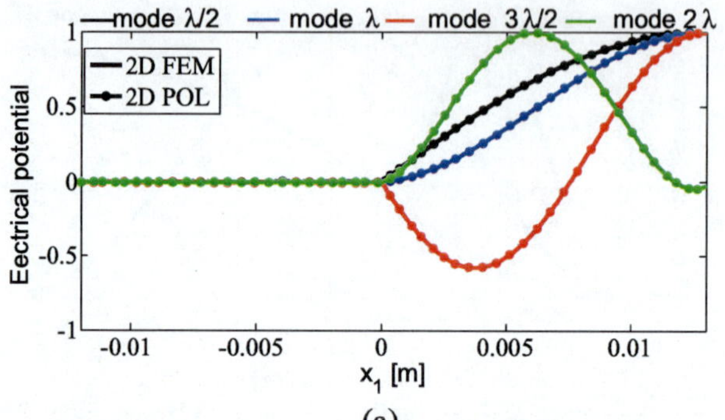

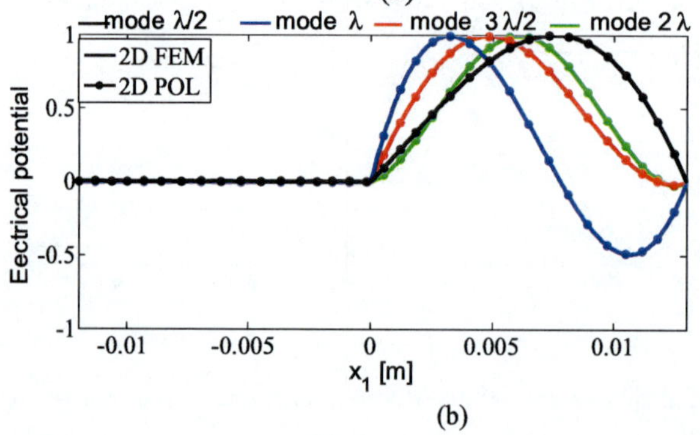

Fig. 3. Electrical potential of the Rosen transformer a) secondary open-circuited b) secondary short circuited

Besides the stress limitation is quite important constraint [18] because of the risk of fracture. The stress is a tensor and it cannot be plotted in an easy manner. As explained in [19] one method to forecast the fracture point of a material is to compare the total mechanical stress of the three principal directions to the Von Mises yield criterion. Indeed, this latest is the point at which the Von Mises stress exceeds its maximum value for a particular material.

It is calculated from the principal stresses, T_1 and T_2, as shown in Eq. 6 [19] and is plotted in Fig. 5:

$$T_{vms} = \sqrt{T_1^2 + T_2^2 + 2T_6^2 - T_1T_2}. \tag{6}$$

Furthermore, the operation of the transformer at one of its resonance modes leads to a mounting strategy. The main aim is to strongly reduce the damping from the mounting of the transformer on its support. Nevertheless, tethering the PT at its nodal points will not only provide an undamped fixing of the PT, but will also provide a method of shunting most of the heat away from the PT. Indeed, as stated in the literature [20], the area of maximum internal stresses inside the piezo structure is responsible for the majority of the heat generated by the PT.

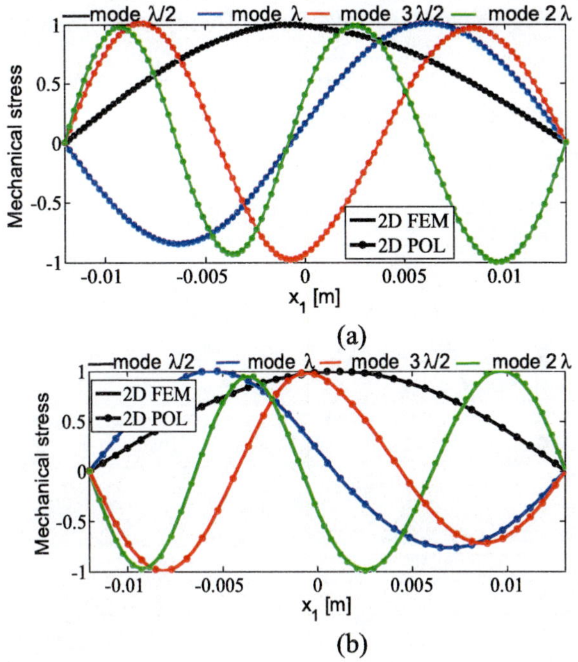

Fig. 4. Mechanical stress of the Rosen transformer a) secondary open-circuited b) secondary short circuited

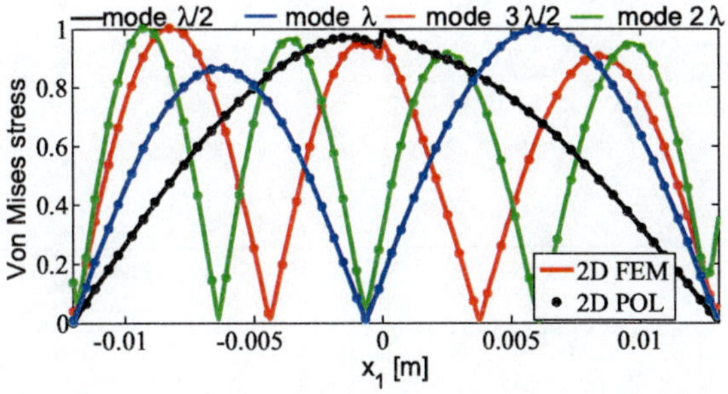

Fig. 5. Von Mises stress of the Rosen transformer.

In addition, the simultaneous operation of two different modes (transverse and longitudinal) deteriorates the transformer, particularly at the primary/secondary interface. Premature destruction of the ceramic coupled with excessive mechanical stresses can lead to a rupture of the structure. In this case, a mode for which the stress at the interface is null is the better choice. For the Rosen PT, the second longitudinal mode is the adequate functioning mode. Near of this mode, the proposed model can calculate the acoustic energy density within the piezoelectric material, the kinetic and the strain energy density and the electrical energy density. These energy densities are presented in Fig. 6. We observe that the maximum acoustic energy density matches to the maximum stress values. This corresponds to the minimum of particle's velocity vibration with null kinetic energy density and mechanical displacement.

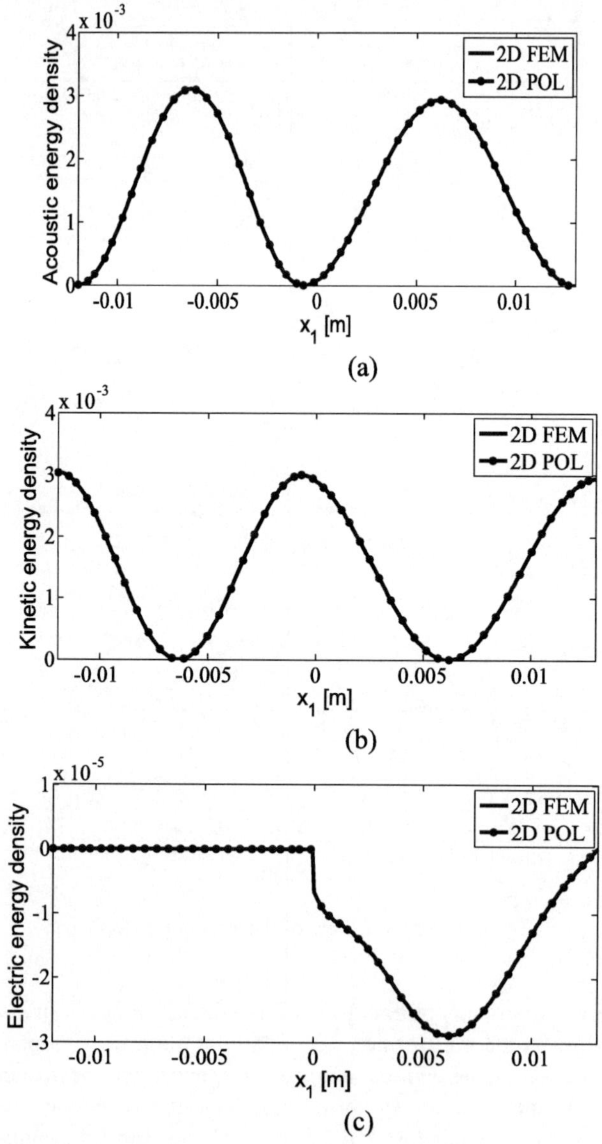

Fig. 6. (a) Acoustic energy density (b) Kinetic energy density (c) Electric energy density of the Rosen transformer at the second longitudinal resonance mode.

4 Conclusion

A PT design is simulated using the polynomial approach with a focus on the modal analysis. The electromechanical behavior of a ceramic piezoelectric transformer is obtained and presented in this article. The specificity of the method is the expansion of the mechanical displacement and the electrical potential in a double series of Legendre polynomial and the direct incorporation of the continuity and boundary conditions into the propagation equations. A very good agreement is obtained. The simulation is carried out with Matlab software. The validity of the method is given and it is shown that the approach is pertinent. The energy density in the piezoelectric transformer is presented.

References

1. Wang, W.C., Wu, L.Y., Chen, L.W., Liu, C.M.: Acoustic energy harvesting by piezoelectric curved beams in the cavity of a sonic crystal. Smart Mater. Struct. **19**, 045016–045022 (2010)
2. Lin, C.Y., Lee, F.C.: Design of a piezoelectric transformer converter and its matching networks. In: Proceedings of 1994 Power Electronics Specialist Conference – PESC 1994, Taipei, Taiwan, pp. 607–612 (1994)
3. Chérif, A., et al.: Improvement of piezoelectric transformer performances using SSHI and SSHI-max methods. Opt. Quant. Electron. **46**(1), 117–131 (2013). https://doi.org/10.1007/s11082-013-9712-2
4. Erhart, J.: Parameters and design optimization of the ring piezoelectric ceramic transformer. J. Adv. Dielectr. **5**(3), 1550022–1550030 (2015)
5. Ekhtiari, M., Zhang, Z., Andersen, M.A.E.: State-of-the-art piezoelectric transformer-based switch mode power supplies. In: Proceedings of IECON, pp. 5072–5078 (2014)
6. Nadal, C., Pigache, F.: Multimodal electromechanical model of piezoelectric transformers by Hamilton's principle. IEEE Trans. Ultrason. Ferroelectr. Freq. Control **56**(11), 2530–2543 (2009)
7. Carazo, A.V.: Ultrasonic piezoelectric transformers for power conversion. In: UIA 42nd Annual Symposium (2013)
8. Emmanuel, S., Vasic, D., François, C.: Transformateurs statiques piézoélectriques. Techniques de l'ingénieur. D 3015 (2012)
9. Thomas, M.: Contribution à l'étude des générateurs piézoélectriques pour la génération des décharges plasmas. Ph.D. dissertation, INP Toulouse (2012)
10. Vasic, D.: Apports des matériaux piézoélectriques pour l'intégration hybride et monolithique des transformateurs. Ph. D. dissertation, ENS de Cachan (2003)
11. Bronstein, S.: Piezoelectric Transformers in Power Electronics. Ph.D. dissertation, Negev (2005)
12. Eddiai, A., Meddad, M., Rguiti, M., Chérif, A., Courtois, C.: Design and construction of a multifunction piezoelectric transformer. J. Aust. Ceram. Soc. **55**(1), 19–24 (2018). https://doi.org/10.1007/s41779-018-0206-3
13. Pigache, F., Nadal, C.: Identification Methodology of Electrical Equivalent Circuit of the Piezoelectric Transformers by FEM. ansys.net
14. Boukazouha, F., Boubenider, F.: Piezoelectric transformer: comparison between a model and an analytical verification. Comput. Struct. **86**, 374–378 (2008)
15. Falimiaramanana, D.J., Ratolojanahary, F.E., Lefebvre, J.E., Elmaimouni, L., Rguiti, M.: 2D modeling of Rosen-type piezoelectric transformer by means of a polynomial approach. IEEE Trans. Ultrason. Ferroelectr. Freq. Control **67**(8), 1701–1714 (2020). https://doi.org/10.1109/TUFFC.2020.2975647

16. Falimiaramanana, D.J., et al.: Modeling of Rosen-type piezoelectric transformer by means of a polynomial approach. In: ICMCS (2018)
17. Lefebvre, J.E., Yu, J.G., Ratolojanahary, F.E., Elmaimouni, L., Xu, W.J., Gryba, T.: Mapped orthogonal functions method applied to acoustic waves-based devices. AIP Adv. **16**, 065307 (2016)
18. Benwell, A.L.: A high voltage piezoelectric transformer for active interrogation. Ph.D. dissertation, University of Missouri-Columbia (2009)
19. Bertram, A., Glüge, R.: Solid Mechanics: Theory, Modeling, and Problems. First German edn. (2013)
20. Marzieh, E., Steenstrup, A.R., Zhang, Z., Andersen, M.A.E.: Power enhancement of piezoelectric transformers for power supplies. In: Proceedings of the 8th IET International Conference on Power Electronics, Machines and Drives (2016)

Civil Engineering and Structures for Sustainable Constructions

Numerical Simulation of Fatigue Crack Propagation of S355 Steel Under Mode I Loading

Jielin Liu[1], Haohui Xin[1](✉), and Ayman Mosallam[2]

[1] Department of Civil Engineering, School of Human Settlements and Civil Engineering, Xi'an Jiaotong University, Xi'an, Shaanxi, People's Republic of China
xjtljl@stu.xjtu.edu.cn, xinhaohui@xjtu.edu.cn

[2] Department of Civil and Environment Engineering, University of California, Irvine, CA, USA
mosallam@uci.edu

Abstract. The in-depth study on the performance of steel structure in the fatigue crack propagation stage is conducive to the rational use of high-strength steel. The load ratio reflecting the average stress effect has a great influence on the fatigue crack growth rate. Therefore, in this paper, Walker equation is used to fit the fatigue crack growth rate. Using Bayesian theory and Weibull distribution, the walker equation coefficients of S355 grade steel at 95% guarantee rate are obtained. Based on Walker equation, two-dimensional finite element models with different element types are used to simulate fatigue crack propagation, and the predicted results of different models are compared with the experimental data. Using the subroutine based on Walker equation, and through the extended finite element method and virtual crack closure method, the fatigue crack growth of S355 grade steel is successfully simulated.

Keywords: Walker equation · Bayesian theory · Weibull distribution model · Extended finite element method · Virtual crack closure technique

1 Introduction

Fatigue phenomenon refers to the cumulative damage in the structure under repeated load. The fatigue process can be divided into two stages: initiation and propagation [1–4]. In 1961, Paris [5] obtained the relationship between stress intensity factor and fatigue crack growth rate through fatigue test, which is called Paris' law. Nowadays, Paris law and its extension [6–10] are widely used in the fatigue performance evaluation of engineering structures. Generally, the crack growth curve is divided into three stages. When the stress intensity factor does not reach a threshold ΔK_{th}, the crack will not propagate in the stage I. The crack growth curve increased linearly in the log-log space in the stage II. In the stage III, the crack growth rate increases sharply.

The fatigue crack propagation not only related to the stress amplitude, but also affected by the stress ratio. However, the original Paris' law did not consider mean stress effects. Walker [11] studied the effect of load ratio on fatigue crack growth rate of

R. Saidi et al. (Eds.): ICATH 2021, LNDECT 110, pp. 333–345, 2022.
https://doi.org/10.1007/978-3-030-94188-8_31

aluminum alloy. The conclusion shows that the fatigue crack growth rate increases with the increase of load ratio. In order to consider the influence of load ratio, an empirical model containing three material parameters is proposed, which is called Walker equation.

Hence, in this paper, Walker equation is used to fit the fatigue crack growth rate. Using Bayesian theory and Weibull distribution, the walker equation coefficients of test results at 95% guarantee rate are obtained. Based on Walker equation, two-dimensional finite element models of different element types are used to simulate fatigue crack propagation. The predicted results of different models are compared with the experimental data. Using the subroutine based on Walker equation, and through the extended finite element method and virtual crack closure method, the fatigue crack growth of S355 grade steel is successfully simulated.

2 Parameters Determination of Fatigue Crack Propagation

2.1 Details of S355 Materials

To evaluate the fatigue properties of S355 steel, it is very important to obtain its material properties. Correia et al. [12] reported the material properties of S355 grade steel. The material properties are listed in Table 1 and Table 2.

Table 1. Material properties of steel S355

Young modulus (E) GPa	Yield strength (f_y) MPa	Tensile strengh (f_u) MPa	Hardness HV10
211.6	367	579	151.28

Table 2. Chemical composition of S355 grade steel

C %	Cu %	Mn %	N %	P %	S %	Si %
0.16	0.2	1.28	0.009	0.03	0.02	0.3

2.2 Experiment

The main purpose of fatigue crack growth rate test is to determine the crack growth rate da/dN as a function of equivalent stress intensity factor ΔK and load ratio R. These tests were performed according to ASTM E647 [13]. In order to study the fatigue properties of S355 steel under uniaxial tension, three groups of fatigue crack growth rate tests (FCGR) were carried out. During loading, the load amplitude of each group is the same, but the load ratio is different, which are 0.1, 0.5 and 0.7 respectively. Compact tension specimens (CT) were used to obtain the fatigue crack propagation rate. The geometry was defined according to ASTM E647 [13] and is shown in Fig. 1. The dimension is listed Table 3. Figure 2 show the two specimens before being tested. Before starting the proper FCGR tests, the specimens had to be pre-cracked in fatigue conditions.

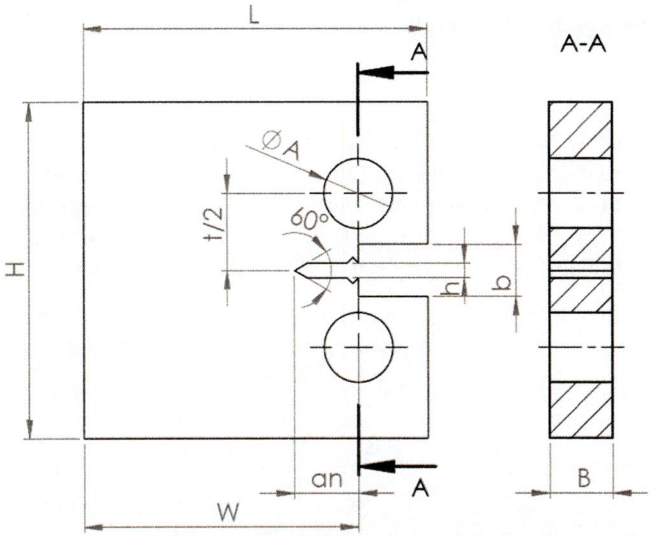

Fig. 1. Geometry of compact tension specimen

Table 3. Dimension of compact tension specimen (mm)

Specimen	R	W	L	H	A	B	b	an	h	t/2
1	0.5	50	62.5	60	12.5	11.3	9.4	11.7	2.75	13.75
2	0.7	49.84	62.3	59.8	12.5	10.27	9.4	11.8	2.94	13.71

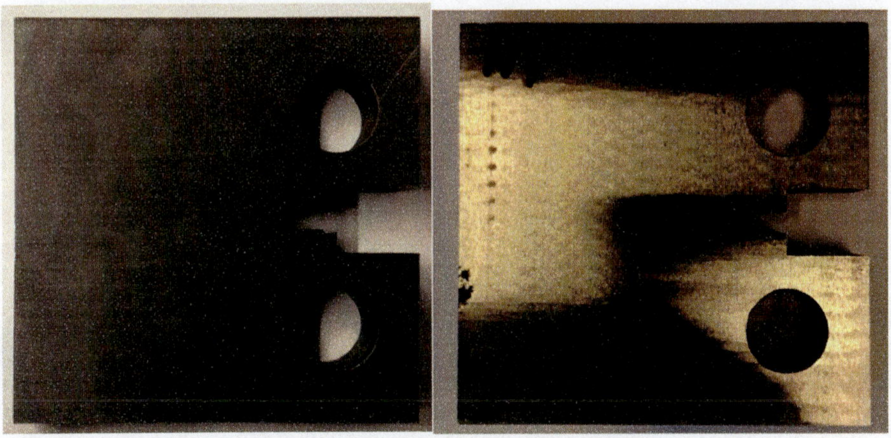

Fig. 2. Compact tensile specimen [14]

2.3 Processing of Experimental Data

Equation Fitting

As shown in Eqs. (1) and (2), the Walker equation illustrates the effect of load ratio on fatigue crack growth rate. As shown in Fig. 3, in order to obtain the material coefficient of Walker equation, after converting the test data to the double logarithmic space, the test data are fitted by using the software MATLAB [15]. Table 4 lists the walker equation fitting coefficients for S355 grade steel and will be used to calculate fatigue crack propagation.

$$\frac{da}{dN} = C_0(\overline{\Delta K})^m \tag{1}$$

$$\overline{\Delta K} = \frac{\Delta K}{(1-R)^{1-\gamma}} \tag{2}$$

where: C_0, m, γ are material coefficient of Walker equation.

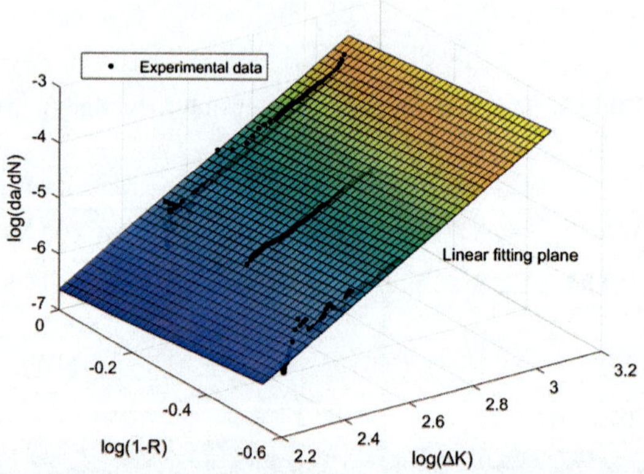

Fig. 3. Comparisons between fitted results and experimental data.

Probability Analysis Based on Experimental Data

Due to the effects of material micro defects, the scatter of fatigue crack propagation rate is relatively larger than general static test results. Therefore, material coefficients of Walker equations considering certain guarantee rate is calculated based on stochastic analysis. Three material parameters in the Walker equation increased the difficulty to predict the probabilistic fatigue life. Hence, a failure probability evaluation method using Markov chain Monte Carlo (MCMC) and Bayesian inference is used to obtain the probabilistic fatigue life with three material parameters. Openbugs [16] is a program for Bayes analysis of complex statistical models using Markov chain Monte Carlo (MCMC) method. It sets all unknown parameters as random variables, and then estimates the posterior distribution of parameters through Gibbs sampling in MCMC method. Hence, we

Table 4. Coefficients of Walker Equation.

Parameters	Statistical values	S355
log (C_0)	Average	−14.34
γ	Average	0.789
m	Average	3.503
R^2		0.9515
RMSE		0.1269
SSE		6.118

obtained the material parameters based on Openbugs software. Table 5 listed material parameters of Walker equation with 95% guarantee rate. Figure 4 compared the fitted probabilistic fatigue crack growth rate estimation with experimental results. The experimental data points are between the upper and lower limits of the guarantee rate under three fixed load ratios R = 0.1, 0.5 and 0.7.

Table 5. Coefficient with 95% guarantee rate

95% guarantee rate	log (C_0)	γ	m
Upper limit	−13.23	3.202	0.7852
Lower limit	−13.42	3.086	0.8398

However, the method of using the software openbugs based on Bayesian theory to calculate the guarantee rate is not validated, and the obtained guarantee rate parameters need to be checked with the existing validated specifications. Therefore, after fixing the load ratio to 0.1, Weibull distribution is used to check the Bayesian method.

Weibull distribution is widely used in data processing of various life tests. Equations (3) gives a three parameter model of failure cycle times and stress amplitude. After replacing the number of failure cycles and stress amplitude in Eqs. (3) with crack growth rate and stress intensity factor, Eqs. (4) is obtained.

$$P_f(N, \Delta\sigma) = 1 - exp\left\{-\left(\frac{(logN - B)(log\Delta\sigma - C) - \lambda}{\delta}\right)^{\beta}\right\}$$
$$(logN - B)(log\Delta\sigma - C) > \lambda \tag{3}$$

$$P_f(dN/da, \Delta K) = 1 - exp\left\{-\left(\frac{(logdN/da - B)(log\Delta K - C) - \lambda}{\delta}\right)^{\beta}\right\}$$
$$(logdN/da - B)(log\Delta K - C) > \lambda \tag{4}$$

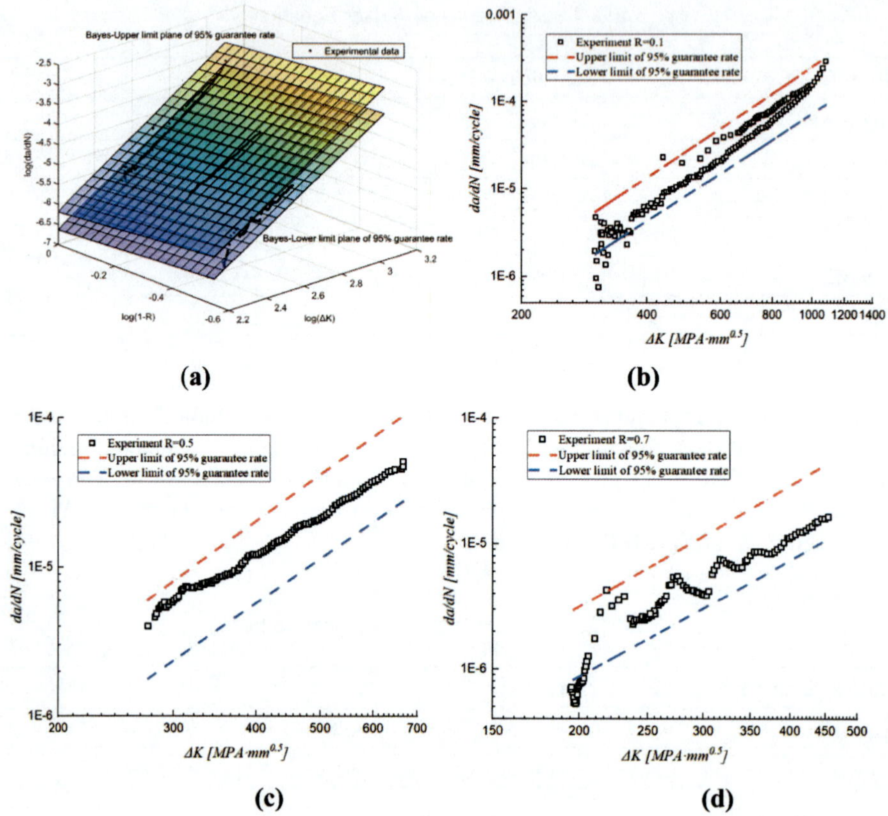

Fig. 4. Comparison between assurance surface and experimental data.

Where: P_f is failure probability for a specimen, N is Number of cycles to failure, $\Delta\sigma$ is stress level, B, C and λ is threshold parameter, δ is scale parameter, β is shape parameter of weibull distribution.

Profatigue [17] is a software that can use Weibull distribution to predict the fatigue failure probability of specimens. By importing the original test data, the program will automatically obtain the values of various parameters in the cumulative distribution function. Table 6 shows the parameters of Weibull distribution obtained by the Profatigue. Figure 5 shows the experimental points and the assurance rate lines obtained by Weibull distribution. Figure 6(a) and Fig. 6(b) shows the difference between the two methods when calculating the upper and lower limits of the guarantee rate. The ordinate in the Fig. 6(b) is the fatigue crack growth rate ratio between Weibull analytical method and Bayesian numerical method. In Fig. 6(b), the red scatter represents the ratio of the results calculated by the two methods when calculating the upper limit of the 95% guarantee interval of the fatigue crack growth rate, and the blue scatter represents the comparison result of the lower limit of the assurance interval. The error of the two methods is less than 5%, and when ΔK is small, the error reaches the peak of 4.7%. Therefore, Bayesian method in predicting fatigue crack probability interval is successfully proved.

Table 6. Weibull distribution parameters

R	C	λ	δ	β
0.1	3.2	28.99	4.61	5.92

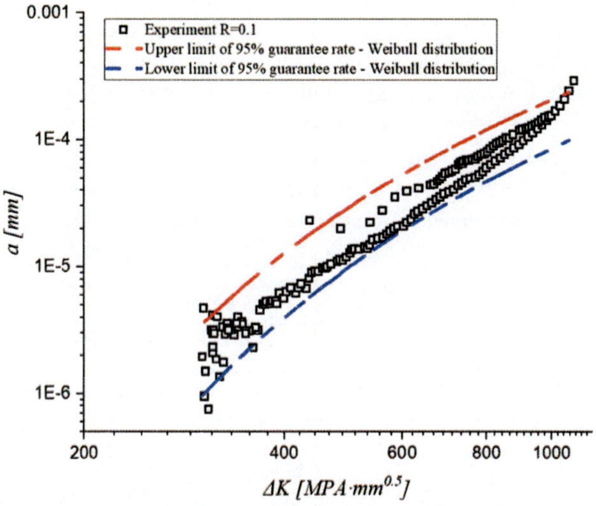

Fig. 5. Comparison between the upper and lower limits of guarantee rate obtained by Weibull distribution and experimental points

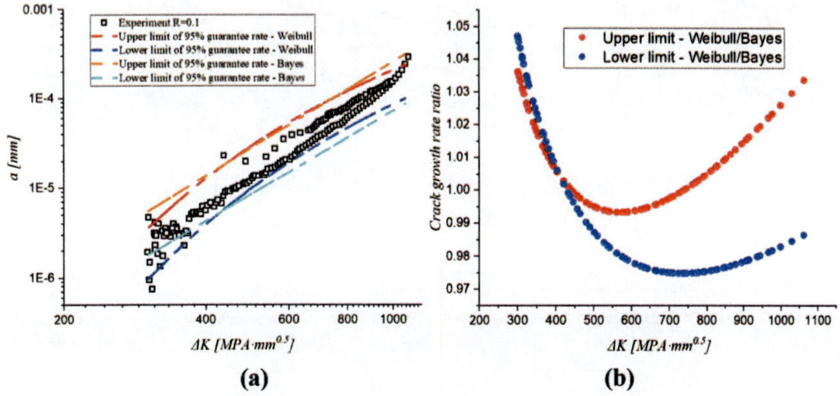

Fig. 6. Computational difference between Weibull distribution and Bayesian theory

3 Numerical Simulation of Pure Mode I Fatigue Crack Propagation

Based on experiments, two groups of numerical simulations have been done. In the first group, the fatigue crack growth of specimens under three different load ratios (R = 0.1, 0.5 and 0.7) is simulated by two-dimensional finite element model, and the element type is assumed to be plane strain. In the second group, except that the element type is assumed to be plane stress, the rest are the same as the numerical simulation of the first group.

XFEM [18] treated the cracks as a enriched function in conjunction with additional degrees of freedom. When the element intersects with the crack, the nodes are enriched by jump function.As shown in Fig. 7, in order to represent the discontinuity of the cracked element, the phantom nodes coincides with the real nodes, and the element is divided into two parts by the level set method (LSM) [19].

In this paper, the virtual crack closure method (VCCT) [20] is used to calculate the strain energy release rate. As shown in Fig. 8, after assuming the material is a linear elastomer, the strain energy release rate of pure mode I [21] fatigue crack can be calculated by Eq. (5).

$$G_I = \frac{1}{2}\frac{v_{1,6}F_{v,2,5}}{bd} \tag{5}$$

where: G_I is the energy release rate, b is the width, d is the length of the elements, $F_{v,2,5}$ is the vertical force between nodes 2 and 5, $v_{1,6}$ is the vertical displacement between nodes 1 and 6.

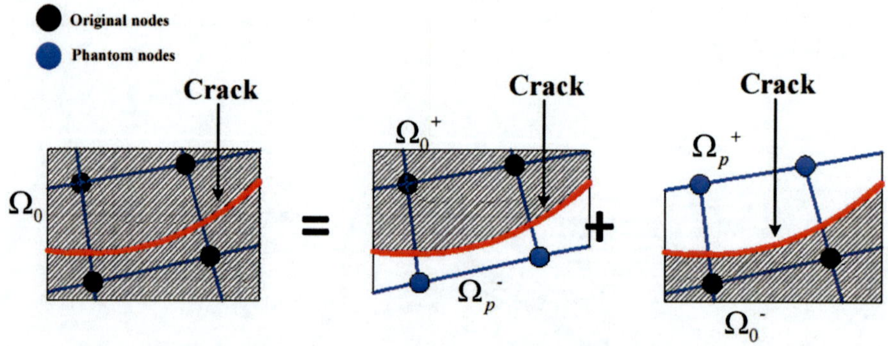

Fig. 7. The schematic of the phantom node method [2, 3]

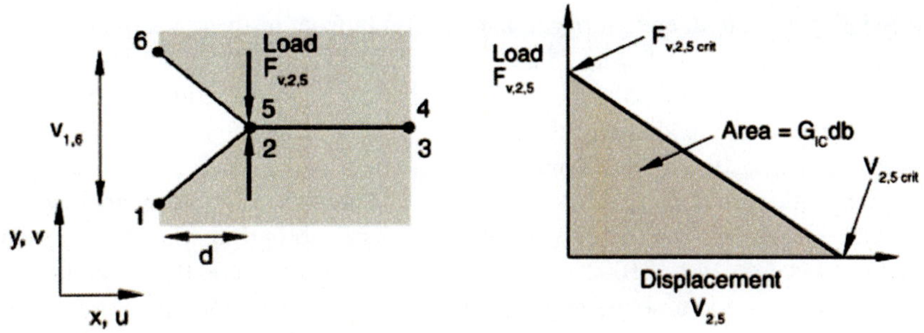

Fig. 8. Schematic of the VCCT method [20]

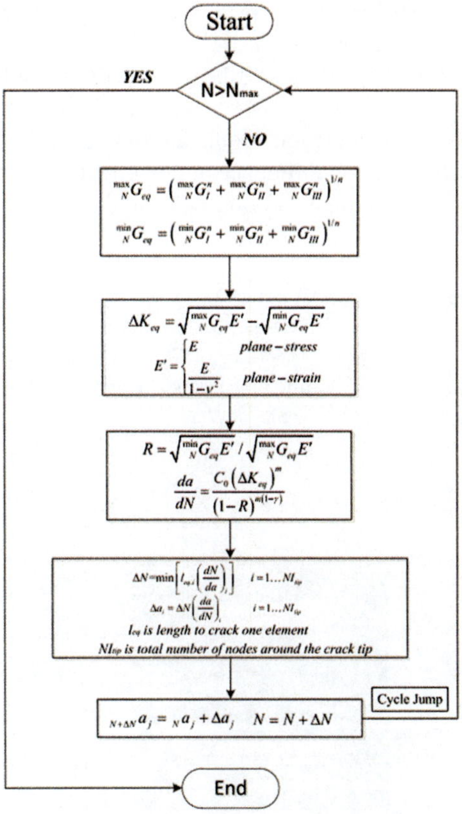

Fig. 9. Implementation flow chart of fatigue crack propagation considering R ratio effects [22]

In order to consider the influence of load ratio on fatigue crack growth rate, Walker equation is implemented by calling subroutine UMIXMODELFATIGUE by software ABAQUS [21]. The calculation method and flow of stress intensity factor are shown in Fig. 9. When the number of loading cycles increases ΔN, the crack length increases Δa forward, and at least one enriched element is broken in this process. Due to the known the enriched element length and propagation direction in front of the crack tip, the number of cycles required for the failure of an enriched element can be calculated by Walker equation. When the enriched element breaks, the crack surface constraint is zero, the stiffness is zero, and the load is redistributed.

In this paper, a two-dimensional finite element model is used to simulate fatigue crack propagation. Among them, XFEM model is established by commercial software ABAQUS, and Walker equation is realized by calling subroutine. The elastic modulus of the model is defined as 210000 MPa, the Poisson's ratio of 0.3, the cyclic load with a period of 1 s, and the time increment step is 0.01 s. The element type of the model is CPE4.

As shown in Fig. 10, the two-dimensional model is fixed through two semicircular surfaces, the contact between them is set as hard contact, and a reference point is set at the lower end of the semicircular surface. The reference point is coupled with the lower end of the semicircular surface. The lower reference point is fixed. The degrees of freedom of the horizontal direction and the rotation direction of the upper reference point are constrained.

After the finite element simulation results are obtained, the fatigue crack growth rate is determined by dividing the difference of crack length by the difference of cycle number. The SIF ranges were computed using the formulation proposed by ASTM E647, as shown in Eq. (6).

$$\Delta K=\frac{\Delta F(2+\frac{a}{W})(0.886+4.46\frac{a}{W}-13.32(\frac{a}{W})^2+14.72(\frac{a}{W})^3-5.6(\frac{a}{W})^4)}{B\sqrt{W}(1-\frac{a}{W})^{1.5}} \tag{6}$$

Where: ΔF is the applied load range, a is crack propagation length, W is the effective width, B is the thickness.

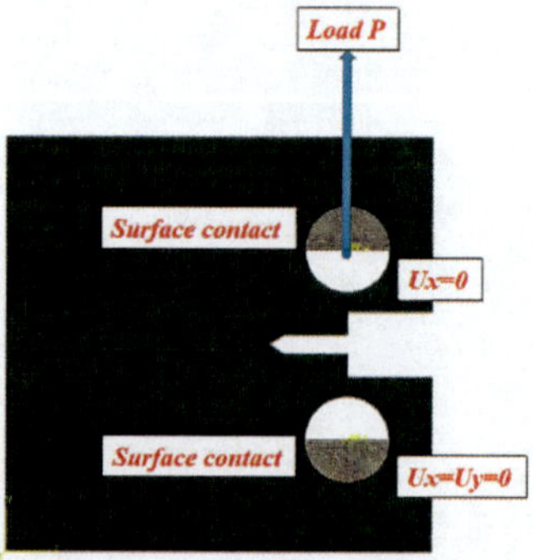

Fig. 10. Boundary conditions of two-dimensional model and three-dimensional model

Figure 11 and Fig. 12 shows the comparison between the finite element calculation results and the experimental data under the assumption that the element type is plane strain and plane stress respectively. Figure 13 compares the simulation results of element type of finite element model using plane stress and plane strain respectively. It is worth mentioning that when using the finite element method to simulate pure mode I crack growth, the two-dimensional model assumed to be plane strain can well show the relationship between crack growth rate and stress intensity factor and can be consistent with the experimental data.

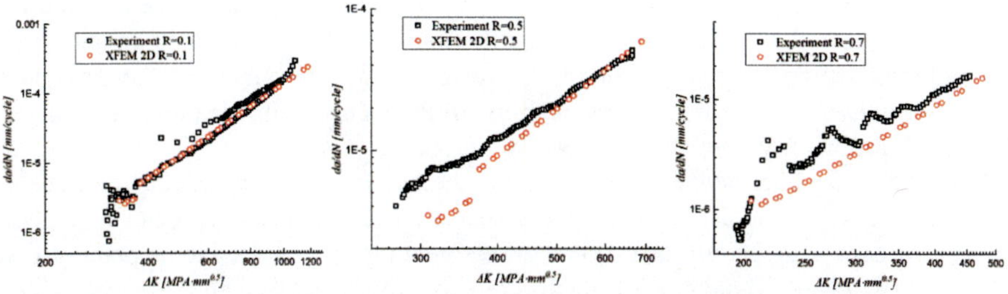

(a) (c) (b)

Fig. 11. Simulation results using plane strain element

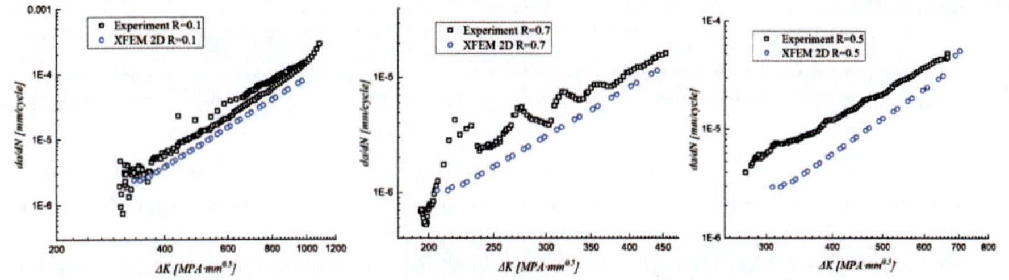

(a) (c) (b)

Fig. 12. Simulation results using plane stress element

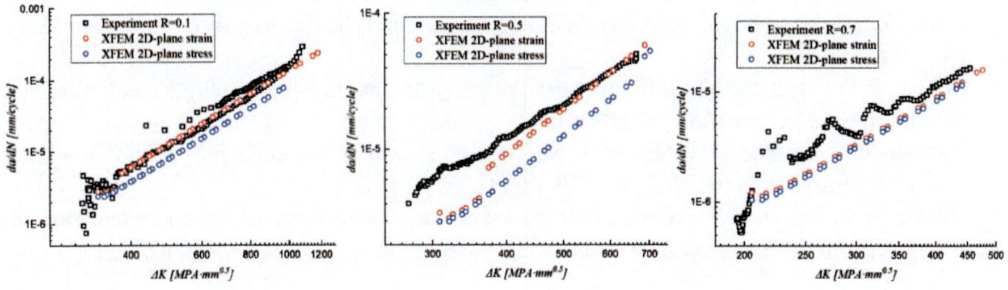

(a) (b) (c)

Fig. 13. Difference of simulation results between two element types

4 Conclusions

In this paper, the coefficient of walker equation with 95% assurance rate is obtained by Bayesian method, and verified by Weibull distribution. The pure mode I fatigue crack propagation is simulated by using the user-defined subroutine. Finally, this paper draws the following conclusions:

(1) Based on Walker equation, the three-dimensional 95% assurance interval of pure mode I fatigue crack growth rate is predicted by Bayesian method. The load ratio is fixed as $R = 0.1$, and then Weibull distribution is used to predict the guarantee rate interval. Compared with the prediction results of Bayesian method, the maximum error is 4.7%. Therefore, the practicability of Bayesian method in predicting fatigue crack probability interval is successfully proved.
(2) Based on the user-defined subroutine and the two-dimensional extended finite element model, when the element type is assumed to be plane strain, it can be in good agreement with the test data. When the element type is assumed to be plane stress, the calculated fatigue crack growth rate is relatively lower than the experimental observations.

References

1. De Jesus, A.M.P., Matos, R., Fontoura, B.F.C., Rebelo, C., Da Silva, L., Veljkovic, M.A.: Comparison of the fatigue behavior between S355 and S690 steel grades. J Constr Steel Res **79**, 140–150 (2012). https://doi.org/10.1016/j.jcsr.2012.07.021
2. Xin, H., Veljkovic, M.: Fatigue crack initiation prediction using phantom nodes-based extended finite element method for S355 and S690 steel grades. Eng. Fract. Mech. **214**, 164–176 (2019)
3. Xin, H., Veljković, M.: Effects of residual stresses on fatigue crack initiation of buttwelded plates made of high strength steel. In: Proceedings of the Seventh International Conference on Structural Engineering, Mechanics and Computation (SEMC 2019), Cape Town, South Africa (n.d.)
4. Xin, H., Veljkovic, M.: Residual stress effects on fatigue crack growth rate of mild steel S355 exposed to air and seawater environments. Mater. Des. 108732 (2020)
5. Paris, P.C., Gomez, M.P., Anderson, W.E.: A rational analytical theory of fatigue. Trend Eng. **13** (1961)
6. Ritchie, R.O.: Near-threshold fatigue crack propagation in ultra-high strength steel: influence of load ratio and cyclic strength (1976)
7. Forman, R.G., Kearney, V.E., Engle, R.M.: Numerical analysis of crack propagation in cyclic-loaded structures. J. Basic Eng. **89**, 459–463 (1967)
8. Suresh, S., Zamiski, G.F., Ritchie, D.R.O.: Oxide-induced crack closure: an explanation for near-threshold corrosion fatigue crack growth behavior. Metall. Mater. Trans. A **12**, 1435–1443 (1981)
9. Elber, W.: The significance of fatigue crack closure. In: Damage Tolerance in Aircraft Structures. ASTM International (1971)
10. Ritchie, R.O.: Mechanisms of fatigue-crack propagation in ductile and brittle solids. Int. J. Fract. **100**, 55–83 (1999)
11. Walker, K.: The effect of stress ratio during crack propagation and fatigue for 2024-T3 and 7075-T6 aluminum. In: Effects of Environment and Complex Load History on Fatigue Life. ASTM International (1970)
12. Correia, J.A., de Jesus, A.M., Fernández-Canteli, A., Calçada, R.A.: Modelling probabilistic fatigue crack propagation rates for a mild structural steel. Fratturaed Integrita Strutturale **31**, 80–96 (2015)

13. ASTM E647, 2015 Edition, May 1, 2015 - Standard Test Method for Measurement of Fatigue Crack Growth Rates
14. Dantas, R.G.R.: Fatigue life estimation of steel half-pipes bolted connections for onshore wind towers applications (2019)
15. Rajasekaran, S.: Related titles (2015). https://doi.org/10.1016/B978-1-78242-253-2.09001-0
16. Carroll, R., Lawson, A.B., Faes, C., Kirby, R.S., Aregay, M., Watjou, K.: Comparing INLA and OpenBUGS for hierarchical Poisson modeling in disease mapping. Spat. Spatio-temporal Epidemiol. **14–15**, 45–54 (2015). https://doi.org/10.1016/j.sste.2015.08.001. ISSN 1877-5845
17. Seitl, S., et al.: Evaluation of fatigue properties of S355 J0 steel using ProFatigue and ProPagation software. Procedia Struct. Integrity **13**, 1494–1501 (2018). https://doi.org/10.1016/j.prostr.2018.12.307. ISSN 2452-3216
18. Moës, N., Dolbow, J., Belytschko, T.: A finite element method for crack growth without remeshing. Int. J. Numer. Methods Eng. **46**, 131–150 (1999)
19. Osher, S., Sethian, J.A.: Fronts propagating with curvature-dependent speed: algorithms based on Hamilton-Jacobi formulations. J. Comput. Phys. **79**, 12–49 (1988)
20. Rybicki, E.F., Kanninen, M.F.: A finite element calculation of stress intensity factors by a modified crack closure integral. Eng. Fract. Mech. **9**, 931–938 (1977)
21. Abaqus, V.: Documentation. Dassault Syst Simulia Corp 2019, 651 (2019)
22. Xin, H., Correia, J.A., Veljkovic, M.: Three-dimensional fatigue crack propagation simulation using extended fifinite element methods for steel grades s355 and s690 considering mean stress effects. Eng. Struct. (2020). https://doi.org/10.1016/j.engstruct.2020.111414vol.Inpress

The Screening Efficiency of Open and Infill Trenches: A Review

Hinde Laghfiri(✉) and Nouzha Lamdouar

GC Laboratory, Mohammadia School of Engineering, Mohammed V University in Rabat, Rabat, Morocco

hinde_laghfiri@um5.ac.ma, lamdouar@emi.ac.ma

Abstract. Open and infill trenches are used as a countermeasure to reduce the ground vibration due to heavy traffic (e.g. road, railway, subway, etc.), machines and pile driving, etc. in order to protect structures alongside the railway tracks and equipment that can be damaged or affected by these vibrations, and to reduce the noise that can impact nearby inhabitants. Furthermore, the trenches are the most widely used vibration attenuation methods due to their lower cost as compared to other methods. Thenceforward, this paper presents an overview of the screening efficiency of open trench and infill wave barriers and the influence of particular parameters that affect the performance of the trench such us depth, width, the distance between the trenches and the source of vibration using different methods of studying the induced vibration (experimental study, numerical study, centrifuge study). This study shall help the stakeholders to choose the most suitable geometry and infill materials of the trenches.

Keywords: Wave barriers · Open trenches · Infill trenches · Screening efficiency · Ground vibration

1 Introduction

Constructing an augmenting number of railway infrastructures of all types has been apparent in the last few years. Satisfying the users' needs, the rail network has got densely packed; this can eventually lead to annoying people, destroying or spoiling buildings and affecting technical processes and sensitive equipment. Hence emerged the importance of using wave barriers as mitigation measures for train-induced vibrations. In this context, studying the effectiveness of the trenches as a countermeasure of vibration has been the focus of many researchers. This latter can be considered as an active technique when it is located close to or inside the track, or as a passive method when it is placed in locations close to vibration-sensitive sites rather than close to the track. The current research presents a review of some of these studies with different modeling methods, namely experimental study, numerical approaches and centrifuge modeling to give a general idea on the efficiency of the trenches as well as the influence of some parameters on the performance of vibration isolation. Table 1 presents a list of references that have been reviewed in the present study.

R. Saidi et al. (Eds.): ICATH 2021, LNDECT 110, pp. 346–351, 2022.
https://doi.org/10.1007/978-3-030-94188-8_32

Table 1. Major reviewed references

References	Attenuation measures	Modeling method
[1]	Geofoam trenches	Centrifuge modeling
[2]	Open trenches	Semi-analytical model
[3]	Open and infill trenches	3D FEM (Finite Element Method)
[4]	Open and infill trenches	Full-scale experimental study
[5]	Open and infill trenches	2.5D dynamic substructuring finite element model
[6]	Open and infill trenches	Experimental study

2 Modeling Methods

The purpose of this study is to explore the effectiveness of trenches as vibration attenuation measures by analyzing different sources which have implemented different methods. Murillo et al. [1] conducted a centrifuge parametric study, however, the barrier length was not considered in order to reduce the geometry of the problem to a 2D problem. Cao et al. [2] proposed an analytical study in which the ground is modeled like a saturated poroelastic half-space, considering the coupling between water and the soil skeleton. Ekanayake et al. [3] developed a 3D FEM model using the ABAQUS/Explicit commercial program to reveal the effectiveness of the open trench barrier, water filled trench and geofoam filled trench for scattering ground vibration. The model was first validated trough results obtained from full scale field experiments using EPS (Expanded Poly Styrene) geofoam as a fill material. Ulgen and Toygar [4] investigated the screening efficiency of open, filled water and filled geofoam trenches, where a vibration plate compactor is being used to generate dynamic loading. to simulate the traffic and the construction vibrations whose frequencies range is 10–100 Hz. Zou et al. [5] studied the efficiency of the open and infill (water and geofoam) trenches in isolating vibration from over-track structures in a metro depot using the concept of dynamic substructuring to develop a numerical model that consists of two sub-structures. The first substructure is a 2D coupled system composed of two component train and track with dynamic model in the time domain considering the vertical dynamic responses solely. The second sub-structure is a 3D model track-soil-building developed using ABAQUS program. Wenbo et al. [6] used a series of model tests to examine the efficiency of open and infill (expanded Polystyrene and Duxseal) trenches in eliminating the ground-borne vibration from tunnels, in which a shaker applied sinusoidal sweep loads and train loads to the model tunnel invert.

3 Screening Efficiency

Based on the results of the aforementioned studies, both open and infill trenches perform well by isolating the ground vibration. In fact, Ekanayake et al. [3] confirmed that for both frequencies 40 Hz and 50 Hz, open trench overpasses the isolation of geofoam and water infill barriers, and that EPS geofoam trench surpasses the water infill trench. Similar results were reported in Ref. [4], and it was shown that the difference in the performance of reducing vibration among the three types of trenches decreases with respect to the distance from the trench. According to Cao et al. [2], for the case where loading speed approaches or exceeds the critical speed, the trench isolation efficiency was considerably underrated by the single-phase elastic soil model. As maintained in Ref. [5], the open trench outperforms the infill trenches in terms of attenuating the vertical floor vibration responses more than the horizontal responses within the over-track buildings. Furthermore, in the frequency range (16 Hz–40 Hz) which corresponds to the vibrations induced by the train, the geofoam barrier outperforms the lightweight aggregate concrete (LAC) infill trench, and the reverse is true for the higher frequency range of 50–80 Hz. Wenbo et al. [6] affirmed that the open trench and the Duxseal trench barrier perform better than the EPS geofoam filled barrier. Otherwise, low impedance ratio results in better isolation performance.

The Impedance ratio is defined as follows [7]:

$$IR = \sqrt{\rho_b G_b / \rho_s G_s} \quad (1)$$

It should be noted that ρs and Gs are the density and the shear modulus of the soil, respectively; and ρb and Gb are the density and the shear modulus of the fill material, respectively. Table 2 depicts the properties of the filling materials of the wave barriers used in the aforementioned studies.

Table 2. The properties of fill materials studied in the references reviewed

References	Fill material	Density	IR
[1]	EPS	15 kg/m^3	0.023
[3]	EPS	20 kg/m^3	<0.015
[4]	EPS	16 kg/m^3	0.01
[5]	Geofoam	80 kg/m^3	0.02
	LAC	1800 kg/m^3	9
[6]	EPS	80 kg/m3	0.119
	Duxseal	860 kg/m^3	0.049
	Air	1.29 kg/m^3	0

4 Parametric Study

In this section we will discuss the influence of certain parameters on the effectiveness of open and fill trenches based on the already mentioned studies. The influence of geometric parameters (see in Fig. 1) is analyzed with respect to their Rayleigh wavelength. Because, the Rayleigh waves transmit the plurality of displacement energy to the ground.

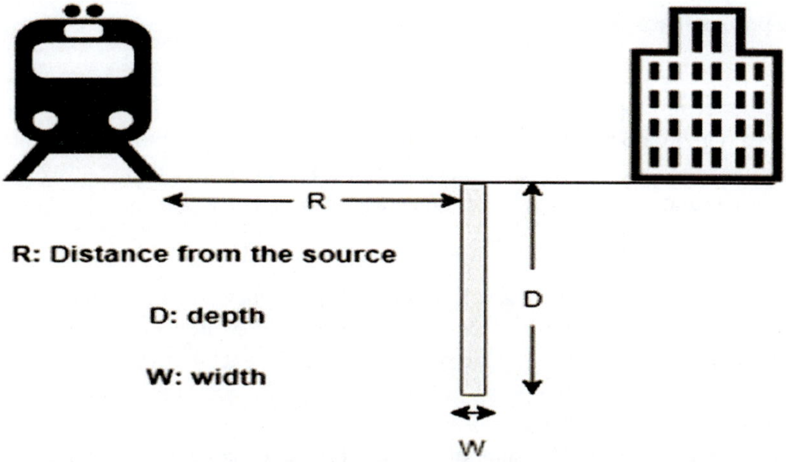

Fig. 1. Geometric parameters of trenches.

4.1 Open trench

The Influence of Depth

For the train-induced vibrations' main frequencies, the isolation vibration is performed for all floors once the normalized depth is almost equal to the Rayleigh wavelength of the soil layer [5]. Moreover, according to the results of the study carried out in Ref. [6], the increase in normalized depth leads to an increase in the reduction of the vibration whether in the frequency-domain, or in the time-domain, especially when the trench is deeper than the tunnel invert.

The Influence of Width

Generally, the effect of width is negligible compared to the effect of depth on vibration attenuation. Indeed, [5, 6] have affirmed that the width of open trench has a negligible effect on the vibration attenuation. Therefore, based on an economic criterion it is better to adopt thin open trenches.

The Influence of Trench Location

Cao et al. [2] proved that increasing the distance between trench and moving load involves an increase on the screening effectiveness. It is noteworthy that the same results are obtained in [5], especially in the 2nd, 3rd and 4th floor in the range frequency 16–40 Hz. Moreover, Wenbo et al. [6] stated that the location impacts the screening efficiency especially in the high frequency range from 63 Hz to 250 Hz. The isolation performance increases by increasing the tunnel-trench distance, while the effect of location is less than that of the depth.

The Influence of Other Parameters

For the high load speed in a saturated soil medium the permeability coefficient has an obvious effect on the trench isolation performance, and increasing the permeability coefficient behind the trench leads to a decrease in the reduction velocity level [2].

4.2 EPS Geofoam Barrier

The Influence of Depth
Deeper EPS geofoam trenches are more effective in isolating ground vibration [1, 3, 4]. However, for the case of low normalized depths ($D < 1$) and the trench away from the source of vibration, the reduction ratio increases significantly [1].

The Influence of Width
The width of the EPS geofoam trench does not affect deeper trenches. while for shallow barriers the acceleration amplifies for thinner barriers [1]. Moreover, the wave isolation increases slightly with increasing the width of the EPS trench barrier [3].

Influence of Trench Location
The EPS barrier as a passive isolation is more efficient than active isolation. In other words, the EPS barrier away from the vibration source is more efficient in attenuating ground vibrations [3]. However, Murillo et al. [1] stated that there is not any observable effect of the distance between source and barrier on vibration isolation.

5 Conclusions

This article aims to provide an overview of the attenuation efficiency of open trenches and infill trenches and the influence of geometric parameters (depth, width, distance trench-source of vibration) on attenuating the vibration using different research methods (analytic, experimental, etc.). The main results are:

1. Open trenches perform better than infill trenches in attenuating vibration for over-track buildings and the ground vibration in which increasing the depth leads to more performance. However, the width has no influence on reducing vibration and passive isolation is more efficient than active isolation. In accordance with the existence of groundwater and soil stability, it is recommended to use the infill trenches.
2. The Duxseal-filled trench outperforms the EPS barrier. This latter outperforms the water barrier and the lightweight aggregate concrete barrier for the frequency range of 16–40 Hz.
3. Deeper trenches are not affected by the EPS geofoam trench width.

All in all, the findings of these studies are of an effectively leading role in designing the geometries and fill materials of the trench.

Acknowledgments. The authors are thankful to the editor and the reviewers for their valuable remarks.

Conflict of Interest Statement. The authors declare that there is no conflict of interest.

References

1. Murillo, C., Thorel, L., Caicedo, B.: Ground vibration isolation with geofoam barriers: centrifuge modeling. Geotext. Geomembr. **27**, 423–434 (2009). https://doi.org/10.1016/j.geotexmem.2009.03.006
2. Cao, Z., Cai, Y., Boström, A., Zheng, J.: Semi-analytical analysis of the isolation to moving-load induced ground vibrations by trenches on a poroelastic half-space. J. Sound Vib. **331**, 947–961 (2012). https://doi.org/10.1016/j.jsv.2011.09.009
3. Ekanayake, S.D., Liyanapathirana, D.S., Leo, C.J.: Attenuation of ground vibrations using in-filled wave barriers. Soil Dyn. Earthq. Eng. **67**, 290–300 (2014). https://doi.org/10.1016/j.soildyn.2014.10.004
4. Ulgen, D., Toygar, O.: Screening effectiveness of open and in-filled wave barriers: a full-scale experimental study. Constr. Build. Mater. **86**, 12–20 (2015). https://doi.org/10.1016/j.conbuildmat.2015.03.098
5. Zou, C., Wang, Y., Zhang, X., Tao, Z.: Vibration isolation of over-track buildings in a metro depot by using trackside wave barriers. J. Build. Eng. **30**, 101270 (2020). https://doi.org/10.1016/j.jobe.2020.101270
6. Wenbo, Y., et al.: An experimental study of the isolation performance of trenches on ground-borne vibration from underground tunnels. Earthq. Eng. Eng. Vib. **19**(3), 779–790 (2020). https://doi.org/10.1007/s11803-020-0595-5
7. Yang, Y.B., Ge, P., Li, Q., Liang, X., Wu, Y.: 2.5D vibration of railway-side buildings mitigated by open or infilled trenches considering rail irregularity. Soil Dyn. Earthq. Eng. **106**, 204–214 (2018). https://doi.org/10.1016/j.soildyn.2017.12.027

Relationship Between Eight-Fold Star and Other Tiles in Traditional Method 'Tastir'

Zouhair Ouazene[1(✉)], Aziz Khamjane[2], and Rachid Benslimane[1]

[1] Laboratory of Innovative Technologies, Sidi Mohamed Ben Abdellah University, Fez, Morocco
zouhair.ouazene@usmba.ac.ma

[2] Laboratory of Applied Sciences, Abdelmalek Essaadi University, Tetouan, Morocco

Abstract. This article concerns the traditional geometric patterns from Morocco and Andalusia. It shows the relationship between the 8-pointed star and other geometric tiles according to the traditional method called "Tastir". It also presents the symmetry groups characterizing these geometric patterns.

Keywords: Star · Rosette · Islamic art · Tiles · Tastir · Geometric patterns · Symmetry groups · Computer design

1 Introduction

Morocco is one of the western Islamic countries best known for its rich diversity of Islamic geometric patterns. These geometric patterns are generally executed on Zellij, plaster, and wood. *Zellij* is a mosaic made by assembling geometric and colored tiles. It is found in the private or the public as well as in the old or modern places. For example, in the city of Fez, Zellij exists everywhere in the old Medina, ranging from old places like Jamiaa Qarawiyine, Moulay Driss Mauselium to more recent ones like Jnan-Sbil park and public baths (Hamams). In such places, Zellij is used to decorate walls, floors, ceilings, fountains, tables, and sinks. Zellij is found everywhere in Morocco, but the artisans of Fez and Meknes are the leading producers of Zellij.

The eight-pointed star/seal of Solomon is ubiquitous in Zellij. This star is constituted by two squares, which are offset each other by 45°. The square's side constitutes a unit measure that defines the width of all the tesserae in the pattern. This unit of measurement is called Qasama by artisans of Morocco.

We will analyze the traditional method called "Tastir" which only uses a compass and a ruler to draw circles and lines. These two simple elements form the basis of all compositions. It is important to emphasize that the traditional method is based on the symmetry of geometric patterns. The craftsman only designs the fundamental pattern from which the whole pattern is created by applying isometric transformations of its symmetry group. For the traditional method, this basic pattern is usually designed in a triangular-shaped fundamental region.

The remainder of this modest work is arranged in the following way: Sect. 2 deals with related works of the construction methods of Islamic geometric patterns. Section 3

R. Saidi et al. (Eds.): ICATH 2021, LNDECT 110, pp. 352–358, 2022.
https://doi.org/10.1007/978-3-030-94188-8_33

provides principles of the symmetry group theory. Section 4 analyzes a pattern and establishes the relationship between the eight-point star and other tiles. Section 5 gives a database of the tiles extracted from the eight-point star. Finally, we conclude the paper in Sect. 6.

2 Related Works

Islamic art has received significant interest from several scientists and historians over the world. Therefore, many works have been conducted to investigate this art to discover

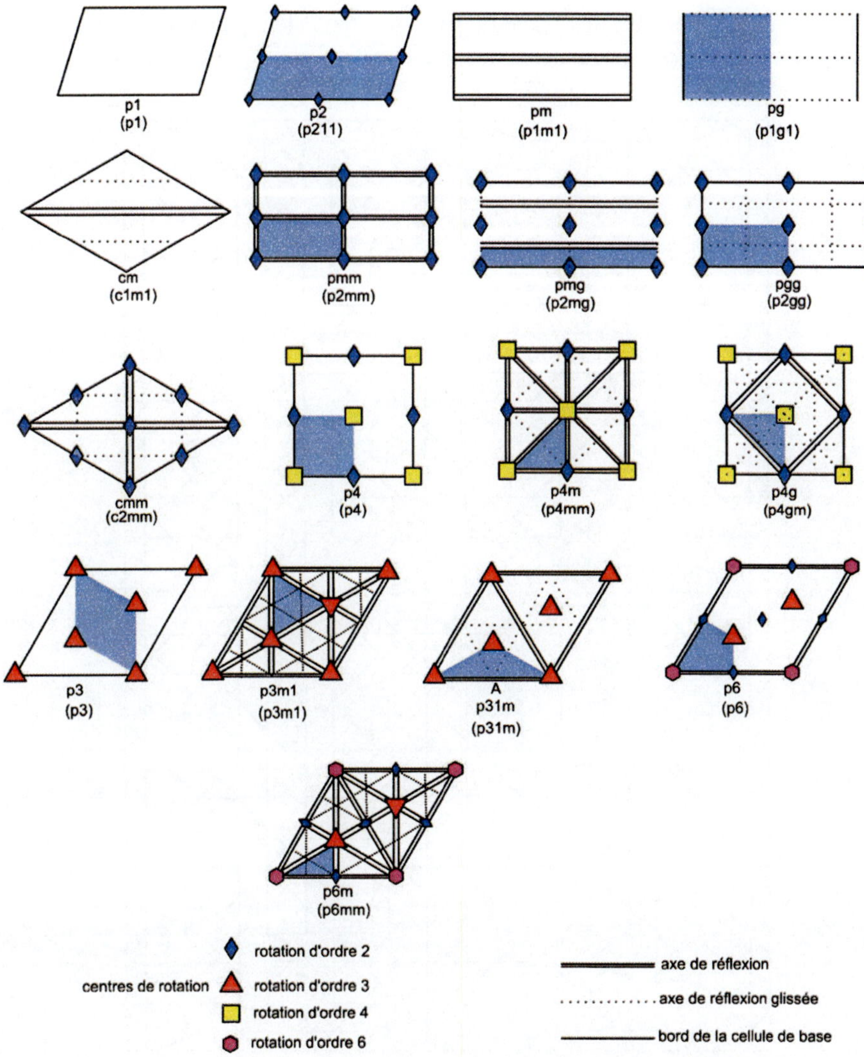

Fig. 1. Fundamental regions and basic cells of the 17 groups

the secret of constructing its patterns. To this end, many researchers have been focusing on a tiling-based technique also known as polygons in contact [1–6]. In the same line of references, Bonner provided a systematic and non-systematic geometric construction method of Islamic geometric patterns [7, 8]. Other works provide mathematical tools for analyzing and constructing Islamic geometric patterns [9–12]. On the other hand, several works have focused on the pattern generation process [13], the symmetry group theory [14], and the quasi periodicity [15–17]. This work aims to present the relationship between the eight-point star and the other derived tiles.

3 Symmetry Groups

The artisans use the notions of symmetry in the decoration to organize the forms according to a certain regularity. *Symmetry* is a mathematical concept that can be applied in

Fig. 2. Example of periodic pattern

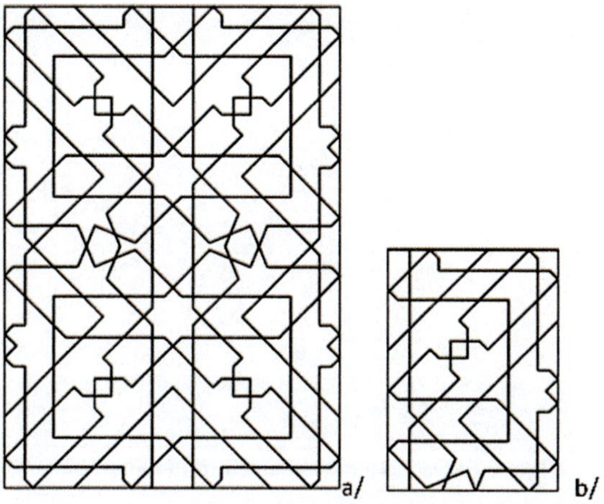

Fig. 3. Basic pattern and its corresponding fundamental motif

various geometric patterns. A geometric pattern is defined by its symmetry group, identified by a set of isometric transformations. There are five types of isometries of the plane: identity, rotations, translations, reflections, and glide reflections.

Figure 1 illustrates the basic cells of the 17 symmetry groups with indications of the isometries and the fundamental region associated with each symmetry group.

4 Definition of Traditional Method 'Tastir' and Relationship Between Tiles

The traditional design geometric patterns method called Tastir constitutes an essential component of Moroccan crafts with strong cultural content. It is also the origin of the world fame of certain historical cities such as Fez, Meknes, Marrakech, and Rabat.

This method applies only to the drawing of the basic repeating pattern, which is called in the Moroccan tradition by Karhet. Moreover, the symmetries of this basic pattern identify the fundamental region, which is the designer creativity area. The periodic pattern, shown in Fig. 2, results from the horizontal and vertical repetition of the basic pattern shown in Fig. 3a. Furthermore, this basic pattern presents two orthogonal axes of symmetry that identify the fundamental motif shown in Fig. 3b.

This fundamental motif is composed of several geometric forms that are generated from the 8-pointed star. Figure 4 shows examples of tesserae extracted from the 8-pointed star.

It can also be seen that the angles in all these tiles are equal to 45 degrees or their multiples or divisions. Therefore, the relative relationship between the tesserae combines length, perimeter, area and angles.

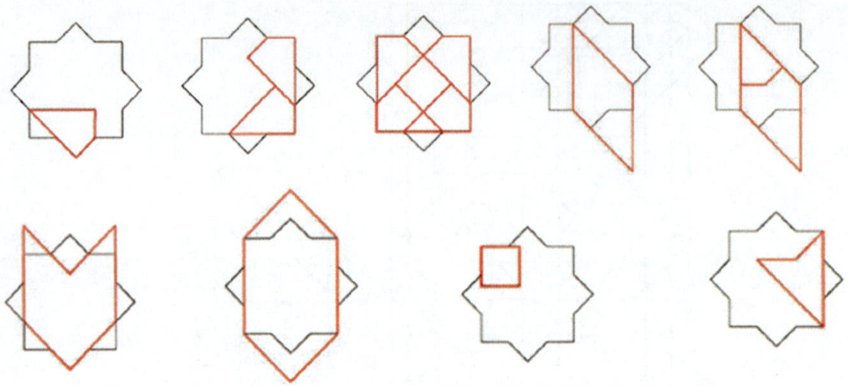

Fig. 4. Examples of tesserae extracted from the 8-pointed star

5 Database of the Tesserae Extracted from the 8-Pointed Star

The Table 1 below presents a database of tesserae extracted from the 8-folded star. Each tessera (elementary shape) is accompanied by its traditional name used by Moroccan artisans.

Table 1. Different tesserae extracted from the 8-folded star

KHATEM SLIMANI	TAJE	SFET BENDRISSIYA	CHOUKA
LKTEF MHLOUL	KHATEM BDLOU	LQFIA	LQFA
MHERRA	MRBAA	LOUZAT LMTEMEN	TALET
QUENDIL BAOULA	QUENDIL TARBIA	LFERMA LMEGRSA	LQUERKBA
LAARJA	NEJIMA	TOUYER	KEF
LFERMA LMGRSA	LQFA	LKTEF	METWA LGRAA
KHTMKHMSYA	SAFET	LAAKDA	SAFET LBAOULA
LKTF MHLOUL	LKEF	KEF MTRKA SRIR	TBAL
TBL	SEFT SGHIR	OKDAT LKOFI	KORA KHMSYA
KHATEM TERA	KHATEM DLOU	KHATEM RKEN	KHATM SDASYA
OKDAT LKOFI	ROUYSE	KORA SDASYA	KORA TMANYA
LMTWA MHDODA	LKEF SGHIR	LQFA	LMTWA

6 Conclusion

This article shows that the traditional "Tastir" method of designing periodic geometric patterns is based on the mathematical approach of symmetry groups. Indeed, the design of a periodic pattern is reduced to its basic pattern, which is itself reconstructed from its fundamental motif and isometric transformations of its symmetry group. We have also shown that the fundamental motif is composed of elementary geometric shapes or tesserae extracted from the 8-pointed star. Finally, we have presented the database of the tesserae, which are accompanied by their traditional names.

References

1. Hankin, E.H.: Examples of methods of drawing geometrical arabesque patterns. Math Gaz **12**, 370 (1925). https://doi.org/10.2307/3604213
2. Hankin, E.H.: The Drawings of Geometric Patterns in Saracenic Art (microfiche) (1925)
3. Hankin, E.H.: Some difficult saracenic designs. II. A pattern containing seven-rayed stars. Math Gaz **18**, 165 (1934). https://doi.org/10.2307/3606813
4. Hankin, E.H.: Some difficult saracenic designs. III. A pattern containing fifteen-rayed stars. Math Gaz **20**, 318 (1936). https://doi.org/10.2307/3607312
5. Kaplan, C.S., Salesin, D.H.: Islamic star patterns in absolute geometry. ACM Trans. Graph **23**, 97–119 (2004). https://doi.org/10.1145/990002.990003
6. Kaplan, C.S.: Islamic star patterns from polygons in contact. In: Proceedings - Graphics Interface, pp. 177–185 (2005)
7. Bonner, J.: The historical use of polygonal systems to create Islamic geometric patterns. In: Fondation de la mosquée Hassan ɪɪ de casablanca (ed.) Les Tracés De L'Arabesque Géométrique, Casablanca, pp. 85–94 (2014)
8. Bonner, J.: 3 polygonal design methodology. In: Bonner, J. (ed.) Islamic Geometric Patterns, pp. 221–548. Springer, New York (2017). https://doi.org/10.1007/978-1-4419-0217-7_3
9. Bodner, B.L.: Bridges 2006: mathematical connections in art, music, and science. In: Nexus Network Journal, pp. 145–150 (2007)
10. Bodner, B.L.: Bridges 2006: mathematical connections in art, music, and science. In: Nexus Network Journal, Birkhäuser Basel, Basel, pp. 145–149 (2008)
11. Bodner, B.L.: The planar space groups of Mamluk patterns. In: Proceedings Bridge 2014 Mathmatics Music Art, Architecture Culture, pp. 35–42 (2008)
12. Cromwell, P.R.: Islamic geometric designs from the Topkap Scroll I: unusual arrangements of stars. J. Math. Arts **4**, 73–85 (2010). https://doi.org/10.1080/17513470903311669
13. Jowers, I., Prats, M., Eissa, H., Lee, J.H.: A study of emergence in the generation of Islamic geometric patterns. In: New Frontiers - Proceedings of the 15th International Conference on Computer-Aided Architectural Design in Asia, CAADRIA 2010, pp. 39–48 (2010)
14. Khamjane, A., Benslimane, R.: A computerized method for generating Islamic star patterns. CAD Comput. Aided Des. **97**, 15–26 (2018). https://doi.org/10.1016/j.cad.2017.11.002
15. Khamjane, A., Benslimane, R.: Generating Islamic quasi-periodic patterns: a new method. J. Comput. Cult. Herit. **11**, 13:1–13:18. https://doi.org/10.1145/3127090
16. Khamjane, A., Taime, A., Ouazene, Z., Benslimane, R.: Computer graphics for generating Islamic geometric periodic and quasi-periodic patterns. In: Proceedings - 2019 International Conference on Intelligent Systems and Advanced Computing Sciences, ISACS 2019. IEEE, pp. 391–400 (2019)
17. Khamjane, A., Benslimane, R., Ouazene, Z.: Method of construction of decagonal self-similar patterns. Nexus Netw. J. **22**(2), 507–520 (2019). https://doi.org/10.1007/s00004-019-00461-4

Periodic Structures as a Countermeasure of Traffic Vibration and Earthquake: A Review

Hinde Laghfiri(✉) and Nouzha Lamdouar

GC Laboratory, Mohammadia School of Engineering, Mohammed V University in Rabat, Rabat, Morocco
hinde_laghfiri@um5.ac.ma, lamdouar@emi.ac.ma

Abstract. This article represents an overview of a new approach in civil and seismic engineering sector, based on the concept used in the solid-state physics field which revealed the manipulation of the dynamic property of the band gap frequency created by "periodic materials/structure" in which the wave cannot propagate if its frequency exists within the range of the band gap frequency of the periodic structure. In fact, the aim of this paper is to propose solutions of periodic structures with a band gap frequency within the range of the main frequency of traffic vibration and earthquake. Herein the research progress of periodic structure in terms of governing equation, analysis of the dispersion curve, eigenvalue problem solving, and application is introduced. Then, the general results on the effect of the mechanical parameters are addressed on the basis of the parametric studies presented in the articles reviewed. Finally, we will provide some recommendations for future research to improve the research framework of periodic structures and develop innovative solutions for civil engineering.

Keywords: Solid-state physics · Band gap · Periodic structures · Traffic vibration · Earthquake

1 Introduction

The attenuation of ground vibrations induced by heavy traffic (e.g., road, railway, subway, etc.), earthquakes, and blasting is one of the most interesting subjects in the world. In order to reduce the harmful effects of these vibrations, such as damaging buildings, disturbing people, and affecting technical processes and sensitive equipment, there are a lot of techniques that may attenuate these vibrations; the most popular among them is wave barriers. This latter could be open or infill trenches, or even rows of piles. There is a wide scope of research that investigates the screening efficiency of wave barriers. There are studies that evaluated the efficiency of trenches as an anti-vibration system through in-situ measurement [1–3]. Some researchers reveal the effectiveness of the open and infill trenches for scattering ground vibration, using the boundary element method (BEM) or the finite element method (FEM) [4–7]. Depending on the stability of the groundwater and soil, it is suggested to use infill trenches or rows of piles rather than open trenches. In this context, there are a lot of studies that investigated the isolation performance of piles

R. Saidi et al. (Eds.): ICATH 2021, LNDECT 110, pp. 359–373, 2022.
https://doi.org/10.1007/978-3-030-94188-8_34

[8, 9]. However, using the attenuation measures mentioned above the vibration can be isolated for certain frequency ranges as it can be amplified for certain frequency ranges. Hence lies the use of periodic structures as a countermeasure for ground vibration. Since 2006, extensive studies have been conducted on the mitigation of ground vibration, using periodic structures and the research is still thriving yearly.

The notion of seismic metamaterials is based on the concept of phononic crystals. Due to the periodicity of the structure the incident wave can be blocked on the attenuation zone. Gei [10] investigated the position of pass and stop bands on the dispersion curves of quasiperiodic infinite beams. Lu and He [11] found that resonance occurs at the bounding frequencies for the pass-bands of a periodic viaduct exposed to a harmonic wave, taking into consideration the interaction between soil and pile using the wave number domain boundary element method (WDBEM). Colombi et al. [12] presented a series of theoretical and experimental researches to reveal the role of the forests as a natural seismic metamaterial. Gao and Shi [13] studied the isolation of anti-plane elastic waves, using a wave guided barrier composed of two materials. Li et al. [14] presented an analytical model to obtain the frequency band gap of a quasi-one-dimensional periodic structure composed of two complex configuration. Chebli et al. [15] reported the use of the Floquet transforms to analyze the response of periodic structures that has been exposed to a moving loads calculated using a generic cell in the frequency-wavenumber domain. Yang et al. [16] employed both the wave and finite element approach to predict the vibroacoustic properties of periodic structures using the space-harmonic assumption. Peng et al. [17] have studied the prediction of railway vibration considering the track as a periodic structure. In addition, Hu et al. [18] Compared the isolation efficiency of a new type of periodic layered slab track with a floating slab track.

A number of researchers reported the isolation efficiency of periodic wave barriers; for example, Pu et al. [19] conducted a laboratory scale study to reveal the screening performance of periodic pile and periodic trench. Palermo et al. [20] presented a seismic metabarrier used to transform the seismic Rayleigh waves within shear bulk waves. Zhang et al. [21] exhibited the role of periodic pile on attenuating the vibration induced by subway and protecting the buildings alongside the subway. Ma et al. [22] conducted a laboratory experiment to explore the dynamic response of three distinct cases of a soil-periodic piles system subjected to a hand hammer impact, as well as to confirm the theoretical frequency band gaps with experimental ones. In this context, the aim of this paper is to present a review of periodic structures using to attenuate the traffic vibration and earthquake, the relevance of the research of the references studied in this article was governed by certain criteria which are indicated in Table 1. Thus, it is useful to focus the review in the following parts: (2) the governing equation using different hypothesis; (3) the most used methods to analyze the dispersion curve; (4) the two methods used to solve the eigenvalue problem; (5) Vibration band gap mechanism of periodic structure; (6) the band gap of the different solution proposed in the literature; and (7) the influence mechanical parameters on the band gap.

Table 1. Selection criteria of the reviewed articles

Parameters	Criteria
Duration	Research papers published from 2006 to 2021
Analysis	Research focuses on vibration attenuation
Application	Band gap frequency within the range of the main frequency of traffic vibration and earthquake
Language	English papers

2 Governing Equation

The majority of researchers have based on the wave propagation given by the Navier-Cauchy equations under the assumptions of homogeneous elastic medium, and without considering body force and damping [23]:

$$\rho \frac{\partial^2 u_j}{\partial t^2} = \frac{\partial}{\partial j}(\lambda \nabla . u) + \nabla .(\mu \left(\nabla u_j + \frac{\partial u}{\partial j} \right)) \quad (1)$$

With ρ is the mass density of the material, and u is the displacement vector, for the case of Cartesian coordinate system j corresponds to x, y or z; λ and μ are the Lamé constants and ∇ represents the nabla operator.

In the case of a poroelastic medium containing compressible Newtonian fluid, the governing equation of elastic waves can be expressed as follow using Biot's theory [24, 25]:

$$\nabla .[\mu(u\nabla + \nabla u)] + \nabla[\lambda_c \nabla .u + \alpha M \nabla .w] = \rho \frac{\partial^2 u}{\partial t^2} + \rho_f \frac{\partial^2 w}{\partial t^2} \quad (2)$$

$$\nabla(\alpha M \nabla .u + M \nabla .w) = \rho_f \frac{\partial^2 u}{\partial t^2} + m \frac{\partial^2 w}{\partial t^2} + \frac{\eta}{\kappa} \frac{\partial w}{\partial t} \quad (3)$$

where u denotes the displacement of the soil frame, w represents the displacement of the fluid relative to the soil skeleton; λc depends on the Lamé constant of soil frame λ in which $\lambda_c = \lambda + \alpha^2 M$, with α and M are the Biot's parameters; $\rho = (1 - \phi)\rho_s + \phi\rho_f$ corresponding to bulk density of porous material, with ϕ, ρ_s, and ρ_f denote the porosity, the density of soil grains and the density of fluid, respectively; $m = \rho_f \tau / \phi$, with τ is the tortuosity; $\frac{\eta}{k} = \frac{\eta}{k_o} = F(\omega)$, η represents the fluid viscosity, and k_o is the permeability and $F(\omega) = \left(1 - k\frac{i \cdot \omega}{w_T}\right)^{1/2}$ denote the dynamic correction factor, with $k = \frac{4\tau k_0}{\Lambda^2 \phi}$ is the geometrical factor, Λ a parameter characterizing the porous medium, and $\omega_T = \frac{\eta\phi}{\rho_f \tau k_0}$ represents the transition frequency.

As mentioned in [26], we can consider the viscoelasticity of the materials, utilizing the complex damping model: $E_d = E(1 + 2i \times \xi)$, ξ represents the damping ratio. Therefore, the Lamé constants associated with E_d in Eq. (1) will have a complex form.

2.1 Bloch Theory

The displacement uj(r, t) may be stated as follows according to the Bloch theory [23]:

$$u_j(r, t) = u_{jK}(r)e^{i(Kr-\omega t)} \tag{4}$$

r is the position vector; $K = (k_x, k_y, k_z)$ is the wave vector limited to the first Brillouin zone; and ω is the angular frequency, where $u_{jK}(r)$ is a periodic function having the same periodicity as the medium (the typical cell), with j corresponding to x, y, or z, we have:

$$u_{jK}(r + R) = u_{jK}(r) \tag{5}$$

where R is the lattice vector for a three-dimensional medium. Instead of examining the complete periodic structure, we may undertake the calculations using just one of the unit cells, and according to the two equations above (4) and (5), the boundary conditions of the elementary cell are as follows:

$$u_j(r + R, t) = u_{jK}(r + R)e^{i(K(r+R)-\omega t)} = u_j(r, t)e^{iKR} \tag{6}$$

Given the periodicity of the structure and applying Bloch-Floquet theory while factoring out the exp(iωt) at cell boundary following the directions of periodicity, we have the following relation:

$$u_j(r + a) = u_j(r) \tag{7}$$

Let us a donate the periodic constant.

3 Dispersion Curves

In order to evaluate the dispersion curves of periodic structures, a great number of researchers have extensively employed the FEM or the PWE (plane wave expansion) method. According to Fig. 1 we notice that FEM is used more than PWE method. This may be due to the fact that the FEM has the advantages of compatibility, convergence, accuracy, and efficiency.

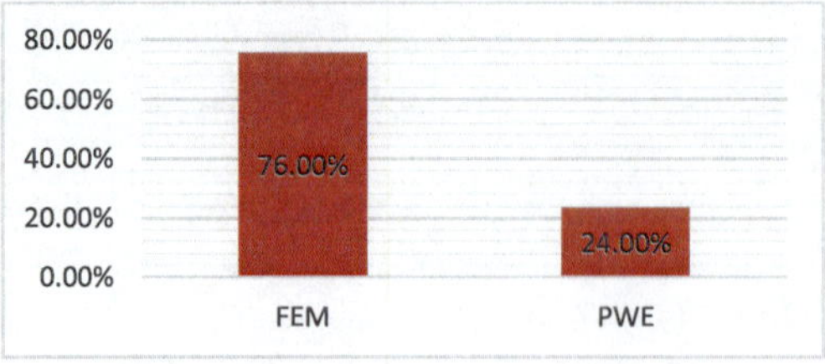

Fig. 1. Number of papers using FEM and PWE methods.

3.1 Finite Element Method

In order to evaluate the dispersion curves of periodic structures, a great number of researchers have employed the FEM. According to the equations above, the eigenvalue problem can be written as [27]:

$$\left(\overline{K} - \overline{M}\omega^2\right)u = 0 \tag{8}$$

where $\overline{\mathrm{K}}$ and $\overline{\mathrm{M}}$ are the stiffness and mass matrices of the unit cell, respectively, and u is the displacement at all the nodes. The commercial program COMSOL 3.5 may be used to solve the eigenvalue problem.

3.2 Plane Wave Expansion Method

Considering the periodicity of the periodic structures, the density ρ(r) and the Lamé constants μ(r) and λ(r) can be written as a Fourier series [23]:

$$f(r) = \sum_{G} f(G)e^{iG.r} \tag{9}$$

Where f (r) = ρ(r), μ(r) or λ(r) is considered, r is the general position vector and i represents the imaginary unit number. Taking the sum over all the integers that give each G, and the function f(G) can be expressed as:

$$f(\mathrm{G}) = \frac{1}{V}\int_{V} f(r)e^{-iG.r}dV \tag{10}$$

with V is the volume of the integration area

Hence is the result:

$$u_j(r,t) = e^{-i\omega t}\sum_{G} u_{jK}(G)e^{i(K+G).r} \tag{11}$$

The eigenvalue formula is obtained by substituting Eqs. (10) and (11) into the displacement Eq. (1). The solution of the eigenproblem is constraining in terms of time required and convergence, especially for the cases of the system with a large elastic mismatch, or the system with either very low or very high filling ratios, owing to the adopted unsuitable formulation of the eigenproblem. Hence lies the development of a new formulation of eigenproblem termed as the improved plane wave expansion method IPWE. Thereby, the only difference between them is the way where the elastic parameter functions are utilized in the equations [28].

4 Eigenvalue Problem Solution

In the literature, there are two methods to solve the eigenvalue problem:

- Traditional approach is known under the name of the ω (K) method. This latter is the most popular and related solely for the real band diagrams, which uses a given real wavenumber K within the first Brillouin zone to obtain the corresponding real eigenfrequency ω [29].
- Contrariwise K(ω) method is applied for the complex band diagrams to find the wave vector based on the incidence angle of wave θ and the given frequency ω [26].

We observe in Fig 2 that the ω (K) method is the most used; this is maybe due to the fact that we simplify the analysis. Besides, the field of application can be widened using K(ω) method because the calculation is possible even if we consider a viscoelastic material, and through this method, we can explain the decay of amplitude within the bandgap. Strictly speaking, the non-zero complex part of the wave vector informs us about the evanescent modes in which the vibration will be attenuated.

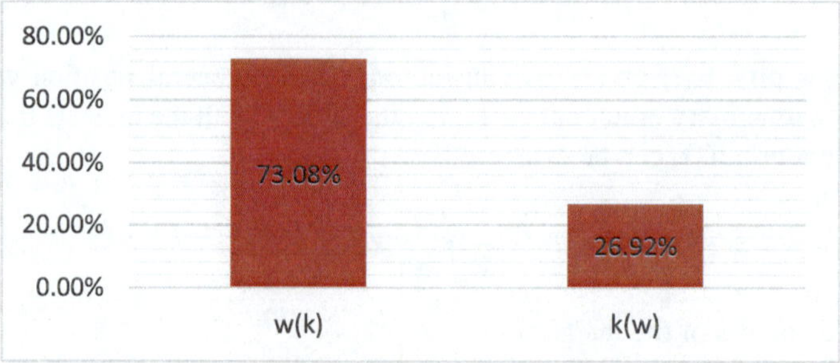

Fig. 2. Number of articles using ω(K) and K(ω) methods.

5 Vibration Band Gap Mechanism of Periodic Structure

Two separate mechanisms, Bragg scattering and local resonant, are responsible for the presence of the band gap. Innumerable works have been devoted to studying the existence of AZs by the Bragg scattering mechanism in different periodic systems. This mechanism explains that the elastic waves are unable to pass the periodic structure due to the reflection of vibration wave layer by layer because of changing material properties at the interface, and thus the band gaps occur with a wavelength similar to the periodicity of the unit cell [26]. According to Huang and Shi [30] AZs created by the Bragg scattering mechanism exists in the middle-frequency band, exhibiting the vibration at the whole typical cell composed of pile and soil; also, the periodic pile barrier in-filled by a hard material has a wide Bragg scattering AZs.

However, and in some specific periodic structures, the existence of the bandgap can be caused by a completely different mechanism, namely the mechanism of local resonance. Here, the bandgap may exist for waves having a wavelength much greater than the periodicity of the structure. Indeed, the resonator behaves like a spring system linked to a mass [26]. As mentioned by Huang and Shi [30] the AZs caused by the local resonant mechanism correspond to the range of low-frequency. Additionally, this mechanism is more promising in acquiring narrow AZs for a periodic pile barrier in-filled by a soft material with a vibration localized in the soft material.

6 Attenuation Zones and Screening Efficiency

Huang and Shi [31] affirm that the vibration frequencies induced by train ranging from 10 Hz to 40 Hz. Pu and Shi [33] reported that the nominal frequency of the train-induced vibrations is 30–80 Hz. As mentioned in [21] the main ground surface frequency of vibration induced by subway is 30–80Hz. According to Huang and Shi [31] the vibration frequencies of bus, Truck and automobile are under the range of 7–27 Hz, 10–35 Hz and 3–30 Hz respectively. In addition, the frequencies of seismic vibration next to Nuclear Power Plants are 0–10 Hz [32], also Ruiz and Saragoni mentioned that the main frequency of Mexico earthquake 1985 is from 0 to 1.5 Hz [33].

Table 2. Vibration frequencies of earthquake and traffic wave.

Domain	Frequency (Hz)
Train	10–80
Subway	30–80
Bus	7–27
Truck	10–35
Automobile	3–30
Earthquakes	0–10

In this part, we summarize the different first attenuation zones (FAZs), within the frequency range of traffic and seismic vibration, found in the most recent references as shown in Table 3. Notice that LBF means the Lower Bound Frequency; UBF means the Upper Bound Frequency; and WFAZ the Width of First Attenuation Zone. Based on the frequency band within the AZs (Table 3) and the range frequency of traffic vibration and earthquakes summarized in Table 2. We defined the field of application for each solution studied in the references reviewed.

Table 3. The AZs mentioned in the reviewed paper.

Reference	LBF (Hz)	UBF (Hz)	WFAZ (Hz)	Application domain
Periodic pile				
[34]	0	4.5	4.5	• Automobile • Earthquake
[35]	22	27	5	• Train • Truck • Automobile • Bus
[24]	36.6	43.2	6.6*	• Train • Subway
[36]	0	≈5	≈5	• Automobile • Earthquake
[37]	0	2.87	2.87	• Earthquakes
[31]	33.321	37.335	4.014	• Train • Truck • Subway
[30]	34.64	36.95	2.31	• Train • Subway
[38]	14.24	27.86	13.62*	• Train • Truck • Automobile • Bus
[26]	13	18.5	5.5	• Train • Truck • Automobile • Bus
[39]	67.45	159.50	92.05	• Train • Subway
[40]	6	16	10	• Train • Truck • Automobile • Earthquakes • Bus
[41]	12.677	13.78	1.103	• Train • Truck • Automobile • Bus

(continued)

Table 3. (*continued*)

Reference	LBF (Hz)	UBF (Hz)	WFAZ (Hz)	Application domain
[42]	16.4		1.8	• Train • Truck • Automobile • Bus
Periodic trench barriers				
[29]	≈45	≈62	≈17	• Train • Subway
[43]	45	62	17	• Train • Subway
Periodic foundation				
[44]	2.5	15	12.5	• Truck • Train • Automobile • Earthquakes • Bus
[45]	9.6	16.3	6.7	• Train • Truck • Automobile • Bus
[46]	7.22	10.16	2.94	• Automobile • Earthquakes • Bus
[47]	8.79	15.72	6.93	• Train • Truck • Automobile • Earthquakes • Bus
[48]	32.9	35.6	2.7	• Train • Truck • Subway
[49]	2.15	15.01	12.86	• Train • Truck • Automobile • Earthquakes • Bus

(*continued*)

Table 3. (*continued*)

Reference	LBF (Hz)	UBF (Hz)	WFAZ (Hz)	Application domain
[50]	8.58	16.49	7.91	• Train • Truck • Automobile • Earthquakes • Bus
[51]	41.2	58.7	17.5	• Train • Subway
[52]	9.9	16.3	6.4	• Train • Truck • Automobile • Bus
[53]	6.03	8.73	2.7	• Automobile • Earthquakes • Bus

*Directional attenuation zone in **ΓX** direction

7 Parametric Study

We observe from Table 3 that the attenuation zone strongly depends on certain mechanical parameters, so we can rely on these parameters to have the desired band gap which corresponds to the desired field of application. In this context, some authors have studied the influence of the mechanical characteristics of the soil on the pile-soil system. As shown in Table 4 the increase of soil elastic modulus generates the increase of LBF, UBF and WFAZ. From Table 5 we cannot conclude that there is a significant effect of the density on the 2 parameters, namely UBF and WFAZ, while the LBF decreases with the increase of soil density. Also, the soil's Poisson ratio has no discernible impact on the band gap [24]. The soil porosity has a small effect, in which the LBF and the UBF increase as porosity increase, but permeability decreases, resulting in an increase in the width of the FAZs [25].

Table 4. The influence of soil elastic modulus on the FAZs of pile-soil system.

References	Soil elastic modulus (MPa)	LBF (Hz)	UBF (Hz)	WFAZ (Hz)
[54]	1–30	8–27.5	10.5–34	2.5–6.5
[24]	1–1000	12.5–143.75	12.5–200	0–56.5
[31]	1–60	7.3–57	8.2–65	0.9–8
[30]	1–1000	7.6–17	8.35–18	0.75–1
[25]	1–30	6–20	8–26	2–6

Table 5. The influence of soil density on the FAZs of pile-soil system.

References	Soil density (Kg/m^3)	LBF (Hz)	UBF (Hz)	WFAZ (Hz)
[54]	1700–2200	23–22	23.75–25	0.75–3
[24]	1000–2500	61–40	76–50	15–10
[37]	1000–1400	0	2.89–2.84	2.89–2.84
[31]	1700–2200	33.5–32	39–34.5	5.5–2.5

It should be noted that very few studies have investigated the influence of mechanical parameters on the FAZs for the other solutions such as the periodic foundation, the periodic trench, etc. In this regard, we can hardly give remarks about the influence of these parameters on the band gap.

8 Conclusion and Recommendation

Given the significant damage and the harmful consequences of vibrations induced by the different modes of traffic (e.g. road, rail, etc.) or earthquakes, targeted protection is required. In this perspective, countless research has studied the effectiveness of periodic structures by attenuating or isolating the aforementioned vibrations. The critical overview of the scientific papers considered in this article highlights the potential applications of the periodic structures among others periodic foundations, periodic pile-soil systems, periodic barriers, and periodic trenches, aiming to help the researchers work on attenuating the vibrations to provide innovative and efficient solutions for civil and seismic engineering.

Through this paper, we have been able to formulate the following recommendations:

- The majority of existing studies are based on theoretical results using finite element simulation while experimental studies are limited to the laboratory scale. Thus, the experimental study on 3D periodic structures for real cases with real data of traffic and seismic vibration must be conducted.
- The researchers should carry out the optimization studies to have solutions with a band gap applicable for both traffic and seismic vibration, considering the soil-structure interaction.
- The researchers should adopt specific periodic structures for the bridges in order to face the vibrations caused by traffic or earthquake.
- There is a paucity or lack of parametric studies for some periodic structures. Therefore, it is necessary to carry out several studies to have general results, taking into account the different parameters influencing the forbidden band.
- There is a scarcity of comparison research between periodic structures and traditional solutions in order to reveal the advantages and disadvantages of each solution, or the combination of two solutions in an optimal way to obtain relevant results.

Acknowledgments. The authors are thankful to the editor and the reviewers for their valuable remarks.

Conflict of Interest Statement. The authors declare that there is no conflict of interest.

References

1. Çelebi, E., Firat, S., Beyhan, G., Çankaya, I., Vural, I., Kirtel, O.: Field experiments on wave propagation and vibration isolation by using wave barriers. Soil Dyn. Earthq. Eng. **29**, 824–833 (2009). https://doi.org/10.1016/j.soildyn.2008.08.007
2. Garinei, A., Risitano, G., Scappaticci, L.: Experimental evaluation of the efficiency of trenches for the mitigation of train-induced vibrations. Transp. Res. Part D Transp. Environ. **32**, 303–315 (2014). https://doi.org/10.1016/j.trd.2014.08.016
3. Ulgen, D., Toygar, O.: Screening effectiveness of open and in-filled wave barriers: a full-scale experimental study. Constr. Build. Mater. **86**, 12–20 (2015). https://doi.org/10.1016/j.conbuildmat.2015.03.098
4. Adam, M., Von Estorff, O.: Reduction of train-induced building vibrations by using open and filled trenches. Comput. Struct. **83**, 11–24 (2005). https://doi.org/10.1016/j.compstruc.2004.08.010
5. Ekanayake, S.D., Liyanapathirana, D.S., Leo, C.J.: Attenuation of ground vibrations using in-filled wave barriers. Soil Dyn. Earthq. Eng. **67**, 290–300 (2014). https://doi.org/10.1016/j.soildyn.2014.10.004
6. Coulier, P., Cuéllar, V., Degrande, G., Lombaert, G.: Experimental and numerical evaluation of the effectiveness of a stiff wave barrier in the soil. Soil Dyn. Earthq. Eng. **77**, 238–253 (2015). https://doi.org/10.1016/j.soildyn.2015.04.007
7. Yarmohammadi, F., Rafiee-Dehkharghani, R., Behnia, C., Aref, A.J.: Design of wave barriers for mitigation of train–induced vibrations using a coupled genetic-algorithm/finite-element methodology. Soil Dyn. Earthq. Eng. **121**, 262–275 (2019). https://doi.org/10.1016/j.soildyn.2019.03.007
8. Dijckmans, A., Ekblad, A., Smekal, A., Degrande, G., Lombaert, G.: Efficacy of a sheet pile wall as a wave barrier for railway induced ground vibration. Soil Dyn. Earthq. Eng. **84**, 55–69 (2016). https://doi.org/10.1016/j.soildyn.2016.02.001
9. Gao, G.Y., Li, Z.Y., Qiu, C., Yue, Z.Q.: Three-dimensional analysis of rows of piles as passive barriers for ground vibration isolation. Soil Dyn. Earthq. Eng. **26**, 1015–1027 (2006). https://doi.org/10.1016/j.soildyn.2006.02.005
10. Gei, M.: Wave propagation in quasiperiodic structures: stop/pass band distribution and pre-stress effects. Int. J. Solids Struct. **47**, 3067–3075 (2010). https://doi.org/10.1016/j.ijsolstr.2010.07.008
11. Lu, J.F., He, C.Y.: A model for the seismic analysis of a periodic viaduct when considering the pile-soil-structure interaction. Soil Dyn. Earthq. Eng. **56**, 13–27 (2014). https://doi.org/10.1016/j.soildyn.2013.09.018
12. Colombi, A., Roux, P., Guenneau, S., Gueguen, P., Craster, R.V.: Forests as a natural seismic metamaterial: Rayleigh wave bandgaps induced by local resonances. Sci. Rep. **6** (2016). https://doi.org/10.1038/srep19238
13. Gao, M., Shi, Z.: A wave guided barrier to isolate antiplane elastic waves. J. Sound Vib. **443**, 155–166 (2019). https://doi.org/10.1016/j.jsv.2018.11.042
14. Li, Y., Wang, X., Yan, G.: Configuration effect and bandgap mechanism of quasi-one-dimensional periodic lattice structure. Int. J. Mech. Sci. **190**, 106017 (2021). https://doi.org/10.1016/j.ijmecsci.2020.106017

15. Chebli, H., Othman, R., Clouteau, D.: Response of periodic structures due to moving loads. Comptes Rendus - Mec. **334**, 347–352 (2006). https://doi.org/10.1016/j.crme.2006.04.001
16. Yang, Y., Mace, B.R., Kingan, M.J.: Vibroacoustic analysis of periodic structures using a wave and finite element method. J. Sound Vib. **457**, 333–353 (2019). https://doi.org/10.1016/j.jsv.2019.06.009
17. Peng, Y., Sheng, X., Cheng, G.: Modelling track and ground vibrations for a slab ballastless track as an infinitely long periodic structure subject to a moving harmonic load. J. Sound Vib. **489**, 115760 (2020). https://doi.org/10.1016/j.jsv.2020.115760
18. Hu, Y., Cheng, Z., Shi, Z.: Vibration reduction performance of a periodic layered slab track. In: Tutumluer, E., Chen, X., Xiao, Y. (eds.) Advances in Environmental Vibration and Transportation Geodynamics. LNCE, vol. 66, pp. 779–791. Springer, Singapore (2020). https://doi.org/10.1007/978-981-15-2349-6_50
19. Pu, X., Meng, Q., Shi, Z.: Experimental studies on surface-wave isolation by periodic wave barriers. Soil Dyn. Earthq. Eng. **130**, 106000 (2020). https://doi.org/10.1016/j.soildyn.2019.106000
20. Palermo, A., Krödel, S., Marzani, A., Daraio, C.: Engineered metabarrier as shield from seismic surface waves. Sci. Rep. **6**, 1 (2016). https://doi.org/10.1038/srep39356
21. Zhang, S., Wang, W., Wu, Z., Dai, Y., Li, Y., Dai, H.: Effect of periodic pile row in subway vibration Isolation. Procedia Eng. **199**, 302–309 (2017). https://doi.org/10.1016/j.proeng.2017.09.043
22. Ma, M., Jiang, B., Gao, J., Liu, W.: Experimental study on attenuation zone of soil-periodic piles system. Soil Dyn. Earthq. Eng. **126**, 105738 (2019). https://doi.org/10.1016/j.soildyn.2019.105738
23. Poggetto, V.F.D., Serpa, A.L.: Elastic wave band gaps in a three-dimensional periodic metamaterial using the plane wave expansion method. Int. J. Mech. Sci. **184**, 105841 (2020). https://doi.org/10.1016/j.ijmecsci.2020.105841
24. Pu, X., Shi, Z.: Periodic pile barriers for Rayleigh wave isolation in a poroelastic half-space. Soil Dyn. Earthq. Eng. **121**, 75–86 (2019). https://doi.org/10.1016/j.soildyn.2019.02.029
25. Meng, L., Cheng, Z., Shi, Z.: Vibration mitigation in saturated soil by periodic pile barriers. Comput. Geotech. **117**, 103251 (2020). https://doi.org/10.1016/j.compgeo.2019.103251
26. Jiang, Y., et al.: Vibration attenuation analysis of periodic underground barriers using complex band diagrams. Comput. Geotech. **128**, 103821 (2020). https://doi.org/10.1016/j.compgeo.2020.103821
27. Huang, J., Ruzzene, M., Chen, S.: Analysis of in-plane wave propagation in periodic structures with Sierpinski-carpet unit cells. J. Sound Vib. **395**, 127–141 (2017). https://doi.org/10.1016/j.jsv.2017.02.020
28. Cao, Y., Hou, Z., Liu, Y.: Convergence problem of plane-wave expansion method for phononic crystals. Phys. Lett. Sect. Gen. Solid State Phys. **327**, 247–253 (2004). https://doi.org/10.1016/j.physleta.2004.05.030
29. Pu, X., Shi, Z.: Broadband surface wave attenuation in periodic trench barriers. J. Sound Vib. **468**, 115130 (2020). https://doi.org/10.1016/j.jsv.2019.115130
30. Huang, J., Shi, Z.: Vibration reduction of plane waves using periodic in-filled pile barriers. J. Geotech. Geoenviron. Eng. **141**, 04015018 (2015). https://doi.org/10.1061/(asce)gt.1943-5606.0001301
31. Huang, J., Shi, Z.: Application of periodic theory to rows of piles for horizontal vibration attenuation. Int. J. Geomech. **13**, 132–142 (2013). https://doi.org/10.1061/(asce)gm.1943-5622.0000193
32. Arif, S.J., Imdadullah, Asghar, M.S.J.: Accurate measurement of velocity and acceleration of seismic vibrations near nuclear power plants. Phys. Procedia. **37**, 43–50 (2012). https://doi.org/10.1016/j.phpro.2012.03.704

33. Ruiz, S., Saragoni, G.R.: Free vibration of soils during large earthquakes. Soil Dyn. Earthq. Eng. **29**, 1–16 (2009). https://doi.org/10.1016/j.soildyn.2008.01.005
34. Li, T., Su, Q., Kaewunruen, S.: Seismic metamaterial barriers for ground vibration mitigation in railways considering the train-track-soil dynamic interactions. Constr. Build. Mater. **260**, 119936 (2020). https://doi.org/10.1016/j.conbuildmat.2020.119936
35. Meng, L., Cheng, Z., Shi, Z.: Filtering property of periodic pile barriers under moving loads. Comput. Geotech. **136**, 104244 (2021). https://doi.org/10.1016/j.compgeo.2021.104244
36. Chen, Y., Feng, Q., Scarpa, F., Zuo, L., Zhuang, X.: Harnessing multi-layered soil to design seismic metamaterials with ultralow frequency band gaps. Mater. Des. **175**, 107813 (2019). https://doi.org/10.1016/j.matdes.2019.107813
37. Mandal, P., Somala, S.N.: Periodic pile-soil system as a barrier for seismic surface waves. SN Appl. Sci. **2**(7), 1–8 (2020). https://doi.org/10.1007/s42452-020-2969-8
38. Liu, X., Shi, Z., Xiang, H., Mo, Y.L.: Attenuation zones of periodic pile barriers with initial stress. Soil Dyn. Earthq. Eng. **77**, 381–390 (2015). https://doi.org/10.1016/j.soildyn.2015.06.010
39. Liang, F., Yang, X.-D.: Wave properties and band gap analysis of deploying pipes conveying fluid with periodic varying parameters. Appl. Math. Model. **77**, 522–538 (2020). https://doi.org/10.1016/j.apm.2019.07.064
40. Zhang, K., Luo, J., Hong, F., Deng, Z.: Seismic metamaterials with cross-like and square steel sections for low-frequency wide band gaps. Eng. Struct. **232**, 111870 (2021). https://doi.org/10.1016/j.engstruct.2021.111870
41. Meng, L., Cheng, Z., Shi, Z.: Vibration mitigation in saturated soil by periodic in-filled pipe pile barriers. Comput. Geotech. **124**, 103633 (2020). https://doi.org/10.1016/j.compgeo.2020.103633
42. Meng, Q., Shi, Z.: Vibration isolation of plane waves by periodic pipe pile barriers in saturated soil. J. Aerosp. Eng. **32**, 04018114 (2019). https://doi.org/10.1061/(asce)as.1943-5525.0000938
43. Pu, X., Shi, Z., Xiang, H.: Feasibility of ambient vibration screening by periodic geofoam-filled trenches. Soil Dyn. Earthq. Eng. **104**, 228–235 (2018). https://doi.org/10.1016/j.soildyn.2017.10.022
44. Cheng, Z., Shi, Z., Palermo, A., Xiang, H., Guo, W., Marzani, A.: Seismic vibrations attenuation via damped layered periodic foundations. Eng. Struct. **211**, 110427 (2020). https://doi.org/10.1016/j.engstruct.2020.110427
45. Bao, J., Shi, Z., Xiang, H.: Dynamic responses of a structure with periodic foundations. J. Eng. Mech. **138**, 761–769 (2012). https://doi.org/10.1061/(asce)em.1943-7889.0000383
46. Cheng, Z., Shi, Z.: Novel composite periodic structures with attenuation zones. Eng. Struct. **56**, 1271–1282 (2013). https://doi.org/10.1016/j.engstruct.2013.07.003
47. Shi, Z., Huang, J.: Feasibility of reducing three-dimensional wave energy by introducing periodic foundations. Soil Dyn. Earthq. Eng. **50**, 204–212 (2013). https://doi.org/10.1016/j.soildyn.2013.03.009
48. Yan, Y., Cheng, Z., Menq, F., Mo, Y.L., Tang, Y., Shi, Z.: Three dimensional periodic foundations for base seismic isolation. Smart Mater. Struct. **24** (2015). https://doi.org/10.1088/0964-1726/24/7/075006
49. Shi, Z., Cheng, Z., Xiang, H.: Seismic isolation foundations with effective attenuation zones. Soil Dyn. Earthq. Eng. **57**, 143–151 (2014). https://doi.org/10.1016/j.soildyn.2013.11.009
50. Cheng, Z.B., Shi, Z.F.: Composite periodic foundation and its application for seismic isolation. Earthq. Eng. Struct. Dyn. **47**, 925–944 (2018). https://doi.org/10.1002/eqe.2999
51. Yan, Y., et al.: Seismic isolation of two dimensional periodic foundations. J. Appl. Phys. **116** (2014). https://doi.org/10.1063/1.4891837
52. Xiong, C., Shi, Z., Xiang, H.: Attenuation of building vibration using periodic foundations. Adv. Struct. Eng. **15**, 1375–1388 (2012). https://doi.org/10.1260/1369-4332.15.8.1375

53. Jia, G., Shi, Z.: A new seismic isolation system and its feasibility study. Earthq. Eng. Eng. Vib. **9**, 75–82 (2010). https://doi.org/10.1007/s11803-010-8159-8
54. Huang, J., Shi, Z.: Attenuation zones of periodic pile barriers and its application in vibration reduction for plane waves. J. Sound Vib. **332**, 4423–4439 (2013). https://doi.org/10.1016/j.jsv.2013.03.028

Numerical Evaluation of Channel-Beam Railway Bridge with Hollow Section Concrete Deck

Shaohua He[1](✉), Luoquan Zou[1], and Ayman Mosallam[2]

[1] School of Civil and Transportation Engineering, Guangdong University of Technology, Guangzhou, Guangdong, China
hesh@gdut.edu.cn

[2] Department of Civil and Environmental Engineering, University of California, Irvine, CA, USA

Abstract. Prestressed concrete (PC) channel-beam bridge, consisting of side girders, solid concrete deck and end transoms, has the advantages of small height, large under-bridge net spacing and excellent sound-insulated noise reduction. However, due to the enormous structural weight of the thick solid track-bed deck, the load-resisting efficiency of the material is comparatively low. To optimize the mechanical behavior of the channel-beam bridge and increase the span-ability, this paper proposes a hollow cross-section track-bed deck for the replacement of solid decks. A nonlinear numerical model to simulate the structural behavior of the Xifu River Railway Bridge that is a 32.6 m-long channel-beam bridge in China was established. Based on the numerical results, three structural forms of hollow cross-section track-bed decks were developed. Numerical results indicate that over 69.7% of the concrete stresses in solid-deck channel-beam bridges were caused by self-weight. The small stresses in components of the solid-deck channel-beam bridge make it possible to further optimize section of the solid deck. Results also demonstrate that favorable mechanical performance of the hollow-deck channel-beam was achieved, and the use of serrated cross section reduced the self-weight of the original solid deck by 15.3%. Comparing to the solid-deck channel-beam, the maximum axial stress, lateral stress and shear stress of the hollow-deck channel-beam under most serve load combinations increased by 40.06%, 28.65%, and 25.8%, respectively. The outcomes of this study provide a reference for the application of hollow track-bed deck in channel-beam bridges.

Keywords: Channel-beam · Railway bridge · Numerical simulation · Track-bed deck

1 Introduction

Prestressed concrete (PC) railway channel-beam (hereafter referred as channel-beam), consisting of main girders, track-bed deck and end transoms, is a type of under-bearing beam that has low structure height and large under-bridge headroom. During the past several decades, the channel-beam bridge has been extensively used in highway bridges, railroad bridges and urban rail transportation [1]. Especially, for the proposed bridges

R. Saidi et al. (Eds.): ICATH 2021, LNDECT 110, pp. 374–386, 2022.
https://doi.org/10.1007/978-3-030-94188-8_35

adjacent to existing highways or railways, the utilization of channel-beam can prevent the restrictions of existing facilities that occupied large land resources, reduce bridge approach fills and save constructing costs [2–4]. Figure 1 shows the arrangement for a typical channel-beam railway bridge.

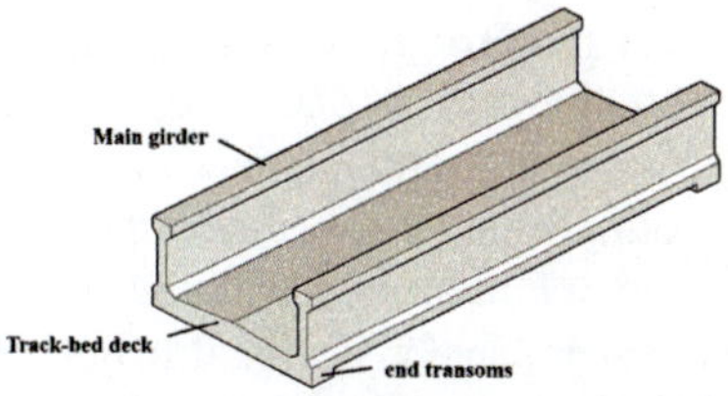

Fig. 1. Arrangement of channel-beam

According to bridge inventory, more than 50% of the stresses in small- and medium-span concrete beam bridges were generated by its self-weight, and the percentage goes up to 90% for large-span concrete beam bridges [5]. The most commonly used track-bed deck in channel-beam bridges is the thick solid flat panel. The average thickness of track-bed decks in previously established channel-beam bridges are between 0.20 m and 1.30 m [6–8]. Due to the heavy self-weight of solid track-bed deck, the efficiency of material strength in resisting the stresses triggered by vehicle loads for channel-beams is relatively low, which hampered the possibility of forming large-span channel-beam railway bridges [9]. Recently, several experiments and finite element (FE) simulations focusing on optimizing the track-bed decks in channel-beam bridges have been reported by researchers around the world [3–12]. It was found that the stresses in solid track-bed decks were minimal, as compared to the allowable stress of the constitutive material [10–12]. The high residual material strength of the solid track-bed deck allows further optimizing the cross section of solid track-bed decks.

Introducing hollow cross-section to track-bed decks of channel-beams is a potential way to solve the problem. Under the precondition of sufficient cross-sectional characteristics and load-bearing resistance, removing some sectional area from the solid deck can reduce self-weight effects. Unfortunately, the existing studies are mainly concentrated on channel-beams using solid track-bed decks, and few studies have been reported on the hollow-deck channel-beams [13–15]. Considering that the mechanic properties of channel-beams with hollow-decks may differ from that of solid concrete deck, it is necessary to explore the structural forms of hollow track-bed decks for the optimization of channel-beam bridges.

The recently constructed Xifu River Bridge, which is a 32.0 m-long simple-supported channel-beam bridge located in Guangdong China, has been selected as the prototype bridge for the study. In this paper, the structural performance of channel-beam bridges with hollow track-bed deck was evaluated using *Abaqus* software. A nonlinear numerical model was established to simulate the structural behavior of the background bridge. Based on the numerical results, three types of hollow cross-section concrete decks were proposed to replace the original solid-deck. Discussions on the feasibility and rationality of the proposed hollow track-bed decks were conducted and presented. Finally, a practical form of hollow track-bed deck is recommended for the channel-beam bridges.

2 Prototype Bridge Descriptions

The recently constructed Xifu River Bridge is a simple-supported channel-beam bridge, which belongs to a new double-line railway that locates in Guangdong province, China. Figure 2 shows the configuration and arrangement of the Xifu River Bridge. As can be seen in the figure, the prototype channel-beam bridge has a total length of 32.6 m, an overall top width of 15.3 m, a bottom width of 13.2 m, and a net width of 11.4 m. The two inclined-wall box girders have a sectional height of 3.15 m at mid-span and 3.75 m at beam supports. The thickness of track-bed deck at the locations of mid-span and supports are 0.7 m and 1.3 m, respectively. The lateral slop of the track-bed deck is 2.0%.

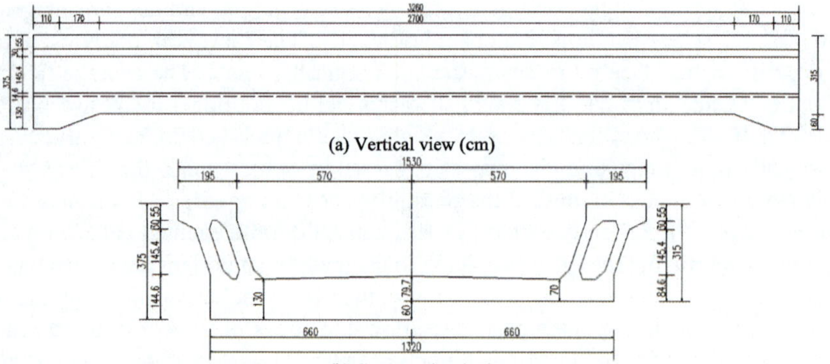

(a) Vertical view (cm)

(b) Cross section (cm)

Fig. 2. Configuration of Xifu River Bridge

The C60 grade concrete, with a target compressive strength of 60.0 MPa, was used for the prototype channel-beam. The lateral and longitudinal prestressing tendons are made of low relaxation steel strands, which have a standard tensile strength $f_{pk} = 1860$ MPa, nominal diameter of 15.2 mm, modulus of elasticity 1.95×10^5 MPa, and thermal expansion coefficient of 1.2×10^{-5}. The PSB830 threaded steel bars with standard strength $f_{pk} = 830$ MPa and 25 mm in diameter were used as vertical prestressing bolts for the side box girder. The other steel reinforcements in concrete side girders and track-bed decks are made of HRB400 (with nominal yielding strength of 400 MPa) and HPB300 (with nominal yielding strength of 300 MPa).

The Xifu River Bridge belongs to a passenger and freight combined double-line railway, which has a central-line spacing of 5.3 m and a designed speed of 160 km/h. According to the Chinese code (TB10002-2017) [16], the vertical load coefficient for longitudinal calculation is 1.19, and that for the horizontal calculation is 1.29.

3 Channel-Beam Bridge with Solid Decks

3.1 Finite Element Model

Elements and Material Modeling. Compacted finite element package Abaqus was used for numerical analysis of the Xifu River Bridge. The main components of the channel-beam bridge consisted of two side concrete box girders, a track-bed concrete deck, prestressing steel strands, and steel reinforcements. Considering the vertical symmetry, 1/2 of the channel-beam was assumed. Figure 3 shows the FE model of the channel-beam bridge.

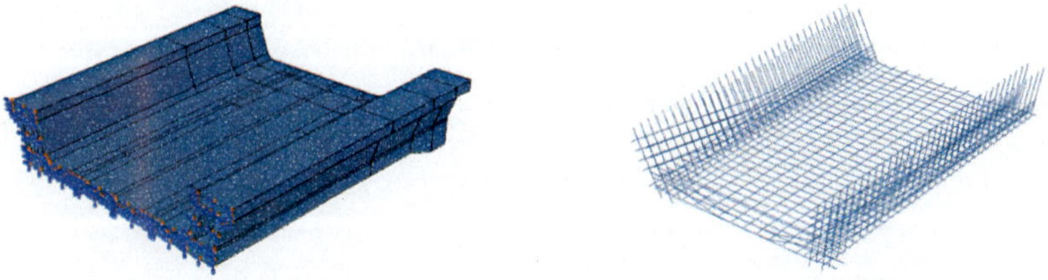

(a) Channel-beam model (b) Pre-stressed tendons

Fig. 3. FE model for Channel-beam with solid concrete deck

Abaqus software agency provides various elements for simulating steel and concrete members. All concrete members were simulated using eight-node solid elements (C3D8R), while the prestressing steel strands and reinforcements were simulated using two-node truss elements (T3D2).

The constitutive model proposed by Ding and Yu [17, 18] was adopted for concrete material. Expressions to determine the compressive stress-strain relationship of the concrete are:

$$y = \begin{cases} \dfrac{A_1 x - x^2}{1 + (A_1 - 2)x} & x \leq 1 \\ \dfrac{x}{\alpha_1 (x-1)^2 + x} & x > 1 \end{cases} \tag{1}$$

$$y = \frac{\sigma}{f_c} \tag{2}$$

$$x = \frac{\varepsilon}{\varepsilon_0} \tag{3}$$

$$A_1 = 9.1 f_{cu}^{-\frac{4}{9}} \tag{4}$$

$$\alpha_1 = 2.5 \times 10^{-5} f_{cu}^3 \tag{5}$$

The tensile stress-strain relationship adopted for the concrete is as follows:

$$y = \begin{cases} \dfrac{Ax - x^2}{1 + (A-2)x} & x \le 1 \\ \dfrac{x}{\alpha_1 (x-1)^{1.7} + x} & x > 1 \end{cases} \tag{6}$$

$$y = \frac{\sigma}{f_t} \tag{7}$$

$$x = \frac{\varepsilon}{\varepsilon_{tp}} \tag{8}$$

$$A_1 = 1.306 \tag{9}$$

$$\alpha_1 = 1 + 0.025 f_t^3 \tag{10}$$

The ideal elastic-plastic model was used for all steel and can be calculated by:

$$\sigma = \begin{cases} E_s \varepsilon & 0 < \varepsilon < \varepsilon_y \\ f_y & \varepsilon \ge \varepsilon_y \end{cases} \tag{11}$$

where, σ is the compressive stress; f_c is the prismal compressive strength of concrete; f_{cu} is the cubic compressive strength of concrete; ε is the compressive strain; ε_0 is the compressive strain correponding to f_c; f_t is the tensile strength of concrete; ε_{tp} is the tensile strain correponding to f_t; E_s is the Youngs modulum of concrete; ε_y is the yielding strain of steel; f_y is the yielding strength of steel.

Displacement and Load Boundaries. Prestressing steel strands were embedded into the concrete, and the relative slip between strands and concrete was ignored. The symmetry boundary condition was applied to the symmetric section of the channel-beam model, while means all nodes located at the symmetric section were prevented from translating in the X-direction and rotating in Y- and Z- directions. As the channel-beam is simply-supported, the displacement in Z-direction at the ends of the beam is restricted.

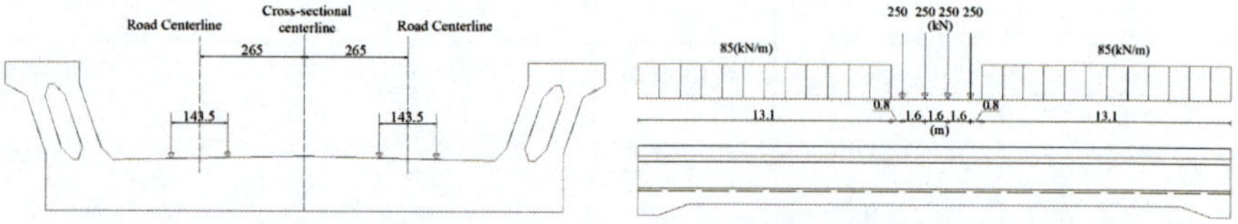

(a) Distribution of track loads in lateral direction (b) Distribution of track loads in axial direction

Fig. 4. Load applied to the FE model

Self-weight of the structure is automatically calculated using the mass density and gravitational acceleration. Force applied to the prestressing steel strands is realized using temperature reduction method. The thermal expansion coefficient of the steel strands is defined as 1.2×10^{-5}. The double-line railway loads applied to the FE model is defined according to the Chinese code [16], as shown in Fig. 4.

3.2 Results and Discussion

Structural performance of the channel-beam subjected to self-weight loading and combined actions of self-weight and double-line railway loads, respectively, were achieved from FE models. In the following sections, discussions are performed based on these numerical results.

Self-weight Loading Case. The calculated stress distribution in channel-beam subjected to self-weight loads is shown in Fig. 5. As can be seen, the maximum longitudinal and lateral compressive stresses in the side main girders are −10.12 MPa and −0.98 MPa, respectively, while the corresponding compressive stresses in the track-bed deck are −6.89 MPa and −3.79 MPa, respectively. The maximum shear stress in the concrete components of the channel-beam model is 3.8 MPa. Apparently, stresses in the channel-beam caused by structural weight are comparatively small and the channel-beam exhibit large safety margin in resisting static loads. The maximum vertical displacement occurred at the mid-span of the beam, with a displacement value of −3.90 mm.

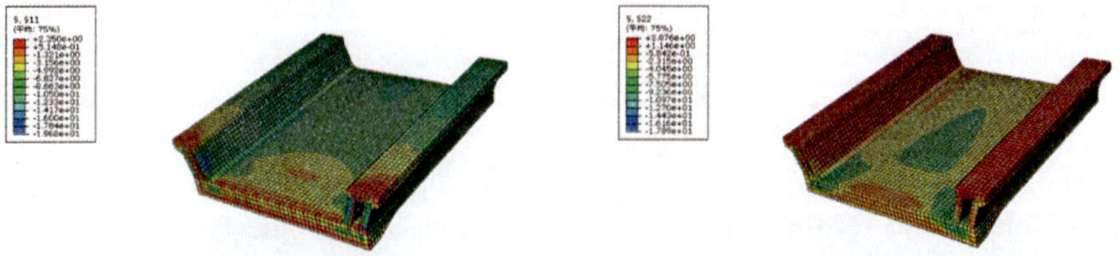

(a) Longitudinal normal stress (b) Transverse normal stress

Fig. 5. Stress of solid-deck channel-beam (MPa)

Most Critical Loading Case. Figure 6 shows the stress nephogram for the channel-beam under the most critical loading case. As can be seen in Fig. 6(a), all members in the channel-beam are in compression. The maximum longitudinal and lateral compressive stresses in the side main girders are −14.52 MPa and −0.83 MPa, respectively, while the corresponding compressive stresses in the track-bed deck are −7.27 MPa and −5.82 MPa, respectively, as shown in Fig. 6(b). By comparing the maximum compressive stresses at the top of side girder, it can be seen that approximately 69.7% of the concrete stresses in the solid-deck channel-beam bridge is generated by the self-weight effect of the bridge.

Under the most critical load combination, the maximum deflection of the track-bed deck was occurred at the mid-span, showing a maximum displacement value of −11.92 mm. The maximum deflection of the side main girder is −9.30 mm. According to the Chinese code (TB10002-2017) [16], the allowable deflection for the simple-supported double-line railway beam is L/1000, equalling to −32.0 mm for the present Xifu River Bridge. Obviously, the prototype channel-beam exhibites sufficient bending stiffness under the most critical combination of static and live loads.

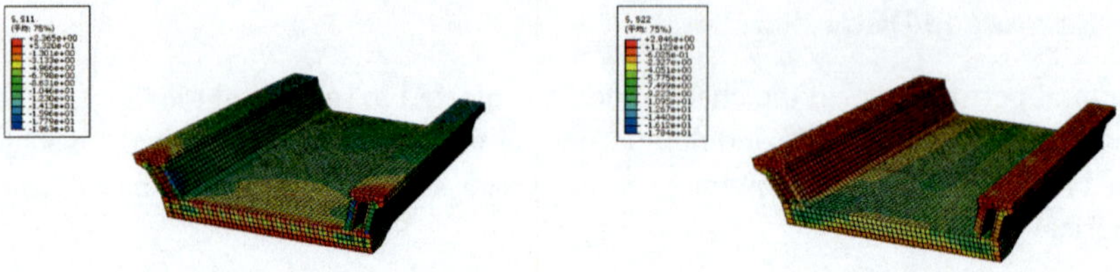

(a) Longitudinal stress (b) Lateral stress

Fig. 6. Stress distribution in channel-beam under the most critical load case (MPa)

The small stresses in solid-deck demonstrated the large safety margin of the track-bed deck, which restricts the development of large-span channel-beam bridges. Accordingly, it is practical and rational to introduce hollow-deck to channel-beam railway bridges to reduce the structural self-weight and improve the load-bearing efficiency of the bridge.

4 Channel-Beam Bridge with Hollow Section Decks

4.1 Design of Hollow Cross-Section Decks

The thickness of the original solid track-bed deck is 700 mm, and large numbers of longitudinal and lateral prestressing strands are arranged in the bottom of the deck. According to the required protection thickness for a prestressed concrete structure, the required thicknesses for the reinforcements in bottom- and top decks are 300–400 mm and 200 mm, respectively. Accordingly, three structural forms of hollow cross-sections track-bed decks, as shown in Fig. 7, are proposed and summarized in Table 1. The hollowed-out cross-sections are box section (HS-I), box-serrated section (HS-II) and serrated section (HS-III). The vacant percentage of HS-I, II and III are 6.71%, 12.08%, and 10.07%, respectively. As the minimal plate thickness of the bottom-deck of HS-II is only 30 cm, which is thinner than the required protection thickness for prestressing strands. Therefore, the channel-beam only uses HS-II at the locations without lateral strand, and use HS-I at other locations. Comparing to the original solid track-bed deck, the self-weight of HS-I, II, and III decreased by 12.32%, 15.33% and 15.44%, respectively.

4.2 Numerical Evaluation of Channel-Beam with Hollow Decks

FE Models. The mechanical properties of the hollow-deck channel-beam are analyzed using the *Abaqus* software. The execution of element selection, material modelling, and boundary definitions are same to the aforementioned solid-deck channel-beam model. Since the load and structure of the channel-beam are symmetrical to the mid-span section, only 1/2 of the channel-beam is assumed for modeling the hollow-deck channel-beam. Figure 8 shows the FE models for the three types of hollow-deck channel-beams.

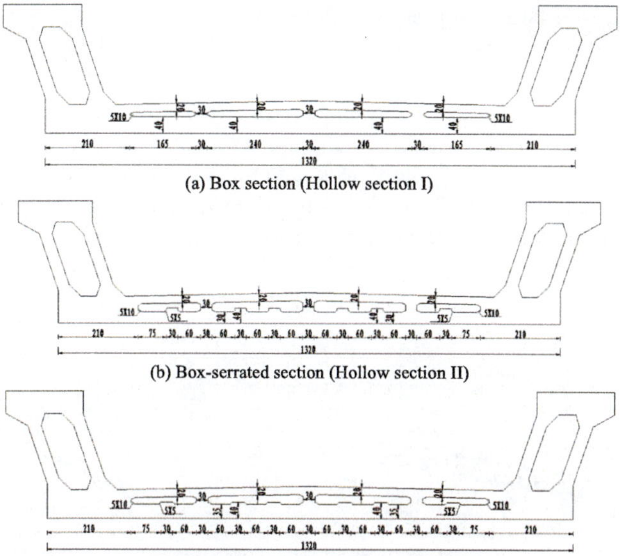

(a) Box section (Hollow section I)

(b) Box-serrated section (Hollow section II)

(c) Serrated section (Hollow section III)

Fig. 7. Hollow sections of concrete deck

Table 1. Characteristics of proposed hollow section track-bed decks

Type of sections	Area ($\times 10^5$ cm^2)	Distance from section center to bottom (cm)	Bending moment of inertia I_{yy} ($\times 10^9$ cm^4)	Volume of track-bed deck ($\times 10^9$ cm^3)
Solid section	1.49	99.36	1.44	259.39
Box section (HS-I)	1.39	103.82	1.40	227.44
Box-serrated section (HS-II)	1.31	107.04	1.37	219.62
Serrated section (HS-III)	1.34	105.30	1.39	219.34

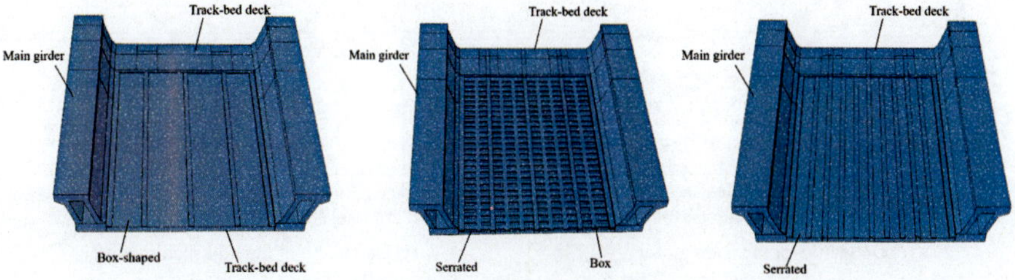

(a) Box deck (HS-I) (b) Box-serrated deck (HS-II) (c) Serrated deck (HS-III)

Fig. 8. Sectional view of hollow deck channel-beam

Results and Discussion. In order to evaluate the structural performance of the hollow-deck channel-beams, the most critical loading cases for longitudinal bending and lateral twisting are selected. The calculated numerical results obtained from the models under designated load combinations are presented in the following sections.

Most Critical Loading Case for Longitudinal Bend. Stresses at the bottom edge of side girders, bottom center of hollow-deck, and mid-span cross-section of the channel-beam are obtained from the FE models. Figure 9 shows the calculated results for the three structural forms of the hollow-deck channel-beam.

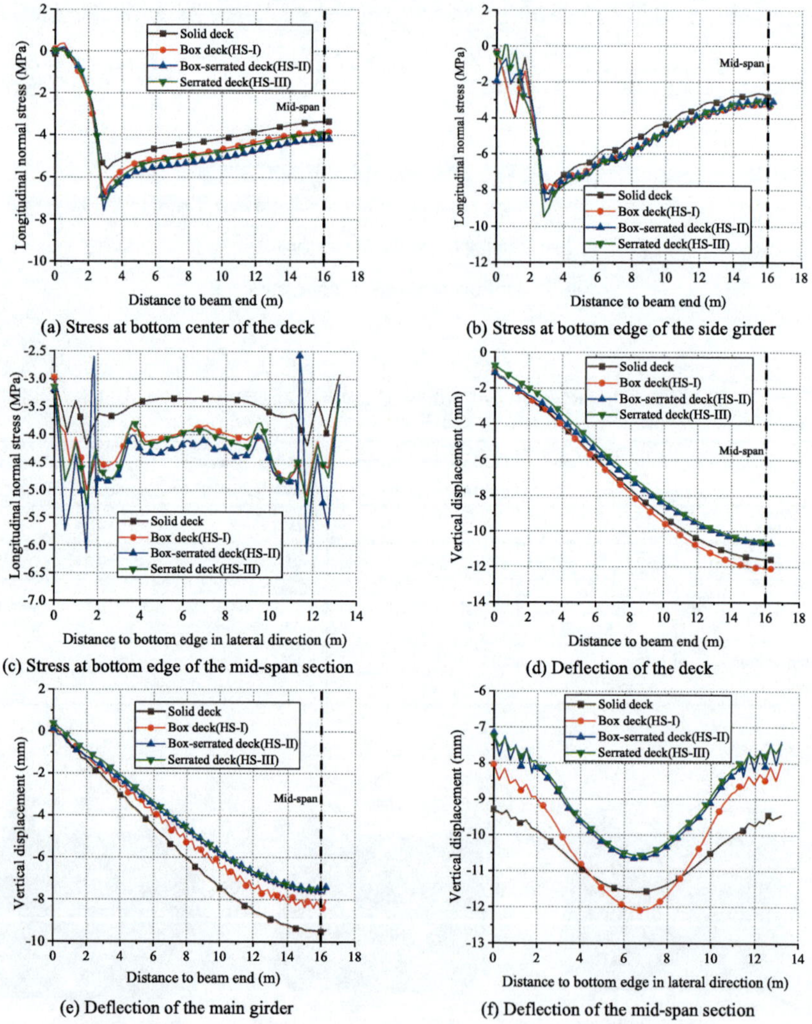

Fig. 9. Longitudinal normal stress and vertical displacement of hollow deck channel-beams with different structural forms

As can be seen in Figs. 9(a)–(c), under the most critical loading case of longitudinal bending, the compressive stress obtained from the hollow-deck bottom is larger than that of the solid-deck. Specifically, the box-serrated deck (HS-II) gave the largest compressive stress, which is followed by the serrated deck (HS-III), and the channel-beam using box deck (HS-I) gave the smallest compressive strength. With respect to the compressive stress at the side girder bottom, the hollow-deck channel-beam with different sections exhibited similar stress results, which are larger than that of the solid-deck channel-beam. The maximum compressive stress achieved from the deck bottom of HS-II is 40.60% larger than that of the original solid-deck.

Referring to Figs. 9(d)–(f), it can be seen that the deflection at the side main girder bottom of HS-II channel-beam is smaller than that of the solid-deck channel-beam, and the former gave a 23.1% smaller girder deflection as compared to that of the solid-deck channel-beam. Concerning to the deflection at the deck bottom, it is observed that the hollow-deck channel-beams presented 19.03% smaller deck deflection than that of the solid-deck channel-beam. For the channel-beam with different hollow-sections, the HS-II channel beam gave the smallest deflection, indicating that the channel-beam using HS-II track-bed deck has the largest longitudinal bending stiffness.

Most Critical Loading Case for Lateral Twist. For the channel-beam subjected to lateral torsion, the side girders would undergo torsional deformation and generate shear stresses. In order to analyze the channel-beam subjected to torsion, the torsional shear stresses obtained from the cross-sections at the beam supports are collected and summarized in Table 2.

Table 2. Shear stress in cross-section of supports (MPa)

Type of sections	Girder web	Beam-deck corner	Girder bottom
Solid deck	−3.01	−5.19	0.89
Box deck (HS-I)	−2.15	−5.43	1.24
Box-serrated deck (HS-II)	−2.22	−5.40	0.86
Serrated deck(HS-III)	−2.52	−4.68	1.12

The shear stress nephogram for cross-section at beam supports are plotted in Fig. 10. As can be seen, the maximum shear stress of the hollow-deck channel-beam occured at the side girder to track-bed corner. The shear stresses at the girder to deck corner of HS-I and HS-II are 4.62% and 4.05% larger than that of the solid-deck. In contrast, the calculated shear stress at the girder to deck corner of HS-III is 9.83% smaller than that of the solid-deck, which indicates that the use of serrated deck effectively optimized the stress in the corner zone.

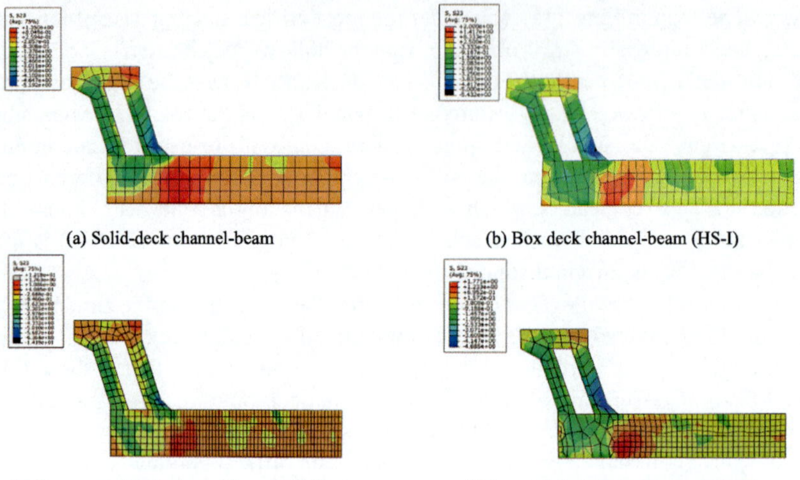

(a) Solid-deck channel-beam

(b) Box deck channel-beam (HS-I)

(c) Box-serrated deck channel-beam (HS-II)

(d) Serrated deck channel-beam (HS-III)

Fig. 10. Shear stresses at cross section of beam supports

By analyzing the calculated results of channel-beams with proposed hollow track-bed decks, it can be concluded that the channel-beam using box-serrated deck (HS-II) gave the smallest stress and mid-span deflection, and in addition of the highest sectional hollowing ratio. However, the combination of box and serrated cross-section of HS-II brings difficulties in constructing the structure. The box deck (HS-I) is easy to fabricate, but the stress and deflection is relatively large and the hollow ratio is the low. The serrated deck (HS-III) has the merits of large volume hollowing ratio, easy construction, small stress and mid-span deflection, can be used as an ideal hollow section form for channel-beams.

5 Conclusions

This paper numerically investigates the structural performance of channel-beam bridge with hollow track-bed deck. With the help of *Abaqus* software, a nonlinear numerical model to simulate the structural behavior of the Xifu River Railway Bridge that is a 32.0 m-long channel-beam bridge located in China was established. The flexural performance of the prototype bridge under the most critical loading case is analyzed. Based on the numerical results, three types of hollow cross-section concrete decks were proposed and numerically evaluated. The achieved conclusions are as followings:

(1) Under the most critical loading combination, there is no tensile stress found in the Xifu River Bridge, and the maximum compressive stresses at the deck bottom of mid-span in longitudinal and lateral directions are −7.27 MPa and −5.82 MPa, respectively. The solid-deck used by the prototype channel-beam bridge has very large safety margin and can be potentially optimized, in order to improve the load-bearing efficiency of the channel-beam.

(2) Three types of hollow-decks are proposed for channel-beams: box section (HS-I), box-serrated section and serrated section (HS-III). The vacant percentage of HS-I, II and III are 6.71%, 12.08%, and 10.07%, respectively. Comparing to the original solid-deck, the self-weight of the hollow-decks of HS-I, II, and III were decreased by 12.32%, 15.33% and 15.44%, respectively.

(3) A comprehensive comparison among the channel-beam using different track-bed decks was conducted and presented. The serrated deck (HS-III) has the merits of large volume hollowing ratio, easy construction, small stress and mid-span deflection, can be used as an ideal hollow section form for channel-beams.

Acknowledgments. The authors express their sincere gratitude for the financial supports provided by the Science and Technology Project of Guangzhou, China (Grant # 202102020652).

References

1. Zhang, X., Li, X.Z., Li, Y.: Dynamic characteristics study of U-beam applied in rail transit. Adv. Mater. Res. **243–249**, 2021–2026 (2011). https://doi.org/10.4028/www.scientific.net/AMR.243-249.2021
2. Zhou, W.: Construction technology of (16+2×20+16) m prestressed concrete continuous channel girder. Railw. Constr. Technol. **11**, 8–11 (2014). (in Chinese)
3. Staquet, S., Rigot, G., Detandt, H., Espion, B.: Innovative composite precast prestressed precambered u-shape concrete deck for Belgium's high speed railway trains. PCI J. **6**, 94–113 (2004)
4. Wang, J.: Design and construction technology of U-beam abutments for Mecca Metro. Railw. Constr. Technol. **S1**, 43–47 (2010). (in Chinese)
5. Wang, C., Li, H., Ren, G.: Innovative design and applications of U-Shaped girder in bridge engineering. In: International Conference on Advances in Steel Structures, pp. 187–193 (2012)
6. Li, Q., Wu, D.: Test and evaluation of high frequency vibration of a U-shaped girder under moving trains. In: International Conference on Structural Dynamics, pp. 1119–1124 (2014)
7. Heymsfield, E., Durham, S.A.: Retrofitting short-span precast channel beam bridges constructed without shear reinforcement. J. Bridg. Eng. **16**(3), 445–452 (2011). https://doi.org/10.1061/(ASCE)BE.1943-5592.0000167
8. Yu, Z., Liang, J., Jiang, S., Li, S., Zou, C.: Safety study of cast-in-situ support for double track railway on wide channel beam bridge. Vibroengineering PROCEDIA **28**, 206–210 (2019). https://doi.org/10.21595/vp.2019.21041
9. Torabian, S., Zheng, B., Schafer, B.W.: Experimental response of cold-formed steel lipped channel beam-columns. Thin-Walled Struct. **89**, 152–168 (2015). https://doi.org/10.1016/j.tws.2014.12.003
10. Cusens, A.R., Rounds, J.L.: Tests of a U-beam bridge deck. Struct. Eng. **1**, 1–14 (1973)
11. Krpan, P., Collins, M.P.: Testing thin-walled open RC structure in torsion. J. Struct. Divis. **107**(6), 1129–1140 (1981). https://doi.org/10.1061/JSDEAG.0005724
12. Raju, V., Menon, D.: Longitudinal analysis of concrete U-girder bridge decks. In: Proceedings of the Institution of Civil Engineers. Bridge Engineering, vol. 167, no. BE2, pp. 99–110 (2014). https://doi.org/10.1680/bren.11.00003
13. Wei, Y.: Design of 48m simply supported U-shaped beam of double-track railway. High Speed Rail. Technol. **8**(02), 41–44 (2017). (in Chinese)

14. Wu, D.: Research on the design of (40+70+70+40) m continuous trough girder of Ji-Qing high-speed railway. High Speed Railw. Technol. **9**(2), 53–56 (2018). (in Chinese)
15. Zhang, Y.: Construction monitoring and research on continuous channel girder arch bridge. Shijiazhuang Tiedao University (2018)
16. TB 10002-2017: Code for design on railway bridge and culvert (2017). (in Chinese)
17. Ding, X., Yu, Z.: Unified calculation method of mechanical properties of concrete in tension. J. Huazhong Univ. Sci. Technol. (Nat. Sci.) **3**, 29–34 (2004). (in Chinese)
18. Yu, Z., Ding, X.: Unified calculation method of compressive mechanical properties of concrete. J. Build. Struct. **4**, 41–46 (2003). https://doi.org/10.1007/s11769-003-0044-1. (in Chinese)

Seismic and Energy Upgrading of Existing RC Buildings: Methodological Aspects and Application to a Case-Study on the Italian Experience

Luciano Feo[1](✉), Enzo Martinelli[1,2], Rosa Penna[1], and Marco Pepe[1,2]

[1] Department of Civil Engineering, University of Salerno, Fisciano, SA, Italy
lfeo@unisa.it
[2] TESIS Srl, Fisciano, SA, Italy

Abstract. In most of the European countries, a significant part of the built stock is made of Reinforced Concrete (RC) buildings realized during the first decades after WWII. Therefore, those buildings do not generally comply with the safety standards implied by the modern codes and standards, both in terms of structural safety and energy efficiency. Therefore, they need to be upgraded with the aim to raise the safety and efficiency deficit with respect to new constructions. However, upgrading existing structures requires huge resources, which cannot generally be sustained by owners, especially in the case of private buildings. This paper presents some preliminary results obtained by following a systemic approach to both seismic and thermal upgrading of existing RC buildings. Specifically, after a general presentation of relevant methodological aspects, it summarizes the results obtained on a residential RC building lying in Italy, where a significant fiscal incentive programme has been launched by the government to foster upgrading operation on the existing built stock.

Keywords: Existing buildings · Reinforced concrete structures · Seismic upgrading · Energy efficiency · Integrated retrofitting

1 Introduction

The catastrophic seismic events occurred during the recent decades and the high consumption of energy make Italian existing built stock one of the most impacting assets in terms of both consequences on the environment and socio-economic implications [1]. In fact, in most of the existing buildings both structural members and installations are obsolete, often due to both a lack of ordinary maintenance and the fact that they are close to the end of service life. Therefore, combined seismic-thermal upgrading becomes one of the most current and most interesting issues, in response to which the current regulations aim at achieving appropriate levels of performance, through appropriate interventions and tax relieves [2].

R. Saidi et al. (Eds.): ICATH 2021, LNDECT 110, pp. 387–403, 2022.
https://doi.org/10.1007/978-3-030-94188-8_36

The Italian built stock was realized in relatively recent times, by means of different construction techniques and with reference to code provisions that were in force at the time of construction and are now obsolete [3].

The 2011 Italian census data (ISTAT – 2011 [4]) show that 12.8% of the total 10.5 million buildings were built between 1919 and 1945, and 4.1% of them is very bad preserved. Moreover, 56.7% of the entire residential stock was built from 1946 to 1980; while, from 1981 to 2011, the total buildings amounted to only 30.5% of the total [4]. The largest number of RC buildings (40.3%) were built in the period 1961–1980, when the regulations were incomplete and lacking in terms of earthquake safety provisions. Therefore, most of those buildings would not be able to withstand major seismic events [5]. The current share of RC buildings with more than 40 years, a time threshold beyond which substantial maintenance is essential, is growing progressively [6].

Therefore, the possible damage or collapse induced by natural events like earthquakes can lead to possible losses of human life and financial assets [7].

Moreover, beyond the aforementioned structural deficiencies, existing buildings presents envelopes with performances well below the currently required limitations, as a results of the significant deficiencies in terms of thermal insulation that would be required by modern standards [8, 9] and make even more relevant as a result of commitments deriving by the adoption of the Kyoto agreement [10].

On the other hand, the performance targets aimed at a sustainable development perspective are also linked to the fact that the construction sector is responsible for the greatest environmental impacts in terms of energy (40%) and raw material consumption (50%), harmful emissions (40%) and waste production (33%) in Europe. Therefore, enhancing the thermal and energy efficiency of existing building may have significantly positive consequences in terms of environmental sustainability and make human settlements safer, more resilient and more sustainable [11].

Therefore, it is necessary to implement upgrading policies of the current building stock, which guarantee the fulfillment of combined and integrated seismic and energy performance objectives, as required for new buildings: this would also minimize the gap between the existing and the newly built building stock. In the last decades, the hopeful implementation of structural and non-structural upgrading measures found a significant barrier in the scarcity of financial resources made available for this aim.

More recently, the raised awareness of the relevance for the aforementioned issues led National Governments in Europe, in coordination with the EU institutions, to settle adequate financial resources on implementing ambitious and widespread programmes aimed at reducing energy demand and raised seismic safety of existing buildings. In Italy, unprecedented fiscal incentive programmes, known as *Ecobonus* and *Sismabonus*, were launched with the aim to trigger and foster the private initiative of people interested in raising on energy efficiency and seismic safety of residential buildings and factories. More recently, in July 2020, the Italian Government established further and stronger fiscal incentives, known as *Decreto Rilancio* [12], as part of a set the urgent measures intended at relaunch economical activities after the Covid-19 pandemics.

However, the possibly uncoupled implementation of seismic and energy intervention strategies has raised serious limitations [11]. Traditional renovation projects carried out so far mostly ignored the multiple and complex needs of a building considered as a whole, as well as the possible interferences and synergies between the different types of intervention (e.g. energy and structural). Moreover, buildings that have only undergone structural upgrading, according to the recommendations of the regulations currently in force, could have no advantage in terms of energy demand. Similarly, in the case of energy upgrading only, a nearly zero-energy building, may paradoxically be characterized by a higher seismic risk induced by a rise in asset value (and, hence, exposition) and a basically unchanged vulnerability.

Therefore, an uncoupled approach to building upgrading or retrofitting is intrinsically ineffective in promoting a sustainable transformation of the built stock. Another aspect that is worthy of mention is that, during the interventions, the main issues to be resolved are linked to *(i)* the possible need for relocation of the inhabitants, *(ii)* the long duration of the renovation works and the downtime of the buildings during the intervention and *(iii)* the high initial construction costs. Therefore, in the perspective of an uncoupled intervention, these problems would increase exponentially.

In this context, the present work aims to summarize the fundamental aspects of the technical regulatory framework of interest for existing RC buildings in terms of seismic-energy efficiency, so as to define its application to a realistic case study, on which to carry out an integrated seismic-energy upgrading intervention. As already mentioned, the work is framed within the Italian context, as a result of the recently released fiscal incentive programmes supporting upgrading projects of existing buildings. However, under a methodological standpoint, it is relevant also to other contexts.

2 Methodological Aspects for Seismic and Energy Analysis

For the definition of intervention measures aimed at reducing energy demand and raising seismic safety in existing building, the first step is the classification of buildings with respect to the two aforementioned aspects.

2.1 Conventional Method for the Definition of the Seismic Class

The seismic classification of buildings is based upon the definition of 8 risk classes ranging from A+ (minimum seismic risk) to G (maximum seismic risk), which can be determined as a function of the two following parameters:

- PAM (*Perdita Annua Media*, Average Annual Loss);
- IS-V (*Indice di Sicurezza*, Safety Index).

These parameters can be determined by comparing the value structural capacity, quantified in terms Peak Ground Acceleration PGA_C that leads the structure to achieve the generic limit state of relevance, and the corresponding demand parameter PGA_D defined, for the same Limit State, with reference for a given return time T_{rD} (e.g., 475 years in the case of Limit State of Life Safety for ordinary buildings).

PAM - Average Annual Loss. Once the structure has been analyzed and the ground acceleration in PGA_C capacity defined, the return periods T_{rC} are calculated from the following formulation:

$$T_{rC} = T_{rD} \cdot \left(\frac{PGA_C}{PGA_D}\right)^{\eta} \tag{1}$$

where:

- $\eta = 1/0.41$ for the whole national territory;
- PGA_C is the peak ground acceleration at which the SLV is reached;
- PGA_D is the peak ground acceleration defined by the Italian regulation [13].

For each return time T_{rC}, the annual average frequency of overrun can be determined as follows:

$$\lambda = \frac{1}{T_{rD}} \tag{2}$$

Once the analysis has been conducted for both SLV and SLD limit states, the following are defined for the SLO and SLC:

$$\begin{aligned} \lambda_{SLO} &= 1.67 \cdot \lambda_{SLD} \\ \lambda_{SLC} &= 0.49 \cdot \lambda_{SLV} \end{aligned} \tag{3}$$

from which it is possible to calculate the direct economic loss CR (%), as the annual average frequency of exceedance (λ) varies [14]. Table 1 shows the percentages of variation of CR as the limit states considered: it is evident that the minimum economic loss (0%) is associated with the limit state of initial damage (SLID) and the state of reconstruction (SLR) is associated with the maximum (100%).

Table 1. Valutazione del CR (%) in funzione degli stati limite [14]

Limite state		CR [%]
Economic convenience limit of demolition and reconstruction	SLR	100
Limit Stat of Collapse	SLC	80
Limit State of Life Safety	SLV	50
Limit State of Damage (structural and non-structural elements)	SLD	15
Limit State of Damage (non-structural elements) – Operation	SLO	7
Limit State of beginning of Damage for non-structural elements	SLID	0

The PAM is the area below the curve of economic losses related to the average annual frequency (λ) of exceeding ($1/T_R$) of seismic events that lead to the achievement of a specific limit state (Fig. 1).

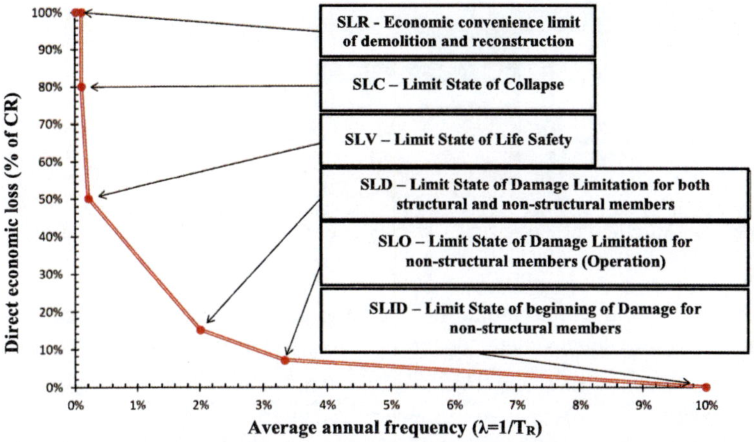

Fig. 1. Curve of direct economic losses defining PAM (adapted from [2]).

Therefore, this parameter is directly related to structural and non-structural damage and is considered in percentage terms of the construction costs (CR) of the building:

$$PAM = \sum_{i=2}^{5} (\lambda_{SLi-1} - \lambda_{SLi}) \cdot \frac{CR_{SLi-1} + CR_{SLi}}{2} + \lambda_{SLC} \cdot CR_{SLR} \quad (4)$$

The seismic classification as a function of the PAM is obtained as defined in Fig. 2.

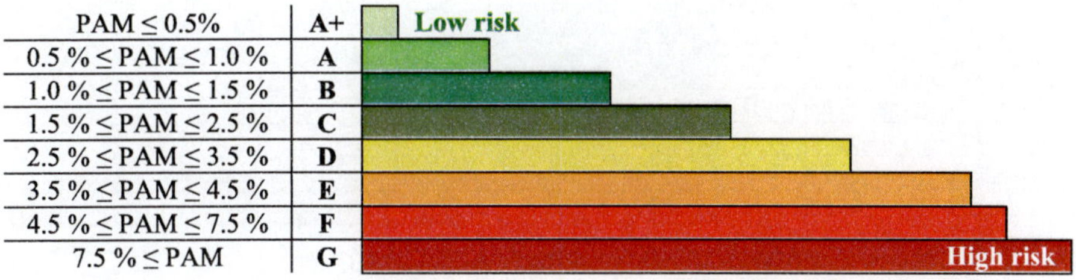

Fig. 2. Definition of the seismic risk classes as a function of the PAM values [14].

Safety Index. The IS-V safety index for the life-saving limit state (SLV), which defines the classification in Fig. 3, is given by the ratio between the construction capacity PGA and the construction site demand PGA:

$$IS - V = \frac{PGA_C}{PGA_D} \quad (5)$$

2.2 Conventional Method for Defining the Energy Class

The determination of the energy classes, generally reported in a document referred to as APE (*Attestato di Prestazione Energetica,* Energy Performance Certificate), aims at

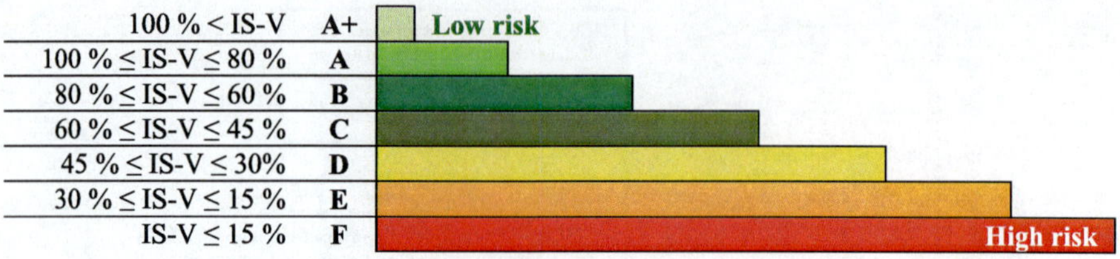

Fig. 3. Definition of the seismic risk classes as a function of the IS-V values [14].

measuring the global annual requirement of non-renewable primary energy consumed in reference to one square meter of the building under consideration. Therefore, this requirement ($EP_{gl,\ nren,\ ref,\ standard}$) is expressed in kWh/m^{2}year and is calculated according to the results obtained at monthly intervals. The factors that come into play for the evaluation of the energy class of a building are many and are mainly related to heating, cooling, ventilation, the production of domestic hot water, lighting and the operation of any lifts or escalators, where present, as well as the insulating quality of the windows and the thermo-technical characteristics of the envelope. As part of the combined interventions, particular attention is paid to the technical characteristics of the envelope and windows. Figure 4 shows the energy classification to be evaluated in the *APE*, defined on the basis of the energy performance guaranteed by each plant.

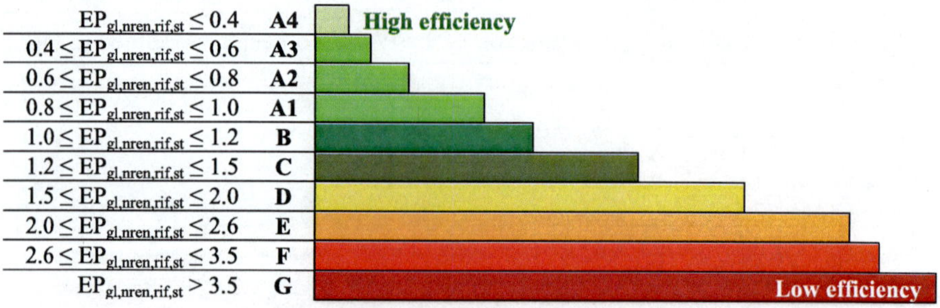

Fig. 4. Energy - Efficiency classification.

3 Case Study: Presentation

3.1 Analysis of the Current State of the Seismic Behavior

For an existing building, it is of fundamental importance to reach an appropriate level of knowledge, so as to be able to define an analysis that is as consistent as possible with the current state of the building. Specifically, designers are requested to survey geometry, construction details, and material properties.

Geometric and Structural Characterization. The structure under consideration is a residential building, located in Campania Region (Italy). The building, intended for residential use, refers to a realistic case study, typical of the '80s, on which to carry out a simulated design project. It has a regular elevation development but is characterized by an irregular plan conformation. The structure, entirely above ground, stands on a ground floor with an inter-story height of 2.65 m, followed by three floors in elevation, each with an inter-story height of 3.35 m, reaching a total height of 12.70 m. The building has frames made entirely of reinforced concrete and horizontals in brick-concrete cast on site, with vertical connections between the various floors guaranteed by appropriate stairwells. The main frame of the floors is in the longitudinal direction, parallel to the main development of the building, as shown in Fig. 5. In this direction there is a frame with three spans in the central area and with four spans to the north and south of the building. In the transversal direction there is a development mainly on three spans both in the more perimeter and central areas, while, where the stairwells are also inserted, there are frames with a single span.

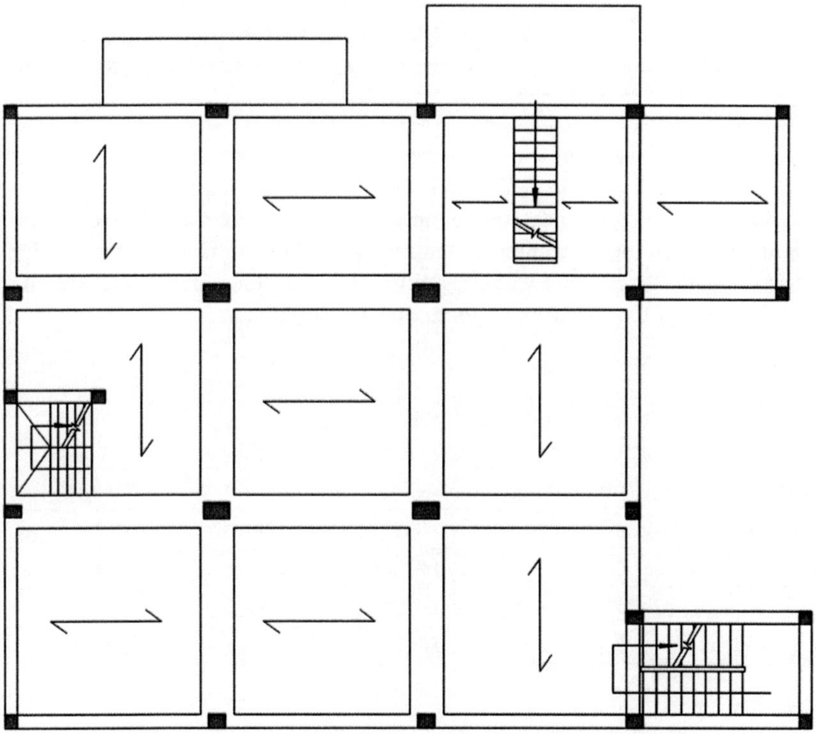

Fig. 5. Structural scheme in plant for the proposed case study.

Characterization of Materials. An adequate level of knowledge of the materials used in the construction of a building can be pursued, not only through accurate and in-depth surveys on site, but also through a critical study of the state of the art "crystallized" in the technical literature, where it is a case study realistic, as in the case in question, and not an actual project.

A recent research has shown that the concrete structures realized in 1980s between the province of Naples and Caserta were made by using concrete mixtures characterized by a compressive strength ranging from 15 MPa to 30 MPa [15]. The indications found herein, which mostly highlight the adoption of C20/25 concrete, comply with the mechanical characterization of the materials used for the structure under investigation. From the technical literature it also emerges that, in the time window that goes from 1974 to 1982, the bars mainly adopted in the province of Naples and in the neighboring areas were ribbed and of the FeB38k type, with a percentage of use of about 50% compared to the other types [16].

Design Criteria: Admissible Stress Method. The design criterion adopted for the construction of the analyzed building is the method of allowable stresses, according to which the structural elements are considered with a behavior typical of the linear-elastic range, both for concrete and for steel. The permanent and variable actions are assumed with characteristic values and the analysis performed is of the linear elastic type. Using this method, the design stresses (σ_d and τ_d) are determined, to be compared with the maximum allowable stresses, by means of point checks.

Seismic Characterization of the Reference Site. Having obtained the main parameters that come into play in the definition of the seismic characterization of the site [13] (see Table 2) it is possible to determine the values of the maximum horizontal acceleration at the site (a_g), the maximum value of the amplification factor of the acceleration spectrum (F_o) and the reference value T_C* for determining the start period of the constant velocity segment of the spectrum in horizontal acceleration.

Table 2. Seismic parameters.

Limit state	T_{rD}[years]	λ	S_S	PGA_D[g]
SLO	30	0.033	1.50	0.068
SLD	50	0.020	1.50	0.081
SLV	475	0.002	1.50	0.195
SLC	975	0.001	1.45	0.238

The elastic response spectrum obtained for the different limit states is shown in Fig. 6.

Modeling and Analysis of Seismic Vulnerability

The structural elements that come into play are mainly beams and columns and, for the sake of simplicity, a lumped-plasticity approach is followed to describe the nonlinear behavior of elements.

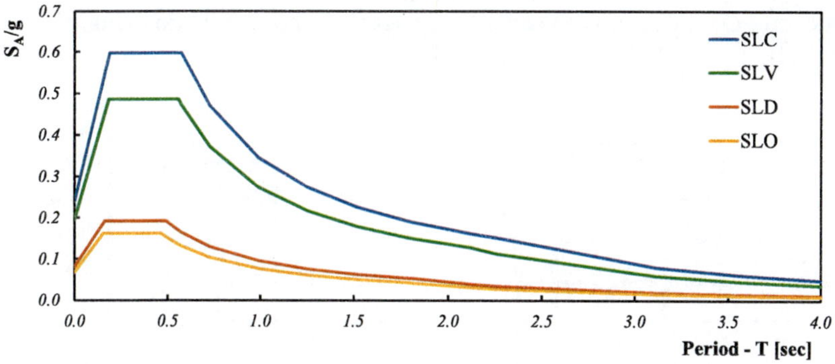

Fig. 6. Elastic spectrum demand for the site of the case study.

The load cases are then defined by including both gravitational loads and a system of horizontal forces equivalent seismic actions.

The software adopted for structural modeling is SAP 2000, thanks to which it was possible to create the structural model shown in Fig. 7.

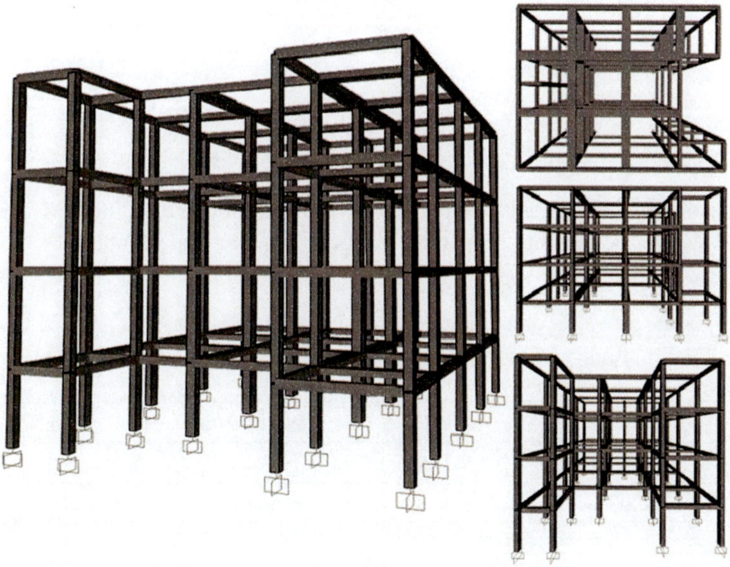

Fig. 7. Structural modelling of the building under investigation with SAP 2000.

Nonlinear static (pushover) analysis and the well-known N2 Method [17] are employed to obtain seismic demand on the structure for the various relevant Limit States. During the pushover analysis, gravitational loads and horizontal forces are applied at each deck, the latter in the direction of the considered seismic action. In the case in question, we have the combination of pushover analyzes different from each other in terms of direction (x, y), direction (positive and negative) and distribution (Fig. 8) (uniform and modal) for a total of 8 non-linear analyzes.

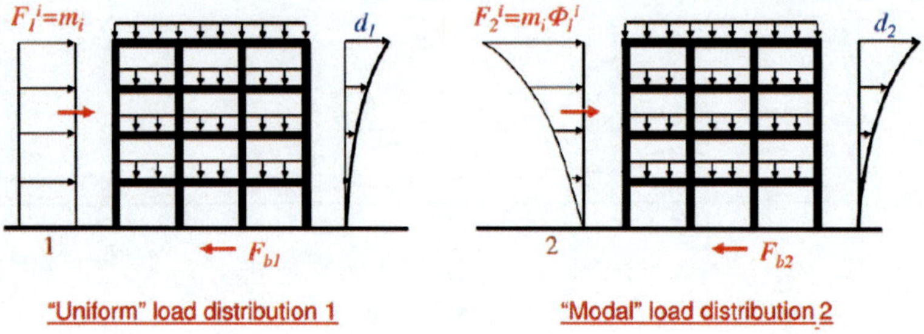

Fig. 8. Uniform and modal distribution for the pushover analysis.

The analyses with uniform distribution are obtained by applying a uniform acceleration to the floor masses, while the "modal" distribution sees a pattern of forces applied to each deck and obtained as follows:

$$F_i = \frac{F_h \cdot z_i \cdot W_i}{\sum_j z_i \cdot W_i} \tag{6}$$

where:

- $F_h = 1$ kN;
- z_i and z_j are the heights, with respect to the foundation plane, of the masses i and j;
- W_i and W_j are the weights, respectively, of the masses i and j (Table 3).

Table 3. Modal force distribution for pushover analysis.

Level	z_i[m]	W_i[kN]	$W_i{\cdot}z_i$[kNm]	F_i[kN]
0	2.65	3243	8593	75
1	6.00	3274	19646	171
2	9.35	3182	29749	258
3	12.70	3190	40519	352
Total		12889	98507	855.63

In the present case study, having launched the analyses for each load case above defined, it was possible to obtain the pushover curves: on each of them the step was defined in correspondence with which the limit states of interest are reached, depending on the formation of the hinges plastics, which occurred during the analysis. Then, moving from the performance curve of the MDOF system to the capacity curve of the equivalent SDOF system, and then proceed to its bilinearization.

Figure 9 shows the curves obtained in the case of uniform load, with analysis launched in x direction with negative direction.

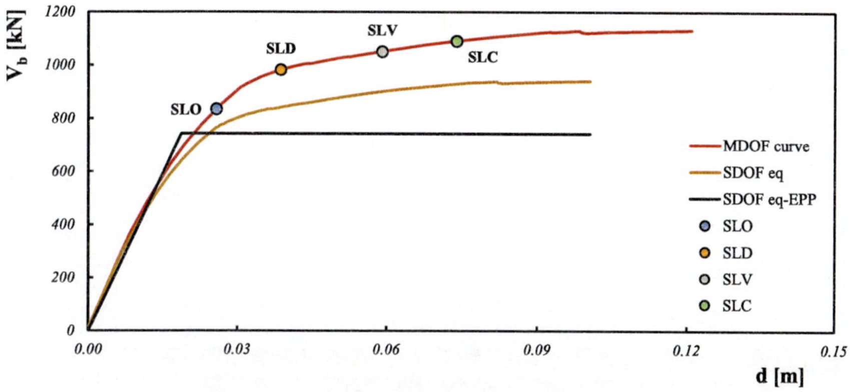

Fig. 9. N2 method: representative pushover curve.

Seismic Safety Classification. The procedural definition of the N2 method allowed a performance comparison of the structure, evaluating both the capacity limits and the demand requirements. In these terms and in accordance with the conventional method for defining the seismic class of the structures, it was possible to define the seismic class of the building under consideration. The results obtained show that the building belongs to the seismic class G (Table 4). Specifically, Table 4 defines in which direction, versus and type of distribution is present the greatest structural deficiency. In the case under investigation, the most critical conditions derive from analyses in y direction and with particular reference to SLD, as PAM is generally characterized by values lower than the corresponding IV-S.

3.2 Analysis of the Actual State of Energy Behavior

Climatic Characterization of the Reference Site. The climatic classification of Italian municipalities was introduced to regulate the operation and period of operation of the thermal systems of buildings, in order to contain energy consumption. Therefore, it is possible to proceed to the climatic characterization of the area in which the building under analysis is located, relevant for the energy assessments. The site in question is characterized by an average temperature ranging from an annual minimum of 8.7 °C to a maximum of 24.3 °C, with winters with higher rainfall than summer and humidity higher than 70% throughout year (Table 5).

The average daily temperature increases necessary to reach the threshold of 20 °C, a temperature that allows you to maintain a comfortable climate in homes, is approximately

Table 4. Seismic classification of the building under analysis.

Load case			Parameter			
Distribution	Direction	Versus	PAM		IS-V	
			Value [%]	Class	Value [%]	Class
Uniform	X	-	6.87	F	27.04	E
		+	6.86	F	27.27	E
	Y	-	852	G	25.07	E
		+	7.98	G	25.62	E
Modal	X	-	9.34	G	23.58	E
		+	9.79	G	23.46	E
	Y	-	14.22	G	21.53	E
		+	12.01	G	22.06	E
Seismic classification			14.22	G	21.53	E
			G			

Table 5. Climatic input data for the energy analysis.

Parameter	Jan	Feb	Mar	Apr	May	Jun	Jul	Aug	Sep	Oct	Nov	Dec
T_{mean}[°C]	8.7	8.7	10.7	13.4	17.1	21.2	23.8	24.8	20.8	17.5	13.7	10.2
T_{min}[°C]	6.6	6.4	8.0	10.5	13.9	17.9	20.6	21.2	18.2	15.1	11.6	8.1
T_{max}[°C]	10.7	10.9	13.2	16.0	19.6	23.7	26.4	27.0	23.3	19.9	15.8	12.1
Rainfall [mm]	128	109	107	94	62	30	18	25	91	139	190	150
R.U. [%]	75	73	75	76	76	74	71	70	70	75	75	74
Rain days [d]	9	8	8	9	6	4	3	3	7	8	10	10

1.18 °C per day: from this it emerges that the site where the building is located refers to a climatic zone "C" [18, 19].

Energy Diagnosis. The building complex under analysis is a residential building consisting of three real estate units to be subjected to energy improvement. It has a heated area of 279.21 m^2 per building unit and a gross heated volume of 1316.73 m^3. The opaque vertical surfaces are not effectively isolated, as they are characterised by high values of U transmittance, generating excessive heat losses due to transmission: in fact, U is equal to 0.86 W/m^2year, while the standards require that the transmittances in energy class C vertical opaque closures must not exceed 0.36 W/m^2K. The same performance shortcomings emerged from the characterization of the transparent surfaces: the windows have transmittances higher than the limits of 2.20 W/m^{2}K for climatic zone C, highlighting excessive dispersions.

From the energy audit it emerged that the inappropriate characteristics of the roof slab are also particularly affecting the thermo-energy performance.

Moreover, the periodic thermal transmittance exceeds the regulatory limit values equal to 0.18 W/m^{2}K for horizontal walls and equal to 0.10 W/m^{2}K for vertical ones. In addition, the analyzes carried out revealed problems regarding the formation of thermal bridges, compromising the health conditions of the building. Therefore, significant criticalities were found concentrated on passive components, with greater relevance for masonry and window frames.

In terms of plant engineering, the existing building is currently characterized by systems that guaranteed both winter air conditioning and the production of domestic hot water. The winter air conditioning was carried out by means of fan coils powered by a 48 kW GPL condensing boiler, the latter also capable of guaranteeing the production of domestic hot water, while a solar thermal system, with a nominal power of 1.84 kW, contributed to the production of energy from renewable sources.

The energy sources adopted, in this case, are responsible for a non-renewable energy performance index equal to 165.35 kWh/m^{2}year, while the renewable one amounts to 5.79 kWh/m^{2}year, resulting in a total of 36.11 kg/m^{2}year of emissions of CO_2. These performances mean that the building under consideration is defined in energy class F: the worst performances occur in winter, while in summer they are mediocre. The defined APE also shows that a new building, similar to the present case study and located in the same area, would be defined in class A1, with a consumption of approximately 52.01 kW/m^2 per year.

4 Case Study: Upgrading Solution

4.1 Structural Reinforcement: Concrete Jacketing

The critical aspects that emerged from the structural analysis phase revealed the need to carry out seismic upgrading. The choice of the latter focused on the main objective of raising the seismic characterization of the building complex by two classes, in accordance with the regulatory requirements in the field of seismic improvement.

The approach adopted involved a structural reinforcement implemented thanks to the concrete jacketing technique, achieved through the reinforcement of the sections of some pillars, obtained by adopting a C25/30 concrete and the use of longitudinal reinforcements and B450C steel brackets. This intervention is particularly advantageous for existing buildings given the extent of the benefits it offers. In fact, it is easy to implement, it is not invasive, it is not particularly expensive, it guarantees an increase in stiffness, an increase in resistance to normal stress, to bending and shear, it increases the ductility, as it turns the structural response from a local- to a global-mechanism.

In the case under consideration, it was decided to work on the corner columns, which limited the impact of the strengthening operation on the existing structural members and finishing. Specifically, the intervention in question concerned overall the 6 corner columns highlighted in Fig. 10, reaching a 70x70 section with the use of a Ø20/10 bar register.

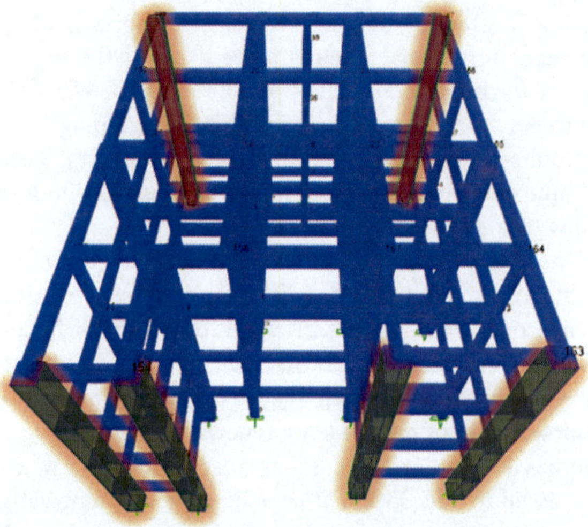

Fig. 10. Concrete jacketing for seismic retrofitting of the case study.

The comparison between the pushover analyses carried out on both the *as-built* and *upgraded* structures highlight a significant improvement in seismic performances (Fig. 11), as the building becomes capable of coping with cutting at the base and increased displacements.

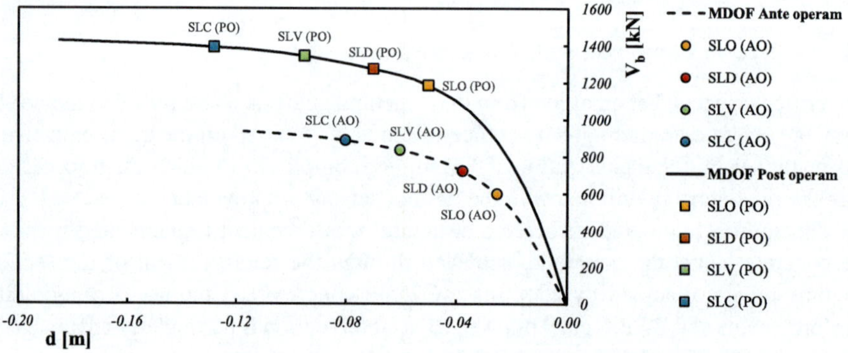

Fig. 11. Representative pushover curves (*as-built and* upgraded).

The designed upgrading intervention results in gaining two seismic classes, letting the building to move from class G (Table 4) of the as-built configuration to the class E of the project state (Table 6).

Table 6. Seismic classification of the case study after the concrete jacketing application.

<table>
<tr><th colspan="3">Load case</th><th colspan="4">Parameter</th></tr>
<tr><th rowspan="2">Distribution</th><th rowspan="2">Direction</th><th rowspan="2">Versus</th><th colspan="2">PAM</th><th colspan="2">IS-V</th></tr>
<tr><th>Value [%]</th><th>Class</th><th>Value [%]</th><th>Class</th></tr>
<tr><td rowspan="4">Uniform</td><td rowspan="2">X</td><td>-</td><td>2.53</td><td>D</td><td>39.68</td><td>D</td></tr>
<tr><td>+</td><td>2.16</td><td>D</td><td>42.41</td><td>D</td></tr>
<tr><td rowspan="2">Y</td><td>-</td><td>2.10</td><td>C</td><td>43.06</td><td>D</td></tr>
<tr><td>+</td><td>2.08</td><td>C</td><td>43.01</td><td>D</td></tr>
<tr><td rowspan="4">Modal</td><td rowspan="2">X</td><td>-</td><td>3.58</td><td>E</td><td>34.61</td><td>D</td></tr>
<tr><td>+</td><td>3.62</td><td>E</td><td>34.29</td><td>D</td></tr>
<tr><td rowspan="2">Y</td><td>-</td><td>3.49</td><td>D</td><td>34.64</td><td>D</td></tr>
<tr><td>+</td><td>3.61</td><td>E</td><td>34.62</td><td>D</td></tr>
<tr><td colspan="3" rowspan="2">Seismic classification</td><td>3.62</td><td>E</td><td>34.29</td><td>D</td></tr>
<tr><td colspan="4">E</td></tr>
</table>

4.2 Energy Upgrading

The building complex was subjected to an intervention scenario aimed at improving the performance of both passive and active components. The main characteristics of the upgrades defined in the project phase are outlined below.

Opaque and Transparent Surfaces. The pre-existing opaque vertical surfaces were coupled to a thermal insulation system in EPS and graphite with a thickness of 8 cm, with high performance thermal insulation properties. The materials are suitable: in fact, even the verification against the risk of interstitial condensation and the formation of mold, in the context of thermal bridges, has returned a positive result. At the same time, the replacement of the existing fixtures with more performing fixtures was defined: the transmittance values for the frame are equal to $U_f = 1.2\ W/m^2K$. These windows are equipped with a low-emission double glazing with transmittance values equal to $U_W = 1.3\ W/m^2K$ for double-leaf windows, in compliance with the regulatory limits that vary according to the climatic zone to which they belong.

The horizontal opaque surface subject to upgrading concerned the roofing surface, which was equipped with thermal insulation by laying an 8 cm thick EPS thermal insulator, to which a layer of bitumen with a thickness of 4 was subjected. mm and a double layer of bitumen sheets, each 5 mm thick, was superimposed.

Installations. Given the ineffectiveness and high consumption of the plant systems currently present, it was decided to intervene by replacing the winter air conditioning and domestic hot water production services with new generation technologies, while at the same time extending the services present, by means of the addition of summer air conditioning and mechanical ventilation systems. The electricity supply is no longer gas, as in the current state of affairs, but of the electric type. In this case, an electrical system with air-to-air exchangers with a nominal power of 37.50 kW were installed to guarantee winter air conditioning, an electrical system with air-air exchangers with a nominal power of 33.50 kW for summer air conditioning, a electrical system with air-water exchangers with 3.83 kW nominal power for the production of domestic hot water and fans to ensure mechanical ventilation. In addition, plants have been set up with the production of energy from renewable sources, distinguishing a photovoltaic system with a nominal power of 7.50 kW, replacing the solar thermal system currently present, and a heat pump of approximately 42 kW. From the analysis of the post-intervention performance it emerged that there is a non-renewable energy performance index of 23.67 kWh/m^2 year. Furthermore, the renewable energy performance index of 41.34 kWh/m^2year is reached. In terms of emissions it moves from a state of fact that causes 36.11 kg/m^2 year of CO_2 emissions to a project state in which emissions are equal to 5.26 kg/m^2 year: the reduction is 85%. From the post-intervention energy classification it emerged that the case-study reaches class A4+, obtaining the performance of a Zen building with almost zero energy and at the same time satisfying the requirements of the standards, as similar buildings would have reached class A3: from the state of fact to the state of project the class jump is equal to 8.

5 Conclusion

Carrying out integrated seismic-energy improvement interventions for existing structures has multiple purposes and objectives, among which to raise seismic safety, reduce energy consumption due to obsolete installation technologies, guarantee healthy conditions, and reduce environmental impacts. Given the extents of these purposes, the construction sector, to date, has developed a market whose range of technologies adopted for integrated seismic-energy efficiency is rich and varied. In the seismic field, both more innovative interventions, to reduce demand, and typically traditional interventions, which make it possible to increase the capacity of the structures, stand out.

Having defined the critical and general overview above, this work has mainly focused its attention on an existing case study, subjecting it to a structural reinforcement by means of concrete jacketing, concentrated only on the perimeter pillars: the logic adopted in the choice of sections subject to reinforcement, in principle, was aimed at carrying out a seismic upgrading integrated with the thermo-energy one, acting mainly on the passive perimeter surfaces of the residential building in question, and then providing additional replacement of the existing systems.

The results that it was possible to achieve made it possible to validate the efficiency of the proposed intervention techniques, reaching significant performance standards. In fact, following the interventions, the seismic-energetic response of the structure confirmed a considerable global improvement. The seismic intervention proposed herein guarantees the resolution of about 80% of the criticalities present at the current state, as well as the jump of two classes, intervening on a negligible number of structural elements. At the same time, the thermo-energy interventions, designed for the residential building being analyzed, have validated the achievement of performance requirements mainly compliant with regulatory requirements.

References

1. Sassu, M., Stochino, F., Mistretta, F.: Assessment method for combined structural and energy retrofitting in masonry buildings. Buildings **7**(3), 71 (2017). https://doi.org/10.3390/buildings7030071

2. Lima, C., Martinelli, E., Pepe, M., Faella, C.: Towards an integrated approach to seismic and energy retrofitting of existing RC frame buildings. ACI Special Publ. SP **326**, 1–10 (2018)
3. Biagini, S.: Vulnerabilità sismica e miglioramento di un edificio ospedaliero in c.a. (in Italian) Master thesis, Alma Mater Studiorum Università di Bologna, Italy (2019)
4. ISTAT: Dati Istat sulle costruzioni Decade 2011 (2011)
5. Menna, C., Del Vecchio, C., Di Ludovico, M., Mauro, G.M., Ascione, F., Prota, A.: Conceptual design of integrated seismic and energy retrofit interventions. J. Build. Eng. **38**, 102190 (2021). https://doi.org/10.1016/j.jobe.2021.102190
6. Berto, L., Vitaliani, R., Saetta, A., Simioni, P.: Seismic assessment of existing RC structures affected by degradation phenomena. Struct. Saf. **31**(4), 284–297 (2009). https://doi.org/10.1016/j.strusafe.2008.09.006
7. Pertile, V., Stella, A., De Stefani, L., Scotta, R.: Seismic and energy integrated retrofitting of existing buildings with an innovative ICF-based system: design principles and case studies. Sustainability **13**(16), 9363 (2021). https://doi.org/10.3390/su13169363
8. Martinelli, L., Matzarakis, A.: Influence of height/width proportions on the thermal comfort of courtyard typology for Italian climate zones. Sustain. Cities Soc. **29**, 97–106 (2017). https://doi.org/10.1016/j.scs.2016.12.004
9. Di Giuseppe, E., Jona Lasinio, G., Esposito, S., Pasqui, M.: Functional clustering for Italian climate zones identification. Theoret. Appl. Climatol. **114**(1–2), 39–54 (2012). https://doi.org/10.1007/s00704-012-0801-0
10. Huovila, P.: Buildings and climate change: status, challenges, and opportunities. UNEP/Earthprint (2007)
11. Passoni, C., Marini, A., Belleri, A., Menna, C.: Redefining the concept of sustainable renovation of buildings: state of the art and an LCT-based design framework. Sustain. Cities Soc. **64**, 1–45 (2021). https://doi.org/10.1016/j.scs.2020.102519
12. www.gazzettaufficiale.it/eli/id/2020/07/18/20A03914/sg. (in Italian)
13. M.II.TT: Nuove Norme Tecniche per le Costruzioni, Ministerial Decree of 17 January 2018. https://www.gazzettaufficiale.it/eli/id/2018/02/20/18A00716/sg. Accessed 14 Oct 2021. (in Italian)
14. M.II.TT.: Sisma Bonus - Linee guida per la classificazione del rischio sismico delle costruzioni e i relativi allegati. Modifiche all'articolo 3 del Decreto Ministeriale numero 58 del 28/02/2017, Italian Ministry of Infrastracture and Transportation, Ministerial Decree n.65 of 7 March 2017. http://www.mit.gov.it/normativa/decreto-ministeriale-numero-65-del-07032017. Accessed 14 Oct 2021. (in Italian)
15. Petrungaro, G.F., Basile, A., Brandonisio, G.: Come sono evolute le resistenze del calcestruzzo dagli anni '30 ad oggi (2020). https://www.ingenio-web.it/28708-come-sono-evolute-le-resistenze-del-calcestruzzo-dagli-anni-30-ad-oggi. Accessed 14 Oct 2021. (in Italian)
16. Verderame, G.G.M., Ricci, P., Esposito, M., Manfredi, G.: STIL software v1.0 (2012). https://www.reluis.it/it/progettazione/software/stil.html. Accessed 14 Oct 2021. (in Italian)
17. Martinelli, E., Lima, C., De Stefano, G.: A simplified procedure for nonlinear static analysis of masonry infilled RC frames. Eng. Struct. **101**, 591–608 (2015). https://doi.org/10.1016/j.engstruct.2015.07.023
18. UNI 10349 - Riscaldamento e raffrescamento degli edifici - Dati climatici, CTI (2016). (in Italian)
19. UNI/TS 10300 – Prestazioni energetiche degli edifici, CTI (2016). in Italian

Impact of Coronavirus Pandemic Crisis on Construction Control Processes in Egypt

Nora Magdy Essa[1], Hassan Mohamed Ibrahim[2], and Ibrahim Mahmoud Mahdi[3](✉)

[1] MSC Candidate Faculty of Engineering, Port Said University, Port Fuad, Egypt
[2] Faculty of Engineering, Port Said University, Port Fuad, Egypt
[3] Construction Management, Faculty of Engineering, Future University in Egypt, New Cairo, Egypt
ibrahim.mahdy@fue.edu.eg

Abstract. In January 2020, the World Health Organization (WHO) declared the outbreak of COVID-19, a public health emergency of international worry. On March 11, 2020, the World Health Organization officially declared COVID-19 a pandemic. On March 16, 2020, the Egyptian Prime Minister began issuing decisions as preventive measures within the county's comprehensive plan to deal with any possible repercussions of the Coronavirus. This article objectives is to analyze the impact of the coronavirus pandemic on the controlling processes of construction projects in Egypt. This paper presents theoretical and practical findings; a 36-question questionnaire was used to collect data from 147 engineerings and construction (E&C) professionals and a cases studies approach analyzing the impact of the crisis on the controlling processes on construction projects. In addition, this paper discusses the results of the last section of the survey, where respondents indicated the most critical impacts on controlling processes and the work options strategies.

Keywords: Controlling process · Update · Covid-19 · Crisis · Construction projects

1 Introduction

The success of construction projects is crucial for both stakeholders and a country's economic and social development (Sohu et al. 2018). Construction projects produce employment and generate revenue at both the national and local levels. One crucial concern in construction projects is the planning and controlling projects to lessen the probability of potentially devastating consequences of risks on project performance. Effective planning leads to the success of the project and its completion on time at the lowest cost and highest quality according to scope, as the project management team develops the master plan of the project and the planning process develops the components of the project management

R. Saidi et al. (Eds.): ICATH 2021, LNDECT 110, pp. 404–427, 2022.
https://doi.org/10.1007/978-3-030-94188-8_37

plan and establishes reference lines that are monitored and controlled during the implementation process, as well as a study of the risks that may be exposed to The project and when any changes occur during implementation, one or more planning processes are reconsidered as it is a continuous process throughout the life of the project (Mirghani 2016). The control process monitors the project's performance with the project plan and reference lines to determine the variance so that it can take corrective measures to achieve the objectives of the project and its success (PMI 2017). (Figure 1 & Fig. 2) present the planning and control processes group Adapted from (PMI 2017).

Since the first reported infections in Wuhan, China, in December 2019, COVID-19 has had a significant toll on human health and life. Over a million people globally have been infected, and many thousands have died. In addition, billions have been affected by government efforts worldwide to slow the spread of the virus through stay-at-home/lockdowns orders, quarantines, travel restrictions, heightened border scrutiny, and other measures, which have been far-ranging and change almost daily. The pandemic and government measures are taken in response have had a significant impact, both direct and indirect, on businesses and commercial relations, as worker shortages have occurred; supply chains have been disrupted; facilities and production have been suspended and shut down; and demand for and supply of certain goods and services has dramatically fallen (Rothan and Byvydy 2020).

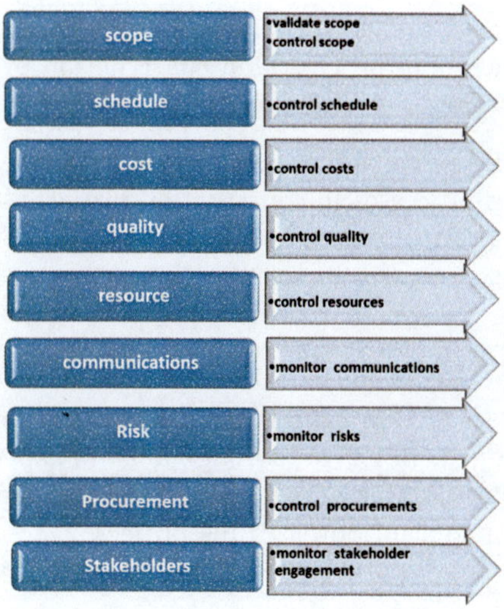

Fig. 1. Present the planning processes group adapted from (PMI 2017).

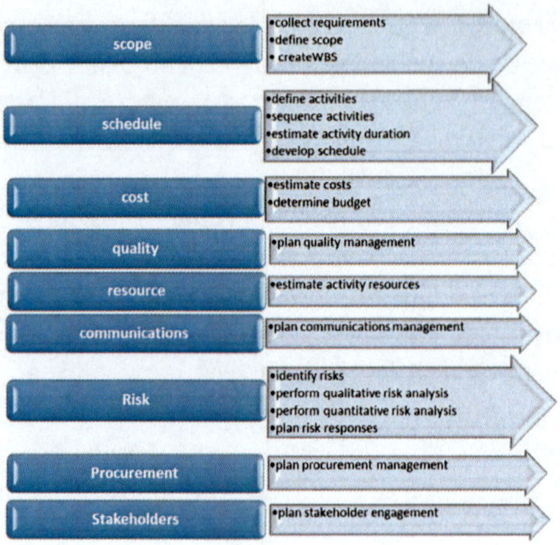

Fig. 2. Present the control processes group adapted from (PMI 2017).

1.1 Problem Statement and Hypotheses

The precautionary measures issued by the states have affected some industries such as construction, shipping, tourism, mining, gas, and aviation (Trenor and Lim 2020). The rapid spread of MERS-CoV from person to person makes labor-intensive industries such as the construction industry more vulnerable (Rothan and Byvydy 2020). Hence the question: what extent does the Corona virus pandemic crisis affect controlling processes on construction projects in Egypt?

The following hypotheses guided the study:

H0. There is no statistically significant effect of the Coronavirus pandemic crisis on controlling processes in construction projects.
Ha. There is a statistically significant effect of the Coronavirus pandemic crisis on controlling processes in construction projects.

1.2 Research Objectives

1. Measuring the impact of Coronavirus pandemic crisis on controlling processes on construction projects in Egypt, through a list of aspects of influence as the researcher read from the literature is presented, and the defendant requested a statement of the extent of power in his projects.
2. Assessment of the work option strategies under the coronavirus pandemic crisis.

2 Literature Review

2.1 Scope

Project scope is one of the most essential constraints to any project and is one of the factors that influence project performance (Shirazi et al. 2017). Scope planning processes include gathering and documenting stakeholder requirements, defining the project scope and work breakdown structure WBS to ensure that the project consists of only the actions required for project completion and success (Khan 2006).

Project requirements must be adequately defined and compliant with stakeholder requirements to confirm the success of the project and maintain the schedule and cost (Luiz Lampa et al. 2017). The input needed for collecting requirements is the project charter, contract documents, and the list of stakeholders. Interviews, questionnaires, surveys, data analysis, group decision-making methods, and prototypes are used to document requirements, develop a management plan and prepare a matrix to track them (PMI 2017). The baseline for the scope is the main output from the WBS process that will be controlled throughout the project's life and updated in the changes required to ensure the project's closure and success (Mirza et al. 2013).

The scope verification process in which applicable regulations verify all project deliveries, design panels, and quality standards continue throughout the project's life (Khan 2006). Project scope control significantly affects project performance and controls time and cost (Corvello et al. 2017). Follow-up the scope of the project and review the site results in the reference line of the scope prevents the abandonment of the scope, which is a common event in construction projects (Koke and Moehler 2019). The reasons for the scope creep are the instructions issued without realizing the size of the scope and oral instructions (Khan 2006).

2.2 Schedule

Good planning of the project's schedule and the planned closure of the project at the scheduled time is a benchmark for ensuring the project's success (Doleeb 2016). Project schedule planning includes the identification of project activities, the process of sequencing of project activities, the assessment of activities' resources, the estimation of their periods, and the development of the project schedule (PMI 2017).

Activity sequencing process using the Schedule Diagram Method PDM, Leads and Lags and Schedule Network Templates to prepare schedule network charts and consequent updates in project documents (Hebert and Deckro 2011).

When scheduling a project with resource constraints, the main challenge for managers is to develop a program that reduces project delivery time by reducing compliance with resource sequences and resource constraints. This problem is identified as resource-constrained project scheduling problem (RCPSP) and is usually solved using flexible resource management (Lima et al. 2019). The schedule baseline is the main output from the development schedule process, controlled throughout the project (PMI 2017).

The input needed for Control Schedule is the project management plan, project documents, and the organization's process assets. Data analysis, leads, and lags, and schedule compression are used to identify schedule forecasts, change requests, work performance information, and project management plan updates

(PMI 2017). In addition, emergency reserve analysis mitigates project Schedule risks (Douglas 2001).

2.3 Costs

The accuracy of cost estimates is very important for all parties involved in the construction project, so the factors affecting the cost estimate such as the complication of the project, the construction method, the volume and scope of the project, the limitations of the project site, the market condition and the duration of the project must be taken into account (Akintoye 2000). Contingency reserves are estimated at the planning stage and are included in the project's core budget (De Marco et al. 2015). A budget is set up to respond to risks and maintain the project budget (Touran and Liu 2015). Failure to estimate contingency reserves could lead to project delays and cost overruns (Salah and Moselhi 2015a).

The project budget is based on the scope of the project and the requirements of the owner. It includes the cost of design, management reserve, the cost of quality control, the cost of financing, the interest, equipment, and the cost of the land, as it should include contingency reserve to allow flexibility in making decisions during the design as the construction cost is estimated at 65% to 70% of the project budget (Yates and Eskander 2002).

In the case of external financing of the project, the time needed to obtain project financing, the cost of financing, and the change of interest rates can affect the overall budget and the project schedule, full knowledge of the financial markets, and many potential sources of financing for the project (Yates and Eskander 2002). The main output from the determined budget is the cost baseline, which is used to compare actual results during implementation and does not include management reserve (PMI 2017).

The control costs practice monitors project costs to avoid incorrect costs from being included in the cost baseline and to guarantee that the project budget is not exceeded (Heldman 2016). Data analysis includes analysis of earned value, variance, trend, and reserve (PMI 2017). Earned value technology integrates time, cost, and project scope to assess project performance and predict project progress (Moradi et al. 2017). In addition, the analysis of the earned value and variance allows early errors detection and reduces budget and schedule deviations (Araszkie-wicz and Bochenek 2019).

2.4 Quality

Quality management means satisfying the client and meeting the specifications and drawings of the contract within a specific budget and schedule. The most acceptable method for calculating the cost of quality includes the cost of prevention or minimizing errors of implementation defects such as training, appropriate equipment, and time, as well as the cost of evaluation such as testing, inspection, validation, and auditing (Love and Irani 2003).

The factors affecting the cost of quality in the planning phase are identifying quality goals, providing effective leadership, developing the team, identifying project objectives, and testing procedures (Mashwama et al. 2017).

Quality control is a process in which processes are monitored, performance and testing problems resolved, and manufacturers are supervised for final product control (Linda and Jaroslav 2019). The factors affecting the cost of quality in the control phase are safety and health requirements, evaluating the performance of activities on the critical path, measuring the work carried out, resource productivity, and variation in the use of planned and actual measurement of resources (Mashwama et al. 2017).

2.5 Resource

The resource plan is essential to ensure the most effective use of resources during the life of the project and to ensure the success of the construction project (Mesároš and Mandičák 2015, January). The input needed to develop the resources plan is the requirements of the activity resources, enterprise environmental factors influencing the project, and the organization process assets. In addition, organizational charts, job status descriptions, networking, and organizational theory are used to prepare the resources plan (PMI 2017).

The input needed for the control resource process is the project management plan, project documents, agreements, work performance data, and the organization's operational assets. In addition, project performance assessments, personal skills, problem-solving, and alternatives analysis are used to make change requests and updates in the project management plan and work performance information (PMI 2017).

The factors influencing resource management are effective communication between contractors and suppliers, providing a risk-free and secure environment for teamwork, sharing knowledge of skilled workers, the experience of site managers from similar projects, periodic meetings, appropriate planning before the project, selecting an effective team, training team, controlling inequality and reducing adjustment (Fapohunda and Chileshe 2014).

Managing the flow of resources is a difficult process where human beings fail to make the best use of resources so information systems are used to manage resources through the use of information and communication technology ICT to ensure reduced time and cost and focus on customer satisfaction The supply chain goes through several stages of the supplier, manufacturer, distributor, and customer. (Mesároš and Mandičák 2015, January). Failure of resource supply leads to project failure, increased cost, and project delays as factors affecting the availability of project resources include transportation, stakeholder factors, and factors related to poor construction such as fluctuating resource prices and factors related to the operational environment of the project such as a shortage of qualified construction workers (Chang et al. 2012).

2.6 Communications

Effective stakeholder communication plan improves project performance, increases trust among team members, participation in decision-making, knowledge sharing, and learning (Yap and Skitmore 2020). How the employer communicates and the employees affect employee satisfaction and break down barriers between project team members, collaboration and sharing of experiences (Wang and Liu 2009, December).

The input required for the communication planning method is the register of stakeholders, the management strategy of those involved, the environmental factors influencing the project, and the operational assets of the organization. Analysis of communication requirements, technology, models, and communication methods are used to set up a communication management plan and make updates in the project management plan (PMI 2017).

The most effective communication methods are face-to-face meetings, e-mails, brainstorming, and information and communication technology ICT learning (Yap and Skitmore 2020). In addition, ICT between consultants, contractors, and suppliers in different geographical locations reduces communication problems as it reduces time, cost, and travel as it helps to retain and retrieve information when needed (Weippert et al. 2002).

Lack of communication can lead to re-employment and increased requests for change, conflict, and delays, so it is necessary to ensure good communication between the main participants in the project and the exchange of information and ideas (Jafari et al. 2020). The input needed for the communications monitor process is the project management plan, project documents, work performance data, and the organization's operational assets. Communication methods and personal skills are used to set up change requests, project management plan updates, project document updates, and work performance information (PMI 2017).

2.7 Risk

The risk management process is intended to increase the likelihood of positive events and reduce the likelihood of adverse events (Grigore et al. 2018). Therefore, it is an important process for the construction project's success and includes risk identification, qualitative and quantitative risk analysis, risk response plan, and risk control (PMI 2017). For the success of risk management, all stakeholders must have a good project manager to understand and assess risks and make the right decision (Khan and Gul 2017).

The risk identification process in which the sources and types of potential risks are identified and documented in the risk register (Banaitiene et al. 2011). This process should be performed throughout the life of the project. The risk record is updated, and the risk is classified internally and externally (Turnbaugh 2005)-also categorized construction project risks into the following nine classes: construction and economy-related, environmental and force majeure, financial, managerial, partnering, legal and regulatory, political, and technical risks (Alashwal and Al-Sabahi 2018). The tools used to determine the risks of the project are the review of documents, brainstorming sessions, interviews, Delphi technology, and strength-weakness opportunity threat (SWOT) analysis (Salah and Moselhi 2016).

The process of qualitative risk analysis in which the priority of risk is determined in terms of its impact and probability (Agrawal et al. 2016). The risks are classified into high, medium, and low categories depending on the probability of the risk and the severity of its impact (Banaitiene et al. 2011). The tools used in this process are decision-making, Delphi technology, expert consultation, Monte Carlo simulation, impact charts, and probability analysis (Adedokun et al. 2013; Sarvari et al. 2019)

Risk response process in which an emergency plan is put in place to control risks as they occur and options to reduce threats (PMI 2017). There are four risk response strategies: avoiding, transferring, mitigating, and accepting risks (Banaitiene et al. 2011). The risk can be mitigated or avoided by reducing the cause, getting rid of it, or transferring the risk to the other. Some risks are beyond the project manager's control, such as disasters and changes in climatic conditions; in this case, the risks can be accepted and wait for the end of the event or continue the construction process (Turnbaugh 2005).

Risk control process in which the project performance, risk situation, and the implementation of the risk management plan are monitored throughout the project's life (Ning and Mao 2011; Mamoghli et al. 2015). Audits, data analysis, and meetings of the tools used for risk monitoring to examine the effectiveness of the risk response (Shilts 2017). The remaining contingency reserves are compared to the remaining risks in the project to ensure that the remaining reserves are sufficient to cover the risks (PMI 2017).

2.8 Procurement

Good procurement planning achieves the desired results and saves money where purchases are determined by the project's scope and requirements, the method of purchase, and the identification of vendors (Laryea 2019). The basis for choosing a bid should depend on the contractor's reputation, timely delivery of the project, and technical capacity (El-khalek et al. 2019). The bidding process can be done through automatic evaluation to save time (Idrees 2015).

Control procurement A process in which contracts are monitored and ensured to be successfully closed with the required changes made (Laryea 2019). The inputs needed for the procurement control process are project management plan, procurement documents, agreements, work performance data, and approved change requests. In addition, procurement performance reviews, checks, audits, and claims management are used to prepare procurement documents updates, the organization's operational assets, requests for change, project management plan updates, and closed procurements (PMI 2017).

2.9 Stakeholder

The construction industry includes many stakeholders who should be well involved in the project to increase support, optimal use of resources, and the project's success (Srinivasan and Dhivya 2020). Initially, stakeholders are identified, the functions and responsibilities are identified, and an appropriate plan is developed to communicate between them and determine their needs, strength, and impact (Senaratne and Ruwanpura 2016). The input needed for the Monitor stakeholder engagement process is the project management plan, work performance data, and project documents. It is used for information technologies, decision-making, and meetings to produce work performance information, change requests, project management plan updates, and project documents (PMI 2017).

2.10 COVID-19 Outbreak

The novel coronavirus is a highly infectious virus that spreads rapidly among humans and was regarded as responsible for tremendous health issues and economic and social impacts on every facet of the environment (Sohrabi et al. 2020). SARS-CoV-2 is a new strain of viruses not known to humans and made its first appearance in Wuhan City, China, at the end of 2019 and is spreading rapidly among people through the spray. Flu-like symptoms accompanied by difficulty breathing and pneumonia can lead to death (Cirrincione et al. 2020).

The infection is transmitted from person to person online and indirectly with an average incubation period of 4 to 6 days (Lai et al. 2020). Another study indicated that the symptoms of infection appear after the average incubation period of 2 to 5 days and that the incubation period depends on the immune system and the age of the human (Rothan and Byvydy 2020). The corona-virus epidemic has shut down many countries such as Italy, Malaysia, and many major cities such as Matila, Daegu, and Wuhan. As a result of this closure (Rothan and Byvydy 2020).

2.11 The Risk Consequences of Pandemic Coronavirus Crisis on the Construction Projects

Due to the outbreak of the Coronavirus epidemic, the Prime Minister of Egypt declared a state of emergency and the state and public sector companies issued some precautionary measures to protect citizens from any possible repercussions of the new Coronavirus, such as working from home without being at work, working on a rotational system, banning the movement or movement of citizens at times, curfews in some areas and suspension of international air traffic in all Egyptian airports (Egyptian Newspaper and Official Facts 2020). Precautionary measures issued by countries have affected some industries such as construction, shipping, tourism, mining, gas, and aviation (Trenor and Lim 2020). The rapid spread of MERS-CoV from one person to another makes labor-intensive industries such as the construction industry more vulnerable (Rothan and Byvydy 2020).

Increased potential for infection due to work in the construction industry, which needs to be located on-site and hotels near the site and work in remote areas. Which makes it difficult to access health care. (Edwards and Bowen 2019). In addition, continuous movement in the worksites to follow the work leads to the transmission of infection and inability to work, which makes drivers more vulnerable to infection due to the nature of their Work (Hishan et al. 2020).

The Coronavirus pandemic has affected the quantities bill, the project budget, and the schedule, the completion of the project, human resources, and the occurrence of force majeure and affects both sides of the contract and the performance of some contractual obligations (Kabiru and Yahaya 2020). It also affected the supply of building materials, supply chains, and staff and his presence on-site due to the difficulty of moving to work sites, who usually stay in hotels near work sites (Jallow et al. 2020). In addition, CORONA virus pandemic could lead to a request by one of the parties to the contract to adjust the prices or the duration of the project as a result of the contract containing clauses on the rights of delay as a result of certain events or conditions of price adjustment (Trenor and Lim 2020). Therefore, the pandemic, directly and indirectly, affected the construction industry and led to delays, suspensions and the completion of the contract, and bankruptcy (Bleby 2020; Johnson et al. 2020).

The term force majeure is widely used in construction contracts where in the event of force majeure, the contractor has the right to prevent the fulfillment of some of its contractual obligations as a result of the event's out-of-control or the extension of project time due to an unexpected shortage of goods or manpower due to government procedures and epidemics (Hansen 2020).

Force majeure could lead to an extension of the project time, suspension, or complete project termination (Hagedoorn and Hesen 2007; Ezeldin and Helw 2018). In the event of a force event mentioned in the contract and affects the performance of one of the parties from the performance of its contractual obligations, the work on the site is suspended until the end of the event. Then the work is resumed at the end of the event, and in case the circumstances of the project do not allow the disruption the contract is terminated (Amkhan 1991).

If the force leads to the termination of the project, each party to the contract will bear its consequences, and the contractor is compensated for part of the work done before the event (Hagedoorn and Hesen 2007). Not all the provisions of the force majure include the epidemic clause in some private contracts, which shows the conflict if the force clause is interpreted in force (Hansen 2020). Therefore, caution should be exercised in drafting future contracts, the conditions of force majeure, dispute settlement provisions, and expert consultation in the drafting of such clauses (Trenor and Lim 2020).

2.12 Construction Projects Continuity Under Coronavirus Pandemic Crisis

The state and companies face a difficult challenge to the potential consequences of the new CORONA virus by continuing to perform work and provide services within the specified schedules, especially multinational companies, where they have pursued a risk mitigation strategy such as remote working (Belzunegui-Eraso and Erro-Garcés 2020). The use of modern communication methods between project managers and project staff, such as video chat and online meetings, allows the team to manage and work at home (Jallow et al. 2020). Evaluating alternative options such as diversification of sources of supply, transport, employment, and buyers (Trenor and Lim 2020).

The need to work on the site and use physical spaces can be reduced by using intelligent and virtual applications, the internet, and shopping through e-commerce sites (Hishanetal 2020). Use the design of building information modeling and 3D structures to visualize the project in virtual meetings (Jallow et al. 2020).

Shared and distributed risks in the case of public-private partnership, the formulation of profitable incentives to face common risks, and increased cooperation and trust between the two parties (Baxter and Casady 2020). Increase confidence between customers and contractors and increase confidence in the workforce to deliver the project and continue it as agreed (Jallow et al. 2020).

Social spacing and telework reduce traffic and thus reduce the spread of infection (Megahed and Ghoneim 2020). The post-pandemic emphasizes the importance of looking forward to the innovations in construction techniques that speed the creation of emergency architecture. The COVID-19 pandemic represents an unprecedented challenge for health care systems internationally. Medical facilities and their human resources are usually overwhelmed By considering the Solving approach with SWOT analysis and Proactive, the construction companies should solve the crisis in the construction projects with minimal loss by establishing an early warning system (Rajprasad et al. 2018).

If necessary to go to work on-site and continuity of activities, preventive methods should be used to reduce the spread of the Coronavirus such as temperature detection and the use of the rotation system by dividing the team into two groups and in case of symptoms, one group is isolated (Cirrincione et al. 2020). In addition, use face masks and gloves and maintain recommended safety spaces in the presence of an official to verify these procedures (Casanova et al. 2016).

Reduce the number of people involved in trucking and provide hand sanitizers in offices and throughout the site (Jallow et al. 2020). Involve employers and construction contractors to use health education materials in local languages and provide free diagnosis and treatment (Shivalli et al. 2016). Use of photographs, posters, and signage in offices and workplaces to educate employees (Edwards and Bowen 2019).

Companies review their contracts to determine the rights and obligations of contracting parties, assess risks affecting certain contract obligations, and review contract terms that may justify the performance of the work in the event of unforeseen circumstances (Trenor and Lim 2020).

3 Methodology

To meet the aims of this research, a scientific methodology is pursued in:-

Review of any relevant local and international references such as (books, journals, researches, papers, etc.) That is directly relevant to the study and aided by reports and articles from the internet from websites of engineering facilities of relevancy; this represents the secondary data-primary data is formal using open questionnaire and the traditional closed questionnaire forms based on the data and information collected through the theoretical review and the open questionnaire to conclude and examine some concepts concerning the actual study status. This research depending on the questionnaire, must be workable. Thus, it is possible to define the significant effects of the Coronavirus pandemic crisis on controlling construction processes and assess the work option strategies under the coronavirus epidemic crisis using a method or criteria in projects management update to reach this goal. The questionnaire contained 36 questions, and it was tested by a panel of three experts who were asked to review the survey instrument. The experts were two professors in the project management field with knowledge of project management and one research consultant with knowledge of the academic research process, revised, and then distributed to 190 industry construction professionals. In total, 147 Construction industry professionals returned the surveys. The survey results were obtained by ranking the significant effects of the Coronavirus pandemic crisis on controlling processes on construction projects. The respondents also were asked to assess the work option strategies under the coronavirus epidemic crisis.

At the end of all the sections, the respondents were provided with blank spaces to add any other effects and the work option strategies listed in the survey instrument. A cases study approach that analyzes the significant impacts of the Coronavirus pandemic crisis on controlling processes on construction projects in Egypt.

The Likert scale of 1 to 5, where 1 - "Strongly disagree", 2 - "Disagree", 3 - "neutral or unsure", 4 - "Agree" and 5 - "Strongly agree" was used. The Likert scale is a popular format of a questionnaire that is used in education research. The Likert scale is chosen in this study because it allows the respondents to express how much they agree or disagree with certain statements. The Mean Item Score (MIS) is ranked in descending order (from the highest to the lowest). The statement with the highest ranking is the one that was considered to be dominant. The Mean Item Score (MIS) was derived from the following formula (Mashwama et al. 2016)

$$\frac{1n1\ +\ 2n2\ +3n3+\ 4n4+\ 5n5}{\sum N} \tag{1}$$

where

n1 = number of respondents for strongly disagree
n2 = number of respondents for disagree
n3 = number of respondents for neutral
n4 = number of respondents for agree
n5 = number of respondents for strongly agree
N = Total number of respondents

Table 1. A brief description of the case studies

Project	The estimated cost	Project Description
Infrastructure project	(L.E.415 million)	Infrastructure project for social housing (roads - water - sewage - irrigation - electricity) with an area of 202 acres owned by New Urban Communities Authority, Ministry of housing. Financial close in 2021
The essential education school building	(L.E.21 million)	Implementation project of an essential education school building with an area of 3000 m^2 owned by New Urban Communities Authority, Ministry of housing. Financial close in 2021.

3.1 Sampling Procedure

The researcher was keen to ensure that the research sample includes the parties to the work team, namely the employer, the main contractor, and the consultant and that the research sample includes different work sites directly related to the topic of the research, as a random sample was chosen in a scientifically systematic way so that this sample represents the research community in a correct representation.

The questionnaire survey was performed in Egypt; the survey targets the private sector (contractors, engineering consultants for large projects) and the Public Sector (Ministry of Housing, Utilities, and Urban Communities), the number of contractor's works in construction projects with L.E.2.5 million or more is 465, and this number was obtained from Egyptian Federation for Construction & Building Contractors at the year 2013. Therefore, the survey targets are considered a sample of 25 consultant companies and 50 entities affiliated with the Ministry of Housing, Utilities, and Urban Communities.

The total population for the survey is 540 contractors, consultants, and entities affiliated with the Ministry of Housing, Utilities, and Urban Communities.

3.2 Cases Studies

Two different projects in cost and specifications were chosen from the researcher's worksite (New Port Said City - Salam) to verify the results obtained from the questionnaires and know the impact of the Coronavirus pandemic crisis on these projects. The research methodology for these case studies is to Interview consultants and the contractors of these projects to discuss the questions contained in the questionnaire questions to verify the results. Table 1 shows a brief description of the case studies.

4 Results Analysis and Discussion

4.1 Survey Participants

The first section of the survey requested that the participants provide information on the job description, sector of work, experience, etc. This chart (Fig. 3) represents that 25.2% were owners, then 32% of them were consultants, then 38.1% contractors, then 3.4% subcontractors, and 1.4% suppliers. In addition, a group of construction industry professionals was interviewed to obtain in-depth information on the significant effects of the Coronavirus pandemic crisis on controlling processes on construction projects. The 12 elements provided in the survey were ranked by the respondents and then prioritized using statistical weight factors. Then they assessment the work option strategies for mitigating the effects of the coronavirus epidemic crisis.

4.2 Ranking Impact of Coronavirus Pandemic Crisis on Controlling Processes on Construction Projects Based on Survey

This crisis on controlling processes on construction projects based on survey. This section discusses the impact of the Coronavirus pandemic crisis on controlling processes on construction Projects in Egypt that the survey participants ranked in priority order. The effect that was ranked number one by the survey respondents was the cash flows of the project, In addition to being ranked as the most common the factors most affected by the crisis with (MIS = 4.02, R = 1) which is a large percentage As a result of the worldwide economic recession; The project's schedule was ranked second with (MIS = 3.97, R = 2); work completed was ranked third with (MIS = 3.89, R = 3); The performance of certain contractual conditions was ranked fourth with (MIS = 3.88, R = 4); The project cost was ranked as the fifth with (MIS = 3.85, R = 5); the start and finish dates of each activity was ranked as the sixth with (MIS = 3.74, R = 6); the shift system for workers was ranked next was with (MIS = 3.69, R = 7); deliverables status was ranked eighth with (MIS = 3.66, R = 8); stakeholders engagement was ranked ninth with (MIS = 3.6, R = 9); the number of working hours and construction

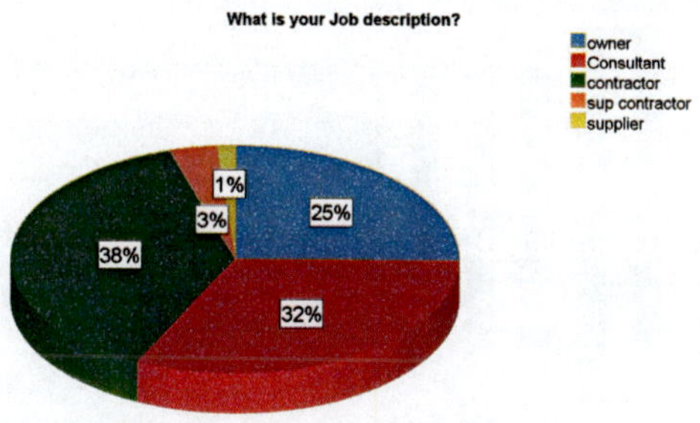

Fig. 3. Present the planning processes group adapted from (PMI 2017).

materials supply were ranked with (MIS = 3.57, R = 10) and finally project team management with (MIS = 3.41, R = 11) (Table 2 and Fig. 4).

Table 2. Ranking impact of Coronavirus pandemic crisis

Controlling processes	MIS	RANK
Cash flow of the project	4.02	1
The project's schedule	3.97	2
Work completed	3.89	3
The performance of certain contractual conditions	3.88	4
Project cost	3.85	5
The start and finish dates of each activity	3.74	6
The worker's shift system	3.69	7
Deliverables status	3.66	8
Stakeholders engagement	3.60	9
The supply of building materials	3.57	10
The number of working hours	3.57	10
Project team management	3.41	11

4.3 Ranking the Impact of Coronavirus Pandemic Crisis on the Contract Conditions

Respondents ranked the potential effects on the contract conditions (some contractual conditions can be waived) highest with (MIS = 3.45, R = 1); compensation of the contractor was ranked second (MIS = 3.42, R = 2); respondents ranked claim of the Alternative prices third with (MIS = 3.35, R = 3) and termination of some contracts was ranked last with (MIS = 3.03, R = 4) (Table 3).

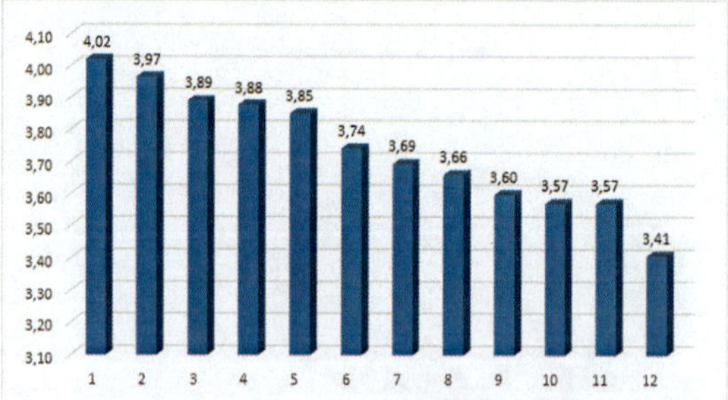

Fig. 4. Organization's categories of respondents

Table 3. The potential effects on the contract conditions

The possible impact on the contract conditions	MIS	RANK
Some contractual conditions can be waived under the Coronavirus epidemic	3.45	1
The contractor can be compensated as a result of the coronavirus epidemic	3.42	2
Alternative prices can be claimed in the face of the Coronavirus epidemic	3.35	3
Some contracts could be terminated under the Coronavirus epidemic	3.03	4

4.4 Ranking of Work Option Strategies Based on Survey

Respondents ranked work option strategies under the crisis of the Coronavirus epidemic (Having clauses in the contract related to the rights of delay or termination) highest with (MIS = 4.27, R = 1); Following precautionary measures was ranked second with (MIS = 4.26, R = 2) Adjusting the duration of activities was ranked third with (MIS = 4.25, R = 3); Settlement of claims and disputes through negotiation was ranked fourth with (MIS = 4.23, R = 4); Diversification of supply sources was ranked as the fifth with (MIS = 4.14, R = 5); The re-analysis of contingency reserves to maintain the project's schedule was ranked next was with (MIS = 4.08, R = 6); The re-analysis of contingency reserves to keep the budget for the project with (MIS = 4.03, R = 7); increase resources was ranked eighth with (MIS = 3.99, R = 8); Using building information modelling and 3D design models to facilitate remote work was ranked ninth with (MIS = 3.93, R = 9); Implement project activities in parallel with (MIS = 3.88, R = 10); Use of modern technology methods for monitoring with (MIS = 3.86, R = 11); Use virtual teams online and work remotely and with (MIS = 3.83, R = 12); finally enhancing the owner's confidence in the contractor to complete the project on time at the specified cost (Table 4 and Fig. 5).

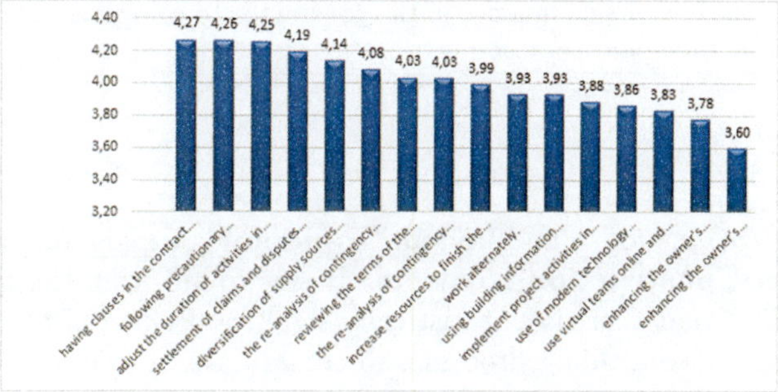

Fig. 5. Ranking of work option strategies.

Table 4. Ranking of work option strategies.

Work option strategies	MIS	RANK
Having clauses in the contract related to the rights of delay or termination of the conditions of price adjustment or conditions of hardship	4.27	1
Following preventive measures by detecting temperature, using a respiratory mask, maintaining safety distances, and using guidelines and health education materials are helpful in case of work on site	4.26	2
Adjust the duration of activities by the constraints related to the coronavirus epidemic crisis	4.25	3
Settlement of claims and disputes through negotiation	4.23	4
Diversification of supply sources and the use of alternative sources	4.14	5
The re-analysis of contingency reserves to maintain the project's schedule	4.08	6
Review the contract's terms to evaluate any provisions in the agreement that justify performance under coronavirus epidemic crisis	4.03	7
The re-analysis of contingency reserves to maintain the budget for the project	4.03	7
Increase resources to finish the project on schedule	3.99	8
Work alternately	3.93	9
Using building information modeling and 3D design models to facilitate remote work	3.93	9
Implement project activities in parallel to maintain the project schedule	3.88	10
Use of modern technology methods for monitoring to follow up on business, such as modern cameras	3.86	11
Use virtual teams online and work remotely	3.83	12
Enhancing the owner's confidence in the contractor to complete the project on time	3.78	13
It enhances the owner's confidence in the contractor to complete the project at the specified cost	3.60	14

4.5 Discussion

Measuring Impact of Coronavirus Pandemic Crisis on Controlling Processes on Construction Projects. It was found from the questionnaire and cases studies that there is a statistically significant effect of the Coronavirus pandemic crisis on controlling processes in construction projects.

- The highest impact of the crisis on controlling processes is the project's cash flows as a result of the worldwide economic recession.

- Deviation in the project schedule resulted from the 50% reduction in the number of workers and repeated infections of workers with the virus.
- The project was not completed on time.
- The crisis and the preventive measures taken by the state affected the performance of some contractual conditions, such as the project's duration, as the company executing the project submitted a request to extend the project period for an additional three months.
- Increasing the project cost due to the expenses related to the precautionary measures, such as masks, continuous sterilization of workplaces, and temperature measuring devices, also affected the wages of laborers.
- The start and finish dates of each activity were affected due to repeated injuries, which hinder follow-up, the rotation system.
- Projects operating in shifts have been affected.
- Activities deliverables affected as a result of poor communication
- Engage stakeholders were involved due to the rotation system and the state's 50% reduction of labor.
- Project resources are affected in the case of imports from abroad.
- The number of working hours has decreased due to the imposition of the ban and the state's preventive measures.
- Difficulty managing a work team as a result of poor communication

Assessment of the Work Option Strategies Under the Coronavirus Epidemic Crisis

- The existence of clauses in the contract related to rights of delay or termination, terms of price adjustment, or hardship conditions amending the duration of the contract.
- Following preventive measures by detecting temperature, using a respiratory mask, maintaining safety distances, and using guidelines and health education materials are helpful in case of work on site.
- Adjust the duration of activities by the constraints related to the coronavirus epidemic crisis.
- Settlement of claims and disputes through negotiation
- Diversification of supply sources and the use of alternative sources.
- The re-analysis of contingency reserves to maintain the project's schedule.
- Review the contract's terms to evaluate any provisions in the agreement that justify performance under coronavirus epidemic crisis.
- The re-analysis of contingency reserves to maintain the budget for the project.
- Increasing the materials if financial resources are available
- Planning and coordination between workers for work alternately.
- Using building information modeling and 3D design models to facilitate remote work.
- Implement project activities in parallel to maintain the project schedule.
- Use of modern technology methods for monitoring to follow up on business, such as modern cameras.

- The use of virtual work teams and remote work in cases that do not need direct supervision
- It enhances the owner's confidence in the contractor to complete the project on time and at the specified cost.

5 Conclusion and Recommendations

5.1 Conclusion

The findings of this study presented the effects on the controlling processes of projects in Egypt due to the Coronavirus epidemic crisis. The result revealed that the main effects are cash flows, non-compliance with the project schedule, non-compliance with the contractual conditions of the project, the additional cost, reducing the number of working hours, and difficulty engaging stake-holders. The work option strategies under the coronavirus epidemic crisis are as follows: preventive measures, adjusting the duration of activities, diversification of supply sources and the use of alternative sources, re-analysis of contingency reserves, and modern technology methods for monitoring and use virtual teams, and work remotely.

5.2 Recommendations

The study revealed the importance of having clauses in the contract related to delay rights, termination, price adjustment terms, or hardship clauses to be negotiated to allocate risks to specific events, include diseases and epidemics in the project risk register, and develop a plan to respond to them if they occur, and identify and document lessons learned from Dealing with the crisis to benefit from it in the future, and training in the use of modern technology methods to monitor and follow up business for ease of working remotely in the event of the spread of diseases and epidemics, spreading awareness and reducing panic at the time of outbreaks.

References

Adedokun, O.A., Ogunsemi, D.R., Aje, I.O., Awodele, O.A., Dairo, D.O.: Evaluation of qualitative risk analysis techniques in selected large construction companies in Nigeria. J. Facil. Manag. **11**(2), 123–135 (2013). https://doi.org/10.1108/14725961311314615

Agrawal, R., Singh, D., Sharma, A.: Prioritizing and optimizing risk factors in agile software development. Presented at the Ninth International Conference on Contemporary Computing (IC3), 1. GLA University, Mathura, India, 20 March (2016). https://doi.org/10.1109/IC3.2016.7880232

Akbiyikli, R., Dikmen, S.U., Eaton, D.: Insurance issues and design and build construction contracts. NWSA Eng. Sci. **7**(1), 203–217 (2012)

Akintoye, A.: Analysis of factors influencing project cost estimating practice. Constr. Manag. Econ. **18**(1), 77–89 (2000). https://doi.org/10.1080/014461900370979

Amkhan, A.: Force majeure and impossibility of performance in Arab contract law. Arab Law Q. **6**(3), 297–308 (1991). https://doi.org/10.1163/157302591X00359

Banaitiene, N., Banaitis, A., Norkus, A.: Risk management in projects: peculiarities of Lithuanian construction companies. Int. J. Strateg. Prop. Manag. **15**(1), 60–73 (2011). https://doi.org/10.3846/1648715X.2011.568675

Baxter, D., Casady, C.B.: A coronavirus (COVID-19) triage framework for (Sub) national public-private partnership (PPP) programs. Sustainability (Switzerland) **12**(13) (2020). https://doi.org/10.3390/su12135253

Belzunegui-Eraso, A., Erro-Garcés, A.: Teleworking in the context of theCovid-19 crisis. Sustainability **12**(9), 3662 (2020). https://doi.org/10.3390/su12093662

Bleby, M.: Construction feels COVID-19 delays in supply chain (2020). https://www.afr.com/property/commercial/construction-feels-covid-19-delays-in-supply-chain-20200304-p546pv. Accessed 4 Mar 2020

Casanova, L., et al.: Assessment of self-contamination during removal of personal protective equipment for Ebola patient care. Infect. Control Hosp. Epidemiol. **37**(10), 1156–1161 (2016). https://doi.org/10.1017/ice.2016.169

Chang, P.H.: Applying resource capability for planning and managing contingency reserves for software and information engineering projects. In: 2012 IEEE International Conference on Electro/Information Technology, p. 1 (2012). https://doi.org/10.1109/EIT.2012.6220715

Chang, Y., Wilkinson, S., Potangaroa, R., Seville, E.: Managing resources in disaster recovery projects. Eng. Constr. Archit. Manag. (2012). https://doi.org/10.1108/09699981211259621

Cirrincione, L., et al.: COVID-19 pandemic: prevention and protection measures to be adopted at the workplace. Sustainability (Switzerland) **12**(9), 1–18 (2020). https://doi.org/10.3390/su12093603

Corvello, V., Javernick-Will, A., Ratta, A.M.L.: Routine project scope management in small construction enterprises. Int. J. Project Organ. Manag. **9**(1), 18–30 (2017). https://doi.org/10.1504/IJPOM.2017.083109

De Marco, A., Rafele, C., Thaheem, M.J.: Dynamic management of risk contingency in complex design-build projects. J. Constr. Eng. Manag. **142**(2), 04015080 (2015). https://doi.org/10.1061/(ASCE)CO.1943-7862.0001052

Doleeb, S.M.M.: The process of planning & scheduling in construction projects in Sudan towards optimum applications (Doctoral dissertation, Sudan University of Science and Technology) (2016). http://repository.sustech.edu/handle/123456789/13016

Douglas, E.E.: Contingency management on DOE projects. AACE Int. Trans. 1–6 (2001)

Edwards, P., Bowen, P.: Language and communication issues in HIV/AIDS intervention management in the South African construction industry. Eng. Constr. Archit. Manag. (2019). https://doi.org/10.1108/ECAM-12-2017-0260

Egyptian Newspaper and Official Facts. Official website for princely printing presses in Arabic (2020). http://www.alamiria.com/a/index.html

El-khalek, H.A., Aziz, R.F., Morgan, E.S.: Identification of construction subcontractor prequalification evaluation criteria and their impact on project success. Alexandria Eng. J. **58**(1), 217–223 (2019). https://doi.org/10.1016/j.aej.2018.11.010

Ezeldin, A.S., Helw, A.A.: Proposed force majeure clause for construction contracts under civil and common laws. J. Leg. Aff. Disput. Resolut. Eng. Constr. **10**(3), 04518005 (2018). https://doi.org/10.1061/(ASCE)LA.1943-4170.0000255

Fapohunda, J.A., Chileshe, N.: Essential factors towards optimal utilisation of construction resources. J. Eng. Des. Technol. (2014). https://doi.org/10.1108/JEDT-02-2013-0016

Grigore, M.C., Ionescu, S., Niculescu, A.: New methods for project monitoring. FAIMA Bus. Manag. J. **6**(1), 35–44 (2018)

Hagedoorn, J., Hesen, G.: Contract law and the governance of inter-firm technology partnerships: an analysis of different modes of partnering and their contractual implications. J. Manag. Stud. **44**(3), 342–366 (2007). https://doi.org/10.1111/j.1467-6486.2006.00679.x

Hansen, S.: Does the COVID-19 outbreak constitute a force majeure event? A pandemic impact on construction contracts. J. Civil Eng. Forum **6**(2), 201–214 (2020). https://doi.org/10.22146/jcef.54997

Hebert, J.E., Deckro, R.F.: Combining contemporary and traditional project management tools to resolve a project scheduling problem. Comput. Oper. Res. **38**(1), 21–32 (2011). https://doi.org/10.1016/j.cor.2009.12.004

Heldman, K.: PMP: Project Management Professional Exam Study Guide, 8th edn. Wiley, Indianapolis (2016)

Hishan, S.S., Ramakrishnan, S., Qureshi, M.I., Khan, N., Al-Kumaim, N.H.S.: Pandemic thoughts, civil infrastructure and sustainable development: five insights from COVID-19 across travel lenses. J. Talent Dev. Excellence **12**(2s), 1690–1696 (2020)

Idrees, A.M.: Towards an automated evaluation approach for e-procurement. In: 2015 13th International Conference on ICT and Knowledge Engineering (ICT & Knowledge Engineering 2015, pp. 67–71. IEEE, November 2015. https://doi.org/10.1109/ICTKE.2015.7368473

Abdul-Rashid, I., Aboul-Haggag, S., Mahdi, I.M., El-Hegazy, H.: Construction performance control in steel structures projects. Ind. Eng. Manag. **5**(4), 2–11 (2016)

Jafari, P., Mohamed, E., Lee, S., Abourizk, S.: Social network analysis of change management processes for communication assessment. Autom. Constr. **118**, 103292 (2020). https://doi.org/10.1016/j.autcon.2020.103292

Jallow, H., Renukappa, S., Suresh, S.: The impact of COVID-19 outbreak on United Kingdom infrastructure sector. Smart and Sustainable Built Environment, Vol. ahead-of-print No. ahead-of-print (2020). https://0810ba82q-1106-y-https-doi-org.mplbci.ekb.eg/10.1108/SASBE-05-2020-0068

Johnson, N., Moore, R., Mitha, Y.: Is COVID-19 likely to be a valid basis for avoiding contractual obligations? (2020). https://www.herbertsmithfreehills.com/latestthinking/is-covid-19-likely-to-be-a-valid-basisfor-avoiding-contractual-obligations. Accessed 14 Mar 2020

Kabiru, J.M., Yahaya, B.H.: Can Covid-19 considered as force majeure event in the Nigeria construction industry? Int. J. Sci. Eng. Sci. **4**(6), 34–39 (2020). ISSN (Online): 2456-736

Khan, A.: Project scope management. Cost Eng. **48**(6), 12–16 (2006). https://doi.org/10.1201/b15011-3

Khan, R.A., Gul, W.: Empirical study of critical risk factors causing delays in construction projects. Presented at the 9th IEEE International Conference on Intelligent Data Acquisition and Advanced Computing Systems: Technology and Applications (IDAACS, 21–23 September 2017), Bucharest, Romania (2017). https://doi.org/10.1109/IDAACS.2017.8095217

Koke, B., Moehler, R.C.: Earned Green Value management for project management: a systematic review. J. Clean. Prod. **230**, 180–197 (2019). https://doi.org/10.1109/CONISOFT.2019.00038

Laryea, S.: Procurement strategy and outcomes of a new universities project in South Africa. Eng. Constr. Archit. Manag. (2019). https://doi.org/10.1108/ECAM-04-2018-0154

Lima, R., Tereso, A., Faria, J.: Project management under uncertainty: resource flexibility visualization in the schedule. Procedia Comput. Sci. **164**, 381–388 (2019). https://doi.org/10.1016/j.procs.2019.12.197

Linda, V., Jaroslav, S.: Quality control in building and construction. In: IOP Conference Series: Materials Science and Engineering, vol. 471, no. 2 (2019). https://doi.org/10.1088/1757-899X/471/2/022013

Love, P.E., Irani, Z.: A project management quality cost information system for the construction industry. Inf. Manag. **40**(7), 649–661 (2003). https://doi.org/10.1016/S0378-7206(02)00094-0

Luiz Lampa, I., de Godoi Contessoto, A., Rici Amorim, A., Francisco Donegá Zafalon, G., Valên-cio, C., Souza, R.: Project scope management: a strategy oriented to the requirements engineering. In: Proceedings of the 19th International Conference on Enterprise Information Systems - Volume 2: ICEIS, pp. 370–378 (2017). https://doi.org/10.5220/0006318603700037. ISBN 978-989-758-248-6

Mamoghli, S., Goepp, V., Botta-Genoulaz, V.: An operational "risk factor-driven" approach for the mitigation and monitoring of the "misalignment risk" in enterprise resource planning projects. Comput. Ind. **70**, 1–12 (2015). https://doi.org/10.1016/j.compind.2015.01.010

Mashwama, N., Aigbavboa, C., Thwala, D.: An assessment of the critical success factor for the reduction of cost of poor quality in construction projects in Swaziland. Procedia Eng. **196**, 447–453 (2017). https://doi.org/10.1016/j.proeng.2017.07.223

Megahed, N.A., Ghoneim, E.M.: Antivirus-built environment: lessons learned from Covid-19 pandemic. Sustain. Cities Soc. **61**(May), 102350 (2020). https://doi.org/10.1016/j.scs.2020.102350

Mesároš, P., Mandičák, T.: Information systems for material flow management in construction processes. In: IOP Conference Series: Materials Science and Engineering, vol. 71, no. 1, p. 012054. IOP Publishing, January 2015. https://doi.org/10.1088/1757-899X/71/1/012054

Mirghani, A.A.E.: Planning and scheduling of construction projects in Sudan (Doctoral dissertation, Sudan University of Science and Technology) (2016). http://repository.sustech.edu/handle/123456789/13914

Mirza, M.N., Pourzolfaghar, Z., Shahnazari, M.: Significance of scope in project success. Procedia Technol. **9**(1) (2013). https://doi.org/10.1016/j.protcy.2013.12.080

Moradi, N., Mousavi, S.M., Vahdani, B.: An earned value model with risk analysis for project management under uncertain conditions. J. Intel. Fuzzy Syst. **32**(1), 97–113 (2017). https://doi.org/10.3233/JIFS-151139

Ning, Y., Mao, Y.: The risk monitoring of coal construction project based on system dynamics model. In: Proceedings of the International Conference on Information Systems for Crisis Response and Management, pp. 330–334. IEEE, New York (2011). https://doi.org/10.1109/ISCRAM.2011.6184127

El-Nawawy, O.A., Mahdi, I.M., Badwy, M., Gamal Al-Deen, A.: Developing methodology for stakeholder management to achieve project success. Int. J. Innov. Res. Sci. Eng. Technol. **4**(11), 10651–10660 (2015). ISSN (Online): 2319-8753

Project Management Institute: A guide to the project management body of knowledge (PMBOK® Guide), 6th edn. Project Management Institute, Newtown Square, PA (2017)

Rajprasad, J., Thamilarasu, V., Mageshwari, N.: Role of crisis management in construction projects. Int. J. Eng. Technol. **7**(2.12), 451–453 (2018)

Rothan, H.A., Byrareddy, S.N.: The epidemiology and pathogenesis of coronavirus disease (COVID-19) outbreak. J. Autoimmunity (2020, in press). https://doi.org/10.1016/j.jaut.2020.102433

Salah, A., Moselhi, O.: Contingency modeling for construction projects using fuzzyset theory. Eng. Constr. Archit. Manag. **22**(2), 214–241 (2015). https://doi.org/10.1108/ECAM-03-2014-0039

Salah, A., Moselhi, O.: Risk identification and assessment for engineering procurement construction management projects using fuzzy set theory. Can. J. Civ. Eng. **43**(5), 429–442 (2016). https://doi.org/10.1139/cjce-2015-0154

Sarvari, H., Valipour, A., Yahya, N., Noor, N., Beer, M.D., Banaitiene, N.: Approaches to risk identification in public-private partnership projects: Malaysian private partners' overview. Adm. Sci. (2076–3387) **9**(1), 17 (2019). https://doi.org/10.3390/admsci9010017

Senaratne, S., Ruwanpura, M.: Communication in construction: a management perspective through case studies in Sri Lanka. Archit. Eng. Des. Manag. **12**(1), 3–18 (2016). https://doi.org/10.1080/17452007.2015.1056721

Shilts, J.: A framework for continuous auditing: why companies don't need to spend big money. J. Account. **223**(3), 1–7 (2017)

Shirazi, F., Kazemipoor, H., Tavakkoli-Moghaddam, R.: Fuzzy decision analysis for project scope change management. Decis. Sci. Lett. **6**(4), 395–406 (2017). https://doi.org/10.5267/j.dsl.2017.1.003

Shivalli, S., Pai, S., Akshaya, K.M., D'Souza, N.: Construction site workers' malaria knowledge and treatment-seeking pattern in a highly endemic urban area of India. Malar. J. **15**(1), 168 (2016). https://doi.org/10.1186/s12936-016-1229-2

Sohrabi, C., et al.: World health organization declares global emergency: a review of the 2019 novel corona-virus (COVID- 19). Int. J. Surg. **76**(February), 71–76 (2020). https://doi.org/10.1016/j.ijsu.2020.02.034

Srinivasan, N.P., Dhivya, S.: An empirical study on stakeholder management in construction projects. Mater. Today: Proc. **21**, 60–62 (2020). https://doi.org/10.1016/j.matpr.2019.05.361

Touran, A., Liu, J.: A method for estimating contingency based on project complexity. Procedia Eng. **123**, 574–580 (2015). https://doi.org/10.1016/j.proeng.2015.10.110

Trenor, J.A., Lim, H.S.: Navigating force majeure clauses and related doctrines in light of the COVID-19 pandemic. Young Arb. Rev. **37**, 13 (2020)

Turnbaugh, L.: Risk management on large capital projects. J. Prof. Issues Eng. Educ. Pract. **131**(4), 275–280 (2005). https://doi.org/10.1061/(ASCE)1052-3928(2005)131:4(275)

Wang, Y., Liu, G.: Research on relationships model of organization communication performance of the construction project based on shared mental model. In: 2009 International Conference on Information Management, Innovation Management and Industrial Engineering, vol. 1, pp. 208–211. IEEE, December 2009. https://doi.org/10.1109/ICIII.2009.57

Weippert, A., Kajewski, S.L., Tilley, P.A.: Internet-based information and communication systems on remote construction projects: a case study analysis. Constr. Innov. **2**(2), 103–116 (2002). https://doi.org/10.1108/14714170210814702

World Health Organization (WHO): WHO Director-General's opening remarks at the media briefing on COVID-19 - 11 March 2020 (2020). www.who.int/dg/speeches/detail/whodirector-general-s-opening-remarks-at-themedia-briefing-on-covid-19-11-march-2020. Accessed 17 Mar 2020

Yap, J.B.H., Skitmore, M.: Ameliorating time and cost control with project learning and communication management. Int. J. Managing Projects Bus. (2020). https://doi.org/10.1108/IJMPB-02-2019-0034

Yates, J.K., Eskander, A.: Construction total project management planning issues. Proj. Manag. J. **33**(1), 37–48 (2002). https://doi.org/10.1177/875697280203300107

Mashwama, X.N., Aigbavboa, C., Thwala, D.: Investigation of construction stakeholder's perception on the effects and cost of construction dispute in Swaziland. Procedia Eng. **164**, 196–205 (2016)

Structural Evaluation of FRP Composite Systems for Repair Upgrade of Reinforced Concrete Beams

Ashraf A-K Mostafa Agwa(✉)

Civil Engineering Department, Valley Higher Institute for Engineering and Technology, Cairo, Egypt
ashraf.abdel.khalek@sva.edu.eg

Abstract. This paper presents results summary of study aims at evaluating the efficiency of U-shaped fiber reinforced polymer (FRP) composite for repairing and structural upgrade of reinforced concrete (RC) beams. Tests were based upon procedures described in AC125 International Code Council Evaluation Service (ICC-ES) criteria. The materials used in this study conform to the AC 85 acceptance criteria. Displacement, strains and loads were continuously monitored and recorded during the tests. Large- scale experimental results showed that the use of FRP composite system has resulted in an appreciable increase of the strength of the retrofitted beams with an average of increase of 400% of the original capacity of the undamaged beams.

Keywords: FRP composites · Repair · Reinforced concrete beams · Experimental mechanics

1 Introduction

For the past three decades or so, fiber reinforced polymeric (FRP) composites have been used extensively in different civil infrastructural applications. One of the attractive utilization of FRP composites in construction is repair and rehabilitation of aging structures that includes steel, reinforced concrete, and wood structural systems. The two important applications are in the area of corrosion and seismic deficiencies of different structural members. The majority of FRP composites strengthening application have been focused on RC building, bridges and tunnels due to the vulnerability of reinforced concrete members to corrosion. Also, composite repair systems can be used in other cases such construction and design faults, overloading, change-of-use that will require addition of live loads and accidental damages. Figure 1 shows some examples of the use of composites in repair and rehabilitation of RC structures. A comprehensive summary of different repair and rehabilitation applications of polymeric composites is reported in Ref [1,2] Panahi, et al. [3] discussed the feasibility of two types of composites repair for RC beam members. The two systems are bonded laminates and near-surface-mounted rebars.

R. Saidi et al. (Eds.): ICATH 2021, LNDECT 110, pp. 428–437, 2022.
https://doi.org/10.1007/978-3-030-94188-8_38

2 Experimental Program

2.1 Description Beam Specimens and Test Apparatus

All beams were fabricated using Grade 60 reinforcing steel from the same mill batches. Tension (bottom) reinforcement consisted of 2#13 steel rebars with 2.5 cm cover. Shear reinforcements consisted of #10 steel stirrups spaced evenly at 10.8 cm along the beam length as shown in Fig. 1. The clear span of all beam specimens was 3.45 m with a cross sectional dimensions of 152.4 mm in width by 254 mm in depth.

The average 28-day compressive strength of concrete used in fabricating all beams determined from standard cylinder tests is 43.0 MPa as shown in Table 1.

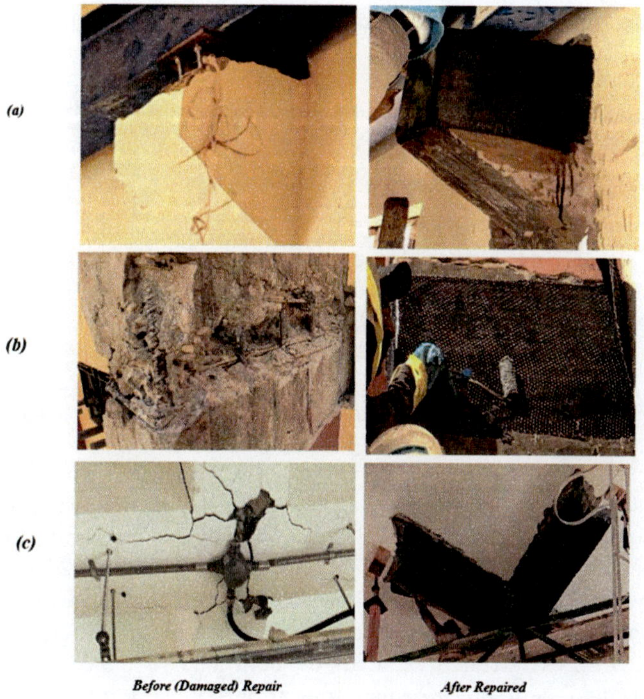

Fig. 1. Examples of Repair of RC Members using FRP Composites: (a) Repair of RC Corbels, (b) Repair of Corroded and Honeycomb RC Columns, and (c) Repair of Cross-Beams

Table 1. Concrete compressive strength

Test	Test ID	Concrete Strength [psi]
1	BC1	42.22
2	BC2	43.64
3	BC3	42.20
4	BC4	43.10
5	BC5	43.36
Average		43.00

2.2 Test Setup

All specimens were tested under 4-point bending protocol as illustrated in Fig. 2.

Fig. 2. Typical four-point load application for beam specimens

The test assembly and the special transfer steel fixture were designed and calibrated to perform this type of full-scale beam evaluation tests. The test actuator used was a calibrated 245-kN MTS hydraulic actuator. A 245-kN calibrated load cell was used for recording the applied load as shown in Fig. 2.

All deflection and strain data were collected using a computerized data acquisition system. Stress/strain σ/ϵ. load/deflection (P/δ curves were developed for each specimen and failure initiation, progress, and ultimate failure were recorded and then analyzed. Figure 3 shows the locations of the different measuring devices for both deflection and strain.

2.3 Lamination Schedule

In this study, E-glass/epoxy polymer (GFRP) composite system was used for strengthening beam specimens. Prior to the application of composite laminates, loose concrete was removed, surface was cleaned and a layer of low-viscosity epoxy-based primer was applied to the treated concrete. After allowing for the epoxy primer curing, the surface was ready for applying the E-glass/epoxy composite laminates.

All application procedures were in accordance to the manufacturer's installation manual procedures. A total of three plies of unidirectional E-glass/epoxy composites were applied to the prepared surface in a U-shape configuration. All laminates were fully saturated using two-part, room cure epoxy. The laminates covered the entire beam span except of 50.80-mm clearance away from each end support. The laminates were extended vertically to cover 178 mm of the depth from the bottom (see Fig. 4).

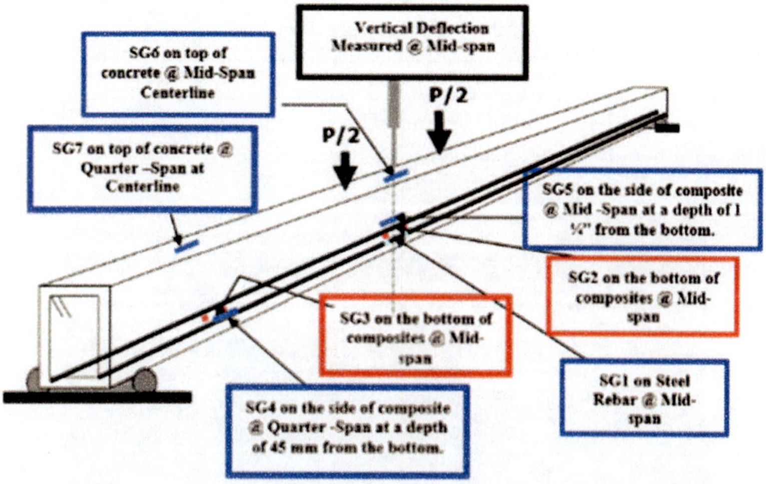

Fig. 3. Locations of deflection and strain measurements for all beam specimens

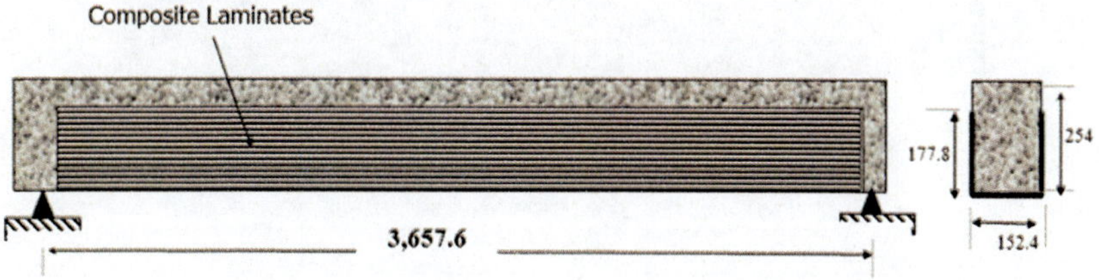

Fig. 4. U-shape lamination details (Units: mm)

3 Experimental Results

Three sets of beams were tested; i) Control Beams, ii) Repaired Beams, and iii) Retrofitted Beams. The following paragraphs summarize the findings of each test group.

3.1 Control Specimens

In general for all beams tested in this study, at low load levels, no cracks were noticed. As a load of about 8.9 kN, fine cracks were developed at the constant moment region (between the two concentrated loads). As the load was increased, the number and sizes of flexural cracks were increased (see Fig. 5), and typical flexural failure of the beams was observed due to crushing of the compression concrete at an average load of 28.7 kN. Figures (6) and (7) present samples of experimental load/deflection (P/δ, and stress/strain (σ/ϵ) curves of the different control beam specimens, respectively.

Fig. 5. Cracking of the control beam specimen

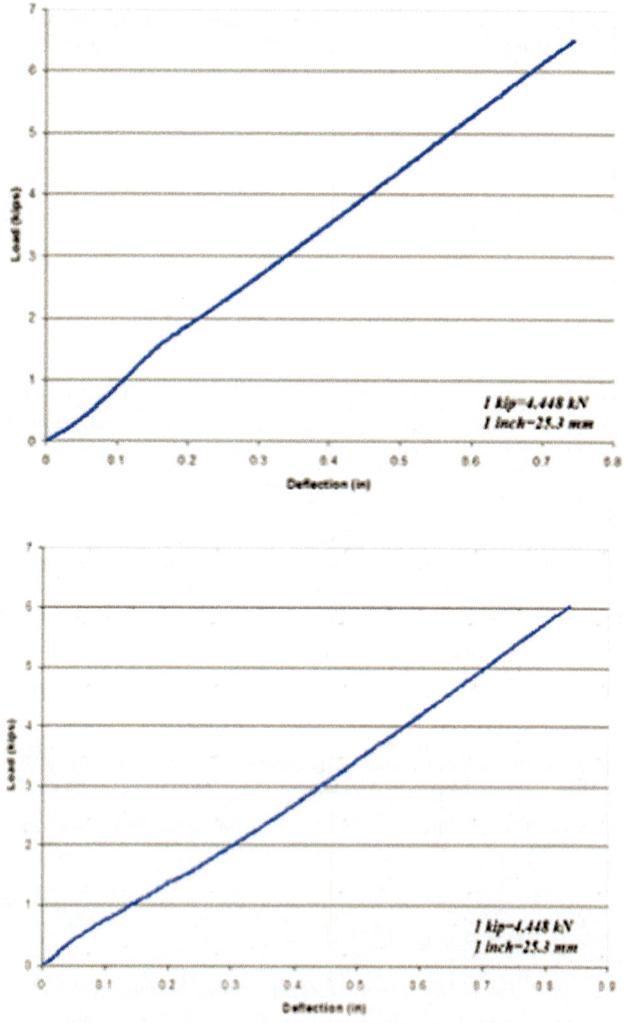

Fig. 6. A sample of Load/Deflection curves for control beam specimens

3.2 Repaired Specimens

As shown in Fig. 5, a U-shaped lamination schedule was used in all repaired specimens. The lamination was made of three layers of E-glass/epoxy unidirectional composites. The U-Shape laminate which covers both the bottom side and the vertical sides of the beam. In order to utilize the full strength properties of the composite repair system, the U-lamination system design was selected (refer to Fig. 4). In this lamination geometry, the ends of the composite laminates were extended vertically to cover 177.8 mm of the beam depth (0.70 d). As test results indicated, this design resulted in a more favorable mode of failure and relatively higher flexural strength.

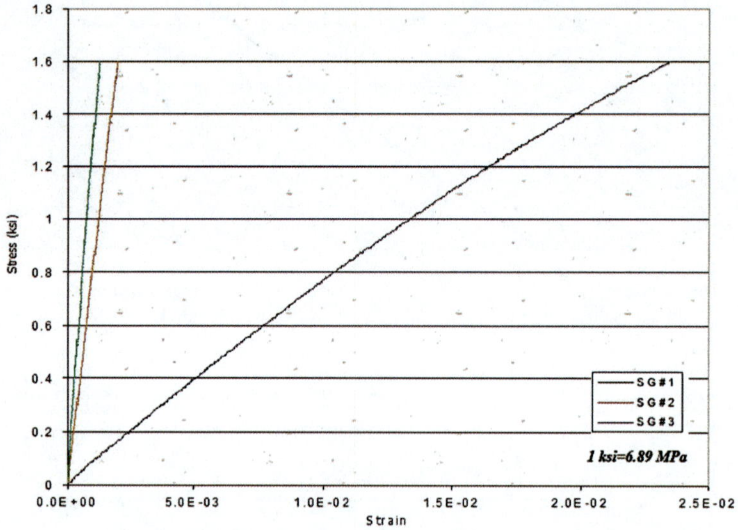

Fig. 7. A sample of Stress/Strain curves for control specimens

The common mode of failure was observed. As shown in Fig. 6, the common mode of failure was primarily concrete crushing at the compression (top) side of the beam that occurred at the constant moment area (between the two applied line loads).

3.3 Retrofitted Specimen RET-E1

Figure 7 shows the load/deflection (P/d) curve measured at mid-span for retrofitted beam specimen RET-E1. As shown in this figure, the behavior was linear up to about 62.3 kN, after which some non- linearity was observed which was continued until failure. It should be noted that the test was halted just before the total collapse to avoid unexpected brittle failure and to maintain the safety of the lab crew and witnesses. The ultimate flexural capacity of this beam specimen was 109.64 kN (corresponds to a flexural stress of 41.64 MPa). This gain in strength is about 394% increase in strength as compared to the average control specimen strength (average ultimate strength was 10.56 MPa). The mid-span deflection at ultimate load was 62.4 mm and the measured tensile strain of the external FRP laminate at the mid-span corresponding to the ultimate load was 0.905% (refer to Fig. 8). This strain is about 41% of the E-glass/epoxy composite' rupture strain of 2.2%. It should be noted that this strain is not the ultimate composite strain of the system since failure did not occur to the laminate, instead, failure was due to local compressive failure of the concrete at the maximum compression zone located between the two line loads (refer to Fig. 9). The compression failure of the concrete at the top was followed by a cohesive failure of the topside laminates which resulted on a local buckling of the side laminate with a symmetrical sine wave with a length of about 300.0 mm from the beam center line.

This beam specimen achieved the highest strength of 113.0 kN (as corresponds to a flexural stress of 42.89 MPa). This gain in strength is about 406% increase in strength as compared to the average control specimen strength. The mid-span deflection at ultimate load was 66.52 mm and the measured tensile strain of the external FRP laminate at the mid-span corresponding to the ultimate load was 1.05% (refer to Fig. 8). This strain is only 48% of the E-glass/epoxy composite' rupture strain of 2.2%. It should be noted that this strain is not the ultimate composite strain of the system since failure did not occur to the laminate, instead the failure was due to local compressive failure of the concrete at the maximum compression zone located between the two line loads. As for the specimen REP-E1, this compression failure of the concrete at the top was followed by a cohesive failure of the topside laminates. The top of the side laminate buckled at approximately the beam centerline as shown in Fig. 9.

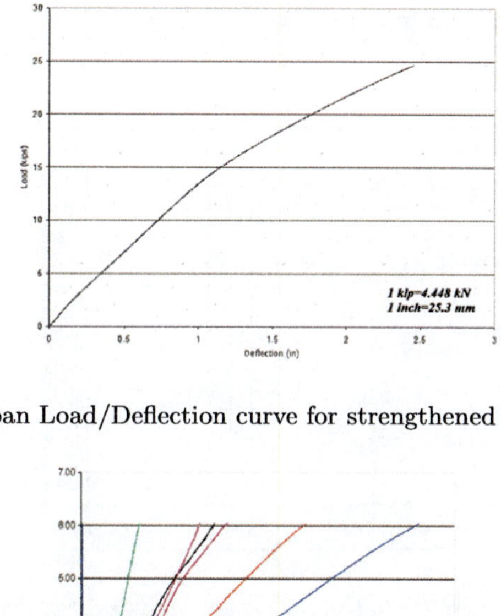

Fig. 8. Mid-Span Load/Deflection curve for strengthened beam specimen

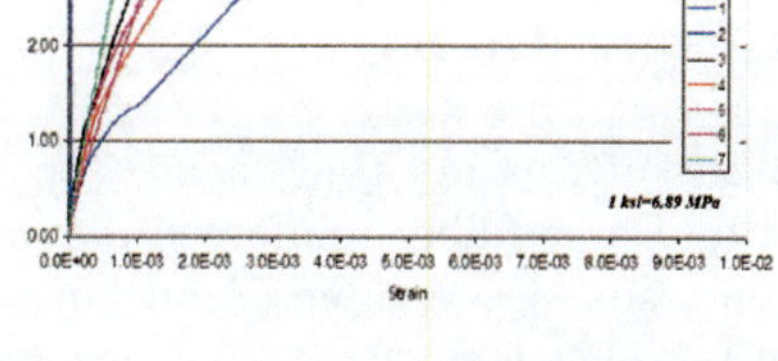

Fig. 9. Stress/Strain curves for strengthened beam specimen

Fig. 10. Local concrete compressive failure

Table 2. Summary results for the E-glass/Epoxy beam retrofitting system

Specimen ID	Ultimate Load [kN]	Deflection @ Ultimate, [mm]	Flexural Stress [MPa]	Composite Ten. Strain @ Ultimate (%)
Control (Average)	27.80	20.0	10.96	0.13*
RET-E1	114.09	62.36	41.65	0.91
RET-E2	112.93	66.55	42.89	1.05

** Concrete Strain (no composites)*

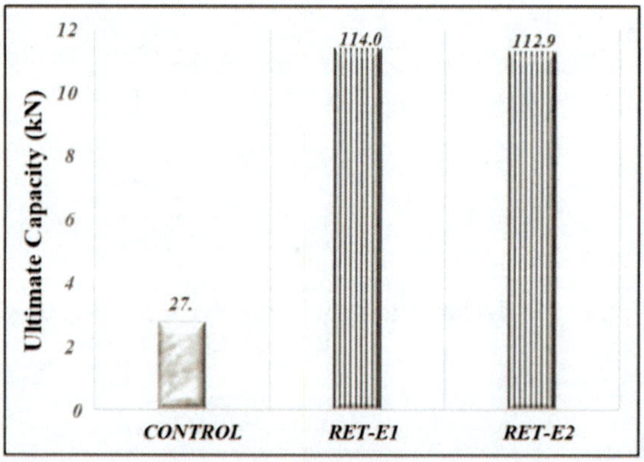

Fig. 11. Comparison between control and E-glass/Epoxy retrofitted beam specimens ultimate strength

4 Summary and Conclusions

This paper presents results of a study that focuses on structural behavior of RC beams strengthened by E-glass fiber reinforced polymeric (GFRP) composites. A series of full-scale tests were conducted on both control (as-built) and strengthened beams with E-glass/epoxy external laminates. Full-scale experimental results indicated that the use of the GFRP composites resulted in an appreciable increase of the strength of the retrofitted beams with an average of 400% as compared to the capacity of unstrengthened (as-built) beam specimens. The failure was preceded with a large deformation (up to 6.6 cm) which provided enough visual warning before ultimate failure. In all tests, no failure occurs to the composite system; instead a localized compression failure of the concrete was the common failure mode. Table 2 and Fig. 10 summarizes the results of this study (Fig. 11).

References

1. Mosallam, A.S., Bayraktar, A., Elmikawi, M., Pul, S., Adanur, S.: Polymer composites in construction: an overview. SOJ Mater Sci. Eng. **2**(1), 25 (2013)
2. Mosallam, A.: Composites: construction materials for the new era. In: Advanced Polymer Composites for Structural Applications in Construction, pp. 45-58. Woodhead Publishing (2004)
3. Panahi, M., Zareei, S.A., Izadic, A.: Flexural Strengthening of Reinforced Concrete Beams through Externally Bonded FRP Sheets and Near Surface Mounted FRP Bars. Case Studies in Construction Materials, p. e00601 (2021)
4. A.C.I. Committee 440: Guide for the design and construction of externally bonded FRP systems for strengthening concrete structures. American Concrete Institute, USA (2017)
5. ICC-ES AC125: Acceptance Criteria for Concrete and Reinforced and Unreinforced Masonry Strengthening Using Fiber-Reinforced Polymer (FRP) Composite Systems. ICC- Evaluation Service Inc., Brea, California, USA (2013)

Materials and Smart Buildings

The Macroscopic Effect of COVID 19 on Flexible Pavement Condition Indicators Based on Analysis of Road Inspection Results

Mohammed Amine Mehdi[1](✉), Toufik Cherradi[1], Said Elkarkouri[2], and Ahmed Qachar[3]

[1] Civil Engineering and Construction Laboratory, Mohammedia School of Engineers, Rabat, Morocco
aminemehdi@reserach.emi.ac.ma

[2] National Center for Road Studies and Research, The Ministry of Equipment, Transport, Logistics and Water, Rabat, Morocco

[3] The Ministry of Equipment, Transport, Logistics and Water, Rabat, Morocco

Abstract. In Morocco, as for all countries worldwide, the COVID-19 pandemic has created a huge disruption in the daily life of modern society. The transportation services have been one of the most severely affected of the various urban systems, specifically the national roads and highways. In collaboration with the National Center for Road Studies and Research, this study deals with the challenges of maintenance and monitoring to ensure a good level of comfort on the one hand, and the variability of traffic levels and its impact on the variability of surface indicators and the proper functioning of a Moroccan National road number 6, linking the city of Meknes and Khemissat over a length of 50 km composed of a flexible pavement from 2018 to 2020, which represents the national road most solicited by traffic in Morocco in recent years. The road managers were hardly prepared for such an event, and were concerned to give priority to the safety of their employees and customers to avoid any service interruption. In this regard, the analysis of past deterioration and variation of structural and surface indicators becomes a necessity in these pandemic circumstances.

Keywords: Moroccan road · Pavement · Road inspection · Road deterioration · COVID-19

1 Introduction

In 2020, the world was surprised by the first global pandemic of the 21st century due to the recently induced Coronavirus, causing a disease called COVID-19, or SARS-Cov2. It first appeared in the Chinese city of Wuhan in late 2019 and rapidly spread worldwide over a few months. As of March 2020, this COID-19 had affected most countries in the world. To control the pandemic, most countries imposes what the IMF called the "Great Lockdown" from March to June of that year. This meant the closure of most economic activities and non-essential businesses (such as hotels, restaurants, and most shops). In

R. Saidi et al. (Eds.): ICATH 2021, LNDECT 110, pp. 441–452, 2022.
https://doi.org/10.1007/978-3-030-94188-8_39

Morocco, a first case of infection with the new Corona virus was confirmed at 9.04 pm on 2 March 2020, after confirmation by the Pasteur Institute on Monday evening. It is a Moroccan from Italian homes.

Since that date, the country has declared a state of health emergency. Automatically, the measures also included a substantial restriction on travel, given the high concentration of people and the increased likelihood of the virus spreading. These quarantine measures (along with self-quarantine for the majority of people and businesses and social distancing) have led most employees to work from home, increasing the percentage of telecommuting to previously unknown levels. Without mentioning the economic crisis that the world is experiencing, the people affected, the road as a development lever has experienced intense variability in statistics relating to traffic, maintenance, comfort, safety and quality offered.

Speaking of the Moroccan situation, the National Center for Studies and Road Research under the supervision of the Ministry of Equipment, Transport, Logistics and Water, like all state bodies was blocked in front of this pandemic situation at the beginning of the year 2020. Road inspections have been postponed until sanitary gives an acceptable indicator for road operation this unplanned stop disrupted several factors related to road management and control. The level of road traffic decreased considerably, in because people had to work practically from home because leisure and travel was prohibited in those months. However, unlike air traffic which fell by 90%, the decline in road traffic was smaller. There are three main reasons for this. The first is that some workers continued to come to work and use the roads. The second reason is logistics and distribution, have required an increase of door-to-door distribution to cope with the exponential growth of purchases in the third reason is the switch from public transport private car use in most.

1.1 Global Context

Despite a recent increase in the use of public transport in previous years, the pandemic led to a substantial decline in this mode of transport. New challenges have arisen, the difficulty of managing the structural comfort of the road system, and at the same time, ensuring the safety and welfare of the works in a sector where face-to-face interaction may be inevitable. All these parameters are interdependent and present a certain problem for the management of road assets in Morocco [1]. This study is presented in the form of a case study methodology, serves to have a theoretical idea based on empirical data on road inspections conducted in 2018 [1] and those conducted after the first wave of the pandemic COVID 19, based on the Moroccan method [2].

In this same, the evolution of surface deterioration in the absence of maintenance, such as potholes, pull-outs and cracks, after the government introduced a strong and wider "lockdown", imposing significant restrictions on peoples mobility is stanalyzed. This case study was used to provide exploratory work and explanatory variables. Using data from inspections conducted in late 2020.

Results of this research showed that the pandemic has led to a focus on providing a short-term response to ensure continuity of service delivery. However, in terms of traffic, the data indicates a significant drop in traffic levels, particularly for light passenger vehicles, but to a lesser extent for trucks and commercial vehicles. In order to understand the impact of COVID-19 on the evolution of the state of road deterioration, we propose results by analyzing the road database relating to post-COVID19 inspections, and a critical discussion of the results towards the end.

1.2 Moroccan Context

Morocco has an extensive road network in Africa, comprising highways, national, provincial, and regional roads. This network is used extensively by a huge amount of traffic, including heavy vehicles, one of the main reasons for the road maintenance policy. The main objective was to offer users a high level of comfort and safety, including the correct maintenance of the roadway, signs and safety equipment, and to preserve the road's pavement patrimony (pavements, bridges, tunnels, retaining walls) [1].

2 Case Study Research

This study concentrates on the flexible pavement of national road number 06 extending over a length of 50 km, beginning at kilometer point KP 0+081 and ending at KP 0+130, joining the city of Meknes and Khemissat [2]. The selection of this section is based on the high traffic rate estimated annually at 3639207 vehicle km/day. This study presents an in-depth analysis of the pavement deterioration [3], of the auscultations and inspections carried out "between" 2018 to 2020, in view of preparing a road database (RDB) that will be used to predict the strategies and model the deterioration of this road.

2.1 Geolocalization

We used a case study methodology to answer fundamental questions during a pandemic. Our first research question is, "How did the road manager react to the pandemic crisis and the requirement to maintain operational activities and health security measures?" This issue is unprecedented, as no emergency plan for highway operators had ever preceded such an event. Confronted with a new, unexpected and disastrous situation with very little time to react effectively. We used this study to determine the degree to which the analysis of roadway databases can facilitate the task of managers (Table 1).

Table 1. Location and coordinates of study section

Kilometer point (National Road number 06)	
Origin	Fin
X: 439617,099607; Y: 358252,362678	X: 478570,08923 Y: 364436,69712

3 Methodology

This article used a qualitative approach [1] for our first research question to assess how this road accident coped with an unexpected and serious new situation requiring a rapid and robust response. The National Center for Road Studies and Research applies the Moroccan method [1] to assess the level of understanding and safety of the Moroccan network at each 2-year level. By the bias of road inspections to define the surface and structural condition, by analyzing the level of deductions in each subset of 1 km. The division including the following elements:

1. Analysis of the surface state: by defining the quantities of deteriorations such as: potholes (Fig. 1), pull-out (Fig. 2), and cracks (Fig. 3).
2. Structural condition analysis (STI): through the deflection calculation method and the evenness calculation [4, 5].
3. Road data reduction and surface indicator evolution analysis [2].

Fig. 1. Potholes

Fig. 2. Pull-out

Fig. 3. Cracks

These indicators characterize 1 km of road and can have 4 values:

A: Good,
B: Acceptable,
C: Poor,
D: Very bad.

3.1 Surface Condition Indicator (SUI)

These indicators, which reflect the state of deterioration of roads, are updated every two years on the basis of a visual survey of pavement distresses. The visual survey concerns the following types of deterioration: cracks (longitudinal and transverse) [3], tears (feathering, peeling, combing, potholes), deformations (rutting, sinking, etc.), and shoulder slopes. The survey is based on the observation at regular 200 m intervals of the existence or absence of each type of deterioration, indicating its severity. The integration of the observations is done by adding the scores attributed to each of the deteriorations (cracks, tears, potholes, slopes) on each 1 km section (Refer to Fig. 4).

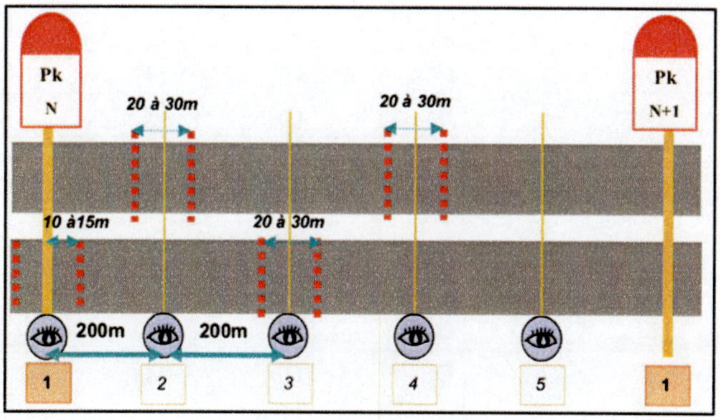

Fig. 4. Visual survey

For each km of pavement, the surface distress condition is classified into four classes for each distress according to the following decision grid, based on the cumulative scores over kilometers (refer to Table 2).

Table 2. Deteriorations repartition class

Deteriorations	Cracks (longitudinal and transversal)	Potholes	Pull-out
A	0	0	0–1
B	1–2	1	2–3
C	3–10	2–4	4–5
D	11–20	5–10	6–10

The pavement surface condition indicator (SUI) for each km of paved road is determined according to the three classes of deterioration status, according to the following matrix (Table 3):

Table 3. Surface deterioration matrix

Potholes	Pull-out	Deteriorations			
		A	B	C	D
A	A	A	B	C	D
	B	A	B	C	D
	C	B	C	C	D
	D	B	C	D	D
B	A	B	B	C	D
	B	B	C	C	D
	C	C	C	D	D
	D	C	D	D	D
C	A	B	C	C	D
	B	C	C	C	D
	C	C	D	D	D
	D	D	D	D	D
D	A	C	D	D	D
	B	D	D	D	D
	C	D	D	D	D
	D	D	D	D	D

The overall condition of the paved road network is assessed by the % of the length of the road with a SUI of class (A) or class (B), known as % (A+B).

3.2 Surface Condition Indicator (SUI)

To deduce the structural indicators, we must refer to the results of the inspection carried out by the techniques represented in the following table (Table 4):

Table 4. Inspection materials

Materials	Role
Lacroix Deflectograph [8]	Used to define the deflection and the bearing capacity of pavements
Longitudinal Profiles Analyzer (APL) [7]	Two-track and single-track for the evaluation of the longitudinal evenness of pavements

This indicator [8] reflects the bearing capacity of the road. It is a function of:

- Deflection [7];
- State of cracking and evenness [6].

The following figures represent the materials used (Figs. 5 and 6):

Fig. 5. Deflection measurement (LACROIX)

Fig. 6. Evenness measurement (APL Method)

This indicator is based on the International Roughness Indicator and defines the road's roughness [6]. It is determined from the table below: (T is the traffic class) (Table 5)

Table 5. Evenness repartition matrix [9]

	International Roughness Indicator IRI en mm/Km			
	T0	T1	T2	T3–T4
A	<2000	<2300	<3000	<3200
B	[2000, 2500]	[2300, 2800]	[3000, 3500]	[3200, 3700]
C	[2500, 3000]	[2800, 3300]	[3500, 4000]	[3700, 4200]
D	≥3000	≥3300	≥4000	≥4200

The deflection class is determined from the deflection value (LACROIX deflectograph [9]) and the traffic class, according to the following table (Table 6):

Table 6. Deflection repartition matrix

Class	Deflection in 1/100 mm			
	T0	T1	T2	T3-T4
A	<80	<100	<120	< 140
B	[80, 120[	[100, 140[	[120, 160[	[140, 180[
C	≥120	≥140	≥160	≥180

The cracking and deflection indicators is determined from the described in Table 7:

Table 7. Cracking and deflection matrix

Cracking and deflection	Class of deflection		
Class of cracks	A	B	C
A	A	B	C
B	B	B	C
C	B	C	C
D	B	C	D

Finally, the STI is determined according to the matrix shown in Table 8:

Table 8. STI matrix

STI	Evenness class			
Class of cracks	A	B	C	D
A	A	B	B	C
B	B	B	B	C
C	B	C	C	D
D	C	C	D	D

4 Results and Discussion

4.1 Deflection Results

Deflection results separate the distribution in two homogeneous parts, the first one is from the original KP 0+081 to KP 0+082 with class C: deterioration in bad condition o (Table 8), the second one from KP 0+082 to KP 0+130 belongs to class A: good conditions (Fig. 7, Table 9).

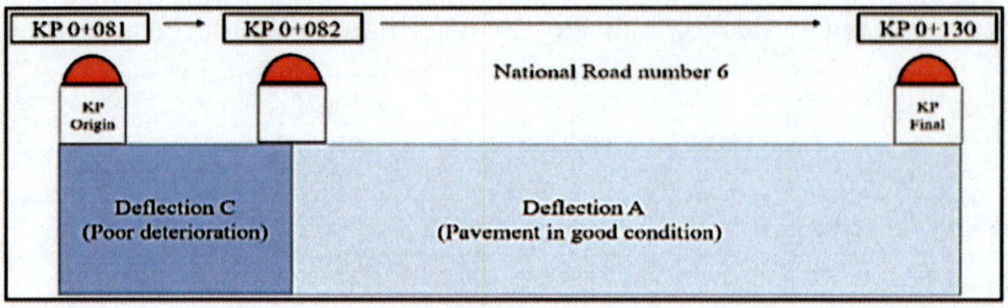

Fig. 7. Deflection repartition

Table 9. Deflection measurement repartition

Kilometer point	Deterioration Class	Traffic	Deflection in 1/100
KP0+081 to KP0+082	C	T1	≥140
KP 0+082 to KP0+124	A	T1	
KP 0+124 to KP 0+130	A	T0	<80

4.2 Evenness Measurement Results

The method of this measurement is mainly based on structural damage, cracks, rutting and subsidence [8]. The results of the evenness share the distribution in three parts, two homogeneous sections rated (A): in a good condition (KP 0+081 at KP 0+124 and KP 0+126 at KP 0+130), and the remained one rated B: heterogeneous of deformation (from KP 0+125 to KP 0+126) class (B). The following figure represent the evenness distribution (Fig. 8):

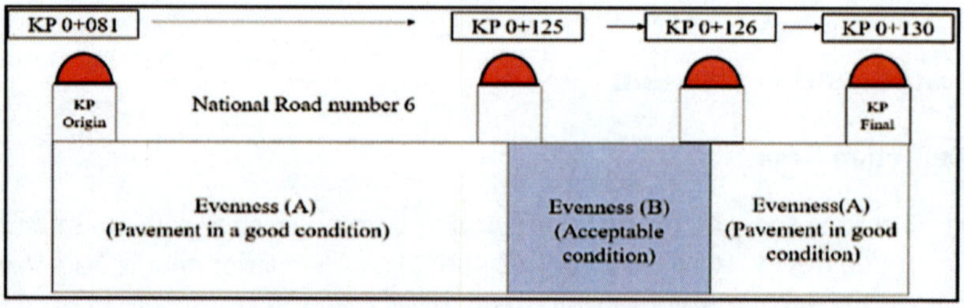

Fig. 8. Evenness distribution

4.3 Analysis of Surface Deterioration Results

The method of this operation is mainly based on surface damage, pullout, potholes and cracks. Taking into account the previous results, crack-type distress are the most dominant on the existing pavement surface. Analyzing the previous results, and taking

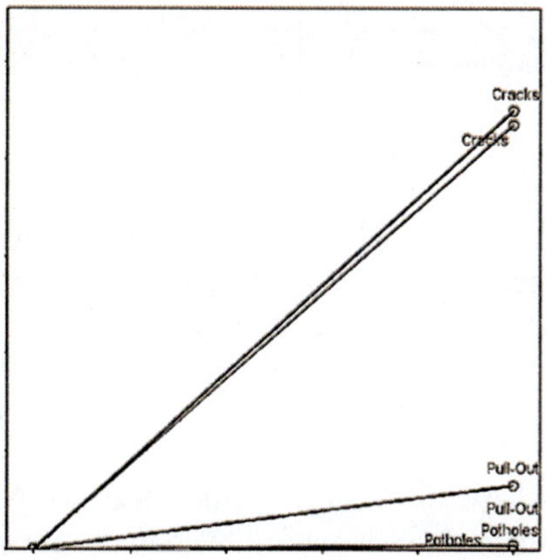

Fig. 9. Distribution of surface deterioration in 2018 and after containment

into account the deteriorations classes fixed by the National Center for Road Studies and Research, the distribution of the surface deteriorations is represented in Fig. 9:

Figure 10 shows the surface indicators distribution indicators (SUI) in 2018 and at the end of 2020.

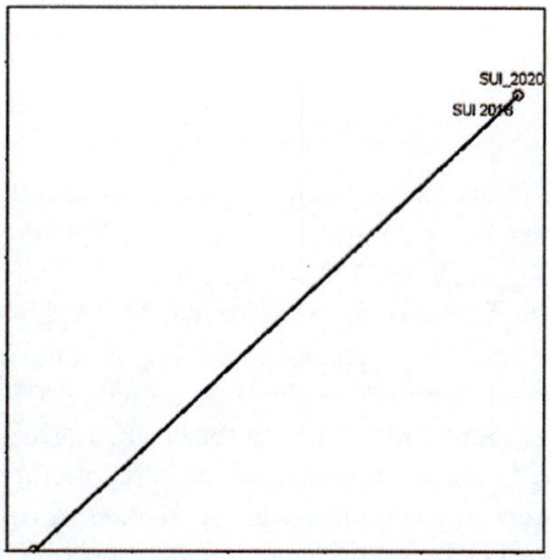

Fig. 10. Distribution of the surface indicators: SUI 2018 & 2020

4.4 Covid-19 Impacts in Surface Condition of the Pavement

Analyzing the results of the environmental inspection carried out on section in 2018, the condition of this link shows a high percentage of potholes, tears and cracks, which increase the level of the surface indicator, with a low percentage of A+B. After containment, the road has not undergone any maintenance and repair interventions, and over 50 km length, the link is divided into sub-assemblies in a state of advanced fatigue equi-quality of level C, which has not allowed the maintenance work carried out to be valued, in particular the tears, cracks of level C and D which have disturbed the stability of the surface condition indicator, especially in the last few kilometers.

5 Conclusion and Perspectives

Although the volume of traffic has decreased, the comfort indicator on this link between 2018 and after the first wave of COVID19 maintained same trend, they are grouped in the same range of the graph, which indicates the strong correlation between the intensity of cracks, potholes, tears, and the deflection and comfort offered by this pavement. COVID 19 has succeeded in reducing the impact of traffic on the pavement, but it has blocked maintenance and repair actions.

All distress have been treated after de-containment and with the required quality, and the existing drainage works have been rehabilitated to pre-serve the pavement however there is still a need to re-plan the interventions by introducing probabilistic prediction of future distress, analyzing the evolution of the database, which is the subject of the present research perspectives, proposing a distress model, a simulation and maintenance and servicing scenarios.

Acknowledgment. I would like to thank my dear professor Mr. Toufik Cherradi for his relevant guidance, the National Centre for Road Studies and Research on behalf of Mr. Said El Karkouri, and the Moroccan Road Directorate on behalf of Mr. Ahmed Qachar for their collaboration and availability.

References

1. Mehdi, M.A., Cherradi, T., Quachar, M., Chigr, A.: Analysis of structural and surface deteriorations of a flexible pavement, based on inspection results using the MCA method. In: Materials Today Proceedings, vol. 45, no. 8, pp. 7538–7546 (2021)
2. Mehdi, M.A., Cherradi, T., El Karkouri, S., Quachar, M.: Applying geographic information systems (GIS) for surface condition indicators modeling of a flexible pavement. In: E3S Web Conference, 1st International Congress on Costal Research (ICCR2020), vol. 298 (2021)
3. Nkomoa, L.S., Desaia, S.A., Seutloalib, K.E., Peerbhaya, K.Y., Dubec, T.: Assessing the surface material quality of unpaved rural roads to understand susceptibility to surface deterioration. A case study of four rural areas in KwaZulu-Natal, South Africa, vol. 112, pp. 3–11 (2019)
4. Hopkins, J., Univ. Pr.: Road Deterioration and maintenance effects: models for planning and management (1987)
5. Harral, C., Faiz, A.: Road deterioration in developing countries: causes and remedies (English). A World Bank policy study. World Bank Group, Washington, D.C. (1988)
6. Vaillancourt, M.: Méthodologie de modélisation de l'uni des chaussées souples et impact des sols d'infrastructure. Thèse de doctorat électronique, Montréal, École de technologie supérieure (2004)
7. Jendryka, W., Pariat, J.C. Robert, B.: Mesure de l'uni longitudinale des chaussées routiéres et aéronautiques. LCPC (Méthode d'Essais LCPC (ME), 46), Paris, version 2.0 (2009)
8. Henia, M.O., Braber, R.: Falling weight deflectometer et Déflectographe Lacroix: Comparaison, Domaine d'Application et Perspectives, Suisse, 6 p. (2008)
9. Nam, B.H., An, J., Kim, M., Murphy, M.R., Zhang, Z.: Improvements to the structural condition index for pavement structural evaluation at network level. Int. J. Pavement Eng. **17**(8), 680–697 (2015)

Effect of Control Partitions on Drag Reduction and Suppression of Vortex Shedding Around a Bluff Body Cylinder

Youssef Admi(✉), Mohammed Amine Moussaoui, and Ahmed Mezrhab

Laboratory of Mechanics and Energetics, Faculty of Sciences, Mohammed First University, 60000 Oujda, Morocco
admi_youssef1718@ump.ac.ma

Abstract. The present work involves a numerical study of the flow of fluid passing around a square block controlled by four partitions. The two downstream partitions are fixed at a distance g = 3.9. While the two upstream partitions are fixed at a distance $g_1 = 0.25$ whose length Lp varies from 1d to 5. Numerical simulations are performed at a fixed Reynolds number (Re = 150) using the multiple relaxation time- lattice Boltzmann method (MRT-LBM). The results are presented as graphs of the velocity contours and the time drag coefficient and show that there is a critical length Lp = 2.25. At this critical length, a maximum percentage reduction of the drag coefficient reaches 38.86% in the case where the two upstream bulkheads are placed at a distance $g_1 = 0.25$. The numerical code has been validated with existing numerical and experimental investigations in the literature for a flow around a square cylinder without and with a single detached partition.

Keywords: Multi-relaxation-time lattice Boltzmann method · Control partitions · Partitions length · Vortex shedding · Drag force · Flow regimes · Control flow

1 Introduction

The flow of fluids in front of the bluff bodies is a major topical subject of several types of research in many practical engineering applications [1–7]. Among these applications, we mention the flows at the level of an airplane, a submarine and an automobile, heat exchanger systems, cooling of electronic components, gas turbine blades and particularly in the field of construction of buildings and high-rise buildings (Civil Engineering). In this domain, the structures become cliff bodies characterized by loads induced by the high wind-structure interaction. These effects can be drastically reduced with new aerodynamic attenuation techniques. Usually, the design of the exterior form and orientation of buildings is unfortunately based on architectural considerations and rather than aerodynamic considerations [7].

Over the past decade, the world has witnessed the biggest wave of tall building construction in history [7]. This type of building (tall buildings) is most sensitive to wind loads and wind-induced stresses which have the potential to reduce structural

R. Saidi et al. (Eds.): ICATH 2021, LNDECT 110, pp. 453–463, 2022.
https://doi.org/10.1007/978-3-030-94188-8_40

integrity and cause discomfort to occupants. These strong wind loads induced by the wind-structure interaction, therefore, represent one of the most critical parameters for the design of buildings. To reduce these wind loads on buildings, it is advisable to use a technique called "aerodynamic attenuation" [7]. This technique effectively uses simple and innovative architectural features to modify the aerodynamic shape of buildings. Aerodynamic changes help either by disrupting the formation of powerful wedge vortices; either by breaking the coherent formation of vortices; or by deflecting the flows in the partition area on the edge of the roof or away from weak elements. And hence, it reduces wind loads on buildings. Aerodynamic mitigation techniques that use modifications to the external form of a building provide a considerable reduction in the response of the building to wind loads. This reduction is due to the modification of the wind flow pattern around the building and can lead to a more economical and comfortable design [8, 9]. For example, changes in angle, variation in form and size of the cross-section along the height of the building, etc.

In this simulation work, we have modelled a building by square block located in a horizontal channel. The fluid entering the channel and passing around the square block actually represents the wind blowing on a building structure. Also, in our work, the control partitions play the role of aerodynamic attenuation techniques. This is a passive type of control where no external energy is required, which makes the process more economical. When the wind blows over a cliff structure, the flow separates and causes a periodic loss of vortices. This periodic vortex shedding exerts transverse forces on the body creating fluctuating pressures. The objective of the implementation of the control partitions is to decrease or suppress these vortices [10–12] and consequently a reduction of the forces applied on the best square block is observed.

2 Description of the Physical Problem and Boundary Conditions

2.1 Description of the Physical Problem

The physical problem studied is illustrated in Fig. 1 where a square cylinder is surrounded by four control partitions in a bi-dimensional channel. The square cylinder of size "d" is located on the axis of symmetry of the channel at an inlet length Lu = 6d, while the outlet length "Ld" is chosen long enough to avoid the outlet effects (Ld = 30d). The blockage ratio of the channel is $\beta = 1/11$ and its total length L = 37d. Each downstream partition has a length of Lp = 1d, whereas the upstream partitions are a variable length. Also, the thickness of each partition is h = 0,02d. All these partitions are positioned symmetrically after and before the block. The gap spacing between the square cylinder and the downstream partitions noted "g" is fixed at 3,9. Likewise, the gap spacing between the square cylinder and the upstream partitions noted "g_1" is fixed at 0,25.

The flow is fully developed with a parabolic velocity profile at the inlet and as well as at the outlet ($u = U_{max}[1 - (y/H)^2$; v = 0), where y is the vertical distance from the centerline, Umax is the maximal entrance velocity of the flow and u and v are the components of velocity vectors.

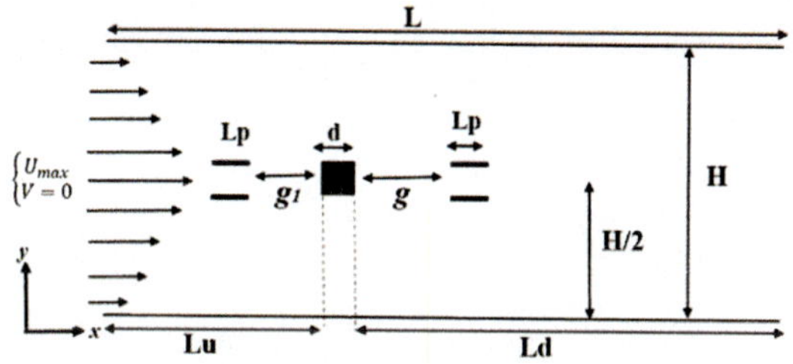

Fig. 1. Schematic of the numerical domain.

2.2 Boundary Condition

The step of the implementation of boundary conditions occupies an important place for the lattice Boltzmann method in the level of the stability and accuracy of the numerical calculations. In fact, in this step, the functions of distributions entering the computational domain (unknown functions) are defining from the functions leaving the computational domain, which are known.

The Bounce-Back boundary conditions [13] are the best-known and the best used boundary conditions in the literature for the LBM approach. The equation used for these conditions is expressed by:

$$f_i(\vec{x}_B, t) = f_{\bar{i}}(\vec{x}_B, t) \tag{1}$$

Where $f_{\bar{i}}(\vec{x}_B, t)$ and $f_i(\vec{x}_B, t)$ are the known and an unknown distribution function at the wall node ($\vec{x}_B$).

Because of its parabolic profile, the inlet and outlet of the computational domain are oriented by the boundary conditions of Zou and He [14].

3 Multi-relaxation-Time Lattice Boltzmann Method

In the last twentieth years, the Lattice Boltzmann method has undergone great development and it has been used by a large number of researchers in various disciplines such as physics, chemistry, health, biology, etc. [15–17]. Two models of LBM are widely used in the literature: The SRT model where all physical quantities relax to a single simple relaxation time, this model is also known as "BGK" [18]. The other model is called the lattice Boltzmann method with multiple relaxation time (LBM-MRT) [17, 19–23] where each physical quantity relaxes to a proper relaxation time. Due to its convergence and stability, the LBM-MRT is the most widely used model in recent times and is used in this work.

The expression of the evolution of the Boltzmann equation with the MRT-LBM model [17, 19, 21] is:

$$fi(x + ci\,\Delta t, t + \Delta t) = fi(x, t) + \Omega(f), i = 0, 1, \ldots., 8 \tag{2}$$

Where f_i designates the distribution function at position x, with velocity ci at the time t. The expression Ω denotes the collision between the fluid particles.

Practically, the collision process is realized in the momentum space by using the following expression:

$$fi(x + ci, t + 1) - fi(x, t) = M^{-1} \times S\left[m(x, t) - m^{eq}(x, t)\right]i = 0, .., 8 \quad (3)$$

Where m and m^{eq} are the moment's vector, and the equilibrium momentum, respectively:

$$m = (m_0, m_1, m_2, , m_n)^T; m^{eq} = \left(\rho, e^{eq}, \varepsilon^{eq}, j_x, q_x^{eq}, j_y, q_y^{eq}, p_{xx}^{eq}, p_{xy}^{eq}\right)^T \quad (4)$$

The correspondence between the velocity and moment spaces denotes by the transformation Matrix M: $m = Mf$ and $f = M^{-1}m$.

The matrix 9×9 given by [17, 19, 21]:

For reasons of stability of the calculation code, the nine values of the relaxation time S must be between 0 and 2 strictly, where the matrix S denotes the diagonal of the relaxation time which is expressed as:

$$S = \mathrm{diag}(s_0, s_1, s_2, s_3, s_4, s_5, s_6, s_7, s_8) \quad (5)$$

The discretization of the computational domain is performed by the D2Q9 model. The choice of this model is based on reasons of stability, accuracy and convergence of the computational code. The numerical scheme of the D2Q9 model is illustrated in the following figure (Fig. 2).

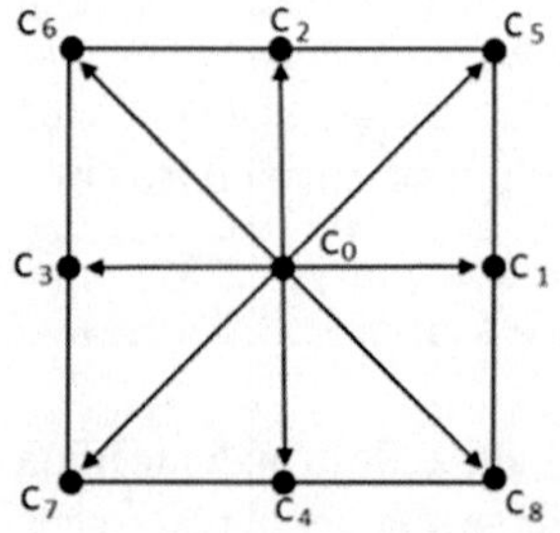

Fig. 2. The D2Q9 model for the fluid flow

The macroscopic quantities such as the density and the movement quantities obtained during the numerical simulation of Eq. (2) are given by:

$$\rho = \sum_{i=0}^{8} f_i, \quad \rho u = \sum_{i=0}^{8} c_i f_i \quad (6)$$

4 Validation of Simulation Code

In this research, the physical problem discussed is the flow around a square cylinder with four control partition placed upstream and downstream of the square cylinder (see

Fig. 1). First, our code is validated against the problem of flow through a square cylinder without a control plate with a blocking ratio of $\beta = 1/8$ and for a range of $20 \leq Re \leq 300$. The results obtained are compared with those found by Breuer et al. (2000) [24]. The graph showing the time-averaged drag coefficient is the only validation parameter with Breuer et al. presented in this work.

First of all, our code is validated with the numerical results found by Breuer et al. (2000) [24] where they treated the physical problem of a fluid flow passing around a square cylinder without a control plate with a blocking ratio of $\beta = 1/8$ and for a range of $20 \leq Re \leq 300$. Several parameters are validated with Breuer et al. but we will suffice to mention in this paper only the curve showing the time-averaged drag coefficient as a function of the Reynolds number. Figure 3-a shows that our LBM results are generally in good agreement with the reference results. Note that in this work, Breuer et al. employed two different methods (LBM and MVF).

In order to ensure the quality and power of our numerical code, a second validation is performed. In this case, the simulation of the flow of a fluid around a square cylinder with a control plate at a blocking ratio $\beta = H/d = 11$ was performed and validated with the work studied by Islam et al. [12]. The plot in Fig. 3-b shows the average values of the drag coefficient for different gap spacing at a fixed Reynolds number (Re = 150). This figure shows that our results obtained using our LBM-MRT code are almost identical to the reference results.

From these two validation studies, we can conclude that our LBM-MRT based code is reliable for studying our present problem.

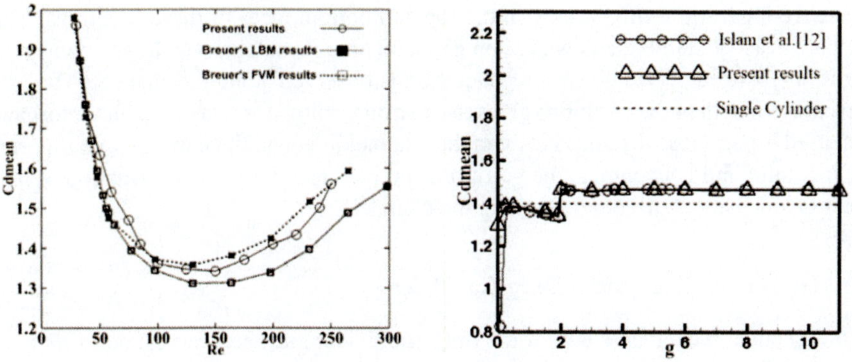

(a) (b)

Fig. 3. Validation of average drag coefficient with previous works for: (a) A single obstacle without control partition. (b) A single obstacle with control partition.

5 Results and Discussion

5.1 Regime of a Fluid Flow

In this section, the velocity contours in the wake of the cylinder in the presence of upstream and downstream partitions for different lengths and for fixed spacing are presented in Fig. 4. We remind that the bi-partition which are placed behind the square obstacle have the same length Lp = 1d and are located at a fixed spacing noted g = 3.9. Whereas, the upstream bi-partition are fixed at a distance $g_1 = 0.25$ and have a variable length. The objective of this work is to study the effect of the length Lp of these upstream partitions on the reduction or suppression of the vortex shedding as well as on the reduction of the coefficient of drag applied on the square cylinder.

In this study, three flow regimes are observed. Firstly, a flow regime called extended body is observed for short lengths where $1d \leq Lp \leq 2.75d$. In this case, the free shear layers delivered by the two upstream control partitions pass beyond the square cylinder and the two downstream control partitions. The incoming fluid considers the upstream and downstream partitions and the square cylinder as one single bluff body. Therefore, alternating vortex shedding is observed only after a distance behind the downstream plates moves along the channel. Also, the number and size of the vortices produced weaken compared to the other cases.

On the other hand, a strong vortex generation behind the two downstream partitions is observed without any fusion or deformation for lengths $3 \leq Lp < 5$. This flow regime is called a periodic flow regime where an alternation of positive and negative vortices is clearly observed (see Fig. 4 d–e). In this regime, the shear layers delivered by the edges of the front partitions are packed on the top and bottom faces of these partitions. Then they are evacuated into the downstream gap, and after its collision with the downstream partitions, a strong periodic vortex generation is observed along the channel. The great increase in length of the partitions gives rise to a disturbing flow regime. This disturbance is caused by the close distance between the channel inlet and the partitions. In this case, the incoming fluid encounters the partitions directly, resulting in the formation of large shear layers that disturb the flow along the channel.

5.2 Time-Trace Analysis of Drag Coefficient

In this section, a detailed study of the drag coefficient has been carried out. Firstly, an investigation of the optimal location of the downstream bi-partition of length Lp = 1d was carried out. This study shows that a percentage reduction of the drag coefficient reaches 12.5% compared to the case without any control partition for an optimal gap spacing g = 3.9. Next, we performed another investigation in which two partitions were added before the square block. In this study, the position of this bi-partition was varied to find out their optimal spacing g_1. Table 1 shows the average values of the drag coefficient for different downstream spacing g_1. This study achieves a maximum percentage reduction in drag coefficient that exceeds 36.33% for an optimal gap spacing $g_1 = 0.25$.

In the present study, we fixed the two downstream partitions at a distance g 3.9 and the two upstream partitions at a distance $g_1 = 0.25$ and we varied the length of these two downstream partitions Lp. Figure 5 shows the graph representing the variation

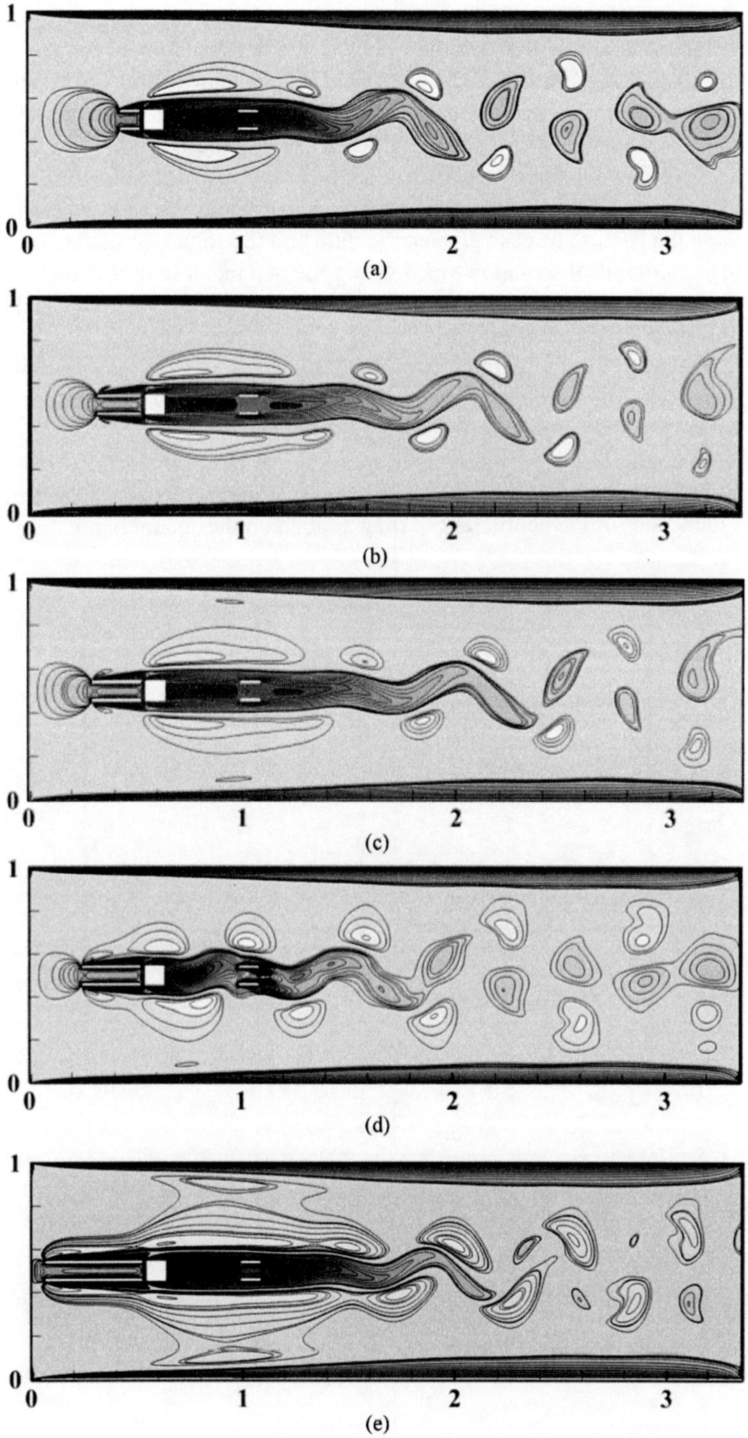

Fig. 4. Visualization of instantaneous flow velocity contours for different lengths of the upstream partitions: (a) Lp = 1d; (b) Lp = 2.25d; (c) Lp = 2.5d; (d) Lp = 3d; (e) Lp = 5d

of the drag coefficient (Cd) as a function of time for different lengths 1d ≤ Lp ≤ 5d. Similarly, Table 2 presents the average values of the drag coefficient for these various lengths (1d ≤ Lp ≤ 5d). From these results, it can be seen that the drag coefficient admits small values for a length range of 1d ≤ Lp ≤ 2.75d. A maximum reduction of the drag coefficient is reached 38.86% for a length Lp = 2.25d. For these lengths, the shear layers are produced by the collision between the incoming fluid and the leading ends of the two upstream partitions. These shear layers pass over and under the square obstacle. This will decrease the friction forces between the fluid and the square obstacle. Therefore, a maximum reduction is observed in this flow regime. An increase in partition length (3d ≤ Lp ≤ 4d) leads to an increase in the values of the drag coefficient. This is due to the fact that the shear layers produced are abutted on the upper and lower sides of the two partitions. They then meet with the square block and are evacuated into the downstream gap. This increases the friction forces applied to the square block and consequently an increase is observed in the values of the drag coefficient.

For long lengths Lp ≥ 5, a perturbed flow regime is observed. This perturbation increases further the contact forces between the fluid and the block. This remarkably increases the values of the coefficient of drag applied on the square block.

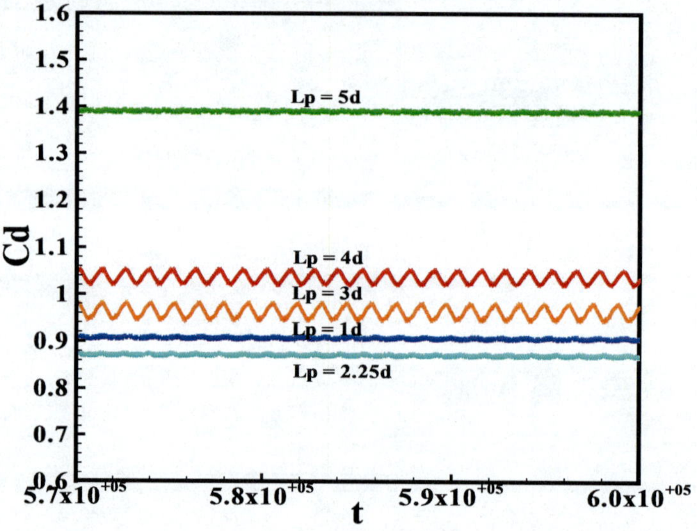

Fig. 5. Time-trace of drag coefficient for different length of upstream partition.

Table 1. The average values of the drag coefficient in case of the implementation of the upstream partition for different gap spacing g_1

g_1	Cd_{mean}
0	**0.9128**
0.25	0.9072
0.5	**1.1075**
0.75	**1.1523**
1	**1.1625**
2	**1.1788**
3	**1.2185**
4	**1.3558**

Table 2. The average values of the drag coefficient various upstream partitions length

Lp	Cd_{mean}	Percentage reduction in drag coefficient
1	**0.9072**	**36.33%**
2	**0.873**	**38.73%**
2.25	0.8712	**38.86%**
2.5	**0,8717**	**38.82%**
2.75	**0,8745**	**38.63%**
3	**0,9638**	**32.36%**
4	**1,0374**	**27.2%**
5	**1,3907**	**2.4%**

6 Conclusion

In this work, the numerical study of fluid flow along a square cylinder controlled by four detached partitions for various lengths of upstream partitions(from $1d \leq Lp \leq 5d$) was performed by using the multi-relaxation-time lattice Boltzmann method at a fixed Reynolds number ($Re = 150$).

The main conclusions drawn from this work are summarized as follows:

* Three different types of flow regimes appear successively: the extended body flow regime ($1d \leq Lp \leq 2.75$), the periodic flow regime ($3d \leq Lp \leq 4d$) and the disturbed flow regime $Lp \geq 5d$.
* A percentage reduction in the drag coefficient reaches 12.5% in the case of the implementation of two downstream partitions only.
* A percentage reduction of the drag coefficient reaches 36.33% in the case of the addition of two upstream partitions of length $Lp = 1d$ at a distance $g_1 = 0.25$.

* A percentage reduction of the drag coefficient reaches 38.86% in the case where the length of two upstream partitions Lp = 2.25d at a distance $g_1 = 0.25$.

References

1. Monat, J.P., Gannon, T.F.: Applying systems thinking to engineering and design. Systems **6**(3), 34 (2018)
2. Wang, X., Chen, J., Zhou, B., Li, Y., Xiang, Q.: Experimental investigation of flow past a confined bluff body: Effects of body shape, blockage ratio and Reynolds number. Ocean Eng. **220**, 108412 (2021)
3. Gilliéron, P.: Flow control applied to the automobile. State of the art. Mech. Ind. **3**(6), 515–524 (2002)
4. Chapain, S., Aly, A.M.: Vibration attenuation in high-rise buildings to achieve system-level performance under multiple hazards. Eng. Struct. **197**, 109352 (2019)
5. Deng, Y., et al.: Effects of different coolants and cooling strategies on the cooling performance of the power lithium ion battery system: a review. Appl. Therm. Eng. **142**, 10–29 (2018)
6. Florides, G., Kalogirou, S.: Ground heat exchangers—a review of systems, models and applications. Renewable Energy **32**(15), 2461–2478 (2007)
7. Mooneghi, M.A., Kargarmoakhar, R.: Aerodynamic mitigation and shape optimization of buildings. J. Build. Eng. **6**, 225–235 (2016)
8. Irwin, P.A.: Bluff body aerodynamics in wind engineering. J. Wind Eng. Ind. Aerodyn. **96**(6–7), 701–712 (2008)
9. Elshaer, A., Bitsuamlak, G., El Damatty, A.: Enhancing wind performance of tall buildings using corner aerodynamic optimization. Eng. Struct. **136**, 133–148 (2017)
10. Rashidi, S., Hayatdavoodi, M., Esfahani, J.A.: Vortex shedding suppression and wake control: A review. Ocean Eng. **126**, 57–80 (2016)
11. Karthik, K., Vishnu, M., Vengadesan, S., Bhattacharyya, S.K.: Optimization of bluff bodies for aerodynamic drag and sound reduction using CFD analysis. J. Wind Eng. Ind. Aerodyn. **174**, 133–140 (2018)
12. Islam, S.U., Rahman, H., Abbasi, W.S., Shahina, T.: Lattice Boltzmann study of wake structures and force statistics for various gap spacings between a square ROD with a detached flat plate. Arab. J. Sci. Eng. **40**, 2169–2182 (2015)
13. Bouzidi, M., Firdaouss, M., Lallemand, P.: Momentum transfer of a Boltzmann lattice fluid with boundaries. Phys. Fluids **13**(11), 3452–3459 (2001)
14. Zou, Q., He, X.: On pressure and velocity boundary conditions for the lattice Boltzmann BGK model. Phys. Fluids **9**(6), 1591–1598 (1997)
15. Noël, R., Ge, F., Zhang, Y., Navarro, L., Courbebaisse, G. Lattice Boltzmann method for modelling of biological phenomena. In: 25th European Signal Processing Conference (EUSIPCO), pp. 2654–265. IEEE (2017)
16. Naughton, N.M., Georgiadis, J.G.: Lattice Boltzmann method for simulation of diffusion magnetic resonance imaging physics in heterogeneous tissue models. arXiv preprint arXiv: 1907.00908 (2019)
17. Admi, Y., Moussaoui, M.A., Mezrhab, A.: Effect of a flat plate on heat transfer and flow past a three side-by-side square cylinders using double MRT-lattice Boltzmann method. In: IEEE 2nd International Conference on Electronics, Control, Optimization and Computer Science (ICECOCS), Kenitra, Morocco, pp. 1–5 (2020)
18. Bhatnagar, P.L., Gross, E.P., Krook, M.: A model for collision processes in gases. I. Small amplitude processes in charged and neutral one-component systems. Phys. Rev. Lett. **94**, 511–525 (1954)

19. Mohamad, A.A.: Lattice Boltzmann Method: Fundamentals and Engineering Applications with Computer Codes. Springer Science & Business Media (2011)
20. Moussaoui, M.A., Jami, M., Mezrhab, A., Naji, H., Bouzidi, M.: Multiple-relaxation-time lattice Boltzmann computation of channel flow past a square cylinder with an upstream control bi-partition. Int. J. Numer. Meth. Fluids **64**, 591–608 (2010)
21. Admi, Y., Moussaoui, M.A., Mezrhab, A., Bottom, V., Henry, D. Convective heat transfer past a three side-by-side square cylinders using double MRT- lattice Boltzmann method. In: 24th French Mechanical Congress (CFM), Brest, French (2019)
22. Moussaoui, M.A., Jami, M., Mezrhab, A., Naji, H.: MRT-Lattice Boltzmann simulation of forced convection in a plane channel with an inclined square cylinder. Int. J. of Therm. Sci. **49**, 131–142 (2010)
23. Moussaoui, M.A., Lahmer, E.B., Admi, Y., Mezrhab, A.: Natural convection heat transfer in a square enclosure with an inside hot block. In: International Conference on Wireless Technologies, Embedded and Intelligent Systems, WITS 2019. IEEE (2019). https://doi.org/10.1109/WITS.2019.8723863
24. Breuer, M., Bernsdorf, J., Zeiser, T., Durst, F.: Accurate computations of the laminar flow past a square cylinder based on two different methods: lattice Boltzmann and finite-volume. Int. J. Heat Fluid Flow **21**(2), 186–196 (2000)

Probabilistic Fatigue Life Analysis of Fiber Reinforced Polymer (FRP) Composites Made of ±45° Layers

Qinglin Gao[1], Haohui Xin[1](✉), and Ayman Mosallam[2]

[1] Department of Civil Engineering, School of Human Settlements and Civil Engineering, Xi'an Jiaotong University, Xi'an, Shaanxi, People's Republic of China
gaoqinglin@stu.xjtu.edu.cn, xinhaohui@xjtu.edu.cn

[2] Department of Civil and Environment Engineering, University of California, Irvine, CA, USA
mosallam@uci.edu

Abstract. The fiber reinforced polymer (FRP) composites are considered to be an ideal material for realizing long-span and high-performance structural system. It is necessary to analyze the reliability of FRP composites based on fatigue failure probability due to the fact that the experimental results of fatigue life are relatively discrete. The relationship between fatigue strength and fatigue life of ±45° ply FRP composites under various failure probabilities is investigated by using the Weibull distribution model in this paper. The cumulative distribution functions under different stress ratios for failure probabilities of 1%, 2.3% and 5% (guarantees of 99%, 97.7% and 95%) are given and the nonlinear probabilistic Weibull *S-N* percentile curves are deemed consistent with the actual situation. The distribution of test data show evidence discreteness when the stress ratio is relatively high and the distance between *S-N* curves with different failure probabilities is also increased. When the maximum stress is used as the evaluation index of stress level, the fatigue life increases with the increase of stress ratio. But when the stress amplitude is used as the evaluation index, the increase of stress ratio will lead to the decrease of fatigue life.

Keywords: Fiber Reinforced Polymer (FRP) · Fatigue life · Weibull distribution model · S-N curve

1 Introduction

Fiber reinforced polymer (FRP) has the advantages of light weight, high strength, corrosion resistance and excellent designability. It has been widely used in aviation, aerospace, automobile, petroleum, shipping, medical equipment and many other fields [1]. With the development of research on FRP composites, more and more attention has been paid to a series of properties that meet the needs of long-span and high-rise structural systems, especially bridge or offshore engineering [1–4].

The fatigue behavior of FRP is obviously different from that of steel and concrete materials [5]. Comprehensive assessment of the probabilistic fatigue life of FRP is

R. Saidi et al. (Eds.): ICATH 2021, LNDECT 110, pp. 464–473, 2022.
https://doi.org/10.1007/978-3-030-94188-8_41

a priority for applications. Previous research [6–8] on the fatigue performance show that fatigue characteristics of FRP has its own complexity and multitude, for example, different directions and stacking sequence of the layers have a significant effect on their fatigue life, the values of fatigue strength or fatigue life are relatively discrete, etc.

Stress levels significantly affect the fatigue life of composites [9]. In order to investigate the relationship between stress level and fatigue life, a fitted equation has been developed, known as *S-N* equation or Wöhler fatigue equation. The stress level, *S*, could be the maximum fatigue stress ($\sigma_{\max}$) [10–14], or stress amplitude ($\Delta\sigma = \sigma_{\max} - \sigma_{\min}$) [15, 16], while *N* represents the fatigue life of the composites.

Many studies have shown that Weibull distribution is an effective method for probabilistic fatigue life prediction of FRP composites. The Weibull distribution is based on the weakest link theory, or the largest flaw theory, which agree with that of the failure of polymer matrix composites. Therefore, the model has been used by many researchers for property modelling and design strength analysis of FRP composites. Alqam et al. [17] and Zureick et al. [18] analyzed the distribution of strength and stiffness of FRP composites by comparing normal, log-normal and Weibull distribution. The results recommended the use of the Weibull distribution to describe the characterization of FRP composites. But two-parameter model was selected considering the computational efficiency. Gomes et al. [19] performed extensive tensile tests, which showed that the Weibull distribution provided the best prediction in the tail region, and therefore is more suitable as the distribution model.

This paper aims to investigate the probabilistic fatigue life of FRP composites made of ±45° Layers using the Weibull distribution model. By using statistical tools, the computational efficiency increases greatly in the process of three parameter analysis. First, the orthogonal regression (OR) methods was used to obtain *S-N* equation for the test results collected from the previous study literature and the results were compared with the fitted equations in literatures. Then the relationship between fatigue strength and fatigue life of ±45° ply FRP composites under various failure probabilities is investigated by using the Weibull distribution model. Finally, the stress levels for the two million cycles and the cumulative distribution functions (CDF) under different stress ratios for failure probabilities of 1%, 2.3% and 5% (guarantees of 99%, 97.7% and 95%) are presented.

2 Experimental Results and Fitted *S-N* Equations

The fatigue tests of GFRP composite laminates with a stacking sequence of $(\pm 45)_{2s}$ are given in References [20, 21]. The tests were conducted for 44 specimens with stress ratio $R = 0.1$ and 0.5. The tests were performed by applying a sinusoidal cyclic loading pattern under load control. The dimension of all specimens is 250 mm × 25 mm × 2.3 mm. Tables 1 list the test results at different stress levels and stress ratios.

In general, the relationship between the maximum stress, $\sigma_{\max}$, and the number of cycles can be expressed as a linear relationship in logarithmic space, as shown in Eq. 1. The fitted regression equation of this linear line can be obtained by two methods of simple linear regression: the least-squares regression (LR) [22] and orthogonal regression (OR) [23] method, as shown in Fig. 1(a) and (b). The fitted equations are shown in Fig. 2. *R*

Table 1. Laboratory fatigue test results [20, 21]

No	R	σ_{max}(MPa)	$\Delta\sigma$(MPa)	N_f	R	σ_{max}(MPa)	$\Delta\sigma$(MPa)	N_f
1	0.1	67.94	61.15	1969	0.5	78.00	39.00	20505
2		67.97	61.17	1007		78.00	39.00	5206
3		67.18	60.46	591		78.00	39.00	1767
4		67.55	60.80	3689		78.00	39.00	1471
5		64.37	57.93	2593		72.00	36.00	60130
6		64.32	57.89	2304		72.00	36.00	61616
7		64.39	57.95	1161		72.00	36.00	36435
8		64.28	57.85	4112		72.00	36.00	6513
9		59.53	53.58	6785		68.00	34.00	5436
10		59.26	53.33	22321		68.00	34.00	112681
11		59.05	53.15	18028		68.00	34.00	119132
12		59.05	53.15	15063		68.00	34.00	16642
13		54.56	49.10	49151		64.20	32.10	76431
14		54.11	48.70	29832		64.20	32.10	228923
15		53.97	48.57	22583		64.20	32.10	345818
16		54.80	49.32	79623		64.20	32.10	108719
17		49.46	44.51	289050		58.20	29.10	2110761
18		49.45	44.51	194115		58.20	29.10	295972
19		49.08	44.17	460664		58.20	29.10	645568
20		49.45	44.51	277227		58.20	29.10	5245384
21		48.04	43.24	533634				
22		48.02	43.22	1198627				
23		47.94	43.15	430211				
24		47.94	43.15	545689				

is the ratio of minimum stress to maximum stress, i.e. $R = \sigma_{min}/\sigma_{max}$.

$$\lg(\sigma_{max}) = A_1 + B_1 \lg(N_f) \tag{1}$$

where A_1 and B_1 are material parameters associated with $\sigma_{max} - N_f$.

Similarly, The fitted regression equation of the relationship between the stress amplitude, $\Delta\sigma = \sigma_{max} - \sigma_{min}$, and the number of cycles, N_f, is shown in Eq. 2. The $\Delta\sigma - N_f$ curves by LR and OR methods are shown in Fig. 3.

$$\lg(\Delta\sigma) = A_2 + B_2 \lg(N_f) \tag{2}$$

where A_2 and B_2 are material parameters associated with $\Delta\sigma - N_f$.

The fitted parameters and equations of *S-N* curves are shown in Table 2. S_{LR} and S_{OR} are the sum of the squares of residuals for LR and OR methods and the Pearson Correlation Coefficient (ρ_{PCC}) is used to measure the degree of correlation between stress level *S* and fatigue life *N*. It can be seen that the PCC for both $R = 0.1$ and 0.5 are larger than 0.8, testifying a strong negative correlation of stress level with the fatigue

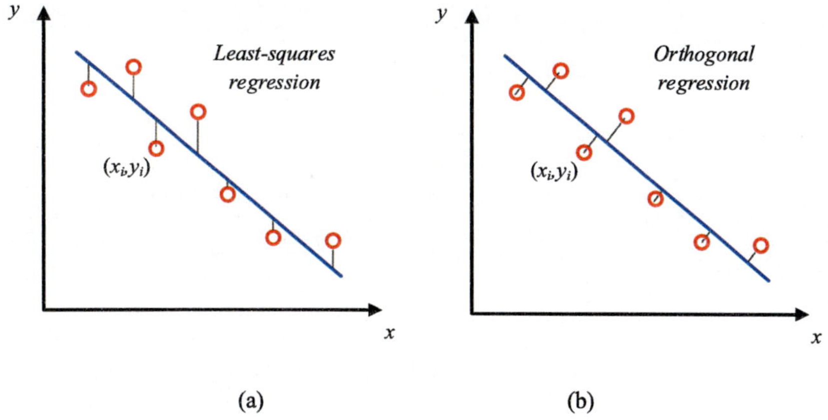

Fig. 1. The least-squares regression (LR) and orthogonal regression (OR) methods

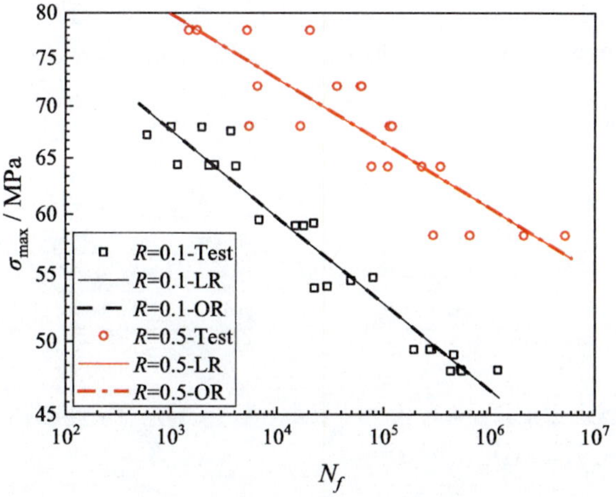

Fig. 2. Fitted $\sigma_{\max} - N_f$ curves and experimental data

life. But the PCC of $R = 0.5$ is 0.8675, obviously lower than that of $R = 0.5$ which is 0.9728, indicated that the distribution of test data show evidence discreteness when the stress ratio is relatively high. In addition, the *S-N* equation and parameters fitted by LR and OR methods are basically the same. In general, the OR method assumes that both independent and dependent variables may be subject to error while least-squares regression (LR) [22] method supposes that the error in the independent variable is negligible. But the parameters and curves fitted by LR and OR methods are very similar in this study. This similarities may result from the uniform distribution of experimental data, causing that the fitting results by LR method are accurate enough while OR method cannot play a greater role in correction.

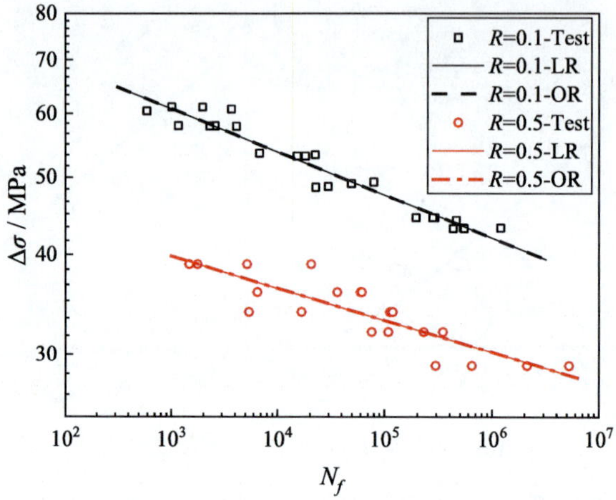

Fig. 3. Fitted $\Delta\sigma - N_f$ curves and experimental data

Table 2. The fitted parameters and equations of *S-N* curves

Relationship	R	Least-squares Regression (LR)		Orthogonal Regression (OR)		ρ_{PCC}
		Fitted equation	S_{OR}	Fitted equation	S_{LR}	
$\sigma_{max} - N_f$	0.1	$\lg(\sigma_{max}) = 1.9915–0.0538\lg(N_f)$	0.0134	$\lg(\sigma_{max}) = 1.9915–0.0538\lg(N_f)$	0.0134	–0.9728
	0.5	$\lg(\sigma_{max}) = 2.0224–0.0401\lg(N_f)$	0.0227	$\lg(\sigma_{max}) = 2.0225–0.0401\lg(N_f)$	0.0226	–0.8675
$\Delta\sigma - N_f$	0.1	$\lg(\Delta\sigma) = 1.9457–0.0538\lg(N_f)$	0.0134	$\lg(\Delta\sigma) = 1.9458–0.0538\lg(N_f)$	0.0134	–0.9728
	0.5	$\lg(\Delta\sigma) = 1.7214–0.0401\lg(N_f)$	0.0227	$\lg(\Delta\sigma) = 1.7215–0.0401\lg(N_f)$	0.0226	–0.8675

3 Failure Probability Analysis Based on the Three-Parameter Weibull Distribution Model

The cumulative distribution functions (CDF) of the three-parameter Weibull distribution, associated with $\sigma_{max} - N_f$ and $\Delta\sigma - N_f$, are expressed in Eqs. 3 and 4.

$$P_f(N_f;\sigma_{max}) = 1 - \exp\left\{-\left[\frac{(\ln N_f - B)(\ln\sigma_{max} - C) - \lambda}{\delta}\right]^{\beta}\right\} \quad (3)$$

$$P_f(N_f;\Delta\sigma) = 1 - \exp\left\{-\left[\frac{(\ln N_f - B)(\ln\Delta\sigma - C) - \lambda}{\delta}\right]^{\beta}\right\} \quad (4)$$

$$V = (\ln N_f - B)(\ln S - C) \quad (5)$$

Where B and C are, respectively, a limit or threshold number of cycles and fatigue limit. It should be noted that exp(B) and exp(C) are the threshold value of N and S. λ, β and δ are the location, shape and scale parameters of the Weibull distribution, respectively. The normalized variable V, as shown in Eq. 5, is introduced to consider damage as the probability of failure, and allows the whole Wöhler filed to be simplified to a cumulative distribution function (CDF).

The three-parameter Weibull distribution model was carried out in a program named ProFatigue [24]. The program is a statistical tool to predict the Wöhler field. When the estimation process has finished, Profatigue offers *S-N* curves and CDF (or the failure probability) graphics to analyze the suitability of the statistical model. Through this program, the values of parameters for $R = 0.1$ and 0.5 and the equations for CDF are shown in Table 3.

Figure 4 shows the *S-N* curves with failure probability of 1%, 2.3% and 5% (guarantees of 99%, 97.7% and 95%). The curves, deemed consistent with the actual situation, allow a reasonable calculation of the FRP composite failure probability at a certain stress level and number of cycles, as well as an estimate of design strength and life expectancy. Figure 5 shows CDF of the normalizing variable V with three-parameter Weibull distribution. The figures indicated that the whole *S-N* filed to be reduced to a CDF is reasonable.

Table 4 presents the corresponding fatigue equations and stress level of $\text{N} = 2 \times 10^6$ cycles for $\sigma_{\max} - N_f$ and $\Delta\sigma - N_f$ with failure probability of 1%, 2.3% and 5% (guarantees of 99%, 97.7% and 95%). The stress level of $N = 2 \times 10^6$ cycles for $\sigma_{\max} - N_f$ and $\Delta\sigma - N_f$ are generally in the range 42–43 MPa and 38–39 with $R = 0.1$ while for $\sigma_{\max} - N_f$ and $\Delta\sigma - N_f$ are generally in the range 51–53.5MPa and 25.3–26.6 with $R = 0.5$. The results show that $R = 0.5$ possesses a higher maximum fatigue stress but a lower stress amplitude than $R = 0.1$ with the same failure probability. In other words, when the maximum stress is used as the evaluation index of stress level, the fatigue life is positively correlated with stress ratio. But when the stress amplitude is used as the evaluation index, the fatigue life is negatively correlated with stress ratio.

Table 3. Three-parameter Weibull distribution function and parameters

Relationship	R	B	C	λ	δ	β	Weibull distribution function
$\sigma_{\max} - N_f$	0.1	0.0	3.4	3.84	2.18	6.36	$P_f(N_f;\sigma_{\max}) = 1 - \exp\left\{-\left[\frac{(\ln N_f)(\ln\sigma_{\max}-3.4)-3.84}{2.18}\right]^{6.36}\right\}$
	0.5	0.0	3.6	0.02	7.05	11.84	$P_f(N_f;\sigma_{\max}) = 1 - \exp\left\{-\left[\frac{(\ln N_f)(\ln\sigma_{\max}-3.6)-0.02}{7.05}\right]^{11.84}\right\}$
$\Delta\sigma - N_f$	0.1	0.0	3.3	3.84	2.18	6.36	$P_f(N_f;\Delta\sigma) = 1 - \exp\left\{-\left[\frac{(\ln N_f)(\ln\Delta\sigma-3.3)-3.84}{2.18}\right]^{6.36}\right\}$
	0.5	0.0	2.9	0.02	7.05	11.84	$P_f(N_f;\Delta\sigma) = 1 - \exp\left\{-\left[\frac{(\ln N_f)(\ln\Delta\sigma-2.9)-0.02}{7.05}\right]^{11.84}\right\}$

Table 4. The stress level when the fatigue failure number N is equal to 2×10^6

Relationship	R	Fatigue equation	Failure probability	The stress level for $N = 2 \times 10^6$
$\sigma_{max} - N_f$	0.1	$P_f(N_f;\sigma_{max}) = 1 - \exp\left\{-\left[\frac{(\ln N_f)(\ln \sigma_{max} - 3.4) - 3.84}{2.18}\right]^{6.36}\right\}$	0.01	42.00
			0.023	42.43
			0.05	42.90
	0.5	$P_f(N_f;\sigma_{max}) = 1 - \exp\left\{-\left[\frac{(\ln N_f)(\ln \sigma_{max} - 3.6) - 0.02}{7.05}\right]^{11.84}\right\}$	0.01	50.95
			0.023	52.20
			0.05	53.49
$\Delta\sigma - N_f$	0.1	$P_f(N_f;\Delta\sigma) = 1 - \exp\left\{-\left[\frac{(\ln N_f)(\ln \Delta\sigma - 3.3) - 3.84}{2.18}\right]^{6.36}\right\}$	0.01	38.00
			0.023	38.39
			0.05	38.82
	0.5	$P_f(N_f;\Delta\sigma) = 1 - \exp\left\{-\left[\frac{(\ln N_f)(\ln \Delta\sigma - 2.9) - 0.02}{7.05}\right]^{11.84}\right\}$	0.01	25.30
			0.023	25.92
			0.05	26.56

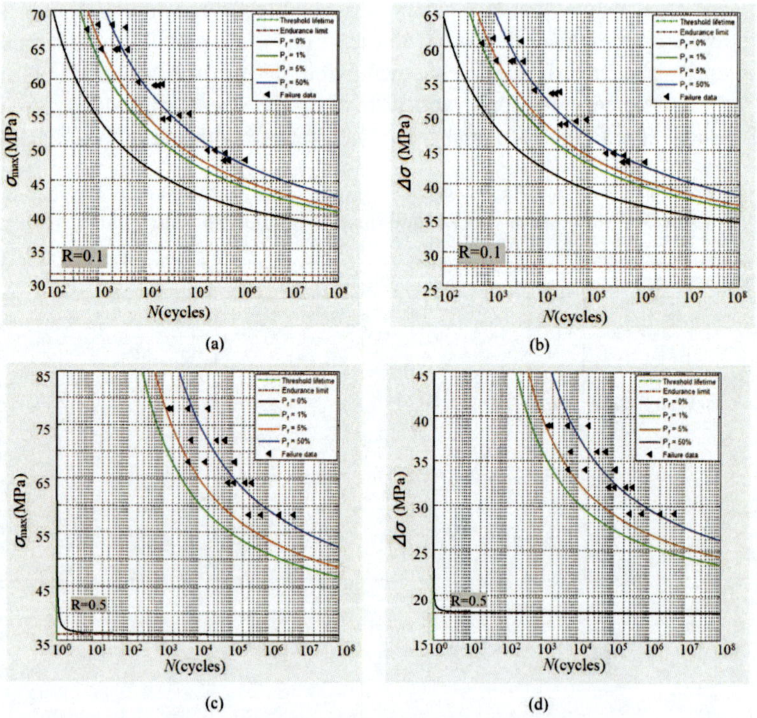

Fig. 4. *S-N* curves with three-parameter Weibull distribution

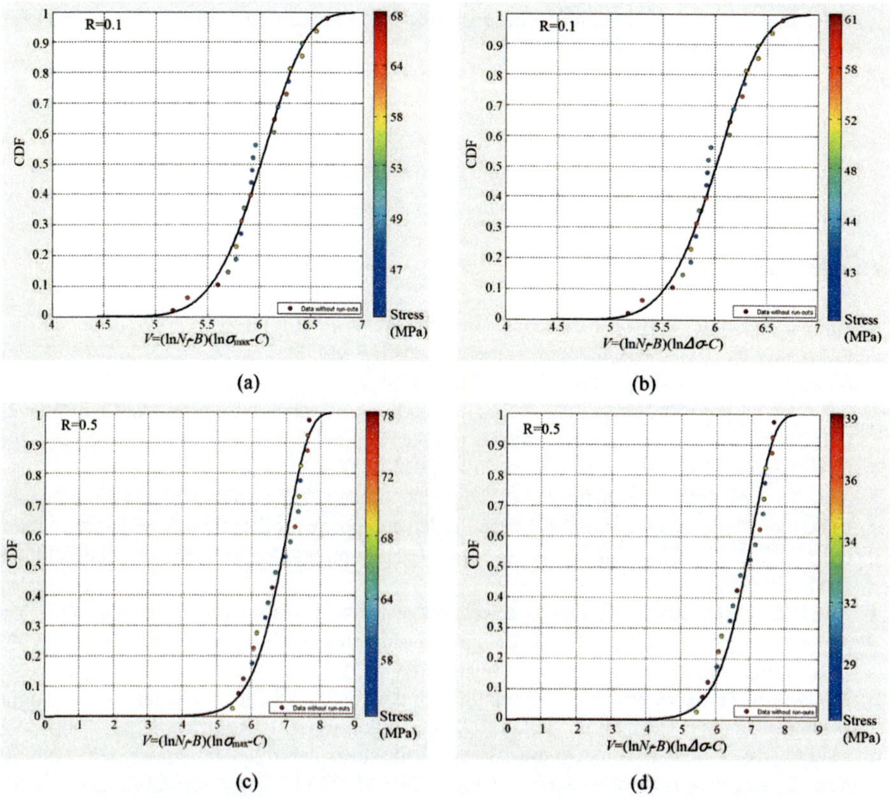

Fig. 5. CDF of the normalizing variable V with three-parameter Weibull distribution

4 Conclusions

The application of fiber reinforced polymer (FRP) benefits long-span and high-rise structural system. More attention to fatigue life is drawn for its application in bridge or offshore structures. The following conclusions can be made in this paper:

(1) The *S-N* equations were fitted by the least-squares regression (LR) and the orthogonal regression (OR) methods. The results testified a strong negative correlation of stress level with the fatigue life. The parameters fitted by two methods are similar, which may result from the uniform distribution of experimental data. The OR method cannot play a greater role in correction.
(2) The relationship between fatigue strength and fatigue life of FRP composites under various failure probabilities is investigated by using the Weibull distribution model. The results indicated the nonlinear curves obtained from the three-parameter model are deemed consistent with the actual situation.
(3) The cumulative distribution functions under different stress ratios for failure probabilities of 1%, 2.3% and 5% are given. When the maximum stress is used as the evaluation index of stress level, the stress ratio is positively correlated with fatigue

life. But when the stress amplitude is used as the evaluation index, the stress ratio is negatively correlated with fatigue life.

Acknowledgments. The authors gratefully acknowledge the financial support provided by the National Natural Science Foundation (Grants #51808398) of the People's Republic of China.

References

1. Ahmed, A., et al.: A review on the tensile behavior of fiber-reinforced polymer composites under varying strain rates and temperatures. Constr. Build. Mater. 294ARTN 123565 (2021). https://doi.org/10.1016/j.conbuildmat.2021.123565
2. Xin, H.H., Liu, Y.Q., Mosallam, A.S., He, J., Du, A.: Evaluation on material behaviors of pultruded glass fiber reinforced polymer (GFRP) laminates. Compos. Struct. **182**, 283–300 (2017). https://doi.org/10.1016/j.compstruct.2017.09.006
3. Xin, H.H., Mosallam, A., Liu, Y.Q., Wang, C.Z., Zhang, Y.Y.: Analytical and experimental evaluation of flexural behavior of FRP pultruded composite profiles for bridge deck structural design. Constr. Build. Mater. **150**, 123–149 (2017). https://doi.org/10.1016/j.conbuildmat.2017.05.212
4. Xin, H.H., et al.: Experimental and numerical investigation on in-plane compression and shear performance of a pultruded GFRP composite bridge deck. Compos. Struct. **180**, 914–932 (2017). https://doi.org/10.1016/j.compstruct.2017.08.066
5. Whitworth, H.A.: A stiffness degradation model for composite laminates under fatigue loading. Compos. Struct. **40**, 95–101 (1997). https://doi.org/10.1016/S0263-8223(97)00142-6
6. Mallick, P.K., Zhou, Y.: Effect of mean stress on the stress-controlled fatigue of a short E-glass fiber reinforced polyamide-6,6. Int. J. Fatigue **26**, 941–946 (2004). https://doi.org/10.1016/j.ijfatigue.2004.02.003
7. Varvani-Farahani, A., Shirazi, A.: A fatigue damage model for (0/90) FRP composites based on stiffness degradation of 0 and 90 composite plies. J. Reinf. Plast. Compos. **26**, 1319–1336 (2007). https://doi.org/10.1177/0731684407079771
8. Gibson, R.F.: A review of recent research on mechanics of multifunctional composite materials and structures. Compos. Struct. **92**, 2793–2810 (2010). https://doi.org/10.1016/j.compstruct.2010.05.003
9. Brunbauer, J., Pinter, G.: Effects of mean stress and fibre volume content on the fatigue-induced damage mechanisms in CFRP. Int. J. Fatigue **75**, 28–38 (2015). https://doi.org/10.1016/j.ijfatigue.2015.01.014
10. Keller, T., Tirelli, T., Zhou, A.: Tensile fatigue performance of pultruded glass fiber reinforced polymer profiles. Compos. Struct. **68**, 235–245 (2005). https://doi.org/10.1016/j.compstruct.2004.03.021
11. Korkiakoski, S., Sarlin, E., Suihkonen, R., Saarela, O.: Internal structure and fatigue performance of quasi-unidirectional non-crimp fabric reinforced laminates. J. Compos. Mater. **51**, 3405–3423 (2017). https://doi.org/10.1177/0021998316683783
12. Mivehchi, H., Varvani-Farahani, A.: The effect of temperature on fatigue strength and cumulative fatigue damage of FRP composites. In: 10th International Fatigue Congress, FATIGUE 2010, 6–11 June 2010, (Procedia Engineering). Elsevier Ltd., Prague, pp. 2011–2020 (2010). https://doi.org/10.1016/j.proeng.2010.03.216
13. Shah, D.U., Schubel, P.J., Clifford, M.J., Licence, P.: Fatigue life evaluation of aligned plant fibre composites through S-N curves and constant-life diagrams. Compos. Sci. Technol. **74**, 139–149 (2013). https://doi.org/10.1016/j.compscitech.2012.10.015

14. Vieira, P.R., Carvalho, E.M.L., Vieira, J.D., Toledo Filho, R.D.: Experimental fatigue behavior of pultruded glass fibre reinforced polymer composite materials. Compos. B Eng. **146**, 69–75 (2018). https://doi.org/10.1016/j.compositesb.2018.03.040
15. Manjunatha, C.M., Jagannathan, N., Padmalatha, K., Kinloch, A.J., Taylor, A.C.: Improved variable-amplitude fatigue behavior of a glass-fiber-reinforced hybrid-toughened epoxy composite. J. Reinf. Plast. Compos. **30**, 1783–1793 (2011). https://doi.org/10.1177/0731684411426202
16. Ospina Cadavid, M., Al-Khudairi, O., Hadavinia, H., Goodwin, D., Liaghat, G.H.: Experimental studies of stiffness degradation and dissipated energy in glass fibre reinforced polymer composite under fatigue loading. Polym. Polym. Compos. **25**, 435–446 (2017). https://doi.org/10.1177/096739111702500602
17. Alqam, M., Bennett, R.M., Zureick, A.-H.: Three-parameter vs. two-parameter Weibull distribution for pultruded composite material properties. Compos. Struct. **58**, 497–503 (2002). https://doi.org/10.1016/S0263-8223(02)00158-7
18. Zureick, A.-H., Bennett, R.M., Ellingwood, B.R.: Statistical characterization of fiber-reinforced polymer composite material properties for structural design. J. Struct. Eng. **132**, 1320–1327 (2006). https://doi.org/10.1061/(ASCE)0733-9445(2006)132:8(1320)
19. Gomes, S., Dias-da-Costa, D., Neves, L.A.C., Hadigheh, S.A., Fernandes, P., Julio, E.: Probabilistic-based characterisation of the mechanical properties of CFRP laminates. Constr. Build. Mater. **169**, 132–141 (2018). https://doi.org/10.1016/j.conbuildmat.2018.02.190
20. Movahedi-Rad, A.V., Keller, T., Vassilopoulos, A.P.: Fatigue damage in angle-ply GFRP laminates under tension-tension fatigue. Int. J. Fatigue **109**, 60–69 (2018). https://doi.org/10.1016/j.ijfatigue.2017.12.015
21. Movahedi-Rad, A.V., Keller, T., Vassilopoulos, A.P.: Stress ratio effect on tension-tension fatigue behavior of angle-ply GFRP laminates. Int. J. Fatigue **126**, 103–111 (2019). https://doi.org/10.1016/j.ijfatigue.2019.04.037
22. Björck, Å.: Numerical Methods for Least Squares Problems SIAM (1996)
23. Carroll, R., Ruppert, D.: The use and misuse of orthogonal regression in linear errors-in-variables models. Am. Stat. **50**, 1–6 (1996). https://doi.org/10.2307/2685035
24. Fernandez-Canteli, A., Przybilla, C., Nogal, M., Aenlle, M.L., Castillo, E.: Profatigue: a software program for probabilistic assessment of experimental fatigue data sets. In: 17th International Colloquium on Mechanical Fatigue of Metals, ICMFM 2014, 25–27 June 2014 (Procedia Engineering). Elsevier Ltd., Verbania, pp. 236–241 (2014). https://doi.org/10.1016/j.proeng.2014.06.255

Evaluation for the Punching Shear Design of Concrete Slabs Subjected to Compressive Forces

A. Deifalla(✉)

Department of Structural Engineering and Construction Management, Future University in Egypt, New Cairo City 11835, Egypt
diffalaf@mcmaster.ca

Abstract. Concrete slabs are subjected to compressive forces due to prestressing or restraint conditions. Using prestressed concrete slabs for long spans is a top choice in many applications, particularly lightweight ones. However, the number of studies tackling the punching shear design of lightweight concrete slabs subjected to compressive forces is minimal. The purpose of this investigation is to assess the punching shear design of lightweight and normal-weight concrete slabs. The available punching shear testing for lightweight and normal-weight concrete slabs is compiled. The punching shear strength is calculated using selected design codes compared with the experimentally measured strength. Concluded remarks were outlined and discussed.

Keywords: Punching shear · Lightweight concrete · Prestressing

1 Introduction

Various design codes provide punching shear provisions for concrete slabs under in-plane compression forces. However, a disagreement exists among those design codes [1–5]. Considered parameters include prestressing compression stress, prestressing eccentricity, prestressing vertical component, aggregate nominal size, concrete density, rotation, control perimeter, and flexure reinforcement ratio. All design methods include the effect of the concrete compressive strength; however, the effect of all other parameters is different from none. This investigation is a part of a project by Deifalla and co-workers to investigate the behavior and design of beams and slabs under one-way shear, punching shear, flexure, and torsion [5–21]. The purpose of this investigation is to evaluate those existing design codes for the case of punching shear combined with in-plane compression forces

2 Selected Design Codes for Punching Shear Combined with In-Plane Compression Forces

Several selected codes and models exist for punching shear design of concrete slabs under compressions forces, including EC2 [1], ACI [2], ECOP-203 [3], and

R. Saidi et al. (Eds.): ICATH 2021, LNDECT 110, pp. 474–481, 2022.
https://doi.org/10.1007/978-3-030-94188-8_42

MC [4]. In this section, a brief of each design code is outlined. Various design codes differ from each other in principle, where some are based on the pure empirical formulation (i.e., EC2, ECOP, and ACI), and some are based on physical models calibrated using experimental databases (i.e., MC), which is based on the critical shear crack theory (CSCT). In addition, there is a disagreement between models on the parameters considered in the strength. For example, the EC2 and MC account for the flexure reinforcement ratio and the maximum aggregate size, while the ACI ignores it. For Lightweight concrete slabs, the MC reduces the aggregate interlock by neglecting the aggregate size. In contrast, the EC2 and ACI reduce the tensile concrete strength using a factor based on the concrete density. For in-plane compression, the EC2, ACI, ECOP use a fraction of the axial compression stresses. At the same time, the MC implemented a three approximation levels approach, which accounts for the decompression moment due to the axial compression stresses.

2.1 EC2

The two-way shear strength for concrete slabs with axial compression using the EC2 is expressed in Eq. (1). An empirical model accounts for size effect, flexure reinforcement ratio, concrete compressive strength, axial compression stresses from prestressing, lightweight concrete effect, and loading eccentricity. It takes the critical section at 2d from the loaded area. Thus, The total two-way shear capacity of the concrete slab in Newton is calculated such that:

$$V = \left[0.15\lambda\lambda_s\left(100\rho_l f_c'\right)^{\frac{1}{3}} + 0.1\sigma\right] b_o d \geq \left(0.028\lambda{\lambda_s}^{\frac{3}{2}}\sqrt{f_c'} + 0.1\sigma\right) b_o d \quad (1)$$

Where λ is the concrete type factor, which is calculated differently for each model and taken as $0.40 + 0.60\gamma_c/2200$, λ_s is the size effect factor, which is calculated differently for each model and taken as $\left(1+\sqrt{\frac{200}{d}}\right) \leq 2.0.$ and ρ the flexure reinforcement ratio and take ≤ 0.02. γ_cThe concrete density in kg/m3. b_o is the control perimeter taken at distance d/2 from the edge of the loaded area, except for EC2 taken at distance 2d. d is the effective depth of the slab in mm. σThe normal stress (positive for compression). f_c' The concrete compressive strength in MPa.

2.2 ACI

The two-way shear strength for concrete slabs with axial compression using the ACI is expressed in Eq. (2). It is an empirical model, which accounts for concrete compressive strength, axial compression stresses from prestressing, and loading eccentricity as well as take the critical section at d/2 from loaded area. However, it did not consider the size effect and flexure reinforcement ratio.

$$V = smallelsestof \begin{cases} \left(0.29\lambda\sqrt{f_c'} + 0.3\sigma\right)\ b_o d \\ \left(0.083\left(1.5 + \frac{\alpha_s d}{b_o}\right)\lambda\sqrt{f_c'} + 0.3\sigma\right)\ b_o d \end{cases} \quad (2)$$

where α_s is the factor for eccentric loading, which is different for each design code used and taken as 40, 30, 20 for the inner, edge, and corner slab-column connection, respectively, λ is the concrete type factor, which is calculated differently for each model and taken as $\frac{3}{6400}\gamma_c \leq 1.0$.

2.3 ECOP-203

The two-way shear strength for concrete slabs with axial compression using the ECOP is expressed in Eq. (3). It is an empirical model, which accounts for concrete compressive strength, axial compression stresses from prestressing, and loading eccentricity as well as take the critical section at d/2 from loaded area. However, it did not consider the size effect and flexure reinforcement ratio.

$$V = smallelsest\ of \begin{cases} \left(0.275\lambda\sqrt{1.25f_c'} + 0.2\sigma\right)\ b_o d \\ \left(0.080\left(1.5 + \frac{\alpha_s d}{b_o}\right)\sqrt{1.25f_c'} + 0.2\sigma\right)\ b_o d \end{cases} \tag{3}$$

Where is α_s eccentricity factor is taken as 4, 3, and 2 for the middle, edge, and corner column.

2.4 MC

The two-way shear strength for concrete slabs with axial compression using the MC is expressed in Eq. (4). It is based on the physical sound model critical shear crack theory (CSCT). In addition, it offers three levels of approximation (L.O.A.) approach for design, which improves the accuracy of strength predictions by progressively refining the values of the physically effective parameters. Such an approach allows for performing quick and easy preliminary design checks (LOA I) and highly accurate and sophisticated strength predictions (LOA II & LOA III). It takes the critical section at d/2 from the loaded area-however, no direct account for the size effect. LOA I was selected to simplify the calculations.

$$V = k_b b_o d\sqrt{f_c'} \tag{4}$$

Where

$$k_b = \frac{1}{1.5 + 0.9\left(\frac{32}{16}\right)\psi d} \leq 0.6$$

and

$$\psi = 1.5\frac{r_s}{d}\frac{f_y}{E_s}.$$

E_s The Young's modulus of flexure reinforcements. f_yThe tensile strength of shear reinforcements. r_sThe distance between the centerline of the loading area and the inflection point.

3 Comparison Between Selected Design Codes Predictions

This evaluation approach is the same implemented for validating models and design codes used in many investigations, where the strength calculated using the selected design codes is compared with that observed during the experimental testing [5–11,14,21]. The results were split into two categories, namely normal weight concrete (NW) and lightweight concrete (LW). Although many researchers may believe that an SF (i.e., the ratio between the measured and calculated strength) less than 1 is unsafe, however, an approach developed by the technical committee for concrete design of the Canadian Society Association and implemented in many investigations [5–11,15,22], where the following six categories were identified: Extremely dangerous, Dangerous, Low safety, Approximate safety, Conservative, Extremely conservative for S.I. values of less than 0.5, 0.5-0.65, 0.65-0.85, 0.85-1.30, 1.30-2.00, and more than 2. The details of the experimental testing database are available in the literature review [6].

3.1 Overall

Table 1 shows the statistical measures for the ratio between the measured and calculated strength (S.F.). It is clear that the EC2 is the most consistent with the lowest coefficient of variation (C.O.V.). At the same time, the MC is the most accurate, with an average closer to unity compared with other design codes. Thus, all design codes are reasonably safe, while the EC2 is overly conservative. Figure 1(a–f) shows the measured strength versus the calculated using the various codes. Again, the scattering of all design codes is reasonable; however, MC (III) is the least scattered compared to other methods (Figs. 2 and 3).

Table 1. Statistical measures using various design codes.

Statistical measures	ECOP	ACI	EC2	MC(I)	MC(II)	MC(III)
Average	1.08	1.06	1.86	1.38	1.27	1.11
C.O.V	79%	78%	41%	63%	59%	59%
Lower 95%	0.96	0.95	1.76	1.26	1.15	1.01

3.2 The Size Effect on the Safety Factor Using Selected Design Codes

The size significantly affects the strength. Thus, the A.C.I. and the EC2 proposed a size effect factor to reduce the strength for slabs with thickness more than 200–250 mm. Although tested slabs are less than 250 mm, it is clear that design codes are overly conservative for slabs less than 100 mm thickness. While, for slabs larger than 100 mm thickness, design codes are consistently reasonable. It can be concluded that the design is less conservative for larger depth.

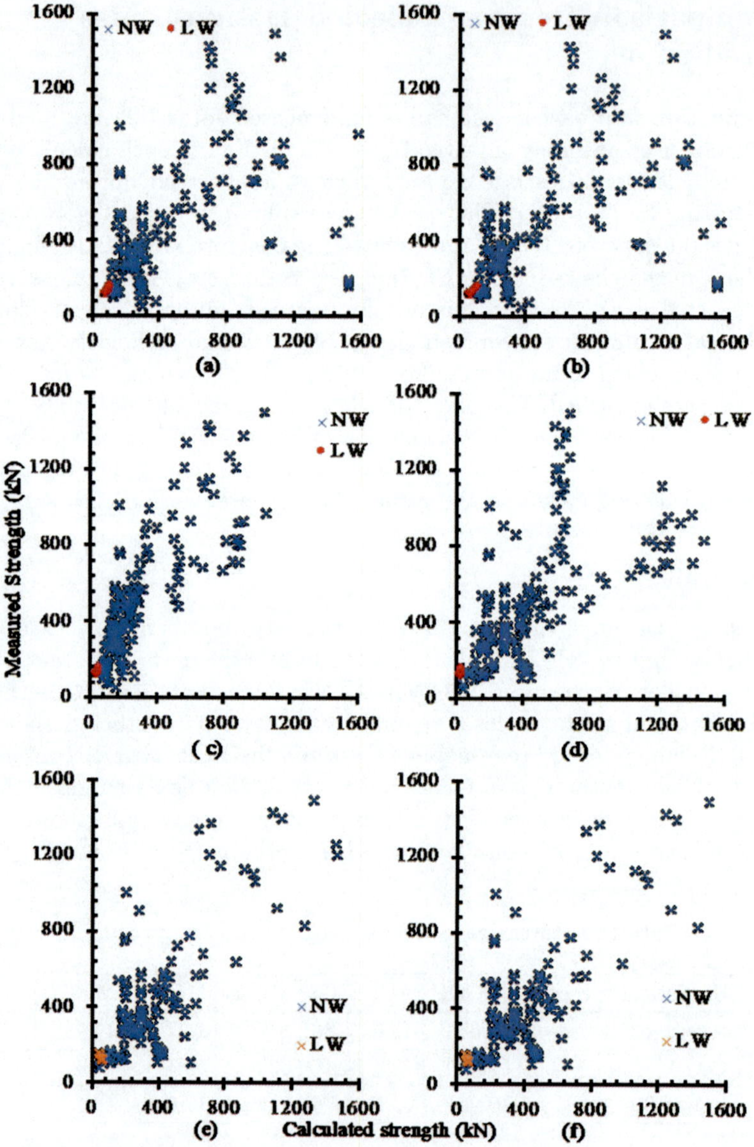

Fig. 1. Measured versus calculated strength using a) ECOP, b) A.C.I., c) EC2, d) MC(I), e) MC(II), and f) MC(III).

3.3 The Effect of the In-Plane Compressive Axial Stress on the Safety Factor Using Selected Design Codes

The in-plane compressive stress increases the punching shear resistance, decreasing ductility and resulting in brittle failure. Furthermore, it is clear that safety decreases with the increase in axial stress. Therefore, it can be concluded that the design is less conservative for larger in-plane axial stress.

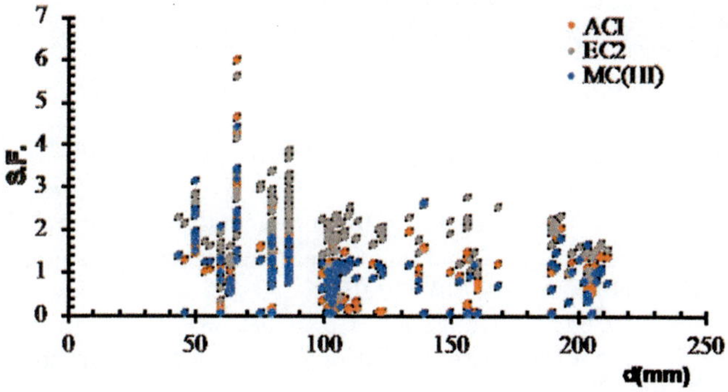

Fig. 2. The effect of size on the ratio between measured and calculated strength (S.F.).

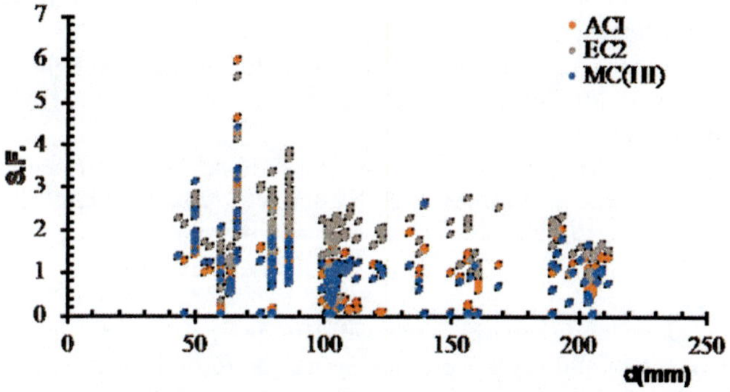

Fig. 3. The effect of in-plane stress on the ratio between measured and calculated strength (S.F.).

4 Summary and Concluding Remarks

Selected methods were used to calculate the punching load of lightweight and normal-weight concrete slabs. As a result, the following conclusions were reached:

- The design using selected design codes is more conservative for lower slabs size and smaller in-plane axial stress.
- The EC2 is the most consistent, while the MC is the most accurate.
- All design codes are reasonably safe, while the EC2 is overly conservative.
- Future work should include artificial intelligence for the optimization of current design support methods [23–26]

Acknowledgments. The author wishes to extend gratitude to the Future University In Egypt for their continuous support and encouragement.

References

1. EC2: EN 1992-1-1:2004: Eurocode 2: Design of concrete structures - Part 1–1: General rules and rules for buildings. Incl. Corrigendum 1: EN 1992-1-1:2004/AC:2008, incl. Corrigendum 2: EN 1992-1-1:2004/AC:2010, incl. Amendment 1: EN 1992-1-1:2004/A1:2014 (2004)
2. ACI: Committee 318. Building Code Requirements for Structural Concrete (ACI 318–19) and Commentary on Building Code Requirements. Farmington Hills (MI): American Concrete Institute; 2019 (2019)
3. ECOP: Egyptian Code for Design and Construction of Concrete Structures, Ministry of housing and infrastructures, National research center for housing and building, pp. 203–300 (2018)
4. MC: Fib model code for concrete structures 2010. Ernst & Sohn (2013)
5. Deifalla, A.: Design of lightweight concrete slabs under two-way shear without shear reinforcements: a comparative study and a new model. Eng. Struct. **222**(2020), 111076 (2020). https://doi.org/10.1016/j.engstruct.2020.111076
6. Deifalla, A.: A strength and deformation model for prestressed lightweight concrete slabs under two-way shear. Adv. Struct. Eng. 1–12 (2021). https://doi.org/10.1177/13694332211020408
7. Deifalla, A.: A mechanical model for concrete slabs subjected to combined punching shear and in-plane tensile forces. Eng. Struct. **231**(2021), 111787 (2021)
8. Deifalla, A.: Strength and ductility of lightweight reinforced concrete slabs under punching shear. Structures **27**, 2329–2345 (2020). https://doi.org/10.1016/j.istruc.2020.08.002
9. Deifalla, A.: Torsion design of lightweight concrete beams without or with fibers: a comparative study and a refined cracking torque formula. Structures **28**, 786–802 (2020). https://doi.org/10.1016/j.istruc.2020.09.004
10. Deifalla, A., Awad, A., AbdElrahman, A.A., Seleem, H.: Investigating the behavior of lightweight foamed concrete t-beams under torsion, shear, and flexure. Eng. Struct. **219**, 110741 (2020). https://doi.org/10.1016/j.engstruct.2020.110741
11. Deifalla, A., Awad, A., Seleem, H., AbdElrahman, A.: Experimental and numerical investigation of the behavior of LWFC L-girders under combined torsion. Structures **26**, 362–77 (2020). https://doi.org/10.1016/j.istruc.2020.03.070
12. Deifalla, A.: Torsional behavior of rectangular and flanged concrete beams with F.R.P. J. Struct. Eng. ASCE. In press since March 2015, STENG-3340R5 (2015). https://doi.org/10.1061/(ASCE)ST.1943-541X.0001322
13. Saleh, A., Fathy, A., Deifalla, A., Nasser, M.: Performance of steel fiber reinforced concrete corbels. Int. Res. J. Innovations Eng. Technol. (IRJIET) **3**(2), 22–27 (2019)
14. Deifalla, A.: Refining the torsion design of fibered concrete beams reinforced with F.R.P. using multi-variable non-linear regression analysis for experimental results. Eng. Struct. 226 (2020). https://doi.org/10.1016/j.engstruct.2020.111394
15. Deifalla, A., Ghobarah, A.: behavior and analysis of inverted T-shaped RC beams under shear and torsion. Eng. Struct. **62**, 776–786 (2014)
16. Deifalla, A., Awad, A., El-Garhy, M.: Effectiveness of externally bonded CFRP strips for strengthening flanged beams under torsion: an experimental study. Eng. Struct. **56**, 2065–2075 (2013)
17. Deifalla, A., Ghobarah, A.: Full torsional behavior of RC beams wrapped with F.R.P.: analytical model. ACSE, Compos. Constr. **14**(3), 289–300 (2010)

18. Deifalla, A., Ghobarah, A.: Strengthening RC t-beams subjected to combined torsion and shear using F.R.P. fabrics - experimental study. ASCE, Compos. Constr. **14**(3), 301–311 (2010)
19. Deifalla, A., Hamed, M., Saleh, A., Ali, T.: Exploring GFRP bars as reinforcement for rectangular and L-shaped beams subjected to significant torsion: an experimental study. Eng. Struct. **59**, 776–786 (2014)
20. Deifalla, A., Khali M.S., Abdelrahman, A.: Simplified model for the torsional strength of concrete beams with GFRP stirrups. Compos. Constr. ASCE, 19(1), 04014032 (2015). http://dx.doi.org/10.1061/(ASCE)CC.1943-5614.0000498
21. ElMeligy, O., El-Nemr, A.M., Deifalla, A.: Re-evaluating the modified shear provision of CAN/CSA S806–12 for concrete beams reinforced with F.R.P. stirrups. A.E.I. (2017). https://doi.org/10.1061/9780784480502.027
22. Hassan, M.M., Deifalla, A.: Evaluating the new CAN/CSA-S806-12 torsion provisions for concrete beams with F.R.P. reinforcements. Mater. Struct. https://doi.org/10.1617/s11527-015-0680-9. Accessed 18 July 2015
23. Chakraborty, D., Elhegazy, H., Elzarka, H., Gutierrez, L.: A novel construction cost prediction model using hybrid natural and light gradient boosting. Adv. Eng. Inf., **46** (2020). https://doi.org/10.1016/j.aei.2020.101201
24. Elhegazy, H.: State-of-the-art review on benefits of applying value engineering for multi-story buildings. Intell. Build. Int. (2021). https://doi.org/10.1080/17508975.2020.1806019
25. Elhegazy, H., Badra, N., Aboul Haggag, S., Abdel Rashid, I.: Implementation of the neural networks for improving the project's performance of steel structures projects. J. Ind. Integr. Manage. (2022). https://doi.org/10.1142/S2424862221500251
26. Zhong, S., Elhegazy, H., Elzarka, H.: Key factors affecting the decision-making process for buildings projects in Egypt. Ain Shams Eng. J. (2021). https://doi.org/10.1016/j.asej.2021.09.024

Random Lattice Modeling of Fracture in Structural Glass-Fiber Reinforced Polymers

Alessandro Fascetti[1], Luciano Feo[2(✉)], Rosa Penna[1], and Yingbo Zhu[2]

[1] Department of Civil and Environmental Engineering, University of Pittsburgh, Pittsburgh, PA 15261, USA
[2] Department of Civil and Environmental Engineering, University of Salerno, 84084 Fisciano, SA, Italy
lfeo@unisa.it

Abstract. Glass Fiber-Reinforced Polymers (GFRP) are an attractive alternative to traditional construction materials, as a result of their strength-to-weight ratio, corrosion resistance and electromagnetic transparency that can be exploited in design for specific structural applications. However, challenges also arise from the relatively low strength and stiffness in the direction orthogonal to the fibers. The increased deformability resulting from these actions can, in fact, result critical in the design of the elements and potentially cause premature failures under service and ultimate limit state conditions. To this end, numerical methods are an effective means for simulating the nonlinear behavior of GFRP elements and assess safety of structural assemblies. This work proposes a review of recent developments in the definition and validation of random lattice modeling approaches for the simulation of fracture in GFRP structural elements. Attention is devoted to the derivation of appropriate rules to scale the stiffness and strength of each lattice element in the network, to describe the transversally orthotropic behavior of the medium. The mathematical derivation of the relevant element-level matrices will be presented first. Results obtained from numerical simulations at different scales will be discussed, and open challenges and directions for future research presented.

Keywords: Glass-Fiber Reinforced Polymers · Random Lattice Modeling · Fracture Mechanics · Numerical Simulations

1 Introduction

Pultruded Glass Fiber-Reinforced Polymer (pGFRP) structural elements have gained increasing attention in structural engineering applications, as a result of their non-traditional features, that make them a viable alternative to classical materials in different scenarios. Such features include an excellent strength-to-weight ratio, electromagnetic transparency and longer durability when compared to traditional materials (particularly with regards to corrosion) [1]. Outside of some pioneering structural engineering applications that mostly interested pedestrian bridges and footways [2], however, the use of pultruded GFRP elements is still restricted to special applications in the field of Civil Engineering. This is mainly due to difficulties deriving from the relatively low stiffness

R. Saidi et al. (Eds.): ICATH 2021, LNDECT 110, pp. 482–491, 2022.
https://doi.org/10.1007/978-3-030-94188-8_43

and strength that pGFRP exhibit in the direction orthogonal to the fibers. This characteristic, in fact, creates challenges in satisfying both Serviceability and Ultimate Limit State conditions, as a result of the relatively high deformability of the members that can induce early failures in areas of the elemental cross-sections that present a lower concentration of fibers [3]. For this reason, several experimental studies concerned the evaluation of mechanical performance of pGFRP structural elements, particularly in the presence of significant tensile actions in the direction orthogonal to the fibers. Xin et al. [2, 4] investigated the in-plane compression and shear performance of pultruded GFRP bridge decks, and concluded that the thickness of the web has a strong influence on the shear behavior of the modular elements. Feo et al. [3] conducted a comprehensive experimental campaign to characterize the ultimate strength of the web-flange junction in pGFRP I beams. The authors evaluated the influence of the type of load applied on the ultimate strength and, based on experimental results and numerical tests, define the so-called "influence zone" to characterize the relationship between the location of the applied load with respect to the beam ends and the measured ultimate strength.

Liu et al. investigated the flexural behavior of pGFRP I beams, with particular attention devoted to the evaluation of their ultimate flexural capacity in presence of local buckling effects occurring in the flange [5] or in the web [6] of the elemental cross-section. The experimental results confirmed that local buckling resistance in the beam flange increased as flange slenderness decreased. The authors also propose predictive equations for the evaluation of the ultimate resistance of pGFRP profiles against local flange buckling and lateral torsional buckling.

1.1 Numerical Models

In view of the aforementioned challenges in the adoption of pGFRP elements in structural applications, several numerical models have been proposed in literature with the aim to investigate their nonlinear behavior. Classical approaches comprise FEM approaches, Virtual Crack Closure Technique (VCCT), Cohesive Zone modeling (CZM). A thorough review of available numerical models for the simulation of pGFRP structural elements is beyond the scope of the present work. The interested reader is referred to a comprehensive literature review recently presented by the authors [7]. In this paper, we present results obtained by means of a random lattice modeling approach for the simulation of pultruded GFRP structural elements. We present results at two different scales (i.e., the material level and the structural element level), to show capabilities of the random lattice modeling approach in simulating failure in pGFRP elements. The main features of the random lattice modeling strategy are first presented, to highlight capabilities of the method. Then, the numerical results are compared with available experimental evidence to demonstrate accuracy of the model in reproducing both the force-displacement response and the crack paths at failure for typical tensile and 3-point bending tests. Conclusions are then drawn, with emphasis on open challenges and future research directions.

2 Random Lattice Modeling Approach

Random Lattice Models are a class of numerical approaches that discretize the behavior of a 3-dimensional body by describing the discrete interactions of rigid, irregular polytopes that compose the domain. Such interactions are resolved on a dense network of 1-dimensional elements, which are called lattice elements. Figure 1 reports a graphical representation of the construction of a typical random lattice modeling simulation. Random Lattice Models have been employed for the solution of different Civil Engineering problems, such as fracture in cementitious composites [8], reinforced concrete structural applications [9], erosion in geomaterials [10], as well as fiber-reinforced polymers [1, 11]. A thorough review of the random lattice modeling technique is beyond the scope of the present work, and the interested reader is referred to available literature reviews on the topic [7, 12].

In the spirit of the random lattice modeling approach, each 1-dimensional element in the dense lattice network (see Fig. 1) inherits its inertial properties from the Voronoi diagram, which is a graph dual to the Delaunay tetrahedralization. By noting that each pair of nodes connected by a lattice element shares a polyhedral facet in the Voronoi diagram (as depicted in Fig. 2), it is possible to compute the inertial properties (i.e., stiffness and strength) of each element in the lattice. To simulate the mechanical behavior, a discrete spring set is inserted at the centroid of each Fig. 2 also represents the 3 reference systems employed in the definition of the model: 1) the global x-y-z coordinate system, 2) the local system of reference for each lattice element defined by the normal direction n, and 3) the local material coordinate system, defined by the direction of the fibers, denoted as 1. For the purpose of what is presented herein, we define θ^e as the angle between the normal direction n and the direction of the fibers in the composite. The global angle between the application of the load in the tensile tests and the fiber direction is denoted as Θ.

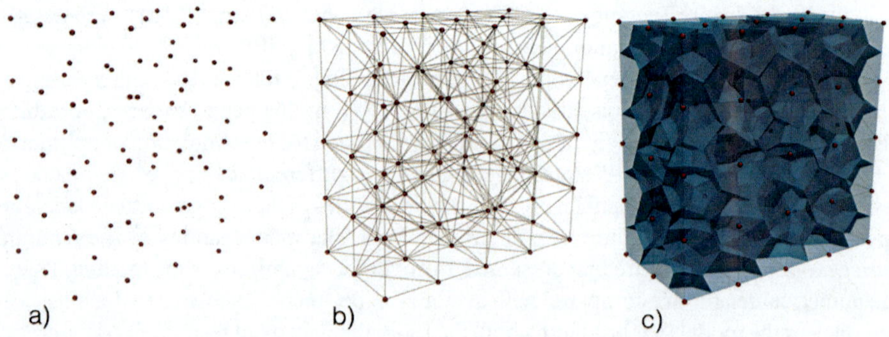

Fig. 1. Graphical representation of the construction of a random lattice modeling simulation. a) random placement of nodal sites, b) Delaunay tetrahedralization and network topology, c) Voronoi diagram

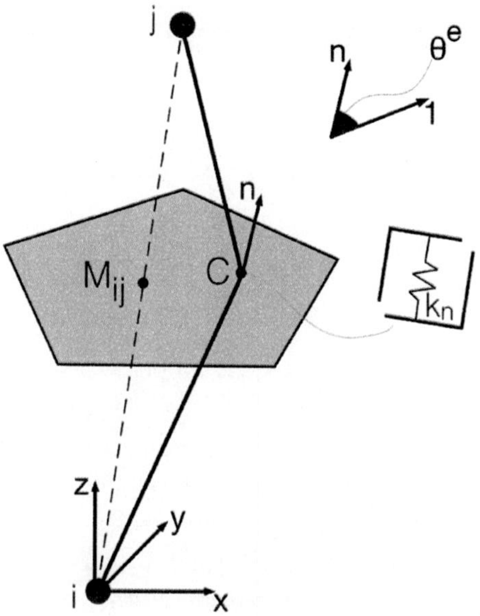

Fig. 2. Definition of a lattice element connecting nodes i and j, resisting area and the normal spring located at the centroid C.

2.1 Transversally Orthotropic Behavior

One of the challenges in deploying random lattice modeling approaches for the simulation of pGFRP structural elements arises from enforcing the transversally orthotropic behavior at the lattice network level. Based on work performed by the authors [1, 11], the stiffness of each element in the network is scaled according to its orientation (defined by the normal vector **n**, see Fig. 2):

$$E^e = \frac{1}{\frac{cos^4\vartheta^e}{E_f} + \frac{sin^4\vartheta^e}{E_m} + \left(\frac{1}{G} - \frac{1}{E_m}\right)cos^2\vartheta^e sin^2\vartheta^e}$$

where E_f and E_m are the macroscopic Young's Moduli in the direction of the fiber and a direction orthogonal to it, respectively, and G is the macroscopic shear modulus.

In a similar manner, the strength of each element is scaled according to its orientation, to reproduce the orthotropic behavior of the material. In the absence of tangential stiffness in the discrete spring set governing the mechanical behavior of each lattice element, such relationship can be expressed as [11]:

$$\sigma^e_{lim} = \frac{1}{\sqrt{\frac{cos^4\vartheta^e}{\sigma_f{}^2} + \frac{sin^4\vartheta^e}{\sigma_m{}^2} + \left(\frac{1}{\tau^2} - \frac{1}{\sigma_m{}^2}\right)cos^2\vartheta^e sin^2\vartheta^e}}$$

where σ_f is the macroscopic strength in the direction of the fibers, σ_m the matrix strength, and τ the macroscopic shear strength.

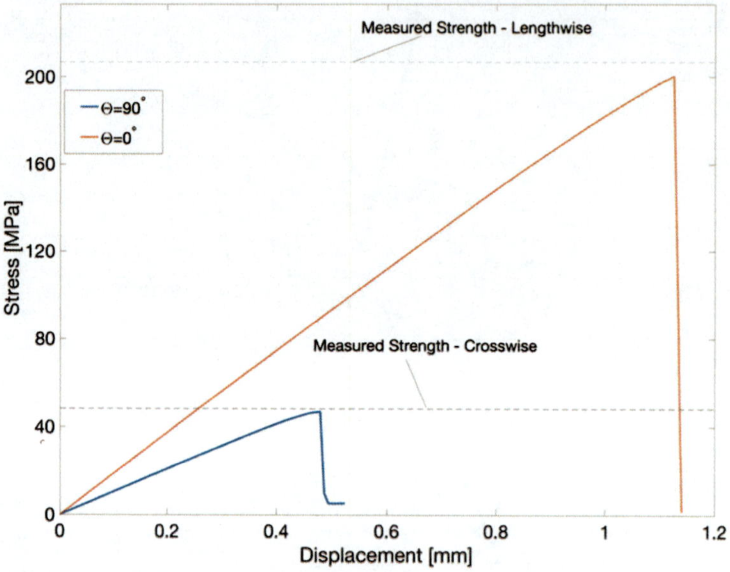

Fig. 3. Numerical stress-displacement curves obtained from the coupon tests for lengthwise and crosswise specimens.

One of the advantages in the adoption of the random lattice modeling approach is that the constitutive law governing the behavior of the spring set is enforced at the 1-dimensional level of the elements, therefore obtaining scalar relationships. This is particularly advantageous for quasi-brittle materials, for which the enforcement of the correct fracture energy dissipation is crucial in the simulation of static and dynamic fracture [8]. The evolution of the stress limit in the direction normal to each lattice element is supposed to be exponentially decaying:

$$\sigma_b = \sigma^e_{lim} e^{K\left(\frac{\tilde{\varepsilon}-\varepsilon}{\varepsilon}\right)}$$

in which $\tilde{\varepsilon}$ is the history-dependent maximum level of strain attained in the element (to model cyclic behavior and irreversibility of damage accumulation) and ε is the current measure of deformation in the element. The value of K can be computed for each lattice element, so that the correct amount of energy is dissipated at the lattice level:

$$K = \begin{cases} -\frac{\sigma^{e\ 2}_{lim} L}{G_I}, & compression\ case \\ 0, & tensile\ case \end{cases}$$

where L is the length of the element and G_I the prescribed Mode I fracture energy.

Moreover, for each lattice element undergoing inelastic deformation, as a result of the discrete kinematics of each lattice element governing the interactions of each couple of Voronoi polytopes, a discrete crack opening can be computed as:

$$w = L\left(\varepsilon - \frac{\sigma}{E^e}\right)$$

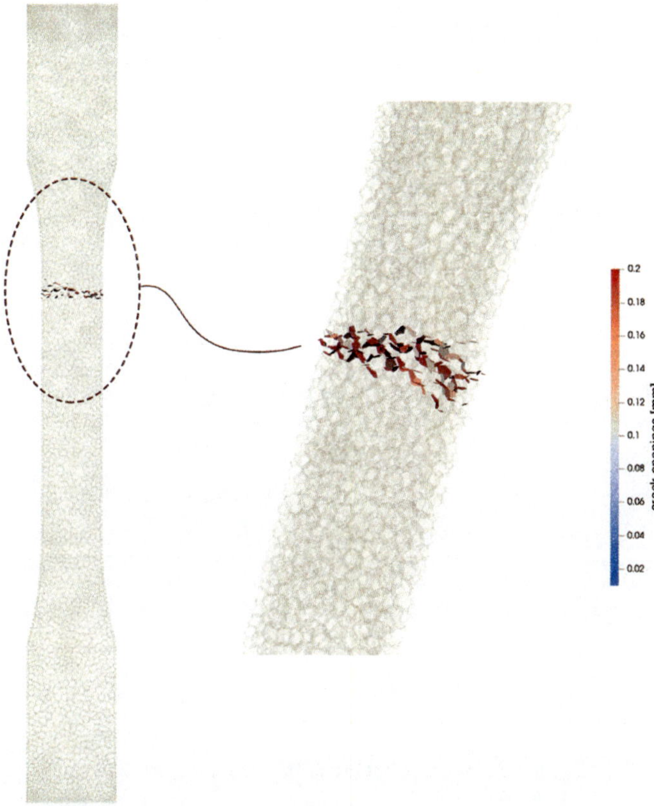

Fig. 4. Numerical crack pattern at failure for the 0° specimen.

3 Numerical Results

The numerical approach presented in the previous was validated at different scales, namely the local-scale of the material and the element-scale of the structural elements. In the first set of numerical tests, typical coupon specimens were simulated for both lengthwise ($\Theta = 0°$) and crosswise ($\Theta = 90°$) specimens. The results are then compared with the corresponding measures of strength obtained from physical specimens. This first set of tests served as calibration of the model. The obtained parameters were then used in the element-scale simulations, to predict the ultimate strength of an I 160 pGFRP beam in a 3-point bending test.

3.1 Local-Scale Validation

In the first step of the validation, the random lattice modeling approach was employed at the material-scale level, to evaluate the mechanical parameters of pultruded GFRP elements in: 1) the direction of the fiber, and 2) in the direction orthogonal to it. To this end, synthetic dog bone specimens were created and tested numerically, according to

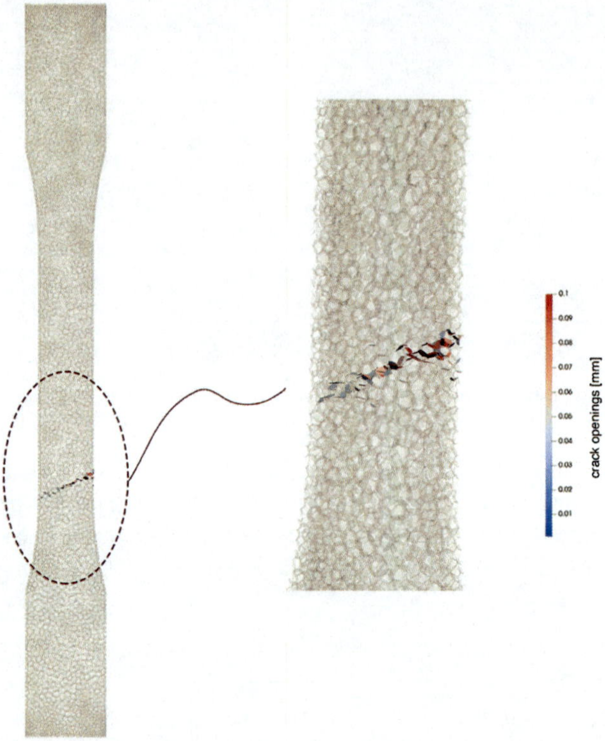

Fig. 5. Numerical crack pattern at failure for the 90° specimen.

ASTM standard D638 (Type I) [13]. Figure 3 reports the results of such tests, which show how the random lattice modeling technique was able to correctly represent the greater stiffness and strength that pGFRP elements exhibit when normal actions are parallel to the application of the load (i.e., $\Theta = 0°$).

Figures 4 and 5 also report the numerical crack patterns at failure for both simulations. The numerical results agree well with experimental evidence obtained from similar tests. The local-scale models were also used to calibrate the values of the lattice-level stiffness and strength values, according to the specifications provided by the manufacturer [14].

3.2 Element-Scale Response

After calibration of the material parameters by means of the previously presented local-scale tests, the proposed random lattice modeling approach was employed in the simulation of failure in pGFRP at the structural element level.

Figure 6 reports the test setup of an I 160 beam subjected to a 3-point bending test. Figure 7 reports the force-displacement curve obtained from the numerical simulation of an I 160 pGFRP beam for a 3-point bending test. The results show a satisfactory agreement with the theoretical maximum force attainable in the beam. Moreover, Fig. 8 reports the crack openings at failure, which qualitatively agree with the experimental evidence [5, 6].

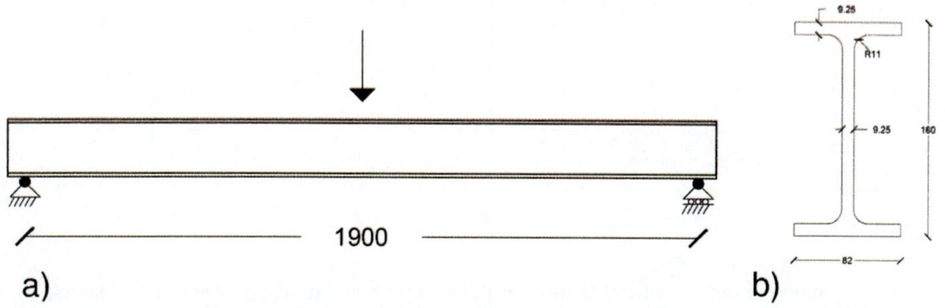

Fig. 6. Schematic representation of the 3-point bending test simulated numerically (all dimensions in mm): a) side view, b) cross-sectional view.

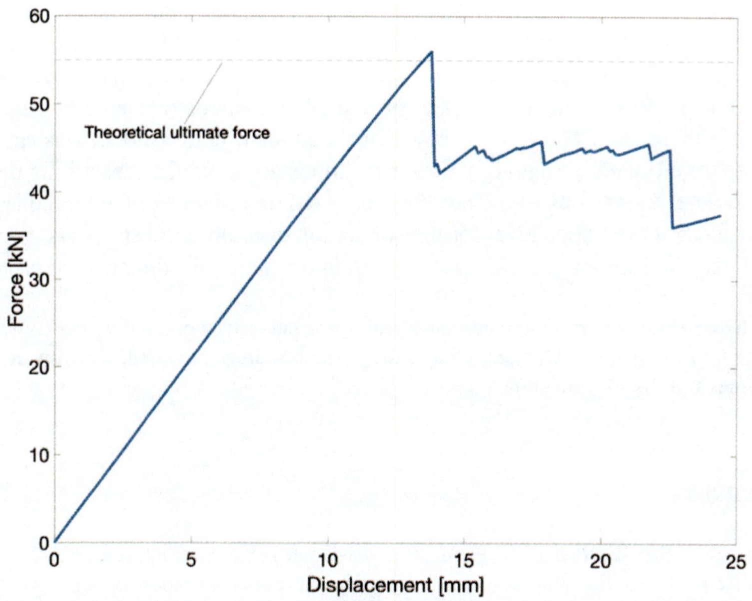

Fig. 7. Numerical response for the 3-point bending test of an I 160 pGRFP beam.

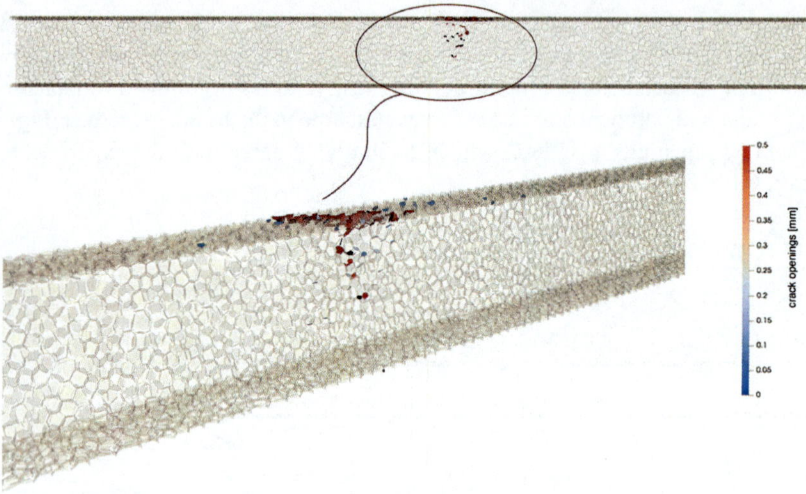

Fig. 8. Numerical crack paths at failure for the 3-point bending test of a pGFRP I 160 beam.

4 Conclusions

In this work, a review of the capabilities granted by the adoption of random lattice modeling techniques in the simulation of failure in pultruded glass-fiber reinforced polymer structural elements has been discussed. The main features of the proposed model were discussed, and a validation at two scales presented, to demonstrate accuracy and applicability of the method. The results show that the random lattice model was capable of predicting the ultimate strength of pultruded composites, as well as describing the crack paths at failure. Extensions of the current model will introduce the tangential and rotational stiffness at the lattice level for the simulation of multi-axial stress states, and the different observed values of the Poisson's ratio in the material's direction of orthotropy.

Acknowledgments. The research presented herein was partially supported by the University of Pittsburgh Center for Research Computing through the resources provided, which were used in the presented numerical simulations.

References

1. Fascetti, A., Feo, L., Nisticò, N., Penna, R.: Web-flange behavior of pultruded GFRP I-beams: a lattice model for the interpretation of experimental results. Compos. B Eng. **100**, 257–269 (2016)
2. Xin, H., et al.: Experimental and numerical investigation on in-plane compression and shear performance of a pultruded GFRP composite bridge deck. Compos. Struct. **180**, 914–932 (2017)
3. Feo, L., Mosallam, A.S., Penna, R.: Mechanical behavior of web–flange junctions of thin-walled pultruded I-profiles: an experimental and numerical evaluation. Compos. B Eng. **48**, 18–39 (2013)

4. Xin, H., Liu, Y., Mosallam, A.S., He, J., Du, A.: Evaluation on material behaviors of pultruded glass fiber reinforced polymer (GFRP) laminates. Compos. Struct. **182**, 283–300 (2017)
5. Liu, T., Vieira, J.D., Harries, K.A.: Predicting flange local buckling capacity of pultruded GFRP I-sections subject to flexure. J. Compos. Constr. **24**(4), 04020025 (2020)
6. Liu, T., Vieira, J.D., Harries, K.A.: Lateral torsional buckling and section distortion of pultruded GFRP I-sections subject to flexure. Compos. Struct. **225**, 111151 (2019)
7. Fascetti, A., Feo, L., Abbaszadeh, H.: A critical review of numerical methods for the simulation of pultruded fiber-reinforced structural elements. Compos. Struct. **273**, 114284 (2021)
8. Fascetti, A., Bolander, J.E., Nisticó, N.: Lattice discrete particle modeling of concrete under compressive loading: multiscale experimental approach for parameter determination. J. Eng. Mech. **144**(8), 04018058 (2018)
9. Bolander, J.E., Hong, G.S., Yoshitake, K.: Structural concrete analysis using rigid-body-spring networks. Comput.-Aided Civil Infrastruct. Eng. **15**(2), 120–133 (2000)
10. Fascetti, A., Oskay, C.: Dual random lattice modeling of backward erosion piping. Comput. Geotech. **105**, 265–276 (2019)
11. Gaetani, A., Fascetti, A., Nistico, N.: Parametric investigation on the tensile response of GFRP elements through a discrete lattice modeling approach. Compos. Part B: Eng. **176**, 107254 (2019)
12. Bolander, J.E., Eliáš, J., Cusatis, G., Nagai, K.: Discrete mechanical models of concrete fracture. Eng. Fract. Mech. **257**, 108030 (2021)
13. ASTM D638–14, Standard Test Method for Tensile Properties of Plastics, ASTM International, West Conshohocken, PA, (2014). www.astm.org. https://doi.org/10.1520/D0638-14
14. Strongwell Corporation. Strongwell Materials Certifications. www.strongwell.com/tools/strongwell-specifications/. Accessed 14 Oct 2021

Industry 4.0 for Smart Factories

Qualitative Functional and Dysfunctional Analysis and Physical Modeling of an Eco-Designed Mechatronics System Using Coloured Petri-nets: Application on a Regenerative Braking System

Imane Mehdi(✉) and El Mostapha Boudi

Mohammadia School of Engineers, Rabat, Morocco

Abstract. Developing robust and high-efficient mechatronics products at a lower cost, requires special control during design the process while verifying that the objectives are met and that the constraints are respected. A study of a mechatronics system performances is generally preceded by functional, dysfunctional and physical modeling. This paper presents a mechatronic system functional analysis using structured analysis and design technique (SADT), dysfunctional analysis based on failure modes and effects analysis methodology (FMEA)/Software Errors and Effects Analysis (SEEL) and physical modeling using coloured Petri nets (CPN).

Keywords: Coloured Petri Net · Mechatronics system · Functional analysis · Dysfunctional analysis · Physical modeling · Design phase · Qualitative modeling

1 Introduction

Mechatronics is the engineering discipline involved in the building of systems integrating mechanical, electronic and computer components [1]. Mechatronics becomes the synergistic incorporation of physical systems with information technology (IT) and sophisticated decision making in the design, manufacture and operations of industrial products and processes [2]. As a matter of fact, the complexity of these systems invites us to evaluate them from their early design stage. Therefore, it is necessary, firstly, to carry out a qualitative analysis in order to highlight all failures affecting the correct function of a system. The critical scenarios leading the system to a "feared" state. Functional analysis precedes generally the qualitative analysis [3].

Model is defined as: "An abstract and simplified representation (i.e. that excludes certain details), of an entity (The phenomenon, The process, The system, etc.) of the real world in order to describe, explain or predict it" [4].

R. Saidi et al. (Eds.): ICATH 2021, LNDECT 110, pp. 495–506, 2022.
https://doi.org/10.1007/978-3-030-94188-8_44

The main advantage of modeling a product in design stage is allowing manufacturers to meet requirements while reducing costs and time. Modeling consists in representing the system by a model which takes into account the physical, functional, behavioral characteristics, etc. in the perspective of studying its performances by simulation.

After introducing a sustainable conceptual design methodology for mechatronics systems in [16].

This paper focuses especially on the a functional, dysfunctional and physical aspects of this mechatronics system. The case study is presented in (2) functional analysis in (3) Dysfunctional analysis in (4) and the regenerative braking system physical model using coloured Petri nets in (5). The global approach is depicted in the Fig. 1:

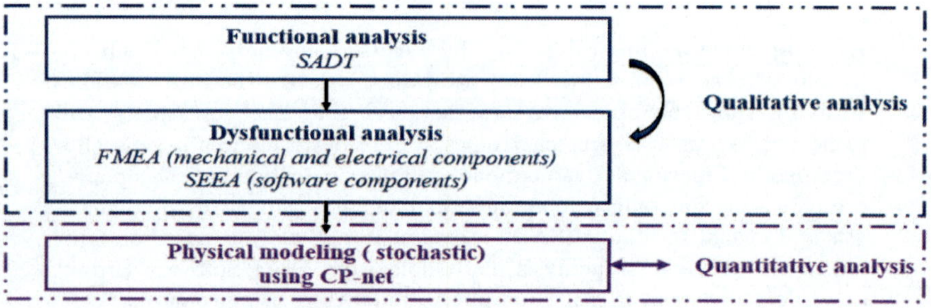

Fig. 1. Proposed approach

Among different models, we have chosen the Coloured Petri Net (CP-net or CPN) implemented in the software called CPN Tools. This choice is motivated by the following reasons:

- The possibility to built a complex model in an easy way. Thanks to the programming language SML implemented in CPN Tools, we can build functions and perform different algorithms with the aim of reducing the size of our model.
- The possibility to build a modular system and to perform Monte Carlo simulations.

2 Case Study

The system under study is a regenerative braking system (Fig. 1). In fact, during regenerative braking kinetic energy of the drive wheels is converted by the electric motor - which is operated as a generator for this purpose - into electrical energy. In this way, some of the energy which is normally lost as frictional heat during braking is fed in the form of electric energy to the battery and Battery and then utilized (Figs. 2 and 8):

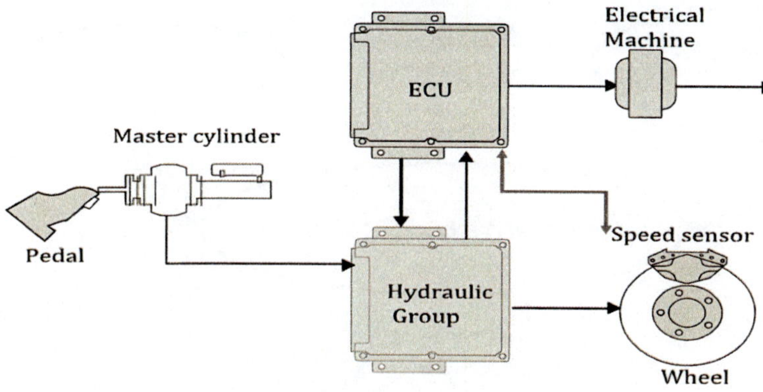

Fig. 2. Regenerative braking system

The introduction of the regenerative braking System has contributed to improve the safety and energy management of modern cars.

3 Functional Analysis

Functional analysis is an approach that consists in searching, ordering, characterizing, prioritizing and/or valuing the system functions expected by the user [5].

Functional analysis allows a synthetic description of the operating modes of a system and the knowledge of the functions to be guaranteed. It systematically and comprehensively sets out the functional interrelation within and outside this system [6–8].

3.1 External Functional Analysis

The aim of external functional analysis is to formalize and validate the requirement analysis. The system is considered as a black box receiving inputs and providing outputs [14]:

Figure 3 represents the External Functional Analysis applied to the Regenerative Braking System. We can deduce that the driver needs the RBS to avoid the locking of the wheel and to recover kinetic energy.

3.2 Internal Functional Analysis

Internal functional analysis allows to identify the technical functions and the technological solutions necessary for the realization of the black box.

The SADT model is comprised of diagrams with boxes to indicate activities. The boxes are linked by arrows, which indicate how each activity influences

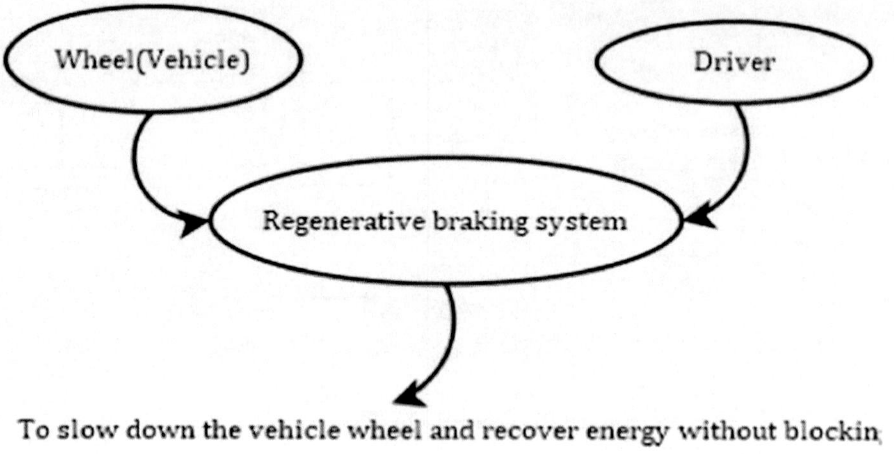

Fig. 3. External functional analysis of a regenerative braking system

the other. The boxes are ordered according to their relative importance. The model captures increasing level of details, in a hierarchical manner, through the decomposition of activities into more detailed diagrams [15]:

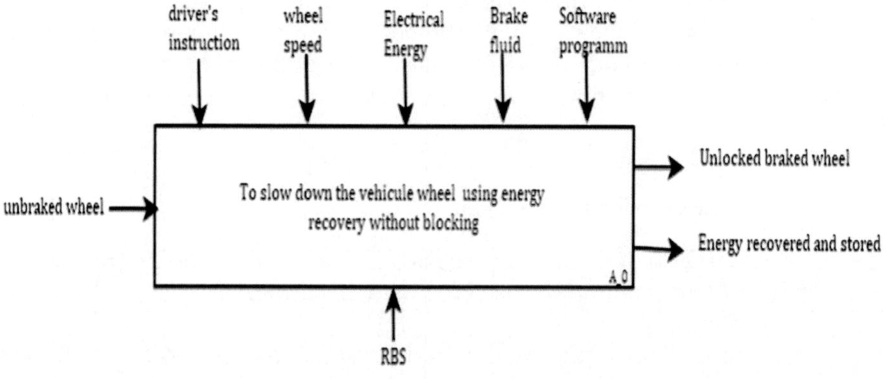

Fig. 4. RBS SADT Level "A_0"

The Fig. 4 shows the principal function of the regenerative braking system, which consists in slowing down the wheel of the vehicle, recovering kinetic energy without blocking. It presents also the relation between the system and its external environment.

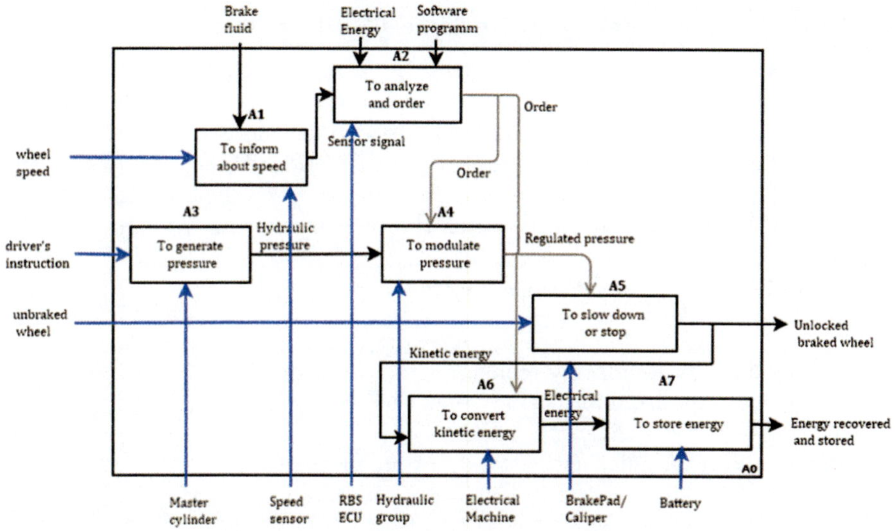

Fig. 5. RBS SADT level A0

Figure 5 represents the level A0 of SADT. It is a decomposition of the level A-0.

The Fig. 6 is the level A5 of SADT. It is the decomposition of the fifth element "to slow down or stop" of the level A0 of SADT.

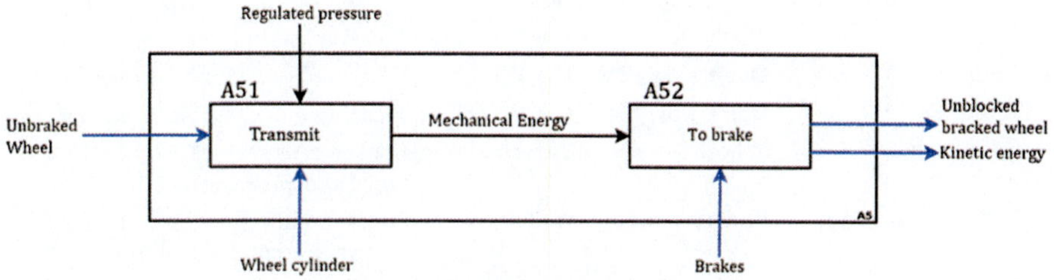

Fig. 6. RBS SADT-level A5

4 Dysfunctional Analysis

The functional analysis previously studied, does not provide any information on the potential failures that can be encountered by a regenerative braking system. For this reason, it is necessary to use dysfunctional analysis permitting to determine the failures causes as well as to specify the different system states.

4.1 Failure Mode and Effects Analysis: FMEA

The FMEA, FMEA was first adopted by the aerospace industry in the 1960s. Unlike other competing failure preventing methods, FMEA was described in universally understandable terms by individuals who had limited technical and/or systems training. This promoted applications that bridged across companies and industries [9]. In particular, the automotive industry brought FMEA into the mainstream by adopting the method as the primary system for error and risk reduction [9].

The FMEA process involves the five following steps [9]:

1. Decide on a process to study;
2. Building a multidisciplinary team;
3. Gather and organize information about the process under study;
4. Carrying out a risk analysis;
5. Develop and implement actions and output measures.

Table 1. RBS failure mode and effects analysis

Component	Failure mode	Cause	Effects
Pedal	Mechanical breakage	Wearing	No breaking
Master cylinder	Missing pressure Over-pressure	Broken piston	No breaking/ Bad braking
Hydraulic Group	Missing pressure Over-pressure	Failed valves	No breaking/ Bad braking
Reservoir	Missing braking fluid	Cracking	No breaking
Electrical machine	No energy conversion	Broken strap	No breaking
Disc	Don't slow down the Wheel	Wearing	No breaking
Caliper	Don't slow down the Wheel	Wearing	No breaking
Brake Pad	Don't slow down the Wheel	Wearing	No breaking
Speed sensor	Wheel speed non-detected	Broken magnet Oxydized magnet	Risk of wheel lock-up
DC-DC Converter	Energy not converted	Short circuit Software error	Risk of wheel lock-up
Battery Lithium-ion	Discharge	Old battery	Risk of wheel lock-up
ECU	No data transmission	Short circuit Software error	Risk of wheel lock-up

Table 1 represents examples of FMEA lines of our system.

4.2 Software Errors and Effects Analysis: SEEA

The SEEA method was first published in 1973 [10] and afterwards referenced in [11], The method is described as "... a systematic evaluation of the effects of errors in a program's performance". The SEEA is an adaptation of the FMEA. Its purpose is to identify critical points detected during the software development phase [12].

Table 2. RBS software errors and effects analysis

Module	Error	Soft effect	System effect
Measure	No information sent to speed sensor from wheels	No data sent to ECU	RBS off
Sent	Bad Connection, false data value,	No data sent to ECU	RBS off
Compute	ECU calculation error	No effect	RBS off
Control	Slip interval error Calcul error	No effect	RBS off
Order	Bad conncetion between ECU and other components	Bug	RBS off

Table 2 below, explores some lines examples of our regenerative braking system SEEA.

5 Coloured Petri Net Model

5.1 CPN Definition

The Coloured Petri Nets (CPN) [13] is a graphical language for building models of concurrent systems and evaluating theirs properties. CPN is a discrete event modeling language, merging the features of Petri nets with those of a high level programming language. The main difference between a classical Petri Net and a CPN is that the CPN tokens can have different color sets representing data types (e.g. Boolean, integer, string) or more complex data structure.

5.2 Timed Coloured Petri Net: Formal Definition

We define formally a timed coloured Petri net as a nine-tuple [13]:

$$CPN = [P, T, A, S, V, C, G, E, I] \tag{1}$$

where:

P: is a finite state of places.
T: is a finite state of transitions such that $P \cap T = \emptyset$
A $\subseteq P \times T \cup T \times P$ is a set of directed arcs.
S: is a finite set of non-empty colour sets. Each colour set is either untimed or timed.

V: V is a finite set of typed variables such that $Type[v] \in S$ for all variables $v \in V$.
C:$P \longrightarrow S$ is a colour set function that assigns a colour set to each place. A place p is timed if C(p) is timed, otherwise p is untimed.
G: $T \longrightarrow EXPR$ is a guard function that assigns a guard to each transition such that Type [G (t)] = Bool.
E: $A \longrightarrow EXPR$ is a function that assigns an arc expression to each arc such that:

- Type[E(a)] $= C(p)_{MS}$ if p is untimed;
- Type[E(a)] $= C(p)_{TMS}$ if p is timed.

Here, p is the place connected to the arc a.

$I : P \longrightarrow EXPR/0$ is an initialization function that assigns an initialization expression to each place p such that:

- Type[I(p)] $= C(p)_{MS}$ if p is untimed;
- Type[I(p)] $= C(p)_{TMS}$ if p is timed.

Timed coloured Petri nets are:

$$CPN_T = (CPN, \Theta_1, \Theta_2, \theta_3) \quad (2)$$

Where CPN stand for coloured petri net, and $\Theta_1, \Theta_2 \Theta_3$ are, respectively, time delays assigned to net places, arcs, and transitions.

5.3 RBS Coloured Petri Net: Specifications Model

In Table 3, we set out the "dysfunctional" distributions associated to the components as well as the ones for repair.

We have adopted the Weibull distribution for mechanical components, the exponential distribution for electronic components and the Putnam model for software components.

For repair, we chose the Erlang distribution.

The Fig. 7 illustrates the physical model of our RBS system, mechanical components are in blue color, electrical in green and software ones in pink.

The place "RBS on hold" contains one token "RBS_Syst" at initial time. The variable "RBS" enables the transition "RBS start-up" by omitting the token from the place "RBS on hold" an adding it to the place "RBS working". Random actions are executed in the transition "RBS startup" considering Weibull, Exponential and Putnam distributions already explained. The is ensured via a transition input() /output() parameters and a function minimum defined to indicate what is the nearest time to failure between all system components. Then, a value of time@wait is associated to the token. In fact, the RBS is working and after the time @wait a failure transition is enabled. The output token corresponds to the "RBS syst", and components failure times. These times are used to enable next transitions.

Table 3. Parameters mean value for reliability and repair distributions.

Component	Distributions	Parameters
Pedal	Weibull	$\eta = 1000$ $\beta = 1{,}5$
Master Cylinder	Weibull	$\eta = 1000$ $\beta = 1{,}5$
Hydraulic Group	Weibull	$\eta = 1000$ $\beta = 1{,}5$
Reservoir	Weibull	$\eta = 1000$ $\beta = 1{,}5$
Electrical Machine	Weibull	$\eta = 1000$ $\beta = 1{,}5$
Disc	Weibull	$\eta = 1000$ $\beta = 1{,}5$
Caliper	Weibull	$\eta = 1000$ $\beta = 1{,}5$
Brake pad	Weibull	$\eta = 1000$ $\beta = 1{,}5$
Tyre	Weibull	$\eta = 1000$ $\beta = 1{,}5$
Speed sensor	Exponential	$\lambda = 3 \times 10^{-4}$
DC-DC converter	Exponential	$\lambda = 3 \times 10^{-4}$
Battery	Exponential	$\lambda = 3 \times 10^{-4}$
ECU Adata-Software	Exponential Exponential-Musa	$\lambda = 3 \times 10^{-4}$ $\lambda = 3 \times 10^{-5}$ $N_0 = 10$
IO -Software	Exponential-Musa	$\lambda = 3 \times 10^{-5}$ $N_0 = 10$
OS -Software	Exponential-Musa	$\lambda = 3 \times 10^{-5}$ $N_0 = 10$
Mechanical components	Erlang	$\mu = 1/0.5$
Electrical components	Erlang	$\mu = 1/144$
Software components	Erlang	$\mu = 1/0.5$

Arcs from "RBS working" place are linked to different failure modes. To select the transition to be enabled, guard conditions functions are defined. A failure transition is enabled if the failure time is smaller then the other components failure times, (it it has the minimum failure time). In our model Failure transitions are precede by the letter F.(For example: FBP means Brake Pad Failure).

A function Erg() was defined in arc inscriptions for reparation times (transitions preceded by R in our model). A Component reparation transition is enabled

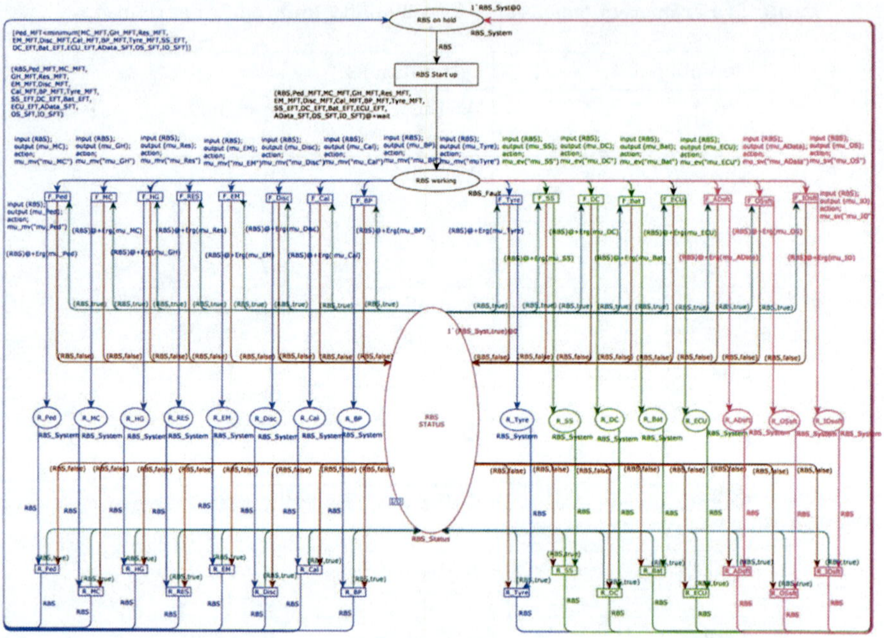

Fig. 7. RBS physical model using timed coloured Petri-Net

only if this component is broken. Otherwise, if a token (RBS, false) was sent to the place "RBS Status".

RBS Status is an important place linking the physical and the logical models of our system. It has one token of type (RBS, BOOL). When the component is working (OK status), the boolean returns a true value, if not it returns a false value.

Fault and reparation transitions are enabled if and so only if all the incoming arcs are connected to a place containing a token satisfying the arc inscriptions. For example: The transition F_Pedal will be enabled if:

- The place RBS working has a token (RBS,Ped_MFT,MC_MFT,GH_MFT, Res_MFT, EM_MFT,Disc_MFT,Cal_MFT,BP_MFT,Tyre_MFT, SS_EFT,DC _EFT,Bat_EFT, ECU_EFT,AData_SFT,OS_SFT,IO_SFT),
- The place RBS status has a token (RBS, true).

Indeed, a fault transition will be enabled only if RBS is working. When a fault transition is fired a correspondent token "(RBS, false)" is placed in place of "(RBS, true)" token in "RBS Status" place. This is used to notify the logical model that a component is broken.

System restart process is instantaneous. In fact, we suppose that after a reparation, our system will restart immediately.

This physical model will be completed by a logical model in order to perform behavioral analysis.

6 Conclusion and Perspectives

In this paper, an approach based on Coloured Petri Net (CPN) has been proposed to cover a regenerative braking system functional and dysfunctional analysis and physical modeling. The proposed approach consists firstly in using SADT methodology to perform functional analysis, FMEA applied to the mechanical and electronic components and its equivalent SEEA applied to the software components for dysfunctional analysis. Quantitative Physical modeling of the system was performed using timed coloured Petri-net. As a perspective of this work, The physical model will be completed by a logical one in order to perform behavioral analysis of the system under study.

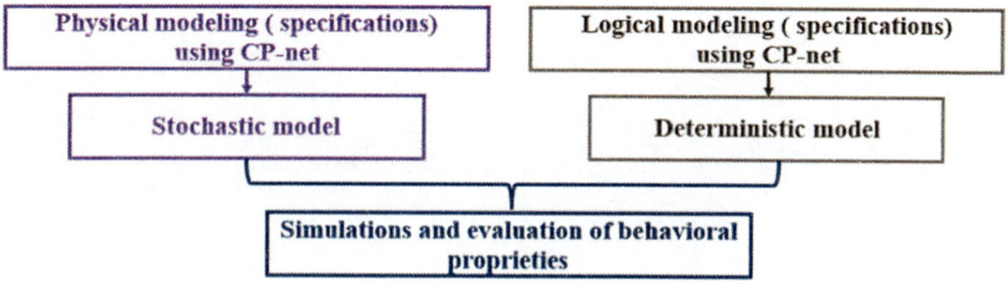

Fig. 8. Perspectives: adding the logical model and behavioral simulations

References

1. Schafer, W., Wehrheim, H.: The challenges of building advanced mechatronic systems. In: Future of Software Engineering (FOSE 2007), pp. 72–88 (2007). https://doi.org/10.1109/FOSE.2007.28.
2. Bishop, R.H.: Mechatronics: An Introduction. CRC Press, Boca Raton (2017)
3. Moncelet, G.: Application des Réseaux de Petri á l'évaluation de la sureté de fonctionnement des systémes mécatroniques du monde automobile. In: Automatique et informatique industriels, Université Paul Sabatier de Toulouse, Octobre 2019
4. Audibert, L.: Introduction á la modélisation objet. http://laurent-audibert.developpez.com/Cours-UML/?page=introduction-modelisation-objet
5. AFNOR. X 50 - 151 analyse de la valeur, analyse fonctionnelle. expression fonctionnelle du besoin et cahier des charges fonctionnel (1991)
6. Conway, J.B.: A Course in Functional Analysis, vol. 96. Springer (2019)
7. Lyonnet, P.: Ingénierie de la fiabilité. In: Lavoisier, tec&doc édition, mars 2006
8. Filali, A.: L'analyse fonctionnelle comme levier de réussite de la décision stratégique de l'entreprise. Revue Organisation et Travail **5.3**, 120-144 (2016)
9. Mikulak, R.J., McDermott, R., Beauregard, M.: The Basics of FMEA. CRC Press, Boca Raton (2017)
10. Chiozza, M., Ponzetti, C.: FMEA: a model for reducing medical errors. Clinica Chimica Acta, **404**(1), 75–78 (2009). ISSN 0009-8981
11. CENELEC EN50128: Railway applications - Communications, signalling and processing systems. In: Software for Railway Control and Protection Systems (2001)

12. Carvalho, P.V.: Mapping the software errors and effects analysis method to iso26262 requirements for software architecture analysis. In: 2014 IEEE International Symposium on Software Reliability Engineering Workshops. IEEE (2014)
13. Jensen, K., Kristensen, L.M.: Coloured Petri Nets. Springer, Heidelberg (2009). https://doi.org/10.1007/b95112
14. Zwingelstein, G.:La maintenance basée sur la fiabilité. Guide pratique d'application de la RCM, Hermes (1996)
15. Yusuf, K.O., Smith, N.J.: Modeling business processes in steel fabrication. Int. J. Project Manage. **14**(6), 367–371 (1996)
16. Mehdi, I., Boudi, E.: Towards a sustainable conceptual design of mechatronic systems application to a regenerative braking system. In: Materials Today: Proceedings, vol. 45, Part 8, pp. 7588–7596 (2021). ISSN 2214-7853

On Permutation Flow Shop Scheduling Problem with Sequence-Independent Setup Time and Total Flow Time

Hajar Sadki[1(✉)], Jabrane Belabid[1], Said Aqil[2], and Karam Allali[1]

[1] Laboratory Mathematics, Computer Science and Applications, University Hassan II of Casablanca, FST, PO Box 146, Mohammedia, Morocco
[2] Ecole Nationale Supérieure des Arts et des Métiers, University Hassan II of Casablanca, Casablanca, Morocco

Abstract. In this work, we study the optimization model for a permutation flow shop scheduling problem with sequence-independent setup time and total flow time. The objective here is to minimize the total completion time for all jobs, namely the total flow time. The constraint of our study is the sequence-independent setup time (SIST) which means that the setup time will depend on the nature of the machine. To solve this problem, we suggest three methods: an exact mixed-integer linear programming (MILP) method; two heuristics, the first one is based on Johnson's rule while the second is based on the NEH algorithm. It was established that the second NEH based heuristic numerical method gives best results than the first based on Johnson.

Keywords: Flow shop scheduling · SIST · MILP · Total flow time · Heuristic

1 Introduction

The numerical methods for solving permutation flow shop scheduling issues (PFSP) have started by the first work of Johnson [1]. Later, many works on this field have been developed taking into account many industrial constraints and objectives. For instance, the setup time may be dependent to the nature of jobs; hence any job may require a certain setup time before it can be processed in the machine. This situation is called flow shop permutation problem with sequence dependent setup time (SDST). PFSP-SDST problem have been resolved using an algorithm based on migratory bird algorithm to minimize the makespan criterion [2]. For the same issue a mixed-integer linear programming (MILP) method was used in [3] to seek the solution in the case of no-wait constraint. We find in [4], the water wave optimization algorithm have been utilized as good algorithm to solve PFSP-SDST. The setup time may be also independent of jobs nature. This case is called sequence independent setup time (SIST) problem. An example of this situation is when the setup time depends of the

R. Saidi et al. (Eds.): ICATH 2021, LNDECT 110, pp. 507–518, 2022.
https://doi.org/10.1007/978-3-030-94188-8_45

nature of the machine instead of jobs. The interest of PFSP-SIST scheduling study is when the manager fix automatically the setup time of each working machine in order to process the job arrivals. The interpretation of this setup time is when the machines need to be cleaned, parts assembled-disassembled and/or maintained. In [5], the authors suggest some MILP models and lower bounds to solve specific cases of PFSP-SIST problem. Recently, the PFSP-SIST scheduling problem was solved in [6] using different numerical approaches and using also MILP method.

It is worthy to notice that the objective function plays an essential role in flow shop problems. It may be under the form of the maximum of job completion time, namely makespan [7], maximum tardiness [8], maximum earliness [9], total flow time (TFT) [10] and several others. This last criterion remains very important in many industrial manufacturies, because minimizing it can reduce significantly the production cost [11]. The importance of this criterion is the completion time of all the jobs by the means of a summation of the jobs completion time in the last arrival machine. The reduction of the production of total flow time also allows the decision makers to minimize the use of resources and avoid any inventory [12]. In this work, we will extend the recent of Belabid et al. [6] by considering TFT as objective function instead of makespan and by taking into account the constraint of sequence independent setup time. In the Fig. 1, we describe the production flow the PFSP-SIST scheduling problem.

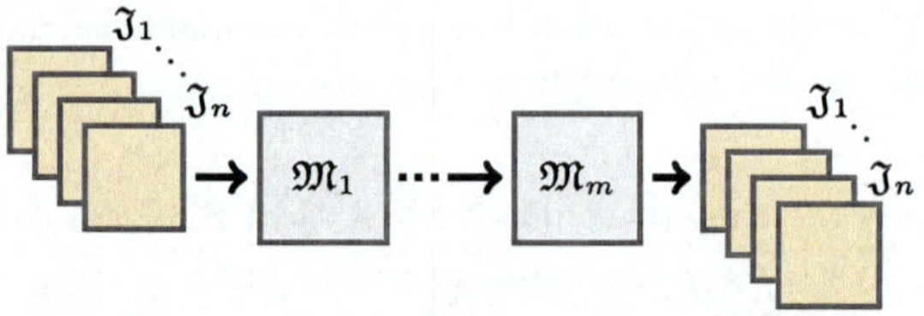

Fig. 1. Flow shop problem with n jobs and m machines.

The present work is organized as follows. In the next section, we will give a brief description of our subject. Next, in Sect. 3, we will give the method for solving PFSP-SIST scheduling problem. Section 4 stands for different numerical results. We conclude in the last section.

2 Problem Statement

The permutation flow shop problem with sequence-independent setup time, is consists of a set of m machines $\mathcal{M} = \{\mathfrak{M}_1, \mathfrak{M}_2, \ldots, \mathfrak{M}_m\}$ arranged in series constituting the production process see Fig. 1. There are a set of n jobs $\mathcal{J} = \{\mathfrak{J}_1, \mathfrak{J}_2, \ldots, \mathfrak{J}_n\}$ at a given time, processing of an operations of a job can be done on only one machine. Likewise, a certain machine can process only one job at one time, the jobs are processed in series of m machines in the same order, and each job $\mathfrak{J}_j$ starts in the first machine and needs to be completed before starting processing on the next machine. The processing time of job $\mathfrak{J}_j$ on machine $\mathfrak{M}_i$ is denoted by $p_{i,j}$, and each machine has a setup time when processing a new job is designated by s_i. The sequence of jobs is defined by $\sigma = \{\sigma_1, \sigma_2, \ldots, \sigma_n\}$ is executed in same order on all machine. The goal then is to minimize the total flow time (TFT) of all job on the last machine m, where $TFT = \sum_{j=1}^{n} CT_{j,m}$ and $CT_{j,m}$ represents the completion time of job $\mathfrak{J}_j$ in machine m. We will give the description of the scheduling equations for the job σ_j in the sequence σ allowing the resolution of our problem.

$$CT_{\sigma_1,1} = p_{\sigma_1,1} + s_1 \tag{1}$$

$$CT_{\sigma_j,1} = CT_{\sigma_{(j-1)},1} + p_{\sigma_j,1} + s_1; j = 2, \ldots, n \tag{2}$$

$$CT_{\sigma_1,i} = \max\left\{s_i, CT_{\sigma_1(i-1)}\right\} + p_{\sigma_1,i}; i = 2, \ldots, m \tag{3}$$

$$CT_{\sigma_j,i} = \max\left\{CT_{\sigma_{(j-1)},i} + s_i, CT_{\sigma_j,(i-1)}\right\} + p_{\sigma_j,i}; \tag{4}$$
$$i = 2, \ldots, m$$
$$j = 2, \ldots, n$$

$$\text{Total Flow Time } TFT = \sum_{j=1}^{n} CT_{\sigma_j,m} \tag{5}$$

Equation (1) is the end date of the first job on the first machine. Equation (2) is the completion time of job σ_j on the first machine. Equation (3) represents the completion time of the first job on the other machines. Equation (4) illustrates the job σ_j finish tome on the other machines. Equation (5) represents the objective function to be minimized.

3 Resolution of the Problem

In this section, we present MILP method and two numerical heuristics to solve the problem $F_m \mid \text{prmu, SIST} \mid \sum_{j=1}^{n} CT_{j,m}$ whose objective is to minimize the total flow time.

3.1 Mixed-Integer Linear Programming

Modeling scheduling problems by PFSP that is of greatest interest of operations research field. Indeed, the formulation of the PFSP scheduling problem is usually adopted by using MILP as exact method with an objective function under constraints according to the industrial studied context. Many models with MILP are suggested to solve different PFSP case study models, a comparative study is described, and also a new MILP model is presented in [13]. Moreover, more models are presented in [14] for the PFSP-SDST problem. We present here a model formulated in MILP for our problem, and we define a set of jobs $\mathcal{J} = \{\mathfrak{J}_1, \mathfrak{J}_2, \ldots, \mathfrak{J}_n\}$ and a set of machines $\mathcal{M} = \{\mathfrak{M}_1, \mathfrak{M}_2, \ldots, \mathfrak{M}_m\}$. In our approach, we define a set of parameters and variables necessary for the modeling of the problem.

$$y_{k,j} = \begin{cases} 1, & \text{if the job j is ordered in the } k^{th} \text{ position in the sequence,} \\ 0, & \text{otherwise,} \end{cases}$$

$$j, k = 1, \ldots, n,$$

$$CT_{k,i} = \text{Completion time of job at position k on machine i,} \tag{6}$$

$$k = 1, \ldots, n,$$

$$i = 1, \ldots, m,$$

The objective is to :

$$\text{minimize } TFT = \sum_{j=1}^{n} CT_{j,m}, \tag{7}$$

The subject is to:

$$\sum_{j=1}^{n} y_{k,j} = 1; k = 1, \ldots, n, \tag{8}$$

$$\sum_{k=1}^{n} y_{k,j} = 1; j = 1, \ldots, n, \tag{9}$$

$$CT_{1,1} \geq s_1 + \sum_{j=1}^{n} y_{j,1} \times p_{1,1}, \tag{10}$$

$$CT_{k,1} \geq s_1 + CT_{(k-1),1} + \sum_{j=1}^{n} y_{k,j} \times p_{j,1}; k = 2, \ldots, n, \tag{11}$$

$$CT_{k,i} \geq CT_{k,(i-1)} + \sum_{j=1}^{n} y_{k,j} \times p_{j,i}; i = 2, \ldots, m; k = 1, \ldots, n \tag{12}$$

$$CT_{k,i} \geq s_i + CT_{(k-1),i} + \sum_{j=1}^{n} y_{k,j} \times p_{j,i}; i = 1, \ldots, m; k = 1, \ldots, n \tag{13}$$

$$CT_{k,i} \geq 0; i = 1, \ldots, m; k = 1, \ldots, n \tag{14}$$

$$y_{k,j} \in \{0, 1\}; j, k = 1, \ldots, n \tag{15}$$

Equation (7) is the objective function to minimize the total flow time. Equation (8) ensure that each position of the sequence is allocated once for all jobs. Equation (9) to ensure that each position is assigned to all positions only once. Equation (10) is used to determine the end date of the job at the first position in the first machine. Equation (11) calculate the end date of the job at position $k, k \geq 2$ in the first machine. Equation (12) ensure that for a job at position k, the end date in machine i is greater than or equal to the end date in the previous machine $i-1$ increasing its processing time on machine i. Equation (13) ensures that for two successive position jobs, the end date in the machine of the one in position k is greater than or equal to that in $k-1$ plus the processing time and setup time. Equation (14) and (15) ensure that the execution end date is positive and that the $y_{k,j}$ assignment variables are binary variables.

Table 1. The processing time of four jobs on three machines.

Job	$\mathfrak{J}_1$	$\mathfrak{J}_2$	$\mathfrak{J}_3$	$\mathfrak{J}_4$	s_i
$\mathfrak{M}_1$	4	6	5	9	1
$\mathfrak{M}_2$	2	2	3	1	2
$\mathfrak{M}_3$	7	4	6	8	3

To illustrate the scheduling issue, we will consider an example with three machines $\mathcal{M} = \{\mathfrak{M}_1, \mathfrak{M}_2, \mathfrak{M}_3\}$ and four jobs $\mathcal{J} = \{\mathfrak{J}_1, \mathfrak{J}_2, \mathfrak{J}_3, \mathfrak{J}_4\}$ in Table 1 with the processing time $p_{i,j}$ and setup time s_i given for all jobs and all machines, the problem will be noted by $F_3 \mid \text{prmu, SIST} \mid \sum_{j=1}^{n} CT_{j,3}$. To solve the problem dealing with the case of PFSP-SIST in MILP model, we will use LINGO 17.3 software as solver. Figure 2 shows the process of scheduling an instance of four jobs in a workshop of three machines. The processing and setup time data for this example are presented in Table 1. By executing our MILP model we have found that the exact jobs scheduling is $(\mathfrak{J}_1, \mathfrak{J}_2, \mathfrak{J}_3, \mathfrak{J}_4)$. Here, we obtain that the optimal objective value is $TFT = 106$.

3.2 Heuristics

PFSP scheduling issues under specific constraints can be resolved by different numerical methods. The heuristics are usually based on some priority rules depending on some jobs characteristics, like processing times, machine setup times and many other characteristics specific to each problem case. In this work, we will present two heuristics, the rule based on Johnson's heuristic [1] and the other one is based on NEH algorithm [15]. These two heuristics are amongst the good approaches to seek the right solution for PFSP manufacturing problems.

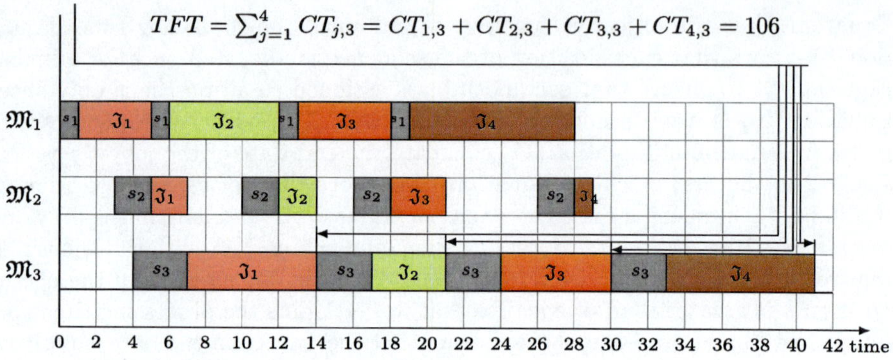

Fig. 2. Gantt chart for four jobs in three machines.

Algorithm 1. Heuristic derived from Johnson's rule

Input: $\sigma_0 = \{\sigma_1, \sigma_2, \ldots, \sigma_n\}$, $\sigma \leftarrow \sigma_0$:%The initial sequence%
$p'_{\sigma_j,1} \leftarrow p_{\sigma_j,1} + s_1$, $p'_{\sigma_j,2} \leftarrow \sum_{k=2}^{m} (p_{\sigma_j,k} + s_k)$,$j = 2, \ldots, n$
$\sigma^* \leftarrow \sigma_0$ $TFT^* \leftarrow TFT(\sigma^*)$
for $i = 2$ to $m - 1$ **do**
$p'_{\sigma_j,1} \leftarrow \sum_{k=1}^{i} (p_{\sigma_j,k} + s_k)$,$j = 1, \ldots, n$
$p'_{\sigma_j,2} \leftarrow \sum_{k=i+1}^{m} (p_{\sigma_j,k} + s_k)$,$j = 1, \ldots, n$
$\sigma'' \leftarrow$ Johnson's rule(σ')
if $TFT(\sigma'') < TFT(\sigma')$ **then**
$\sigma \leftarrow \sigma''$
if $TFT(\sigma) < TFT(\sigma^*)$ **then**
$\sigma^* \leftarrow \sigma$
end if
end if
end for
$TFT^* \leftarrow TFT(\sigma^*)$
Output: σ^*

3.3 Heuristic Derived from Johnson's Rule

The PFSP issue with two machines is a classical problem of many flow shop scheduling problems [1], several studies that have been suggested since the first study refer to the well-known Johnson's rule. The Johnson's algorithm for constructing an optimal schedule may be stated as follows:

- Step 1: from Set-I containing all the job with $p_{j,1} < p_{j,2}$.
- Step 2: from Set-II containing all the job with $p_{j,1} > p_{j,2}$.
- Step 3: from the sequence as follow.

- the jobs in set-I go first in the sequence and they go in increasing order of $p_{j,1}$
- the jobs in set-II follow in decreasing order of $p_{j,2}$.

This type of schedule is referred to as Set-I followed set-II is an optimal solution. In this investigation the Johnson algorithm is adapted to fit more than two machines flow shop problems, We propose a heuristic based on Johnson's rule (HBRJ), which we adopted to solve the case $F_m \mid$ prmu, SIST $\mid \sum_{j=1}^{n} C_{j,m}$. In this heuristic we propose the HBRJ rule modified and adopted with the construction of two virtual machines by assembling the first $1 < k < (m-1)$ machines in a single virtual machine and the $(m-k)$ remaining machines in the second virtual machine. The processing time of a job σ_j on the first virtual machine will be $p'_{j,i}$ which is the sum of the processing times $p_{j,i}$ of the job σ_j on the machines forming the virtual machine.

Algorithm 2. NEH algorithm

Input: A processing time $p_{\sigma_j,1}, ..., p_{\sigma_j,m}$ have the setup time s_i in m machines.
Calculate Sum the total processing time and setup time of each job by: $T(\sigma_j) = \sum_{i=1}^{m} (p_{\sigma_j,i} + s_i)$
Construct the sequence $\sigma = (\sigma_1, ..., \sigma_n)$ by sorting the jobs in descending order of $T(\sigma_j)$
Schedule the first two jobs σ_1,σ_2 and choose the best sequence σ of the first two jobs.
for $j = 3$ to n **do**
Insert σ_j in the all positions of the current sequence built and select the best sequence
with the minimum TFT in the last machine and update σ_{NEH} of the optimal sequence.
end for
Output: σ_{NEH}

3.4 NEH Heuristic

The NEH algorithm is a basic algorithm to solve many PFSP problem with n jobs and m machines. According to several scheduling researchers like Ruiz and Maroto [16], We note that several studies of research as in [17]. Indeed, Ruiz and Maroto compared the NEH algorithm with the most modern heuristic complexes, such as those established in [18,19] showing that the NEH algorithm achieves the best performance of solving problems PFSP. Algorithm (2) presents the steps necessary to establish the sequence of NEH. In order to apply the NEH algorithm to our study case, the setup time is added to the sum of the processing time of each job in the sequence; in the first step of the algorithm we calculate $T(\sigma_j) = \sum_{i=1}^{m} (p_{\sigma_j,i} + s_i)$. In the second step, we classify the jobs in ascending order of $T(\sigma_j)$. We choose the best sequence of the first two jobs $(\sigma_1\sigma_2, \sigma_2\sigma_1)$ with the shortest TFT by testing it two position, thereafter at each iteration we insert the job $\sigma_j, j = 3, \ldots, n$ in the different positions of the constructed sequence and find the shortest TFT by testing it at all possible. Would be selected the best sequence σ_{NEH}, at each iteration, the process is repeated to complete all the jobs.

4 Results

we present here the comparative study based on the calculation of the relative percentage deviation (RPD)

$$RPD_{H_i} = \frac{TFT^{H_i} - TFT^{MILP}}{TFT^{\text{MILP}}} \times 100; i \in \{1, 2\} \tag{16}$$

Where H_1 stands for HBRJ and H_2 stands for the NEH heuristic. We note that TFT^{H_i} represents the solution found by the heuristic H_i and TFT^{MILP} represents the value found by the MILP.

Table 2. Comparison between MILP and heuristics

$n \times m$	***MILP***	***HBJR***	***NEH***	RPD_{H_1}	RPD_{H_2}
5×2	117	130	118	11.11	0.85
8×2	318	373	333	17.29	4.71
11×2	539	600	551	11.31	2.22
14×2	941	1061	977	12.75	3.82
5×3	158	169	165	6.96	4.43
8×3	367	437	373	19.07	1.63
11×3	666	724	671	8.70	0.75
14×3	996	1116	1020	12.04	2.40
5×4	199	219	203	10.05	2.01
8×4	442	487	443	10.18	0.22
11×4	758	825	764	8.83	0.79
14×4	1108	1185	1110	6.94	0.18
5×5	255	280	265	9.80	3.92
8×5	474	546	480	15.18	1.26
11×5	820	945	823	15.24	0.36
14×5	1103	1291	1138	17.04	3.17
5×6	254	282	254	11.02	0
8×6	529	549	541	3.78	2.26
11×6	918	966	927	5.22	0.98
14×6	1213	1418	1233	16.90	1.64

Table 3. Comparison between MILP and heuristics.

$n \times m$	***MILP***	***HBJR***	***NEH***	RPD_{H_1}	RPD_{H_2}
5×7	318	353	323	11.00	1.57
8×7	626	691	642	10.38	2.55
11×7	893	1074	917	20.26	2.68
14×7	1383	1553	1413	12.29	2.16
5×8	346	370	346	6.93	0
8×8	608	689	625	13.32	2.79
11×8	997	1091	1017	9.42	2.00
14×8	1359	1593	1384	17.21	1.83
5×9	370	392	373	5.94	0.81
8×9	691	757	700	9.55	1.30
11×9	1086	1149	1106	5.80	1.84
14×9	1574	1858	1586	18.04	0.76
5×10	407	422	417	3.68	2.45
8×10	747	765	760	2.40	1.74
11×10	1138	1329	1149	16.78	0.96
14×10	1617	1803	1646	11.50	1.79
5×11	445	485	455	8.98	2.24
8×11	733	778	737	6.13	0.54
11×11	1182	1304	1193	10.32	0.93
14×11	1761	2032	1822	15.38	3.46

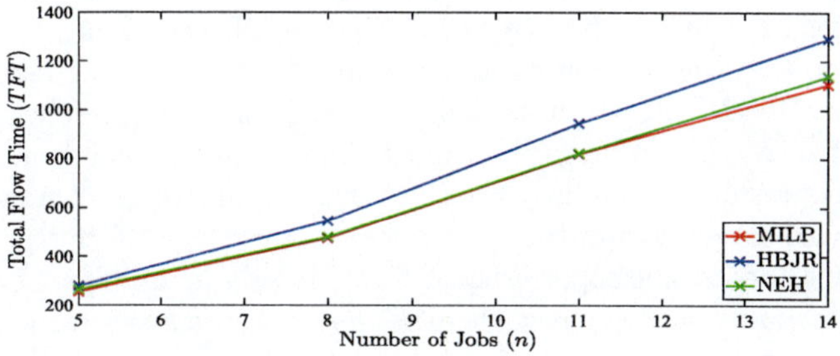

Fig. 3. Total flow time as function of number of jobs for $m = 5$.

Figure 3 shows the evolution of total flow time as function of number of jobs when the number of machines is equal to five ($m = 5$). Our first remark is the total flow time increases when the number of jobs get increased. Secondly, we good performance of NEH algorithm is observed comparing to $HBJR$ heuristic. Therefore NEH can be used as an efficient method to solve our scheduling problem. This result is also illustrated in term of relative percentage deviation

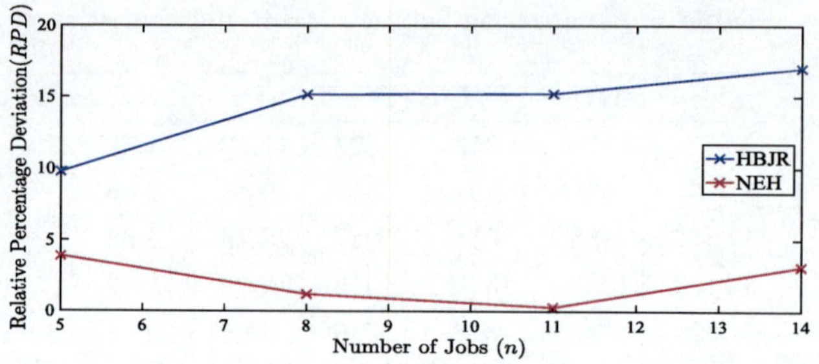

Fig. 4. Relative percentage deviation as function of number of jobs for $m = 5$.

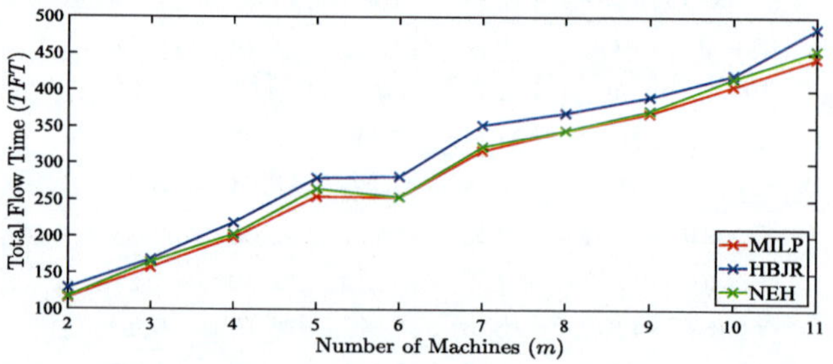

Fig. 5. TFT as function of number of machines for $n = 5$.

(RPD). Indeed, Fig. 4 shows that the RPD_{H_2} given by NEH is smaller than RPD_{H_1} which express the superiority of NEH in solving our flow shop problem. Figure 5 depicts the of total flow time versus the number of machines when the number of jobs is equal to five ($n = 5$). First of all, we remark that the total flow time increases when the number of machines get increased. Also, significant performance of NEH algorithm is well remarked comparing to classical $HBJR$ heuristic. Then NEH can be used as an efficient method to solve our scheduling problem. This result is also illustrated in term of relative percentage deviation (RPD). Indeed, Fig. 6 shows that the RPD_{H_2} given by NEH is smaller than RPD_{H_1} which express the superiority of NEH in solving our flow shop problem. The PFSP scheduling problem with SIST have been solved using an adapted classical algorithm based on Johnson rule and another more efficient algorithm based on NEH algorithm. Our numerical tests have revealed the superiority of the last one compared to the classical HBJR.

We have concluded this fact because the numerical results based on NEH are more closed to the exact results given by MILP method. However, the numerical results with HBJR remain a little bit far from the results given by MILP method. Hence, NEH algorithm can be used as an efficient method to solve PFSP-SIST especially for small to moderate size flow shop instances.

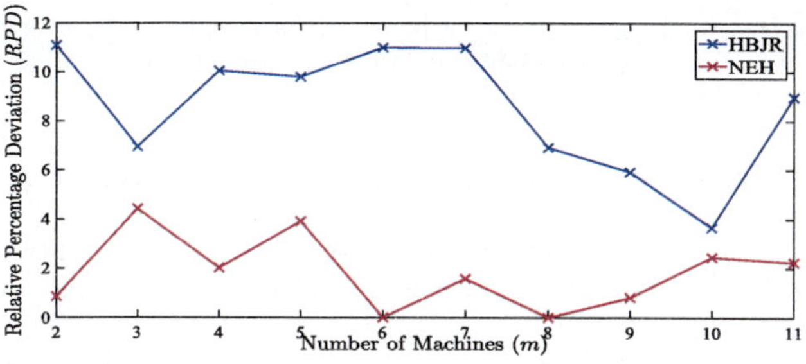

Fig. 6. RPD as function of number of machines for $n = 5$.

5 Conclusion

In this work, we have studied a permutation flow shop scheduling problem with SIST and TFT. To this end, we have used an exact method and two different heuristics. More precisely, we have propose a model in mixed integer linear programming for the case of the flow shop with sequence independent setup time, and for the method approximate, we have applied two heuristics based on HBJR and NEH. The heuristics were compared to the MILP results for all instances tested. The simulations have been given for different size of instances by varying the number of jobs and machines. The numerical results have confirmed that the algorithm based on NEH records best performance than the algorithm based on Johnson.

References

1. Johnson, S.M.: Optimal two and three stage production schedules with setup times included. Nav. res. log. quar. **1**(1), 61–68 (1954)
2. Sioud, A., Gagné, C.: Enhanced migrating birds optimization algorithm for the permutation flow shop problem with sequence dependent setup times. Eur. J. Oper. Res. **264**(1), 66–73 (2018)
3. Cheng, C.Y., Ying, K.C., Li, S.F., Hsieh, Y.C.: Minimizing makespan in mixed no-wait flowshops with sequence-dependent setup times. Comp. Ind. Eng. **130**, 338–347 (2019)
4. Shao, Z., Pi, D., Shao, W.: A novel discrete water wave optimization algorithm for blocking flow-shop scheduling problem with sequence-dependent setup times. Sw. Evol. Comp. **40**, 53–75 (2018)
5. Ladhari, T., Msakni, M.K., Allahverdi, A.: Minimizing the total completion time in a two-machine flowshop with sequence-independent setup times. J. Oper. Res. Soc. **63**(4), 445–459 (2012). https://doi.org/10.1057/jors.2011.37
6. Belabid, J., Aqil, S., Allali, K.: Solving permutation flow shop scheduling problem with sequence-independent setup time. J. Appl. Math. **2020** (2020)
7. Reza Hejazi, S., Saghafian, S.: Flowshop-scheduling problems with makespan criterion: a review. Int. J. Prod. Res. **43**(14), 2895–2929 (2005)

8. Chakravarthy, K., Rajendran, C.: A heuristic for scheduling in a flowshop with the bicriteria of makespan and maximum tardiness minimization. Prod. Plan. Contr. **10**(7), 707–714 (1999)
9. Toktaş, B., Azizoğlu, M., Köksalan, S.K.: Two-machine flow shop scheduling with two criteria: maximum earliness and makespan. Eur. J. Oper. Res. **157**(2), 286–295 (2004)
10. Wang, L., Pan, Q.K., Tasgetiren, M.F.: Minimizing the total flow time in a flow shop with blocking by using hybrid harmony search algorithms. Exp. Sys. Appl. **37**(12), 7929–7936 (2010)
11. Rajendran, C.: Heuristic algorithm for scheduling in a flowshop to minimize total flowtime. Inter. J. Prod. Econ. **29**(1), 65–73 (1993)
12. Rajendran, C., Ziegler, H.: An efficient heuristic for scheduling in a flowshop to minimize total weighted flowtime of jobs. Eur. J. Oper. Res. **103**, 129–138 (1997)
13. Tseng, F.T., Stafford, E.F., Jr.: New MILP models for the permutation flowshop problem. J Oper Res Soc **59**(10), 1373–1386 (2008)
14. Tseng, F.T., Stafford, E.F., Jr.: Two MILP models for the N× M SDST flowshop sequencing problem. Int. J. Prod. Res. **39**(8), 1777–1809 (2001)
15. Nawaz, M., Enscore, E.E., Jr., Ham, I.: A heuristic algorithm for the m-machine, n-job flow-shop sequencing problem. Omega **11**(1), 91–95 (1983)
16. Ruiz, R., Maroto, C.: A comprehensive review and evaluation of permutation flowshop heuristics. Eur. J. Oper. Res. **165**(2), 479–494 (2005)
17. Turner, S., Booth, D.: Comparison of heuristics for flow shop sequencing. Omega **15**(1), 75–78 (1987)
18. Koulamas, C.: A new constructive heuristic for the flowshop scheduling problem. Eur. J. Oper. Res. **105**(1), 66–71 (1998)
19. Suliman, S.M.A.: A two-phase heuristic approach to the permutation flowshop scheduling problem. Inter J prod econ **64**(1–3), 143–152 (2000)

A Novel Hybrid Heuristic Based on Ant Colony Algorithm for Solving Multi-product Inventory Routing Problem

Fadoua Oudouar[1(✉)] and El Miloud Zaoui[2]

[1] Mohammadia School of Engineering, Mohammed V University, Rabat, Morocco
[2] Ecole Nationale Supérieure des Mines de Rabat, Rabat, Morocco
zaoui.elmiloud@enim.ac.ma

Abstract. This work, describes the multi-product inventory routing problem and develops a solution method based on two-phases to solve it. The first one is based on Ant colony algorithm to solve a periodic vehicle routing problem. In the second step, we employ A novel hybrid heuristic that combines an efficient heuristic and two-opt method to improve the initial solution in which we minimizing both the routing and inventory costs. We use the performant heuristic to determine the quantity of products to collect from each supplier at each period, and two-opt method to optimize the routes. The performance of our approach has been tested using a dataset for the inventory routing problem from the literature. The obtained numerical results show the higher performance of the proposed approach to solve the inventory routing problem compared with the best-known algorithms in terms of cost.

Keywords: Inventory Routing · Ant Colony Algorithm · Heuristic · Two-opt

1 Introduction

The Inventory Routing Problem (IRP) combines the two principal decisions in supply chain. It comprises the vehicle routing (VRP) and the inventory management problem. The IRP is one of the most complete problem in the logistic distribution. The goal is to minimize the total cost of VRP and inventory management. The idea of combining both problems, the inventory allocation problem and the VRP was first introduced by [1] which studied the IRP with a single customer. Several varaints of the IRP are described in literature. A classification of the different variants of the IRP can be found in [2–5]. The authors of [6] studied an unique-product IRP where the objective is to deliver a product to a set of clients. [8] solved a multi-product IRP with homogenous fleet types with limited number of vehicles. [7] established an IRP for perishable products. [8] considered a model of IRP with stochastic demand. [9] presented a IRP model, which the horizontal corroboration of suppliers was explicitly considered. The location inventory routing problem [10–12] is another variant of IRP. In this problem, several depots must be located to distribute the products to customers in each period.

R. Saidi et al. (Eds.): ICATH 2021, LNDECT 110, pp. 519–529, 2022.
https://doi.org/10.1007/978-3-030-94188-8_46

Recently, [13] considered a green IRP with time windows. The main idea of this problem is to minimize fuel consumption throughout the distribution. Since the IRP combines two several NP-hard problems, authors have performed large works on it in order to make many techniques. [8] proposed a modified Ant Colony Optimization (ACO) to optimize the difference between the shortage and transportation cost. The authors developed a heuristic method adopted to the IRP with time windows and stochastic demand, which was hybrid two approaches, the Markov Decision Process (MDP) and Clarke-Wright algorithm. Another heuristic method developed permits to solve the IRP with multiple vehicles. They used an approach permits to improve the initial solution, this method is Adaptive Large Neighborhood Search (ALNS), and another optimization technic to determine the optimal solution (Branch-and-Cut algorithm). Another approach used a hybridation between simulated annealing and Tabu search for solving an IRP for perishable goods. The researchers presented a genetic algorithm in order to obtain the solotution of the IRP in an attempt to optimize the sales loss to solve the IRP. An approach integrated green logistics in IRP with a multi-period. They employed an axact method based on Branch-and-Cut to solve the proposed problem. [14] proposed variable neighborhood search and Tabu search algorithms to solve the IRP with location depots.

In this work, we focus on solving the multi-product IRP. The main objective is to optimize the two decisions: the vehicles routing and the inventory management at each period. [15] is one of the first to simulate an IRP with multiple products. They proposed an heuristic method based on a Genetic Algorithm to detrmine the optimal solution for this problem. Also, [16] developed an heuristic based on Variable Neighborhood Search (VNS) for the same objective. Our proposed scheme in this paper containes two levels in order to determine the solution of this problem. The heuristic method named Ant Colony Optimization is used in the first level to solve periodic vehicle routing problem. In the second stage, we develop a hybrid heuristic based on an efficient heuristic to determine the inventory decision and the Two-opt algorithm to improve the total cost.

The structure of this work as follows. We define our problem and present mathematical formulation in [15] in Sect. 2. The Sect. 3 descibes the algorithms used to solve the model. In next section, we give the computational experiments. In the last section, we include some conclusions and discusses opportunities for future research.

2 Problem Description

In this work, we address a multi-product IRP similar to the one proposed by [21]. The problem consists of delevering the products of an assembly plant for each period $s \in S$. A fleet of M homogeneous vehicles with a capacity Q is available. The products are collected from a set of n uppliers. Vehicles are located at the depot. Each supplier can be served by more than one vehicle at each period. Moreover, the inventory costs are calculated just at the manufacturing plant. At each period $s \in S = \{1, \dots, \tau\}$, the demand of the manufacturing plant equals to d_{is} for each product $i \in N = \{1, \dots, n\}$. The corresponding product i defines supplier. At each period, a vehicle k leaves the warehouse 0, visits the suppliers affected to this vehicle, and distributs products to the manufacturing plants n + 1. when a vehicle finishes its tour it returns to the depot. Since IRP aims at simultaneously reducing the fixed cost F of using a vehicle, the traveling cost between node i and node j, c_{ij}, and the inventory cost of product, h_i. The illustration of the problem can be dfined in Fig. 1.

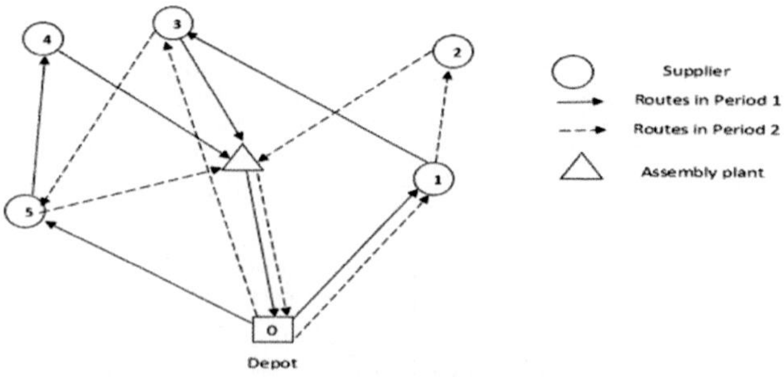

Fig. 1. Example of the IRP

A mathematical formulation based on [15] is given here. We first introduce the following decision variables:

- a_{is}: Total quantity to be collected from each supplier i at each period s.
- I_{is}: Stock level of each product i at the assembly plant at each period s.
- I_{i0}: Initial stock of each product i at the assembly plant.
- q_{ijs}: Quantity of products distributed between the two arcs i and j during the period s.
- x_{ijt}: Number of visits between the two arcs i and j during the period s.

The mathematical formulation states as follows:

$$min \sum_{i\in V} h_i \left(\sum_{s\in S} I_{is}\right) + V\left(\sum_{i\in N} \sum_{\substack{i \in N\cup\{0\} \\ i\neq j}} c_{ij}\left(\sum_{s\in S} x_{ijs}\right) + \sum_{i\in N} c_{i,n+1}\sum_{s\in S} x_{i,n+1,s}\right) + (F + Vc_{n+1,0})\sum_{i\in N}\sum_{s\in S} x_{0is} \quad (1)$$

$$\text{Subject to :}$$

$$I_{is} = I_{i,s-1} + a_{is} - d_{is} \quad \forall i \in N, \forall s \in S \quad (2)$$

$$\sum\nolimits_{\substack{i\in N\cup\{0\} \\ i\neq j}} q_{ijs} + a_{js} = \sum\nolimits_{\substack{i\in N\cup\{n+1\} \\ i\neq j}} q_{ijs} \forall j \in N, \forall s \in S \quad (3)$$

$$\sum\nolimits_{i\in N} q_{i,n+1,s} = \sum\nolimits_{i\in N} a_{is} \quad \forall s \in S \quad (4)$$

$$\sum\nolimits_{\substack{i\in N\cup\{0\} \\ i\neq j}} x_{ijs} = \sum\nolimits_{\substack{i\in N\cup\{n+1\} \\ i\neq j}} x_{jis} \quad \forall j \in N, s \in S \quad (5)$$

$$\sum\nolimits_{j\in N} x_{0js} = \sum\nolimits_{j\in N} x_{j,n+1,t} \quad \forall s \in S \quad (6)$$

$$q_{ijs} \leq cx_{ijs} \qquad \forall i \in N, \forall j \in N \cup \{n+1\}, i \neq j, \forall s \in S \tag{7}$$

$$I_{is} \geq 0 \qquad \forall i \in N, \qquad \forall s \in S \tag{8}$$

$$x_{i,0,s} = x_{n+1,i,s} = 0, \quad \forall i \in N \cup \{0, n+1\}, \quad \forall s \in S \tag{9}$$

$$a_{is} \geq 0 \qquad \forall i \in N, \qquad \forall s \in S \tag{10}$$

$$x_{i,j,s} \geq 0, entier, \quad \forall i, j \in N \cup \{0, n+1\}, \quad \forall s \in S \tag{11}$$

$$q_{ijs} \geq 0, \quad \forall i \in N, j \in N \cup \{n+1\}, \quad \forall s \in S \tag{12}$$

$$q_{0is} = 0, \qquad \forall i \in N, \forall s \in S \tag{13}$$

The objective function (1) aims to minimize both the fixed cost of using vehicles, the routing cost, and the inventory costs. Constraints (2) ensure the inventory level of products at the manufacturing plant. Constraints (3) eliminate all sub tours an assure the flow balance at each supplier. Constraints (4) ensure equality between the quantities of products collected by all vehicles and the demand of the assembly plant at each period. Constraints (5) and (6) guarantee equality between the number of vehicles leaving the depot and the number of vehicles returning to the manufacturing plant. Constraints (7) garantie the respect of the vehicle capacity. Constraints (8) guarantee the satisfaction of the demand at each period. Constraints (9) guarantee the elimination of a direct route from the suppliers to the depot. Finally, constraints (10)–(13) define the decision variables.

3 Solution Presentation and Approach Structures

In this paper, we search to solve the inventory routing problem using a hybrid heuristic based on the Ant Colony algorithm (AC) and an efficient heuristic. Firstly, we use an AC to solve a periodec VRP. More precisely, we must determine the optimum number of routes to collect the number of products from each supplier and transported them to an assembly plant at each period, without considering into the inventory cost. Then we use a combination of two heuristics to minimize the routing and inventory costs simultaneously. In this step, we seek to determine the quantities of products to collect from each supplier while respecting the constraint of storage at the assembly plant. Then, we implement the Two-opt local search to improve the solution.

3.1 Ant Colony Algorithm

The ant colony algorithm was first proposed by [17]: the principle of this heuristic is to simulate the ability of ant colonies to find the shortest path to food. Artificial ant systems have been shown to be effective in solving classic vehicle routing problems [18]. The basic phases of the ACO algorithm are the ant's travel construction and pheromone information (update). In the route construction phase, M simple artificial agents concurrently construct the initial routes starting from the depot selected in the network of N supplier nodes (plus the depot and the assembly plant). At each constructive procedure, the ant k moves from a node i to the next node j by choosing among the nodes that have not been visited yet, the AC algorithm follows the equation to build periodic routes.

$$P_{ij}^{k}(s) = \begin{cases} \frac{(\tau_{ij}(s))^{\alpha}.(\eta_{ij})^{\beta}}{\sum_{l\epsilon N_i^k}((\tau_{il}(s))^{\alpha}.(\eta_{il})^{\beta}}, & j \in N_i^k \\ 0, & \text{otherwise,} \end{cases}$$

- τ_{ij} denotes the level of pheromone on the arc (i, j).
- $\eta_{ij}(s)$ denotes the visibility of the arc (i, j) at planning period s,usually indicates the inverse of the distance/cost on the arc from node i to node j.
- $\alpha \wedge \beta$ denote the relative influence of the pheromone trails.
- N_i^k denotes the ants working memory (i.e. the nodes which are already accessible from node i and not previously choose.

After each passage in the arc (i, j) the pheromone trail is updated locally. The local update is performed during the ant constructive route, and it is calculated according to the following way:

$$\tau_{ij}(s+1) = (1-\rho).\tau_{ij}(s) + \rho.\tau_0 \tag{15}$$

Where $\tau_{ij}(s+1)$ the pheromone level between node i and j after updating, $\tau_{ij}(s)$ is the pheromone level between node i and j before updating, ρ is the constant that optimize the speed of evaporation.

A solution is obtained when all suppliers are visited or no vehicle cannot add a supplier due to capacity of vehicle. The best Ant is the one who visited more suppliers. Once the ants have finished their iterations, the global update is performed using the best solution found. This global pheromone updating guides the following ants in their construction of solutions. The following formula represents the calculation of this update:

$$\tau_{ij}(s+1) = (1-\rho).\tau_{ij}(s) + \rho.\Delta\tau_{ij} \tag{16}$$

Where the arcs (i, j) belong to current optimum route T^+ of length L^+ and $\Delta\tau_{ij} = \frac{1}{L^+}$.

3.2 Efficient Method for the Inventory Management

In this section, we describe the perfermant heuristic used to calculate the amount of products to be assembled. This method is based on a backward technic; it is can generally be divided into two steps: the first determine which suppliers should be visited first in order to collect the demands in each period. The second step determines the order of vehicles that visit a supplier. We note δ_{is} the priority of supplier $i \in K$ at period $s \in S$, it is based on the warehouse cost of its product and on its the distribution period $s' < s$. The following formulation defines this priority:

$$\delta_{is} = h_i\left(s - max\left\{max\left\{s' < s : if\ supplier\ i\ is\ visited\ at\ period\ s'\right\}, 0\right\}\right) \quad (17)$$

The supplier that has the higher value of δ_{is} is the first visited. The list of suppliers is ordered in descending order after the supplier is assigned to a period s. Then, we determine the order of the corresponding allocation of vehicles. We define the second priority rule for this assignment. We start by estimating the average cost. After we control if one unit of product is collected or not by a vehicle v, in period s. This process of estimating is calculated using the formulation as folow:

$$h^*_{vs} = \frac{\sum_{i \in K(s,v)}\left(d_{is} + I_{i,s+1}\right)\delta_{is}}{\sum_{i \in K(s,v)}\left(d_{is} + I_{i,s+1}\right)} \quad (18)$$

A vehicle v with the highest priority at each period s is one who has the smallest γ_{vs}, where:

$$\gamma_{vs} = h^*_{vs} \sum_{i \in N(s,v)} max\left(d_{is} + I_{i,s+1} - C, 0\right) \quad (19)$$

3.3 Two-Opt for Improvement Solution

We present here the Two-opt procedure used to improve the solution. The two-opt local search was proposed by Bertsimas. It is one of the most widely used local search for improving vehicle routes. This algorithm consists of eliminating two arcs and combining the two resulting paths in a different way to obtain a new route. The two-opt procedur is based on exchange edges and search the neighborhood of the current route, that is, perform a local search and provide an optimal route. Figure 2 shows an example of the Two-opt operator.

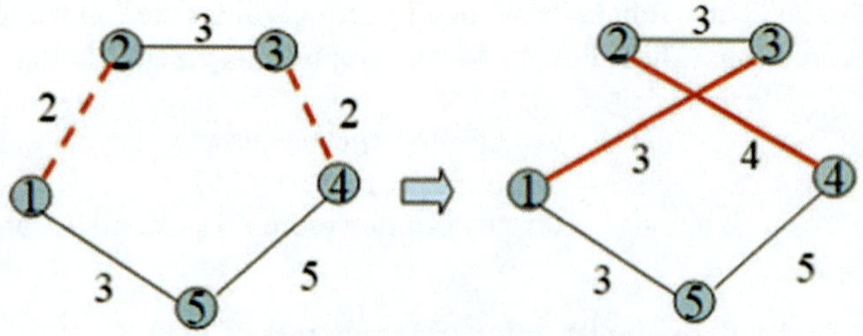

Fig. 2. Example of two-opt algorithm

ALGORITHM 1 : 2-OPT

Input: A tour $k = p_1, p_2, \ldots, p_m$, distance D.
Output: A new shorter tour $k' = p'_1, p'_2, \ldots, p'_m$.
For $i = 1$ to $m - 1$ do
  For $j = i + 1$ to m do
    If $j < m$ then
      If $D(p_i, p_{i+1}) + D(p_j, p_{i+1}) < D(p_j, p_{i+1}) + D(p_j, p_{j+1})$ then
        Swap the position of the two suppliers p_i and p_j;
      End if
    Else
      If $D(p_i, p_l) + D(p_m, p_{i+1}) < D(p_i, p_{i+1}) + D(p_m, p_l)$ then
        Swap the position of the two suppliers p_i and p_m;
      End if
    End if
  End for
End for
Return, the new tour

4 Computational Results

The proposed method was implemented in C++. This section presents the detailed experiments from the proposed approaches. A comparaison between our methods and two algorithms published in the reviws is presented.

We adopt a set of 14 benchmark instances based on the original 4 instances provide in [19], the same ones used in [16]. These four data sets are small instances S12T14 and S20T21 that comprise (12 suppliers, 14 periods) and (20 suppliers, 21 periods), respectively. The medium instances S50T21 with 50 suppliers and 21 periods. The large instances S98T14 with 98 suppliers and 14 periods. The characteristics of the instances are provided in Table 1.

Table 1. Information of instances

Ins.	N. Sup	Fixed cost	T.C/U.D	V.C	R.H.C	R.D.S
S1	12	20	1	10	[3, 27]	[1, 4]
S2	20	20	1	10	[3, 27]	[1, 4]
S3	50	20	1	10	[1, 9]	[0, 9]
S4	98	200	50	400	[1, 44]	[0.04, 393.33]

- N. Sup: Number of suppliers
- T.C/U.D: Travel cost per unit distance
- V.C: Vehicle Capacity
- R.H.C: Rang of Holding Costs

- R.D.S: Range of demand at each supplier

In the first phase, we apply an ACO method to minimize only the routing costs. This method finishes when one of the stopping criteria is satisfied (no improvement). Once it stops, we use an efficient heuristic and the two-opt method, in order to ameliorate the existing solution. Each instance is performed to 10 runs. In Table 2, we report the best result of 10 runs for our proposed algorithm, the result of the Genetic Algorithm (GA) by [15], and the result found by the 2p-VNS method presented by [16]. In the first column, the instances are presented while in the second column the number of suppliers (n) and the number of periods for each instance are given. In columns, 3–6, the best cost of the ten runs and the optimal number of vehicles (op_v) of the solutions of the GA (columns 3 and 4), and the 2p-VNS (columns 5 and 6) are presented, respectively. In columns 7–8, the numerical results of our method (the best cost of the ten runs, the optimal number of vehicles (op_v) are presented. As it can be seen from Table 2.

Table 2. Comparaison results

Ins	(n, T)	AG		2p-VNS		Proposed algorithm	
		Cost	*op_v*	*Cost*	*op_v*	*Cost*	*op-v*
S12T5	(12,5)	2096,75	14	1961,71	14	1989,17	14
S12T10	(12,10)	4333,27	29	4002,85	29	**4000,76**	29
S12T14	(12,14)	6115,19	41	5635,77	41	6045,46	41
S20T5	(20,5)	3143,39	21	2861,26	21	**2859,89**	21
S20T10	(20,10)	6499,4	43	5944,72	43	5995,34	43
S20T14	(20,14)	9208,43	61	8422,02	63	8463,52	63
S20T21	(20,21)	13948,41	92	12700,27	94	12704,62	94
S50T5	(50,5)	5618,09	45	5084	45	5096,43	45
S50T10	(50,10)	11642	95	10694,6	96	10698,23	96
S50T14	(50,14)	16987	135	15384,96	137	**15381,79**	138
S50T21	(50,21)	26448,77	209	23497,4	211	23500,03	211
S98T5	(98,5)	561168,21	57	557287,97	57	**557285,82**	57
S98T10	(98,10)	1124797,57	113	1119767,72	114	1119781,07	114
S98T14	(98,14)	1571652,32	159	1577951,04	161	**1577950,60**	161
AVG	–	**240261,34**	**79,57**	**239371,16**	**80,43**	**239410,91**	**80,5**

It can be seen that our approach produces the new best solutions to six instances out of the 14 instances. The average improvement of the algorithm is about 239410,91.

Table 3 shows a comparison of the numerical result of each method applied on the same dataset. It includes the value of the lower bound presented in [15] (BI), the comparison based on percentage gaps, and shows the computational time (CPU) consumed for obtaining the good result. For the other instances, the quality of the solutions is between 0.4% and 28,38%.

Table 3. Comparison of the two approach and the proposed heuristic

Ins	**BI**	**GA**		**2p-VNS**		**Proposed algorithm**	
		Gap (%)	*CPU*	*Gap (%)*	*CPU*	*Gap (%)*	*CPU*
S12T5	1650	27,08	58,54	18,89	0,1	20,56	0,11
S12T10	3218	34,66	111,3	24,39	0,2	24,32	0,23
S12T14	4709	29,86	120,1	19,68	0,3	28,38	**0,28**
S20T5	2607	20,57	31,81	9,75	27,2	9,70	28,7
S20T10	5227	24,34	126,3	13,73	98,34	14,70	101,2
S20T14	7181	28,23	360,3	17,28	113,36	17,86	114,9
S20T21	10717	30,15	255,8	18,51	185,9	18,55	**182,47**
S50T5	4547	23,56	133,4	11,81	93,03	12,08	**92,6**
S50T10	9285	25,39	226,0	15,18	183,79	15,22	184,23
S50T14	13139	29,29	328,0	17,09	212,07	17,07	**210,22**
S50T21	20185	31,03	496,7	16,41	370,4	16,42	376,1
S98T5	544036	3,15	476,7	2,44	799,6	2,44	**797,1**
S98T10	–	0,45	1307,26	0,03	1128,34	0,4	1130,76
S98T14	–	0,24	1589,71	0,64	1354,8	0.4	1360,31
AVG		**22**	**401,57**	**13,27**	**326,25**	**14,12**	**327,08**

In Table 3, we report the value of the lower bound presented in [18] in column BI. We also give the execution time (CPU) required to find the optimal solution and the deviation from the best solution found (Gap) for each of the three approaches. we denote by g(S) the cost of the solution S qiven by one of the methods. The deviation with respect to the BI is calculated by the following formula:

$$\left(\frac{cost\_of\_the\_solution - BI}{BI}\right) * 100$$

For the last two instances, the lower bound is not defined in [18]. In this case, we calculated the deviation between the value of the solution found by one of the methods and the best value obtained (bestSol) by one of the three algorithms. The objective is to find the best solution found (bestSol) of the two last instances. This formula is presented as follows:

$$\left(\frac{cost\_of\_the\_solution - bestSol}{bestSol}\right) * 100$$

From this table, we can observe that the average of the gaps found vary from 0.4% to 28.38%. From these results, our approach clearly shows its reliability and stability.

5 Conclusion and Future Works

This paper has developed an efficient technic for solvig the inventory routing problem multi-product. The proposed method consists of two phases: the first phase determines an initial solution by solving a periodic vehicle routing problem with Ant colony optimization algorithm. The second stage improves this initial solution. The inventory decisions are optimized at the second phase. For this stage, we combine an efficient heuristic for inventory management to determine the quantities of products to collect from suppliers during each period. Finally, a simple 2-opt is implemented to improve the routes. As the results showed, the proposed method is able to provide better results when compared to the existing approach from the literature, both in terms of solution quality and computational times. We will adopt other approach to this problem using more dataset.

References

1. Bell, W.J., et al.: Improving the distribution of industrial gases with an on-line computerized routing and scheduling optimizer. Informs J. Appl. Analytics **13**, 2–23 (1983)
2. Bertazzi, L., Speranza, M.G.: Inventory routing problems: an introduction. Eur. J. Transp. Logist **1**, 307–326 (2012). https://doi.org/10.1007/s13676-012-0016-7
3. Bertazzi, L., Speranza, M.G.: Inventory routing problems with multiple customers. Eur. J. Transp. Logist **2**, 255–275 (2013)
4. Coelho, L.C., Cordeau, J.-F., Laporte, G.: Thirty years of inventory routing. Transp. Sci **48**, 1–19 (2014)
5. Savelsbergh, M., Song, J.H.: Inventory routing with continuous moves. Comput. Oper. Res. **34**, 1744–1763 (2007)
6. Huang, S.H., Lin, P.C.: A modified ant colony optimization algorithm for multi-item inventory routing problems with demand uncertainty. Transp. Res. Part E: Logistics Transp. Rev. **46**, 598–611 (2010)
7. Solyalı, O., Süral, H.: A branch-and-cut algorithm using a strong formu- lation and an a priori tour-based heuristic for an inventory-routing problem. Transp. Sci. **45**, 335–345 (2011)
8. Sayarshad, H.R., Gao, H.O.: A non-myopic dynamic inventory routing and pricing problem. Transp. Res. Part E **109**, 83–98 (2018)
9. Mahdi, A., Tirkolaee, E.B., Dezaki, Z.K., Hejazi, S.R.: An augmented Tabu search algorithm for the green inventory-routing problem with time windows. Swarm Evol. Comput. **60**, 100802 (2021)
10. Lerhlaly, S., Lebbar, M., Allaoui, H., Ouazar, D., Afifi, S.: An inventory location routing model with environmental considerations. Contemp. Eng. Sci. **9**, 303–314 (2016)
11. Oudouar, F., Lazaar, M., El Fallahi, A.: Self-organizing map and sweep algorithm to solve the capacitated location routing problem. In: Proceedings of 2019 IEEE World Conference on Complex Systems (2019)
12. Oudouar, F., Zaoui, E.M.: An approach based on clustering AH and VND for supply chain management. Int. J. Intell. Eng. Syst. **14**, 306–315 (2020)
13. Franco, C., López-Santana, E.R., Figueroa-García, J.C.: Solving the interval green inventory routing problem using optimization and genetic algorithms. In: Figueroa-García, J.C., López-Santana, E.R., Villa-Ramírez, J.L., Ferro-Escobar, R. (eds.) WEA 2017. CCIS, vol. 742, pp. 556–564. Springer, Cham (2017). https://doi.org/10.1007/978-3-319-66963-2_49
14. Keyvan, F., Saeid, J., Ashkan, H.: An extended robust approach for a cooperative inventory routing problem. Expert Syst. Appl. **116**, 310–327 (2019)

15. Moin, N.H., Salhi, S., Aziz, N.A.B.: An efficient hybrid genetic algorithm for the multiproduct multi-period inventory routing problem. Int. J. Prod. Econ. **133**, 334–343 (2011)
16. Mjirda, A., Jarboui, B., Macedo, R., Hanafi, S., Mladenović, N.: A two phase variable neighborhood search for the multi-product inventory routing problem. Comput. Oper. Res. **52**, 291–299 (2014)
17. Dorigo, M., Maniezzo, V., Colorni, A.: Ant system: optimization by a colony of cooperating agents. IEEE Trans. Syst. Mans, Cybern. **26**, 29–41 (1996)
18. Bulleneimer, B., Hartl, R.F., Strauss, C.: An improved ant system algorithm for the vehicle routing problem. Ann. Oper. Res. **89**, 319–328 (1999). https://doi.org/10.1023/A:1018940026670
19. Li, J., Chu, F., Chen, H.: A solution approach to the inventory routing problem in a three-level distribution system. Eur. J. Oper. Res. **210**, 736–744 (2011)

Check for updates

Overview of Lean Management Within PLM

Nada El Faydy(✉) and Laila El Abbadi

National School of Applied Sciences, Ibn Tofail University, Kenitra, Morocco
Leila.elabbadi@uit.ac.ma

Abstract. Manufacturing is the production of goods through the use of processes, machines and tools, today, many companies want to improve its performance by integrating technology and methods. This performance can be improved by reducing costs, improving quality, and limiting waste. Lean is a method that eliminates waste and improves productivity. The PLM (Product lifecycle management) tool that standardizes all the information throughout the life cycle of a product has for objective: The reduction of the time to market of a new product, the reduction of the costs of conception, excellence in quality. This document presents the major concepts of Lean and PLM and examines the PLM approach related to Lean concepts to limit waste and improve business performance.

Keyword: Lean manufacturing · PLM · Waste

1 Introduction

Companies from all over the world in order to better position themselves on the market, they are always looking for new solutions and innovations to enhance their economic performance and to build a strong brand image [1]. Lean and PLM (Product lifecycle management) help to enhance business performance [2]. Moreover, when talking about improving business performance, it concerns cost, market position, reputation and product design [3]. All of these methods and tools have the common goal of reducing the time and cost of product design and development, and improving quality through more effective collaboration between people [4].

PLM is a business activity that relies on complex networks of stakeholders located remotely, within and across organizational boundaries [2]. Lean has been adopted to describe the Toyota production system, it is an effective method to meet customer requirements in terms of quality and cost while eliminating waste throughout the system [5]. However, information technology has accelerated the implementation of new control, project, and production strategies.

The principles of lean and information technology have combined to create a new concept [6]. In fact, the main idea of Lean is to limit waste. And PLM is an essential tool to exchange wasted time, energy and materials through information [6]. It is important to know that PLM and lean projects are not in competition with each other [7]. PLM can contain waste in its phases such as: Wasted time, extra costs or overstocking [8]. That why, Lean PLM consists in applying lean principles to a PLM implementation to limit waste [9].

R. Saidi et al. (Eds.): ICATH 2021, LNDECT 110, pp. 530–536, 2022.
https://doi.org/10.1007/978-3-030-94188-8_47

This paper focus on lean PLM; the objective of this article is to analyze their complementarities and to show how PLM contributes to waste reduction and improvement of the company's performance. To discuss the issue, and since Lean also contributes to the performance and waste reduction of the company, we analyze the complementarities between Lean and PLM.

2 Lean Management

The word "Lean" was used by John Krafcik, to describe the new production techniques introduced [10]. It is well known as a set of practices aimed at reducing waste and non-value added activities in production [12, 13]. It is not limited to activities that take place in the manufacturing function of a company, but rather activities ranging from product development, procurement and manufacturing to distribution [11].

Lean Management is considered a philosophy that aims to maximize customer value while minimizing waste. Lean is a method that has been used for decades in production, related to the continuous improvement of quality and business performance [14]. It should be noted that Lean in production has become the most relevant and influential new concept in manufacturing in modern industries around the world [15]. The ultimate principle of Lean is to provide the customer with a quality product/service while ensuring no waste in the process. In addition, Lean creates processes that optimize human effort, optimize space while saving capital and time, in order to have products/services that are less expensive and contain fewer defects compared to traditional organizations [2].

Lean consists in creating a good by means of a process which rests on methods as well as the integration of the people, [16] and aims to create value to customers with less resources and waste [17].

3 PLM (Product Lifecycle Management)

Product Lifecycle Management, or PLM, is a software application developed to standardize the various information throughout the life of a product. A product goes through several stages before being put on the market: design, manufacturing, and maintenance. PLM manages product data by centralizing it in a common and secure space to guarantee the traceability of data communicated and traded with the various people who oversee the development of a product [16].

PLM is an approach that helps solve the problem of managing all the information related to product definition, as well as PLM aims to solve the problem of managing all product information by creating this information, managing it throughout its life, disseminating it and using it throughout the product life cycle [18].

The benefits of the PLM approach can be analyzed along 3 axes: [19-9].

Quality: the PLM approach improve the quality, reduces scrap and rework during production/ Improves design engineering productivity.
Delays: With better data management and consistency, PLM can reduce time-to-market. In addition, it can significantly reduce product lifecycles and lead times.
Costs: This approach reduces the costs (manufacturing, transportation and purchasing).

4 Lean Management/PLM (Product Lifecycle Management)

The objective of PLM is to have all the information on the products so as to meet its needs in the different life cycle of the product [19]. PLM provides access to the right information at the right time, with data security [16]. It relies on software tools that centralize and standardize product-related information [20].

PLM contains main tool functions like: Project management and reporting/D2/3D viewer/Data management/WORKFLOW/CAD integration [20]. PLM has a global view on the life cycle of a product, among its main missions is to reduce the development time of products, hence the interest to apply Lean methods. Lean contributes to the improvement of quality, cost and productivity, so the application of Lean throughout the life cycle of a product in the PLM tool will limit the backtracking and errors in the production process and satisfy the customer: we will therefore speak of a Lean-PLM [21].

Lean-PLM can improve company efficiency through [9]:

- Time and cost reduction;
- A more cost-effective execution of a product cycle and deadlines;
- A decrease in scrap;
- Improved productivity.

In addition, a true lean is considered as one of the most effective tools to solve and manage product life cycle issues [5]. Lean PLM is about applying lean principles to a PLM implementation to ensure that there is no waste and minimal change [9].

Organizations in order to satisfy their customers, they seek excellence, this translates into a quality product with less cost and time. To succeed in the markets, organizations must have a method that allows them to succeed, hence the interest in having a blend of Lean and PLM (Fig. 1).

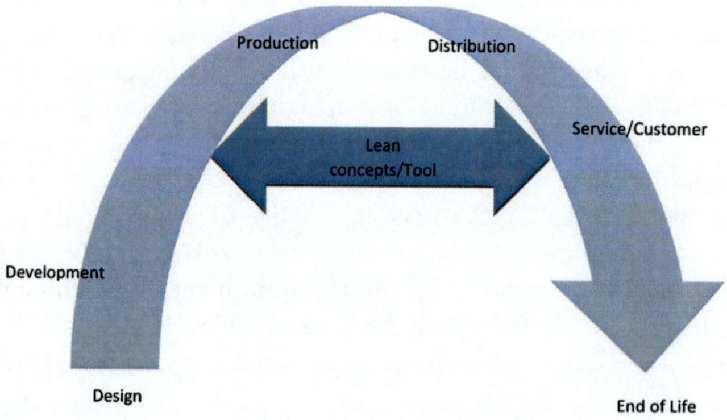

Fig. 1. Lean PLM model

Lean manufacturing is considered one of the safest and most essential tool to manage product life cycle issues [5]. PLM and Lean projects should not be opposed or perceived as being in competition with each other [7]. Lean tools are used to address PLM issues and are described in models that reflect the way these tools are applied [5].

In order to achieve sustainable manufacturing and comply with the Triple Bottom Line (TBL) requirements to have a positive impact on the economy, environment and society throughout a product's life cycle, new manufacturing processes, materials, and the use of the 6Rs (reduce, reuse, recycle, recover, redesign and remanufacture) [5]. The next generation of Lean thinking includes Product Lifecycle Management (PLM) [22]. Lean manufacturing plays a catalytic role in the development of product life cycle, especially in the areas of quality, safety, productivity, cost, human resource management in order to meet customer requirements [5].

5 Wastes in Lean

The Japanese Lean concept aims at improving productivity and business performance as well as reducing hazardous waste [23]. The word waste means an activity that does not add value in the process, the Toyota Production System (TPS) has identified seven wastes: [24].

- Overproduction: (producing more than necessary)
- Defects, rejected/defective finished products: (rejected costs)
- Waiting or idle time: (there is no value to the product).
- Transport: (sometimes causes excess transport)
- Inappropriate treatment, unnecessary processes or procedures: (there is no value to the product).
- Unnecessary movements, poor organization of workstations: (leads to poor ergonomics)
- Unnecessary inventory:(it does not add value).

The seven wastes cited are the basis of Lean [23]. Lean methods help to determine the problems, the causes of the problems as well as to analyze the identified wastes [24], so that companies can reduce their costs and improve the quality of their products [25].

6 Wastes in PLM

Reducing waste in the design phase of the product can be done through the integration of environmental aspects, this is called eco-design [26]. Waste exists throughout the life cycle of a product, the Lean works on their reduction and the PLM which is a tool that allows to avoid, eliminate the losses of time, energy and material through information [6]. In addition, PLM and its components are compatible with lean thinking since they contribute to reducing waste and increasing the revenues of organizations [6].

There is waste coming from all along the life cycle of a product, lost time, extra costs or overstocking can be considered as "waste" [8].

The PLM model includes the following phases: planning, design, build, support and disposal [8].

The Planning Phase: Is a phase that allows the quality of the product to be analyzed before it is launched on the market. Rejections that appear in this phase include excessive inventory and costs, defects [8].

The Design Phase: This phase concerns the physical realization of a product: color, form and size [8]. The design must foresee the recycling of the products in order to avoid waste [27], It is necessary to specify that in this phase the following waste is detected: an inadequate coordination generates waste, which becomes a major cause of waste [28].

The Build Phase: This phase must include all the manufacturing steps, such as the selection of the most qualified plant, equipment, materials and personnel resources. [8]. The wastes of this phase are repetitive and unnecessary steps, rework, waiting time, extra inventory and overproduction [29].

The Support Phase: This phase includes two phases: marketing, distribution and maintenance of the product [8]. The waste identified in this phase is the additional cost [30].

The Disposal Phase: This phase consists in looking for all the tracks allowing, the reuse, the retransformations or the potential recovery of the product [8].

7 Conclusion

The main objective of Lean is to meet the requirements of the customer while minimizing the cost and have a good quality and limit all the waste encountered in the production chain, to do this Lean, has methods that can extract the sources of waste. The manufacturing world has the technology to facilitate the arrival of information at the right time, among these technologies we cite the PLM tool which is very effective in standardizing information throughout the product life cycle and relies on software to have the information at the right time with a secure data. Applying Lean principles in PLM (Product Lifecycle Management) is beneficial for companies, as both converge towards common objectives: Waste reduction, quality improvement and cost reduction. However, during the preparation of this article we found that there is not enough research on Lean PLM. Future research should focus on Lean PLM, the benefits of its implementation as well as the barriers of this implementation.

References

1. Rao, P., Holt, D.: Do green supply chains lead to competitiveness and economic performance? Int. J. Oper. Prod. Manag. **25**(9), 898–916 (2005). https://doi.org/10.1108/01443570510613956
2. Navarro, R., Tiwari, A., Turner, C.: Improving product lifecycle management implementations by applying 'lean' principles. Int. J. Prod. Lifecycle Manag. **6**, 357–380 (2013). https://doi.org/10.1504/IJPLM.2013.063212

3. Wiese, A., Luke, R., Heyns, G., Pisa, N.: The integration of lean, green and best practice business principles. J. Transp. Supp. Chain Manag. **9**, 1–10 (2015). https://doi.org/10.4102/jtscm.v9i1.192
4. Vila, C., Ugarte, D., Ríos, J., Abellán, J.V.: Project-based collaborative engineering learning to develop Industry 4.0 skills within a PLM framework. Procedia Manuf. **13**, 1269–1276 (2017). https://doi.org/10.1016/j.promfg.2017.09.050
5. Maginnis, M.A., Hapuwatte, B.M., Jawahir, I.S.: Implementing total lifecycle product sustainability through true lean thinking. In: Ríos, J., Bernard, A., Bouras, A., Foufou, S. (eds.) PLM 2017. IAICT, vol. 517, pp. 544–553. Springer, Cham (2017). https://doi.org/10.1007/978-3-319-72905-3_48
6. Silva, F., Gamarra, C.J.: Product lifecycle management, digital factory and virtual commissioning: analysis of these concepts as a new tool of lean thinking (2015). https://www.semanticscholar.org/paper/Product-lifecycle-management%2C-digital-factory-and-a-Silva-Gamarra/2cf8e5c9bb037afc9d804c4e01813ea5fb9fb365. Accessed 14 Aug 2021
7. Muriel, P., Braesch, C., Pralus, M., Tabourot, L., Bonnal, P.: Lean and PLM: two complementary pillars for enterprise agility. J. Mod. Proj. Manag. **1** (2013)
8. Gecevska, V., Anisic, Z., Stojanova, T.: Lean product lifecycle management approach. Int. J. Ind. Eng. Manag. 2217-2661, **4**, 207–214 (2013)
9. Gecevska, V., Veza, I., Cus, F., Anisic, Z., Stefanic, N.: Lean PLM -information technology strategy for innovative and sustainable business environment. Int. J. Ind. Eng. Manag. **3** (2012)
10. Bendell, T.: A review and comparison of six sigma and the lean organisations. TQM Mag. **18**(3), 255–262 (2006). https://doi.org/10.1108/09544780610659989
11. Karlsson, C., Åhlström, P.: Assessing changes towards lean production. Int. J. Oper. Prod. Manag. **16**(2), 24–41 (1996). https://doi.org/10.1108/01443579610109820
12. Lasnier, G.: Le lean-manufacturing. La Revue des Sciences de Gestion **223**, 99–107 (2007). https://doi.org/10.1051/larsg:2007009
13. (Mark) Yang, M.G., Hong, P., Modi, S.B.: Impact of lean manufacturing and environmental management on business performance: an empirical study of manufacturing firms. Int. J. Prod. Econ. **129**(2), 251–261 (2011). https://doi.org/10.1016/j.ijpe.2010.10.017
14. Dickson, E.W., Singh, S., Cheung, D.S., Wyatt, C.C., Nugent, A.S.: Application of lean manufacturing techniques in the emergency department. J. Emerg. Med. **37**(2), 177–182 (2009). https://doi.org/10.1016/j.jemermed.2007.11.108
15. Forrester, P., Shimizu, U., Soriano-Meier, H., Garza-Reyes, J.A., Basso, L.: Lean production, market share and value creation in the agricultural machinery sector in Brazil. J. Manuf. Technol. Manag. **21**, 853–871 (2010). https://doi.org/10.1108/17410381011077955
16. Rossi, M., Cattaneo, L., Le Duigou, J., Fugier-Garrel, S., Terzi, S., Eynard, B.: Lean product development and the role of PLM. In: Harik, R., Rivest, L., Bernard, A., Eynard, B., Bouras, A. (eds.) PLM 2016. IAICT, vol. 492, pp. 183–192. Springer, Cham (2016). https://doi.org/10.1007/978-3-319-54660-5_17
17. Duarte, S., Cruz-Machado, V.: Modelling lean and green: a review from business models. Int. J. Lean Six Sigma **4**(3), 228–250 (2013). https://doi.org/10.1108/IJLSS-05-2013-0030
18. Terzi, S., Bouras, A., Dutta, D., Garetti, M., Kiritsis, D.: Product lifecycle management - from its history to its new role. Int. J. Prod. Lifecycle Manag. **4**, 360–389 (2010). https://doi.org/10.1504/IJPLM.2010.036489
19. Donoghue, I., Hannola, L., Papinniemi, J., Mikkola, A.: The benefits and impact of digital twins in product development phase of PLM. In: Chiabert, P., Bouras, A., Noël, F., Ríos, J. (eds.) PLM 2018. IAICT, vol. 540, pp. 432–441. Springer, Cham (2018). https://doi.org/10.1007/978-3-030-01614-2_40

20. Abramovici, M.: Future trends in product lifecycle management (PLM). In: Krause, Frank-Lothar. (ed.) The Future of Product Development, pp. 665–674. Springer, Heidelberg (2007). https://doi.org/10.1007/978-3-540-69820-3_64
21. Hines, P., Francis, M., Found, P.: Towards lean product lifecycle management: a framework for new product development. J. Manuf. Technol. Manag. **17**, 866–887 (2006). https://doi.org/10.1108/17410380610688214
22. Githens, G.: Product lifecycle management: driving the next generation of lean thinking. J. Prod. Innov. Manag. **24**(3), 278–280 (2007). https://doi.org/10.1111/j.1540-5885.2007.00250_2.x
23. Purushothaman, M.B., Seadon, J., Moore, D.: Waste reduction using lean tools in a multicultural environment. J. Clean Prod. **265**, 121681 (2020). https://doi.org/10.1016/j.jclepro.2020.121681
24. Nihlah, Z., Immawan, T.: Lean manufacturing: waste reduction using value stream mapping. In: 3rd International Conference on Energy, Environmental and Information System (icenis 2018), Cedex A, vol. 73, p. 07010 (2018). https://doi.org/10.1051/e3sconf/20187307010
25. Fercoq, A., Lamouri, S., Carbone, V.: Lean/Green integration focused on waste reduction techniques. J. Clean. Prod. **137**, 567–578 (2016). https://doi.org/10.1016/j.jclepro.2016.07.107
26. Popa, I.L., Popa, N.V.: PLM and eco-design of electronic products according with circular economy principles. AMM **657**, 1031–1035 (2014). https://doi.org/10.4028/www.scientific.net/AMM.657.1031
27. Duque Ciceri, N., Garetti, M., Terzi, S.: Product lifecycle management approach for sustainability (2009). http://dspace.lib.cranfield.ac.uk/handle/1826/3762. Accessed 15 Aug 2021
28. Llatas, C., Osmani, M.: Development and validation of a building design waste reduction model. Waste Manag. **56**, 318–336 (2016). https://doi.org/10.1016/j.wasman.2016.05.026
29. Garzaniti, N., Golkar, A., Fortin, C.: Optimization of multi-part 3D printing build strategies for lean product and process development. In: Chiabert, P., Bouras, A., Noël, F., Ríos, J. (eds.) PLM 2018. IAICT, vol. 540, pp. 488–497. Springer, Cham (2018). https://doi.org/10.1007/978-3-030-01614-2_45
30. Gustavo, J.U., et al.: Green marketing in supermarkets: Conventional and digitized marketing alternatives to reduce waste. J. Clean. Prod. **296**, 126531 (2021). https://doi.org/10.1016/j.jclepro.2021.126531

Comparison Study Between CB-SEM and PLS-SEM for Sustainable Supply Chain Innovation Model

Ahmed El Maalmi[1](✉), Kaoutar Jenoui[2], and Laila El Abbadi[1]

[1] National School of Applied Sciences, Ibn Tofail University, Kenitra, Morocco
{ahmed.elmaalmi,laila.elabbadi}@uit.ac.ma

[2] Laboratory Smartilab Sciences, Moroccan School of Engineering Sciences, Rabat, Morocco
k.jenoui@gmail.com

Abstract. Structural equation modeling (SEM) is becoming the most used method when multiple constructs, relationships, and latent and observed variables are involved. Two main statistical analysis techniques of SEM can be applied by researchers: Partial Least Squares SEM (PLS-SEM), and covariance-based (CB-SEM). This paper is presenting a comparison study between the two methods. The comparison is based on a case study of the sustainable supply chain innovation model for the Moroccan industrial field during 2020. It gives a presentation of the conceptual model and the two structural analysis techniques comparison in the literature. Then it explains the methodology. The last sections are consolidating the findings and conclusions revealed where the CB-SEM technique is performed and consistently compared with PLS-SEM.

Keywords: Industry 4.0 technologies · Sustainable supply chain · Innovation · Structural equation model

1 Introduction

SEM is usually used to explain multiple statistical relationships simultaneously through visualization and model validation [1]. It is a powerful and robust multivariate statistical analysis technique. It is enabling researchers to examine the inter-relationships among latent constructs based on theories and their respective observable indicator variables [2]. Complex models can be discussed and analyzed simply through this technique. SEM can be considered an extension of traditional linear modeling techniques, e. g. Analysis of Variance (ANOVA) and multiple regression analysis. Combining at the same time factor analyses and multiple regression analyses [1, 3–5], Scientists' discussions are involving two main approaches: CB-SEM and PLS-SEM, and the discussion are still ongoing regarding the uses and the limitations of each approach. The purpose of this paper is to compare both approaches based on various parameters and results. The two approaches are applied to explore and validate the sustainable supply chain innovation model in the context of the Moroccan industrial field during 2020. The rest of the paper is structured as follows. Section 2 relates important points of SEM history, highlights the various characteristics of SEM. Section 3 presents important commonalities between

R. Saidi et al. (Eds.): ICATH 2021, LNDECT 110, pp. 537–552, 2022.
https://doi.org/10.1007/978-3-030-94188-8_48

the two approaches CB-SEM and PLS-SEM and lists the limitations of each approach basing on the literature review. Section 4 details the methodology and the process of CB-SEM and PLS-SEM in a stepwise manner. Section 5 explains our strategy for data collection. Section 5 presents the results we get from the two methods. In Sects. 6 and 7, all the empirical results from the two methods are discussed and compared. It provides a general discussion, future directions, and limitations. Finally, Sect. 7 concludes this study.

2 Literature Review

2.1 Key Advances in the History of SEM

SEM is becoming a universal tool to study experimental and non-experimental data. Because of its flexibility and generality, this method has become hugely popular across numerous disciplines [1, 6, 7]. A digression of the historical understanding and keys advances involved in the history of SEM is summarized in Fig. 1.

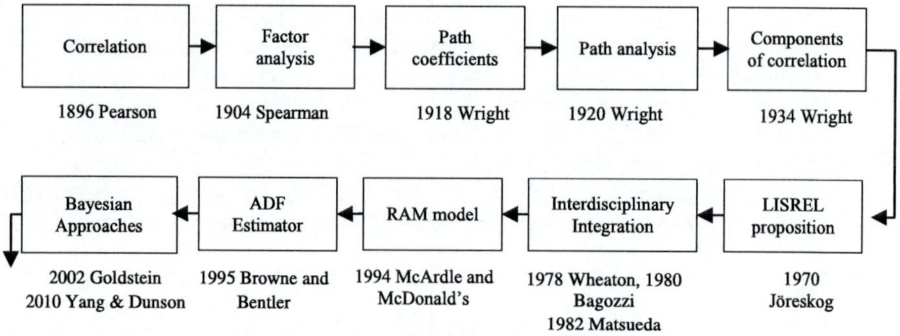

Fig. 1. Key advances in the history of SEM.

The coefficient of correlation is a measure of the relationship between two variables. In 1896, Karl Pearson proposed the new coefficient of correlation, expressed by "r" [8]. It is applied to measure the level of relation between linearly related variables. The early beginnings of SEM models were related especially to the work of Spearman and Wright [9]. Spearman [10] is often cited in the literature as the founding father of factor analysis [11]. Basing on Pearson's correlations, he [12] laid the foundations for SEM by constructing the first-factor model. This became later an important measurement part of the more general SEM analytical [13]. Spearman's work on factor analysis derived from his work on intelligence testing [14, 15]. In 1918, Sewall Wright proposed the method of path coefficients. The method was based on partial correlations. The Path analysis is originally developed by Sewall Wright in the 1920s. The method aims to examine the effects of hypothesized models in phylogenetic studies. Wright's analysis involved writing a system of equations based on the correlations among variables. Especially the ones influencing the outcome. Sewell Wright [14, 16] later developed more general solutions, noting that a mere single external variable is enough if the supply and demand situations were uncorrelated and that, in very complex models with many more shifts than external variables, one could solve for parameters by assuming plausible values for some of the unknowns. For the first time, Jöreskog presented the basis for LISREL during a scientific meeting of 1970 which Matsueda [14] highlighted in series of landmark papers. Jöreskog [16–19] outlined a general approach to covariance analysis and a computer program he called LISREL. Nearly identical models were published by Keesling [20] in his Ph.D. dissertation, and Wiley [21] in the Goldberger-Duncan volume. However, it was Jöreskog's software package that came to dominate the field. The LISREL model incorporates factor analysis, simultaneous equation models, and path analysis into a general covariance structure model [14, 22, 23]. The next two decades were known devastation for many areas of the social sciences, including stratification [14, 24, 25], social psychology [26, 27], marketing [28], mental health [29, 30], sociology of science, criminology, population genetics, and others. Reticular Action Model (RAM) approach rules (1978–1981) were initially presented based on path analysis graphics to simplify and unify structural equation models. The RAM graphic is presented through a series of definitions and axioms that provide a complete and concise isomorphism between algebra and graphics [31]. In 1974, Browne [32] introduced the quadratic form estimator he termed generalized least squares (GLS), which yielded optimal estimates for normally distributed observable variables basing on the covariance matrix of the sample moments. Subsequently, he made a landmark contribution by developing what he termed an "asymptotic distribution-free" (ADF) estimator, by incorporating information about higher-order moments into the weight matrix [14]. Browne's [32] ADF and elliptical estimators first appeared in numerous programs before being introduced to the LISREL program in 2001. Using Markov Chain Monte Carlo (MCMC) algorithm, Bayesian estimation is proving useful for incorporating prior information into confirmatory factor analysis, basically for estimating complex models [33]. It is useful for estimating multilevel factor models in the case of nonlinear latent variable models. When an informative prior is used in s semiparametric estimator, it draws inferences about under-identified parameters from the posterior distribution [14].

2.2 CB-SEM and PLS SEM

SEM, almost from its very beginning, has been divided between covariance-based SEM and composite-based. One of the first CB is PLS. Considering the ultimate need for assurance and reliability, several papers are built to discover similitudes and differences between CB-SEM and PLS-SEM. The purpose of this subsection is to lighten the recent studies involved in this comparison and list the results, conclusion, and limitations of each one. Table 1 is synthesizing the findings.

3 Methods

Figure 2 illustrates the theoretical model for sustainable supply innovation, built based on previous researches in the field such as Abuqada [42] and Surajit [43]. The model aims to enhance sustainable supply chain innovation in the Moroccan industrial context. It is based on six constructs with the mediation role of Industry 4.0 technological

Table 1. Main studies of CB-SEM and PLS-SEM comparison.

Studies and conclusions and recommendations	Limitations
Dash [1]: (1) The item loadings are usually higher in PLS-SEM than CB-SEM. Same conclusion in terms of AVE and CR which indicate better construct reliability and validity. (2) The structural relationship is closer to CB-SEM if a consistent PLS algorithm is undertaken in PLS-SEM. (3) CB-SEM is better in providing model fit indices, whereas PLS-SEM fit indices are still evolving. It is better for factor-based models, whereas composite-based models provide excellent outcomes in PLS-SEM	(1) Simple model used with a small sample size. (2) Only two software packages from each approach were taken. (3) Some factors ignored
Afthanorhan, [34]: (1) Good standardized loading can increase the reliability and validity of construct representation. (2) CB-SEM is particularly yielded valid and unbiased estimation under confirmatory conditions compared with PLS-SEM	(1) No limitation provided in the study
Soo-tai [35]: (1) LISREL and PLS provides the path coefficient, T-value, and R^2, but AMOS does not present the R^2 value. (2) AMOS has the disadvantage that it must be described as T-value instead of R^2 value. If not satisfied, another alternative must be chosen. (3) LISREL has the highest explanatory power of dependent variables than other analytical tools. (4) The path coefficients and T-values presented by the analysis results showed similar results for all three analysis tools. (4) Both advantages and disadvantages exist for each tool	(1) No special superiority found among the three analysis tools of the three representative analytical applications using SEM
Mahadzirah [36]: (1) both approaches produced similar results, but CB-SEM is more appropriate for validation and confirmation. (2) CB-SEM is more appropriate for assessing model fit and provide critical examination to obtain meaningful solution, particularly for decision making. (3) PLS-SEM is more appropriate for prediction and theory development: (for new phenomenon exploration). (4) It allows the use of small sample size	(1) The sample size just satisfying the criteria of required size. It is suitable to have enough data to avoid any bias results

(continued)

Table 1. (*continued*)

Studies and conclusions and recommendations	Limitations
Rigdon [37]: (1) Quality data lead to quality inferences, statistical method turns bad data into good data. It is best to consider the entirety of the research process. (2) The study gives 5 perspectives: a) CBSEM and Pls-SEM are two Different Estimators. b) They are based on different Models. c) The measurement errors must be considered. d) Imperfect World (sample size, few indicators). e) The Long View. (3) Five recommendations are given: a) Focus on the actual phenomenon object of the study. b) Strong focus on building data sets that can credibly lead to insight. c) Use of CB-SEM for a factor model estimation, and PLS-SEM and GSCA to estimate a model of composites. d) PLS-SEM is more recommended for data examination and evaluation of many different configurations. e) More routine exploration of alternative models for explaining the phenomenon under study	(1) No use case is given in this study
Hair [38]: (1) For a well-established theory, and the measurement effectively executed, CB-SEM often works well. (2) CB-SEM assumes normality of data distributions, which is seldom met in social sciences research. (3) PLS-SEM is non-parametric. It works well with non-normal distributions, it has very few restrictions on the use of ordinal and binary scales, when coded properly. (4) Basing on R2 output, for prediction, PLS-SEM is the preferred method because, the variance explained in the dependent variables is substantially higher. It is more appropriate at the theory development stage than CB-SEM is	No limitation provided in the study
Amaro [39]: (1) The both approaches CB-SEM and PLS-SEM produce similar results (2) PLS-SEM produced higher Reliability and Convergent Validity Measures, while CB-SEM achieved higher path coefficients	(1) The model is only a part of Seabra et al.'s model (2014))
Kaufmann [40]: (1) CBSEM is better grounded in statistical theory than PLS and should be the preferred method of choice if all assumptions are met. (2) PLS is an acceptable analytical method and a realistic alternative, if researchers carefully report their reasons for choosing PLS. (3) Recent methodological advances, such as confirmatory tetrad analysis, importance-performance matrix analysis, and the use of the PLSc estimator, equip researchers with more flexibility and allow for a more nuanced testing of theoretical concepts	(1) Only papers of SCM journals are considered in the study. (2) Focus on the drawbacks of PLS and not for CB-SEM. (3) use pf basic criteria of PLS

(*continued*)

Table 1. (*continued*)

Studies and conclusions and recommendations	Limitations
Costa [41]: (1) Sample size requirements of PLS may be smaller than CB. (2) PLS is a useful tool for differential psychology research in scenarios where sample size constrains the use of robust SEM. (3) PLS offers a viable means to explore and text complex models beyond current limits	(1) PLS has some unique aspect that can't be tested with SEM

systems and through two sub-functions which are supply chain innovation and supply chain sustainability. The six constructs are supply chain integration, supply chain orientation, competition capability, business continuity management, sustainable environment strategy, and human and social approach. Each construct is characterized by its main dimensions. An outer model of items is performed based on the inner model illustrated previously in Fig. 2. All items are measured on a five-point Likert-type scale. The database collected from the measurement model will be tested using two different approaches: CB SEM and PLS-SEM.

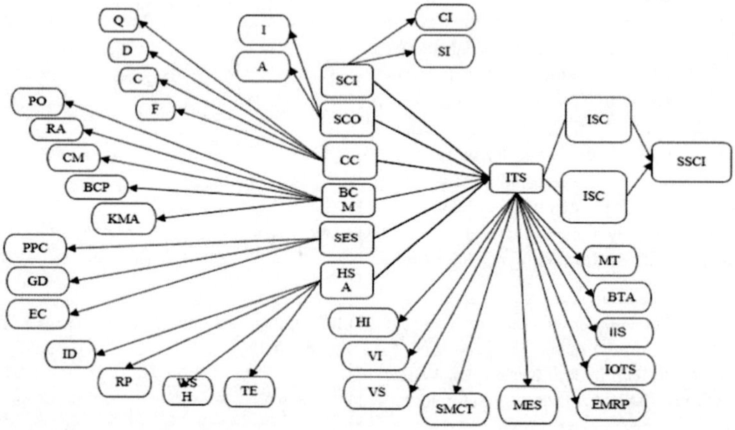

Fig. 2. Theoretical model diagram S. Where: SSCI- Sustainable Supply Chain Innovation; ISC-Innovative Supply Chain; ISC- Sustainable Supply Chain; SCI- Supply Chain Integration; SCO-Supply Chain Orientation; CC- Competition Capability; BCM- Business continuity Management; SES- Sustainable Environment Strategy; HAS- Human and Social Approach; ITS- Industry 4.0 Technological Systems; CI- Customer integration; SI- Supplier integration; A- Attitude; I-Intention; Q- Quality; D- Delivery; C- Cost; F- Flexibility; PO- Policy and objectives; RA- Risk approach; CM- Change management; BCP- Business Continuity Plan; KMA - Knowledge management and Awareness; PPC- Pollution prevention and control; GD- Green Design; EC- Environmental competencies; ID- Information disclosure; RP- Respect for policy; WSH- Worker's safety and health; TE- Trust and Ethics; VS- Virtual store; HI- Horizontal Integration; VI- Vertical Integration; SMCT- Social Media and Collaborative technologies; IIS- Internet Information service; IOTS- Internet of things and systems; BDA- Big data analytics; MT- Mobile Technologies.

3.1 Covariance Based Structural Equation Modeling

CB-SEM is one of the traditional and commun methods used by scientists nowadays. It is increasingly becoming one of the first methods of choice for concept and theory development in supply chain and industry 4.0 research areas. The model framework can be summarized into three matrix equations, two for the measurement model component and one for the path model component [40, 43]. For the measurement model component:

$$x = \Lambda_x \xi + \delta \text{ and } y = \Lambda_y \eta + \varepsilon \quad (1 \text{ and } 2)$$

Where x is the vector of observed exogenous variables, and it is a linear function of exogenous latent variables ξ vector and measurement error δ vector. Λ_x is the matrix of factor loadings relating x to ξ. Similarly, are defined y, ε, Λ_x and η. The two matrices Θ_δ and Θ_ε are defined as variance-covariance matrices. The first among measurement errors δ and the second among measurement errors, ε. The path model component can be written as:

$$\eta = B\eta + \Gamma\xi + \zeta \quad (3)$$

where B is the path coefficients matrix for endogenous latent variables, Γ is the path coefficients matrix for exogenous variables on endogenous variables, and ζ is the errors vector of endogenous variables. Associated with this equation are two variance-covariance matrices: Φ the variance-covariance matrix of latent exogenous variables, and Ψ is the matrix of covariance among errors of endogenous variables.

LISREL, the legendary initial SEM computer program, evolved from a rekindled interest in path analysis during the 1960s for making causal interpretations of nonexperimental data. During the early 1970s, the LISREL model was generalized to accommodate confirmatory factor analysis, multiple-group analysis, and more general models for the analysis of covariance structures. It makes it possible to combine the structural equation and factor analysis, and it can also generate path diagrams for structural equations.

3.2 Part Least Squares Structural Equation Modeling

PLS-SEM, which is one of variance-based SEM methods, is also a useful and increasingly applied approach to examine structural equation models. The formal definition of PLS path models is based on two sets of linear equations: the structural or inner model and the measurement or outer model. The structural model gives the relationships between latent variables, whereas the outer model provides information about the relationships between a latent variable and its observed or manifest variables (indicators) [42, 43]. The structural model relationships can be written as:

$$\xi = B\xi + \varsigma \text{ or simply } (\xi|\xi) = B\xi \quad (4)$$

Where ξ is the latent variables vector, B represents the coefficients matrix of their relationships, and ξ gives the inner model residuals. In the case of a reflexive model, each manifest variable in a certain measurement model is assumed to be generated as a linear function of its latent variables and the residual ε:

$$X_x = \Lambda_x\xi + \varepsilon x \text{ or simply } (X_x|\xi) = \Lambda_x\xi \quad (5)$$

where Λ refers to the loading coefficients matrix. Given that the is no correlation between the outer residuals and the latent variable of the same block. In the formative mode, there is causal relationships from the manifest variables to the latent variable. The linear relationships for those blocks are given as expressed hereafter:

$$\xi = \Pi_x X_x + \varepsilon x \text{ or simply } (\xi | X_x) = \Pi_x X_x \tag{6}$$

The application of PLS is based on five main iterations: initialization of all manifesting variables to a mean equal to zero and a variance equal to 1. Inner approximation by estimating each latent variable LV as a weighted sum of its neighboring LVs. So, the inner estimation is obtained. Outer approximation, the weights are reevaluated based on the weighting scheme. The factor scores calculation and so the outer weights vectors are arranged in an outer weights' matrix. Iteration of the process basing on the previous output. SmartPLS software is a free and user-friendly modeling package for PLS, supported by a community of scholars centered at the University of Hamburg (Germany), School of Business, under the leadership of Prof. Christian M. Ringle. SmartPLS is a successor to the PLS-Path software, which was used in the 1990s [44]. IT is a graphical user interface for variance-based SEM using the PLS path modeling method. The most relevant steps for using Smart PLS and any other software for the PLS-SEM approach are given in Fig. 3.

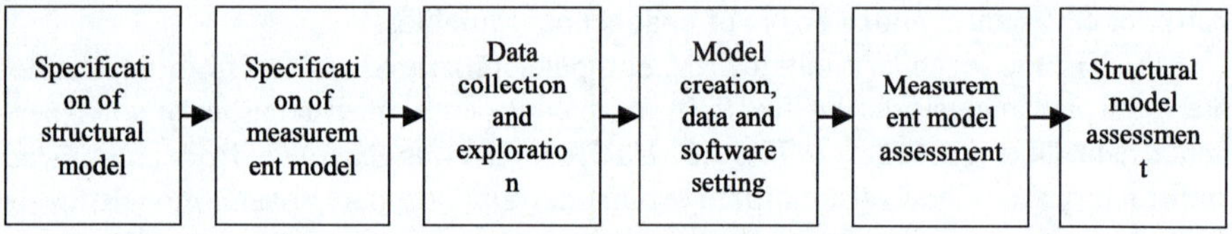

Fig. 3. Flowchart of the most relevant steps of a PLS-SEM analysis and review basing on Russo model [45–47].

4 Data Collection

The database used for the comparison study and model validation concerns the Moroccan manufacturing companies including multinational, Moroccan listed, and medium enterprises during 2020. Mainly are explored the sectors of semiconductors, automotive, aeronautic, and mining. The study is performed based on experts' and managers' feedback regarding internal digitalization strategy, innovation strategy, and sustainability in the supply chain. The questionnaire used in online format includes three sections, the first part is the demographic section, which contains information on gender, academic level, role inside the company, and years of experience, the second section contains information related to firms like the number of employees, activity, annual revenues. The last includes questions and items related to the outer model: each factor is measured based on a Likert scale where the respondent must select an answer according to his agreement: from 1/5 strongly disagree to 5/5 which means strongly agree. Around 300 online questionnaires were emailed, and more than 200 emails were sent on LinkedIn to managerial personnel working in the sampled companies. A total of 239 responses were received, of which 204 were considered accurate for data analysis.

5 Results

5.1 Standardized Loading

The theoretical model consisted of six constructs and one mediator. The result for measurement model assessment between the two SEM methods is reported in Tables 2 and 3. Both CB-SEM and PLS-SEM met their respective indicator loading minimum requirements as illustrated in Table 2. In the case of PLS-SEM, the minimum acceptable requirement for indicator loading in the model is 0.70 and above. In contrast, CB-SEM minimum acceptable indicator loading is 0.50 and above. Results in Table 2 suggest that overall the indicator loadings for PLS-SEM are higher than CB-SEM as PLS-SEM does not give factor loadings, but composite loadings. The findings in Table 2 suggest that all constructs achieved factor loading of more than 0.80 exceeding the recommended level of acceptable factor loading (0.70 for both CB-SEM and PLS-SEM).

Table 2. Indicator loadings obtained using smart PLS and LISREL

	PLS-SEM	CB-SEM		PLS-SEM	CB-SEM		PLS-SEM	CB-SEM
A1	0,975	0,936	F2	0,986	0,946	Q1	0,992	0,952
A2	0,974	0,935	GD1	0,975	0,936	Q2	0,992	0,952
BCM1	0,993	0,953	GD2	0,973	0,935	RA1	0,964	0,925
BCM2	0,992	0,953	HI1	0,956	0,917	RA2	0,931	0,894
BCP1	0,927	0,890	HI2	0,728	0,698	RP1	0,988	0,948
BCP2	0,910	0,873	HSA1	0,974	0,935	RP2	0,987	0,947
BDA1	0,988	0,949	HSA2	0,973	0,934	SCI1	0,992	0,952
BDA2	0,988	0,949	I1	0,967	0,928	SCI2	0,992	0,952
C1	0,992	0,952	I2	0,967	0,928	SCO1	0,981	0,941
C2	0,992	0,952	ID1	0,906	0,869	SCO2	0,978	0,939
CC1	0,978	0,939	ID2	0,945	0,907	SES1	0,989	0,950
CC2	0,981	0,941	IOTS1	0,971	0,932	SES2	0,989	0,949
CCS1	0,977	0,938	IOTS2	0,965	0,926	SI1	0,979	0,939
CCS2	0,972	0,933	ISC1	0,980	0,921	SI2	0,981	0,941
CI1	0,980	0,941	ISC2	0,982	0,923	SMCT1	0,993	0,954
CI2	0,980	0,941	ITS1	0,982	0,923	SMCT2	0,993	0,953

(continued)

Table 2. (*continued*)

	PLS-SEM	CB-SEM		PLS-SEM	CB-SEM		PLS-SEM	CB-SEM
CM1	0,808	0,784	ITS2	0,982	0,923	SSC1	0,971	0,932
CM2	0,915	0,887	KM1	0,976	0,917	SSC2	0,971	0,932
D1	0,981	0,951	KM2	0,978	0,919	SSCI1	0,989	0,950
D2	0,981	0,951	MES1	0,954	0,897	SSCI2	0,989	0,949
DS1	0,990	0,960	MES2	0,961	0,904	TE1	0,961	0,923
DS2	0,990	0,960	MT1	0,991	0,932	TE2	0,967	0,928
EC1	0,959	0,931	MT2	0,990	0,931	VI1	0,957	0,918
EC2	0,963	0,934	P1	0,955	0,897	VI2	0,966	0,927
EMRP1	0,981	0,952	P2	0,950	0,893	VS1	0,998	0,958
EMRP2	0,984	0,945	PPC1	0,930	0,874	VS2	0,998	0,958
F1	0,985	0,946	PPC2	0,943	0,886	WSH1	0,928	0,891
						WSH2	0,958	0,920

5.2 Reliability and Convergent Validity Results

The internal reliability assessment is based on CR (Table 3). The results from Table 3 show that both PLS-SEM and CB-SEM achieved the value of composite reliability threshold except for one dimension of human and social approach construct, which is information disclosure. The convergent validity is assessed using the AVE criterion. As illustrated in Table 3, both PLS-SEM and CB-SEM results satisfied the convergent validity requirement when all values for each construct is greater than 0.50, indicating that the construct explains more than half of the variance of its indicator variables. The AVE values of PLS-SEM are higher than the CB-SEM values. It means that the indicator loadings from PLS-SEM estimates are much higher than the estimates in CB-SEM. In their studies, Rönkkö [48] and Mahadzirah [2] assume that the factor loading produced by PLS-SEM is always biased upwards or overestimated due to the presence of measurement error which could affect the parameter estimates and eventually lead to the committing of type I error.

Discriminant validity criterion was evaluated based on the Fornell & Larcker criterion application for both CB-SEM and PLS-SEM. The results indicate that both methods satisfied discriminant validity requirements. The square root of each construct's AVE is higher than its correlations with any other constructs. Moreover, it also revealed that the values of construct correlations for PLS-SEM is lower than the CB-SEM. This study supported the work of who suggested that the construct correlations from PLS-SEM are always biased downwards or underestimated due to the capitalization on chance correlation existing.

Table 3. Reliability and validity results

	PLS-SEM using SmartPLS				CB-SEM using LISREL software			
	Cronbach's alpha	rho_A	Composite reliability	AVE	Cronbach's alpha	rho_A	Composite reliability	AVE
SSCI	0,995	0,995	0,997	0,995	0,955	0,955	0,957	0,955
SCI	0,991	0,991	0,996	0,991	0,951	0,951	0,956	0,951
SCO	0,941	0,948	0,971	0,944	0,903	0,910	0,932	0,906
CC	0,920	0,934	0,961	0,925	0,883	0,897	0,923	0,888
BCM	0,993	0,994	0,997	0,994	0,953	0,954	0,957	0,954
HAS	0,971	0,971	0,986	0,971	0,932	0,932	0,947	0,932
SES	0,988	0,988	0,994	0,988	0,948	0,948	0,954	0,948
ITS	0,980	0,980	0,990	0,981	0,941	0,941	0,950	0,942
ISC	0,964	0,964	0,982	0,965	0,925	0,925	0,943	0,926
SSC	0,964	0,964	0,982	0,965	0,925	0,925	0,943	0,926
BCP	0,988	0,988	0,994	0,988	0,948	0,948	0,954	0,948
BDA	0,986	0,986	0,993	0,986	0,947	0,947	0,953	0,947
C	0,991	0,991	0,996	0,991	0,951	0,951	0,956	0,951
A	0,972	0,972	0,986	0,973	0,933	0,933	0,947	0,934
CCS	0,763	0,893	0,889	0,800	0,732	0,857	0,853	0,768
CI	0,980	0,980	0,990	0,981	0,941	0,941	0,950	0,942
CM	0,992	0,992	0,996	0,992	0,952	0,952	0,956	0,952
D	0,980	0,980	0,990	0,980	0,941	0,941	0,950	0,941
DS	0,954	0,957	0,977	0,956	0,916	0,919	0,938	0,918
EC	0,892	0,892	0,949	0,902	0,856	0,856	0,911	0,866
EMRP	0,856	0,860	0,933	0,874	0,822	0,826	0,896	0,839
F	0,972	0,972	0,986	0,973	0,933	0,933	0,947	0,934
GD	0,941	0,945	0,971	0,945	0,903	0,907	0,932	0,907
HI	0,903	0,912	0,953	0,911	0,867	0,876	0,915	0,875
I	0,964	0,964	0,982	0,965	0,925	0,925	0,943	0,926
ID	0,352	0,899	0,689	0,563	0,338	0,863	0,661	0,540
IOTS	0,985	0,985	0,992	0,985	0,946	0,946	0,952	0,946
KM	0,989	0,989	0,995	0,989	0,949	0,949	0,955	0,949
MES	0,989	0,989	0,995	0,989	0,949	0,949	0,955	0,949
MT	0,946	0,950	0,974	0,948	0,908	0,912	0,935	0,910
PO	0,991	0,991	0,996	0,991	0,951	0,951	0,956	0,951

(*continued*)

Table 3. *(continued)*

	PLS-SEM using SmartPLS				CB-SEM using LISREL software			
	Cronbach's alpha	rho_A	Composite reliability	AVE	Cronbach's alpha	rho_A	Composite reliability	AVE
PPC	0,881	0,894	0,943	0,893	0,846	0,858	0,905	0,857
Q	0,991	0,991	0,996	0,991	0,955	0,955	0,957	0,955
RA	0,990	0,990	0,995	0,990	0,951	0,951	0,956	0,951
RP	0,988	0,988	0,994	0,988	0,903	0,910	0,932	0,906
SI	0,978	0,979	0,989	0,978	0,883	0,897	0,923	0,888
SMCT	0,992	0,992	0,996	0,992	0,953	0,954	0,957	0,954
TE	0,800	0,800	0,909	0,833	0,932	0,932	0,947	0,932
VI	0,986	0,986	0,993	0,986	0,948	0,948	0,954	0,948
VS	0,919	0,937	0,961	0,924	0,941	0,941	0,950	0,942
WSH	0,926	0,935	0,964	0,930	0,925	0,925	0,943	0,926

6 Analysis and Discussion

Our findings summarize the difference and similarities between PLS-SEM and CB-SEM which could give some clear guidelines for applied researches. Although both approaches produced similar results for our business measurement model, the CB-SEM is more appropriate for validating and confirming the measurement model. This confirm the conclusions of Dash [1], Mahadzirah [2], Hair [3, 4, 50], Afthanorhan [34], Mahadzirah [36], Rigdon [37] and Kaufmann [40]. Keeping in mind that neither of the two SEM approaches is generally superior to the other. On the other hand, PLS-SEM is more appropriate for prediction and theory development as confirmed in the studies of Mahadzirah [36], Rigdon [37] and Hair [38]. Nevertheless, Rönkkö [48], stressed strongly that PLS-SEM is not appropriated for early-stage theory development because, it is unable to identify model specification weaknesses. He added that construct scores and path estimates calculated from an incorrect model are likely to be severely biased. Per design PLS-SEM allows the use of a small sample size but, much care and caution to consider. Otherwise, it could reduce the power of the test and increase the margin of error, which renders the study to be meaningless. This power serves to detect the probability of Type II errors. The best solution is to collect enough data proportionate to the target population. As defined by hair [49] and confirmed by the study of Mahadzirah [36], the generally accepted rule-of-thumb for determining sample size is the sample size should be 5 to 10 times the number of measuring items. The measurement model of the present study consists of 6 constructs, 2 subfunctions, one mediator, and a total of 30 dimensions. The outer model has a total of 60 measurement items. Thus, the appropriate number of cases should be between 300 (5×60) and 600 (10×60). The sample size obtained for this study was 204; this number can be assimilated to the minimum size requirement for employing parametric statistical analysis of which CB-SEM is part. According to Mahadzirah [36], PLS-SEM performs well with non-normal data and employed the non-parametric statistical approach. Nonetheless, if the non-normal data resulted from a biased response where the subjects are responding under fear or pressure, then the data was invalid in the first place. The other possibilities for non-normal data distribution are the use of the wrong target population, the wrong method of sampling, the wrong method of data collection, ambiguous items, biased items, and non-cooperating respondents. The rubbish data due to the wrong methodology cannot be cured by using PLS-SEM. Thus, researchers should observe the methodological discipline properly at every stage of the research process. In our study, we assessed the normality basing on the measure of skewness and kurtosis for every item comparing with the requirements defined by Hair [50]. The findings showed that the required level of −2.58 and 2.58 [51] which confirm that the normality is achieved, and the data are normally distributed. In the case where parametric statistical assumptions are satisfied, the researchers should employ CB-SEM instead of PLS-SEM since the parametric approach is much superior to the non-parametric approach.

7 Conclusion

This study is performed first, to validate for the first time a business model for sustainable supply chain innovation in the Moroccan industrial context. It aims also to perform a comparison study between CB-SEM and PLS-SEM basing on this model. The last versions of software are used in this side to cover limitations of previous comparison studies. Our proposed model is integrating and covering maximum relevant constructs and measurement items and we used a study database that is collected wisely to avoid non-valid data and errors. The results obtained in this study demonstrate that both approaches produce similar results. PLS-SEM leads to higher validation and confirmation indicators values. In fact, PLS-SEM produced higher loadings, reliability, and convergent validity measures, while CB-SEM achieved higher path coefficients. This finding is confirming the conclusion of the studies' selection of the literature review which had pointed out the fact that both approaches achieved comparable results, especially when the models have good measurement proprieties. The wise method for any SEM study is to start the exploration with the PLS-SEM and validate the exploration result with the CB-SEM approach, of course with respect to the minimum sample size. Otherwise, the CB-SEM is more appropriate for validating and confirming the measurement model. The model of sustainable supply chain innovation is validated and confirmed with the two approaches, only one dimension was non-valid in terms of Cronbach's Alpha, CR, and AVE. the model is studied and validated for the first time in the context of the Moroccan industrial field. This can present a valid roadmap with relevant guidelines for the industrial actors and policies makers for enhancing sustainability and innovation in the supply chain in the age of Industry 4.0. For more precisions and details, this model can be extended or simply explored in sections in the future studies.

References

1. Dash, G., Paul, J.: CB-SEM vs PLS-SEM methods for research in social sciences and technology forecasting. Technol. Forecast. Soc. Change **173**, 121092 (2021)

2. Mohamad, M., Afthanorhan, A., Awang, Z., Mohammad, M.: Comparison between CB-SEM and PLS-SEM: testing and confirming the MSQLMM. J. Soc. Sci. Res. **5**(3), 608–614 (2019)
3. Sarstedt, M., Ringle, C., Hair, J.: Partial least squares SEM (2017). https://doi.org/10.1007/978-3-319-05542-8_15-1
4. Hair, J.F., Jr., Matthews, L.M., Matthews, R.L., Sarstedt, M.: PLS-SEM or CB-SEM: updated guidelines on which method to use. Int. J. Multivar. Data Anal. **1**(2), 107–123 (2017)
5. Haenlein, M., Kaplan, A.M.: A beginner's guide to partial least squares analysis, understanding statistics. Stat. Issues Psychol. Soc. Sci. **3**(4), 283–297 (2006)
6. Hancock, G.R., Stapleton, L.M., Mueller, R.O. (eds.): The Reviewer's Guide to Quantitative Methods in the Social Sciences, 2nd edn. Routledge, Milton Park (2018). https://doi.org/10.4324/9781315755649
7. Schinka, J.A., Velicer, W.F., Weiner, I.B.: Handbook of Psychology: Research Methods in Psychology, vol. 2. Wiley, New York (2013)
8. Magnello, M.E.: Karl Pearson and the establishment of mathematical statistics. Int. Stat. Rev./Revue Internationale de Statistique **77**(1), 3–29 (2009)
9. Tarka, P.: An overview of structural equation modeling: its beginnings, historical development, usefulness and controversies in the social sciences. Qual. Quant. **52**(1), 313–354 (2017). https://doi.org/10.1007/s11135-017-0469-8
10. Spearman, C.: 'General intelligence,' objectively determined and measured. Am. J. Psychol. **15**(2), 201–292 (1904)
11. Bartholomew, D.J.: Spearman and the origin and development of factor analysis. Br. J. Math. Stat. Psychol. **48**(2), 211–220 (1995)
12. Spearman, C.: The proof and measurement of association between two things. Am. J. Psychol. **15**, 72–101 (1904)
13. Cudeck, R., Jöreskog, K.G., Sörbom, D., Du Toit, S.: Structural Equation Modeling: Present and Future: A Festschrift in Honor of Karl Jöreskog. Scientific Software International (2001)
14. Matsueda, R.L.: Key advances in the history of structural equation modeling. In: Hoyle, R.H. (ed.) Handbook of Structural Equation Modeling, pp. 17–42. The Guilford Press, New York (2012)
15. Wright, S.: On the nature of size factors. Genetics **3**, 367–374 (1918)
16. Wright, S.: The method of path coefficients. Ann. Math. Stat. **5**(3), 161–215 (1934)
17. Jöreskog, K.G.: A general method for analysis of covariance structures. Biometrika **57**(2), 239–251 (1970)
18. Jöreskog, K.G.: A general method for estimating a linear structural equation system. ETS Res. Bull. Ser. **1970**(2), 1–41 (1970)
19. Jöreskog, K.G.: Structural analysis of covariance and correlation matrices. Psychometrika **43**(4), 443–477 (1978)
20. Keesling Contreras, A.: Quantum simulation and quantum information processing with programmable rydberg atom arrays. Doctoral dissertation, Harvard University Graduate School of Arts and Sciences (2021)
21. Wiley, D.E.: The identification problem for structural equation models with unmeasured variables. In: Goldberger, A., Duncan, O.D. (eds.) Structural Equation Models in the Social Sciences, pp. 69–84. Academic Press, New York (1973)
22. Jöreskog, K.G., Sörbom, D.: LISREL 8 User's Reference Guide. Scientific Software International, Chicago (2001)
23. Jöreskog, K.G., Sörbom, D.: PRELIS 2 User's Reference Guide. Scientific Software International, Chicago (2002)
24. Bielby, W.T., Hauser, R.M.: Structural equation models. Ann. Rev. Sociol. **3**, 137–161 (1977)
25. Bielby, W.T., Hauser, R.M., Featherman, D.L.: Response errors of black and nonblack males in models of the intergenerational transmission of socioeconomic status. Am. J. Sociol. **82**(6), 1242–1288 (1977)

26. Kohn, M., Schooler, C.: Job conditions and personality: a longitudinal assessment of their reciprocal effects. Am. J. Sociol. **87**, 1257–1286 (1982)
27. Bentler, P.M., Weeks, D.G.: Interrelations among models for the analysis of moment structures. Multivar. Behav. Res. **14**(2), 169–186 (1979)
28. Bagozzi, R.P.: Causal Models in Marketing. Wiley, New York (1980)
29. Wheaton, B.: The sociogenesis of psychological disorder. Am. Sociol. Rev. **43**, 383–403 (1978)
30. Wheaton, B.: Models for the stress-buffering functions of coping resources. J. Health Soc. Behav. **26**, 352–364 (1985)
31. McArdle, J.J.: The development of the RAM rules for latent variable structural equation modeling (2005)
32. Browne, M.W.: Generalized least squares estimators in the analysis of covariance structures. South African Stat. J. **8**, 1–24 (1974). In: Aigner, D.J., Goldberger, A.S. (eds.) Latent Variables in Socioeconomic Models, pp. 205–226. Amsterdam: North Holland (1977, reprinted)
33. Lee, S.-Y.: A Bayesian approach to confirmatory factor analysis. Psychometrika **46**(2), 153–160 (1981)
34. Afthanorhan, A., Awang, Z., Aimran, N.: An extensive comparison of CB-SEM and PLS-SEM for reliability and validity. Int. J. Data Netw. Sci. **4**(4), 357–364 (2020)
35. Nam, S., Kim, D., Jin, C.: Comparative analysis between structural equation tools using the same data. J. Korean Inf. Commun. Soc. **22**(7), 978–984 (2018). https://doi.org/10.6109/JKIICE.2018.22.7.978
36. Mohamad, M., Afthanorhan, A., Awang, Z., Mohammad, M.: Comparison between CB-SEM and PLS-SEM: testing and confirming the Maqasid Syariah quality of life measurement model. J. Soc. Sci. Res. **5**(3), 608–614 (2019)
37. Rigdon, E., Sarstedt, M., Ringle, C.: On comparing results from CB-SEM and PLS-SEM: five perspectives and five recommendations. Mark. ZFP 39, 4–16 (2017). https://doi.org/10.15358/0344-1369-2017-3-4
38. Hair, J.F., Jr., Matthews, L.M., Matthews, R.L., Sarstedt, M.: PLS-SEM or CB-SEM: updated guidelines on which method to use. Int. J. Multivar. Data Anal. **1**, 107 (2017). https://doi.org/10.1504/IJMDA.2017.087624
39. Amaro, S., Abrantes, J.L., Seabra, C.: Comparing CB-SEM and PLS-SEM results: an empirical example (2015)
40. Kaufmann, L., Gaeckler, J.: A structured review of partial least squares in supply chain management research. J. Purch. Supply Manag. **21**(4), 259–272 (2015)
41. Willaby, H.W., et al.: Testing complex models with small sample sizes: a historical overview and empirical demonstration of what partial least squares (PLS) can offer differential psychology. Pers. Individ. Differ. **84**, 73–78 (2015)
42. Abudaqa, A., Hilmi, M., Dahalan, N., AlMujaini, H.: Impact of supply chain integration and intelligent information systems in achieving supply chain innovation: a study of retail trading SMEs in Abu Dhabi. UAE J. Uncertain Supply Chain Manag. **8**(4), 3–8, 721–728.B (2020)
43. Surajit, B, Telukdarie, A.: Business logistics optimization using industry 4.0: current status and opportunities. In: 2018 IEEE International Conference on Industrial Engineering and Engineering Management (IEEM), pp. 3–5, 1558–1562 (2018)
44. Grace, J.B.: Structural Equation Modeling and Natural Systems. Cambridge University Press, Cambridge (2006)
45. Khamis, A., Kamarudin, N.K.K., Nor, M.E., Saharan, S., Asrah, N.M.: Covariance based and partial least square structural equation modeling to model job satisfaction among lecturers. Sci. Res. J. **5**(3), 19–28 (2017)
46. Sarstedt, M., Ringle, C.M., Hair, J.F.: Partial least squares structural equation modeling. Handb. Mark. Res. **26**(1), 1–40 (2017)

47. Russo, D., Stol, K.-J.: PLS-SEM for software engineering research: an introduction and survey. ACM Comput. Surv. **54**(4), 38 (2022). Article 78. https://doi.org/10.1145/3447580
48. Rönkkö, M., Evermann, J.: A critical examination of common beliefs about partial least squares path modeling. Organ. Res. Methods **16**(3), 425–448 (2013). https://doi.org/10.1177/1094428112474693
49. Hair, J.F., Ringle, C.M., Sarstedt, M.: PLS-SEM: indeed, a silver bullet. J. Mark. Theory Practic. **19**(2), 139–152 (2011)
50. Hair, J., Sarstedt, M., Hopkins, L., Kuppelwieser, V.: Partial least squares structural equation modeling (PLS-SEM): an emerging tool in business research. Eur. Bus. Rev. **26**(2), 106–121 (2014). https://doi.org/10.1108/EBR-10-2013-0128
51. Hair, J.F., Jr., Hult, G.T.M., Ringle, C.M., Sarstedt, M.: A Primer on Partial Least Squares Structural Equation Modeling PLS-SEM. Sage, Thousand Oaks (2014)

Correction to: Modelling of Cavitation in Transient Flow in Pipe-Water Hammer

Ouafae Rkibi, Nawal Achak, Bennasser Bahrar, and Kamal Gueraoui

Correction to: Chapter "Modelling of Cavitation in Transient Flow in Pipe-Water Hammer" in: R. Saidi et al. (Eds.): *Advanced Technologies for Humanity*, LNDECT 110, https://doi.org/10.1007/978-3-030-94188-8_6

In the original version of the book, the following correction has been incorporated: In Chapter 6, the second author's name has been changed from "Nawal Aachak" to "Nawal Achak". The book and the chapters have been updated with the change.

The updated version of this chapter can be found at
https://doi.org/10.1007/978-3-030-94188-8_6

R. Saidi et al. (Eds.): ICATH 2021, LNDECT 110, p. C1, 2022.
https://doi.org/10.1007/978-3-030-94188-8_49

Author Index

A
Abenna, Said, 96
Abraim, Mounir, 275
Achak, Nawal, 54
Adar, Mustapha, 286
Admi, Youssef, 453
Agwa, Ashraf A-K Mostafa, 428
Akamri, Assia, 173
Ali, Lahcen Oukhoya, 241
Alioui, Abdelmounaim, 286
Allak, Anass, 253
Allali, Karam, 507
Amallah, Mohamed, 253
Amine, Aouatif, 149
Anoun, Houda, 228
Aoulalay, Ayoub, 241
Aqil, Said, 507
Assawab, R., 141
Assawab, Rachida, 241

B
Babay, Mohamed-Amin, 286
Bahrar, Bennasser, 54
Bakdid, Amar, 311
Belabid, Jabrane, 507
Benaya, Nabil, 141, 241
Benelallam, Imade, 253
Benhamou, Jaouad, 106
Bennani, Mohammed, 218
Benslimane, Rachid, 352
Benyacoub, B., 141
Berada, Mohamed, 149
Blanc, Philippe, 275
Boudi, El Mostapha, 495
Bourkhime, Yassine, 300
Bouyghf, Hamid, 96

C
Chafik, Hassan, 149
Chahboun, Noha, 84
Chakir, H. E., 42
Channouf, Salaheddine, 106
Chehaidia, Seif Eddine, 300
Cherradi, Toufik, 441
Chetouani, Elmostafa, 300
Chojaa, Hamid, 300

D
Dahmani, Jamila, 61
Deifalla, A., 474
Derouich, Aziz, 300
Dya, Zitouni, 129

E
Ebobissé Djéné, Yves Frédéric, 182
El Abbadi, Laila, 530, 537
El Alani, Omaima, 275
El Allati, Abderrahim, 141, 241
El Arbaoui, Fatima Ez Zahra, 29
El Bhiri, Brahim, 75, 182
El Faydy, Nada, 530
El Hari, Kaoutar, 29
El Hassouni, Mohammed, 265
EL Idrissi, Mohammed Sbai, 182
El Jbari, Mohamed, 84
El Maalmi, Ahmed, 537
El Massari, Hakim, 193
El Mhouti, Abderrahim, 241

R. Saidi et al. (Eds.): ICATH 2021, LNDECT 110, pp. 553–555, 2022.
https://doi.org/10.1007/978-3-030-94188-8

Elkarkouri, Said, 441
Elkihel, Ali, 311
Elkihel, Yousra, 311
Elmaimouni, L., 320
Elzaar, Abdellah, 141, 241
Ennasar, Mohammed Ali, 75
Erraji, Abdelhak, 206
Essa, Nora Magdy, 404

F
Faieq, Soufiane, 161
Fakhri, Youssef, 182
Falimiaramanana, D. J., 320
Fascetti, Alessandro, 482
Feo, Luciano, 387, 482

G
Gao, Qinglin, 464
Ghennioui, Abdellatif, 275
Ghennioui, Hicham, 275
Gherabi, Noreddine, 193
Gueraoui, Kamal, 54
Gziri, Hassan, 311

H
Habbat, Nassera, 228
Habbouza, Hamdi, 253
Hadmi, Azhar, 10
Hajami, Abdelmajid, 129
Hassouni, Larbi, 228
He, Shaohua, 374

I
Ibrahim, Hassan Mohamed, 404

J
Jami, Mohammed, 106
Jenoui, Kaoutar, 537

K
Kabbaj, Adil, 218
Karfa Bekali, Younes, 75, 182
Khalfi, H., 320
Khalladi, Mohsine, 75
Khamjane, Aziz, 352
Khaouch, Zakaria, 286

L
Laghfiri, Hinde, 346, 359
Lahbas, Aimad, 10
Lahssini, Said, 149
Lamdouar, Nouzha, 346, 359
Legdou, Anass, 149
Liu, Jielin, 333

M
Mabrouki, Mustapha, 286
Magdy, Mahmoud, 19
Mahdi, Ibrahim Mahmoud, 404
Maizate, Abderrahim, 206
Martinelli, Enzo, 387
Massar, Mohammed, 241
Meghni, Billel, 300
Mehanna, Amal S., 19
Mehdi, Imane, 495
Mehdi, Mohammed Amine, 441
Mezrhab, Ahmed, 453
Mhammedi, Sajida, 193
Mosallam, Ayman, 333, 374, 464
Mostafa, Abdallah, 19
Mounir, Fouad, 61
Moussaoui, Mohamed, 84
Moussaoui, Mohammed Amine, 453

N
Naciri, I., 320
Nahid, Mohammed, 96
Naimi, Zakaria, 275
Najih, Yousef, 286
Nasri, Mohammed, 193
Norelyaqine, Abderrahim, 116

O
Ouadi, H., 42
Ouazene, Zouhair, 352
Oudouar, Fadoua, 519
Oularbi, Younes, 61
Oumoussa, Idris, 161
Ozzif, Mohamed, 206

P
Penna, Rosa, 387, 482
Pepe, Marco, 387

Q
Qachar, Ahmed, 441
Qbouche, Khalid, 3

R
Radgui, Amina, 10
Randrianarivelo, J., 320
Ratolojanahary, F. E., 320
Rguibi, Zakaria, 129
Rguiti, M., 320

Rhoulami, Khadija, 3
Rkibi, Ouafae, 54
Roushdy, Ahmed, 19

S
Saadane, Abderrahim, 116
Saadi, Chaimae, 173
Sabri, Mohammed, 265
Sabry, Ahmed, 19
Sadki, Hajar, 507
Saidi, Rajaa, 29, 161
Saint-Drenan, Yves-Marie, 275

T
Taouiri, Souad, 286
Tardif, Pierre-Martin, 182

W
Wagdy, Mahmoud, 19

X
Xin, Haohui, 333, 464

Y
Yessef, Mourad, 300

Z
Zahraoui, F. Z., 42
Zaoui, El Miloud, 519
Zhu, Yingbo, 482
Zou, Luoquan, 374
Zugari, Asmaa, 75